AF740022

NANOTECHNOLOGY IN PHOTOSYNTHESIS

NANOTECHNOLOGY IN PHOTOSYNTHESIS

INNOVATIONS FOR GLOBAL FOOD SUSTAINABILITY

Edited by

VISHNU D. RAJPUT

Academy of Biology and Biotechnology,
Southern Federal University, Russia

MOHAMMAD FAIZAN

Botany Section,
School of Sciences, Maulana Azad National Urdu University,
Hyderabad, Telangana, India

ELSEVIER

ACADEMIC PRESS

An imprint of Elsevier

Academic Press is an imprint of Elsevier
125 London Wall, London EC2Y 5AS, United Kingdom
50 Hampshire Street, 5th Floor, Cambridge, MA 02139, United States

ISBN: 978-0-443-44376-3

For information on all Academic Press publications visit our website at
https://www.elsevier.com/books-and-journals

Publisher: Mara Conner
Acquisitions Editor: Palak Gupta
Editorial Project Manager: Sneha Roy
Production Project Manager: R. Rajeswari
Cover Designer: Mark Rogers

Typeset by TNQ Tech

Contents

5. Nanomaterials for improving photosynthetic efficiency: current innovations

AINA INAM, ASMA AHMAD, MAHAM SHEIKH, SUMERA JAVAD, SUMERA IQBAL AND OZAIR CHAUDHRY

6. Nanoformulated fertilizers and CO_2 delivery systems optimizes photosynthesis in smart agriculture

OMID SOHRABI, ALI SHAMS AND MAHTAB MEHRKISH

7. Nano-formulated fertilizers and CO_2 delivery system for optimizing photosynthesis

PRASANN KUMAR, AMIT RAJ AND PADMANABH DWIVEDI

8. Photocatalytic nanomaterials for enhanced light utilization and energy conversion

PRASANN KUMAR, AMIT RAJ AND PADMANABH DWIVEDI

9. Nanotechnology-enabled chloroplast engineering and gene delivery

NILABJA BANERJEE, ANJALI TIWARI, JOYDEEP ESHORE, VINAI KUMAR, MOHAMMAD FAIZAN, SAJID FIAZ AND DIPTI BISARYA

10. Nanotechnology driven optimization of Photosystem I and Photosystem II functions

SHOAIB AHMAD, ADIBA KHAN SEHRISH AND HAFIZA SADIA

11. Nanoparticle assisted nutrient delivery for enhancing plant photosynthetic performance

YUSHENG LI, SADDAM HUSSAIN, MOHAMMAD NAUMAN KHAN, ZHIYONG AI AND MOHAMMAD FAIZAN

12. Photocatalytic nanomaterials for improving photosynthetic light use efficiency

ADITI, AYUSHI KUMARI, SUBHAM PREETAM AND SUMIRA MALIK

13. Mesoporous silica-based nanotechnology for chloroplast engineering and targeted delivery of double-stranded RNA and ribonucleoproteins

RATHOD SRIDHAR, M.S. PAARSHITHA REDDY, HANUMAN SINGH JATAV, K. ARUNA, T. GOHAIN, AVINASH PILLA AND ANVESH SANKURATRI

14. Life cycle assessment (LCA) of nano-harvest technologies: Evaluating environmental and economic sustainability

MOHAMMAD FAIZAN, S. MAQBOOL AHMED, VISHNU D. RAJPUT AND SHAMSUL HAYAT

15. Nanotechnology for climate-resilient photosynthesis under abiotic stresses

RADWAN KHALIL, AMINA GAMAL AND NESMA ELSAYED

16. Integrating nanotechnology with smart farming and precision photosynthetic management

MEGHNA SARKAR, NILOTPAL DAS, ATIN KUMAR, RIDA BATOOL, SHARAD SACHAN AND MOHAMMAD FAIZAN

17. Policy prespectives: Regulatory pathways for nanomaterials in agriculture

RAJASHREE KRISHNA BHARADWAJ, SUDHIR KUMAR RAJPOOT, ANIMEKH HAZARIKA AND MOHAMMAD FAIZAN

18. Risks and concerns in adopting nanotechnology to tackle hunger: Environmental, health, and ethical assessments

SADIA AFZAL, SUMERA JAVAD, SAIMA BASHIR, AMINA TARIQ AND SAIQA ILYAS

Contributors

Aditi Amity Institute of Biotechnology, Amity University, Ranchi, Jharkhand, India

Sadia Afzal Department of Botany, Lahore College for Women University, Lahore, Pakistan

Asma Ahmad Department of Botany, Lahore College for Women University, Lahore, Pakistan

Shoaib Ahmad State Key Laboratory of Water Pollution Control and Green Resource Recycling, School of the Environment, Nanjing University, Nanjing, Jiangsu, China

S. Maqbool Ahmed Botany Section, School of Sciences, Maulana Azad National Urdu University, Hyderabad, Telangana, India

Zhiyong Ai National Center of Technology Innovation for Saline-Alkali Tolerant Rice in Sanya, Sanya, China

K. Aruna Department of Agronomy, Professor Jayashankar Telangana Agricultural University, Hyderabad, Telangana, India

Nilabja Banerjee School of Agriculture, Lovely Professional University, Phagwara, Punjab, India; Department of Agriculture, Srinath University, Jamshedpur, Jharkhand, India

Saima Bashir Department of Botany, University of Gujrat, Gujrat, Pakistan

Rida Batool Department of Plant Sciences, Faculty of Biological Sciences, Quaid-i-Azam University, Islamabad, Pakistan

Rajashree Krishna Bharadwaj Department of Agronomy (Agroforestry), Institute of Agricultural Sciences, Banaras Hindu University, Varanasi, Uttar Pradesh, India

Aqsa Bibi School of Chinese Medicine, Medical Department, Jianghan University, Wuhan, China

Dipti Bisarya School of Agriculture, Lovely Professional University, Phagwara, Punjab, India

Ozair Chaudhry DECAC, Community Growth & Economic Dev. Department, The Regional Municipality Durham, Durham, ON, Canada

Jingdong Chen Hubei Engineering Research Center for Protection and Utilization of Special Biological Resources in the Hanjiang River Basin, College of Life Sciences, Jianghan University, Wuhan, China

Nilotpal Das Research Scholar, Faculty of Agricultural Sciences, GLA University, Mathura, Uttar Pradesh, India

Priyanka Devi Department of Research and Development Cell, Lovely Professional University, Phagwara, Punjab, India

Padmanabh Dwivedi Department of Plant Physiology, Institute of Agricultural Sciences, Banaras Hindu University, Varanasi, Uttar Pradesh, India

Nesma Elsayed Botany and Microbiology Department, Faculty of Science, Benha University, Benha, Egypt

Joydeep Eshore School of Agriculture, Lovely Professional University, Phagwara, Punjab, India

Mohammad Faizan Botany Section, School of Sciences, Maulana Azad National Urdu University, Hyderabad, Telangana, India

Ahmad Faraz Department of Botany, Shibli National College, Azamgarh, Uttar Pradesh, India

Sajid Fiaz Institute of Molecular Biology and Biotechnology, The University of Lahore, Lahore, Punjab, Pakistan

Amina Gamal Botany and Microbiology Department, Faculty of Science, Benha University, Benha, Egypt

T. Gohain Department of Agronomy, School of Agricultural Sciences, Nagaland University, Medziphema, Nagaland, India

Shamsul Hayat Department of Botany, Aligarh Muslim University, Aligarh, Uttar Pradesh, India

Animekh Hazarika Department of Ecology and Environmental Science, Assam University, Silchar, Assam, India

Saddam Hussain Laboratory of Plant Molecular Genetics & Crop Gene Editing, School of Life Sciences, Linyi University, Linyi, China

Saiqa Ilyas Department of Biotechnology, Lahore College for Women University, Lahore, Pakistan

Aina Inam Department of Botany, Lahore College for Women University, Lahore, Pakistan

Sumera Iqbal Department of Botany, Lahore College for Women University, Lahore, Pakistan

Nayab Ishrat Department of Biotechnology, Era University, Lucknow, Uttar Pradesh, India

Hanuman Singh Jatav Department of Soil Science and Agricultural Chemistry, Sri Karan Narendra Agriculture University, Jobner, Rajasthan, India

Sumera Javad Department of Botany, Lahore College for Women University, Lahore, Pakistan

Uzma Kafeel Asim Siddique Memorial Degree College, Budaun, Uttar Pradesh, India

Radwan Khalil Botany and Microbiology Department, Faculty of Science, Benha University, Benha, Egypt

Mohammad Nauman Khan Laboratory of Plant Molecular Genetics & Crop Gene Editing, School of Life Sciences, Linyi University, Linyi, China

Atin Kumar School of Agriculture, Uttaranchal University, Dehradun, Uttarakhand, India

Prasann Kumar Department of Agronomy, School of Agriculture, Lovely Professional University, Phagwara, Punjab, India

Vinai Kumar Department of Plant Physiology, Govind Ballabh Pant University of Agriculture & Technology, Pantnagar, Uttarakhand, India

Ayushi Kumari Amity Institute of Biotechnology, Amity University, Ranchi, Jharkhand, India

Yusheng Li School of Breeding and Multiplication (Sanya Institute of Breeding and Multiplication), Hainan University, Sanya, China

Sumira Malik Amity Institute of Biotechnology, Amity University, Ranchi, Jharkhand, India; University Centre for Research & Development (UCRD), Chandigarh University, Mohali, Punjab, India; School of Basic and Applied Sciences, K. R. Mangalam University, Gurgaon, Haryana, India

Mahtab Mehrkish Agrotechnology – Crop Plant Physiology, Department of Plant Production and Genetics Engineering, Razi University, Kermanshah, Iran

Amjad Mehsud Key Laboratory of Agro-Forestry Environmental Processes and Ecological Regulation of Hainan Province, School of Environmental Science and Engineering, Hainan University, Haikou, China

M.S. Paarshitha Reddy Department of Agronomy, School of Agricultural Sciences, Nagaland University, Medziphema, Nagaland, India

Avinash Pilla Department of Agriculture, Koneru Lakshmaiah Education Foundation, Vaddeswaram, Andhra Pradesh, India

Subham Preetam Department of Chemical Engineering, School of Engineering, Monash University Malaysia, Bandar Sunway, Selangor, Malaysia

Amit Raj Department of Agronomy, School of Agriculture, Lovely Professional University, Phagwara, Punjab, India

Sudhir Kumar Rajpoot Department of Agronomy, Institute of Agricultural Sciences, Banaras Hindu University, Varanasi, Uttar Pradesh, India

Vishnu D. Rajput Academy of Biology and Biotechnology, Southern Federal University, Rostov-On-Don, Russia

Sharad Sachan Department of Agricultural Economics and Extension, School of Agriculture, Lovely Professional University, Phagwara, Punjab, India

Hafiza Sadia State Key Laboratory of Pharmaceutical Biotechnology, NJU Advanced Institute for Life Sciences, School of Life Sciences, Nanjing University, Nanjing, China

Anvesh Sankuratri Department of Plant pathology, Tennessee State University, Nashville, TN, United States

Meghna Sarkar Department of Soil Science, Assam Agricultural University, Jorhat, Assam, India

Adiba Khan Sehrish State Key Laboratory of Water Pollution Control and Green Resource Recycling, School of the Environment, Nanjing University, Nanjing, Jiangsu, China

Ali Shams Department of Biotechnology, Faculty of Basic Sciences, Al-e-Taha Institute of Higher Education, Tehran, Iran

Maham Sheikh Department of Botany, Lahore College for Women University, Lahore, Pakistan

Monika Sindhu University Centre for Research and Development, Chandigarh University, Mohali, Punjab, India

Omid Sohrabi Department of Horticultural Science and Engineering, Razi University, Kermanshah, Iran

Rathod Sridhar Department of Agronomy, School of Agricultural Sciences, Nagaland University, Medziphema, Nagaland, India

Haider Sultan Hubei Engineering Research Center for Protection and Utilization of Special Biological Resources in the Hanjiang River Basin, College of Life Sciences, Jianghan University, Wuhan, China

Amina Tariq Department of Botany, Lahore College for Women University, Lahore, Pakistan

Anjali Tiwari Department of Agricultural Science, University of Calcutta, Ballygunge, West Bengal, India

Ghafoor Unnisa Botany Section, School of Sciences, Maulana Azad National Urdu University, Hyderabad, Telangana, India

Heping Wan Hubei Engineering Research Center for Protection and Utilization of Special Biological Resources in the Hanjiang River Basin, College of Life Sciences, Jianghan University, Wuhan, China

Changli Zeng Hubei Engineering Research Center for Protection and Utilization of Special Biological Resources in the Hanjiang River Basin, College of Life Sciences, Jianghan University, Wuhan, China

Preface

Photosynthesis is the fundamental biological process that sustains life on Earth, driving plant productivity, food security, atmospheric oxygen balance, and global carbon cycling. However, the increasing pressures of climate change, population growth, soil degradation, water scarcity, and declining agricultural productivity have intensified the urgent need for innovative technologies capable of enhancing photosynthetic performance and crop resilience. In this context, nanotechnology has emerged as a transformative scientific frontier with immense potential to revolutionize agricultural systems and redefine sustainable food production. The book *Nanotechnology in Photosynthesis: Innovations for Global Food Sustainability* presents a comprehensive exploration of the rapidly evolving interface between nanoscience and plant photosynthetic biology. By integrating advances in nanomaterials, plant physiology, molecular biology, biotechnology, and precision agriculture, this volume aims to provide readers with a holistic understanding of how nanoscale technologies can improve photosynthetic efficiency, nutrient utilization, stress tolerance, and crop productivity under changing environmental conditions. This book begins by establishing the scientific foundation of the impact of nanotechnology on photosynthesis, introducing key concepts, opportunities, mechanisms, and emerging innovations in nano-enabled agriculture. Subsequent chapters discuss nanoscale transformations within plant cells, nanoparticle translocation and bioaccumulation, and the mechanisms underlying nanomaterial-mediated enhancement of photosynthesis. Special emphasis is placed on advanced nano-formulated fertilizers, CO_2 delivery systems, photocatalytic nanomaterials, chloroplast engineering, and nano-assisted gene delivery strategies that are shaping the future of photosynthetic optimization. Recognizing the growing demand for climate-smart agriculture, the book further explores the role of nanotechnology in improving photosynthetic resilience under abiotic stresses such as salinity, drought, heat, and oxidative stress. Chapters addressing smart farming, precision photosynthetic management, and life cycle assessment provide important perspectives on sustainability, environmental impact, and techno-economic feasibility. In addition, the volume critically examines regulatory frameworks, biosafety concerns, ethical implications, and environmental risks associated with the agricultural application of nanomaterials. The strength of this book lies in its multidisciplinary approach. It is designed to serve as a valuable resource for researchers, academicians, postgraduate students, policymakers, agricultural technologists, and industry professionals working in plant sciences, nanobiotechnology, sustainable agriculture, environmental sciences, and related disciplines. By combining fundamental principles with recent scientific developments, the book aims to bridge knowledge gaps between nanotechnology and plant photosynthetic research while encouraging future innovations for sustainable crop production. We sincerely hope that this volume will stimulate scientific discussion, inspire collaborative research, and contribute meaningfully to the development of environmentally responsible nano-enabled agricultural technologies. As the world confronts the dual challenge of ensuring food security and environmental sustainability, the integration of nanotechnology with photosynthesis research may open unprecedented opportunities for building resilient agricultural systems capable of feeding future generations. We express our heartfelt gratitude to all contributing authors, reviewers, editors, and publishing staff whose expertise and dedication made this book possible. Their collective efforts have greatly enriched the scientific quality and interdisciplinary value of this work. We hope readers will find this book informative, thought-provoking, and valuable for advancing research and innovation in the emerging field of nano-enabled photosynthetic sustainability.

Vishnu D. Rajput
Mohammad Faizan
(Editor)

Acknowledgments

The editor of the book is sincerely acknowledging the support by Priority 2030 implemented at Southern Federal University, Rostov-on-Don, Russia.

CHAPTER

1

Role of nanotechnology in enhancing photosynthetic efficiency of plants

Uzma Kafeel[1] *and Mohammad Faizan*[2]

[1]**Asim Siddique Memorial Degree College, Budaun, Uttar Pradesh, India;** [2]**Botany Section, School of Sciences, Maulana Azad National Urdu University, Hyderabad, Telangana, India**

1. Introduction

Population expansion has raised the need for productivity per hectare of agricultural land (Kafeel. 2025). This problem has been compounded by agricultural loss, urbanization, soil degradation, and climate change, resulting in uncertainties about enhanced production (Kafeel et al., 2022b). Novel and critical strategies to increase agricultural production in order to fulfill food demand while mitigating the negative impacts of climate change on crop yields (Jahan et al., 2022a). Significant effort has been made to improve agricultural practices and boost crop output in order to offset these possible difficulties. Despite their effectiveness, these strategies' efficacy has visibly plateaued since the so called "green revolution" (Denning 2023). The global population is predicted to reach around 11 billion people by 2100, necessitating an increase in food consumption (Ghosh et al., 2024). There is an urgent need to revolutionize conventional agriculture in order to reach the UN's second sustainable development target of "Zero Hunger" by 2030 (Ghosh et al., 2024). Furthermore, conventional agriculture must be transformed by utilizing a modernized, sustainable, and environmentally friendly strategy. Climate change, which is coupled with abiotic and biotic pressures, has a significant impact on global food production (Denning 2023). Abiotic stress is the most significant impediment to plant growth because it slows physiological responses and obstructs photosynthetic systems (Jahan et al., 2022b, 2023; Kafeel et al., 2022b). Sustainable Development Goal 12, "Responsible Consumption and Production," has become the global economy's driving force for sustainable consumption. This objective promotes sustainable lives by encouraging the use of fewer resources and achieving more with less. Goal 15, "Life on Land," may be met by using novel plant stress management strategies to stop recurrent drought and desertification (Ghosh et al., 2024; Denning 2023).

Plants continue to be the primary source of food for life on Earth due to the intrinsic process of photosynthesis, which employs the specific cellular component chloroplast. Photosynthesis

Nanotechnology in Photosynthesis
https://doi.org/10.1016/B978-0-443-44376-3.00002-1

is required for the creation of oxygen, the regulation of climate, and the powering of biological processes such as agricultural output. Crop output is a by-product of the entire system, not only photosynthesis (Garcia et al., 2023; Dey et al., 2022; Li et al., 2021). Development, leaf and canopy design, source-sink feedback, and other variables all impact the link between crop yield and photosynthesis (Li et al., 2025a, 2025b; Croce et al., 2024; Garcia et al., 2023). Furthermore, unlike labs or greenhouses, the natural environment has a considerable impact on photosynthesis and development. These interactions and feedback have a significant influence on photosynthesis (Araus et al., 2021). The biological activity under abiotic stresses has a significant impact on photosynthetic efficiency. Plants under stress develop reactive oxygen species (ROS), which initiate coordinated cellular oxidative processes. This causes cellular structural denaturation, protein instability, membrane disintegration, and metabolic disturbance. Excessive ROS generation produces severe oxidative stress in plants, negatively reducing photosynthesis and yield (El-Desouki et al., 2024; Zhu et al., 2023; Awad et al., 2022; Yang et al., 2021; Sharma et al., 2020). Traditional agricultural practices decrease stress, yet they are insufficient. Managing abiotic stress is thus the primary objective for increasing crop output when under abiotic stress. As a result, major attempts to solve this issue with imminent technological advancements are required. When exposed to nanoparticles, plant growth hormones are known to participate in a variety of signaling cascades and act as stress-regulating molecules (Wu and Bose, 2024). In the context of a changing environment, nanobiotechnology has gained relevance since it can assist in mitigating the impacts of environmental stresses and is viewed as a viable approach for enhancing agricultural productivity (Zhu et al., 2023; Bora et al., 2022). Nanotechnology is a multidisciplinary discipline in which scientists and engineers work together to build nanoparticles, nanomaterials, and nanocomposites on a nanoscale (Das et al., 2025a, 2025b). Nanotechnology is playing an important role in agriculture. Its intervention is a new crop production approach (Riseh et al., 2024; Patel et al., 2021).

Nanoparticles (NPs) have distinct traits such as a high surface-to-volume ratio, unique physicochemical properties, cargo loading capacity, biodegradability, and sustained release activity. Furthermore, due to fast population expansion and environmental deterioration, traditional agriculture is facing unprecedented problems (Khatri et al., 2024). For many years, fertilizers and insecticides have been critical components of maintaining agricultural productivity in order to satisfy food production demands. However, excessive pesticide usage impedes sustainable agricultural growth (Li et al., 2025a, 2025b; Popp et al., 2013). Plant nanobiotechnology, which combines nanotechnology with plant biotechnology, increases agricultural productivity. Examples of how nanoparticles and nanomaterials are used to control different agricultural factors on farms include soil improvement mediators, nano-fertilizers, nanopesticides, growth stimulators, nutrient absorption enhancers, nanosensors, abiotic stress management, pathogen detection, plant hormone regulation, environmental pollution reduction, and stress amelioration (Singh et al., 2025; Ahmad and Ahmad, 2024). Additionally, it is known that plant growth hormones act as stress-regulating molecules and participate in a number of signaling cascades when exposed to nanoparticles (Kafeel, 2025; Kafeel et al. 2022a, 2022b). Under abiotic stressors, photosynthetic efficiency is very important for biological activity. By accelerating growth, development, and the rate of active photosynthesis, NPs can increase plant productivity. These applications help to increase agricultural output (Sharma et al., 2024; Nayeri et al., 2023; Mony et al., 2022; Rai-Kalal and Jajoo, 2021). Additionally, as nanotechnology becomes more widely used, agriculture may experience profound changes. Innovative methods to enhance sustainable agriculture and satisfy food demand through nano-enabled solar energy harvesting by plants could potentially transform agriculture in the future (Nayeri et al., 2023; Mony et al., 2022; Rai-Kalal and Jajoo, 2021). Regarding NPs' impact on plant photosynthesis, this highlights the beneficial effects of NPs/nanomaterials on plants as well as their method of operation and strategy for overcoming a variety of abiotic difficulties (Li et al., 2025a, 2025b; Sharma et al., 2024; Zhu et al., 2023; Mony et al., 2022; Yang et al., 2021). This chapter aims to explore how functional nanomaterials enhance plant photosynthesis and antioxidant defense systems under stress. It highlights their potential to protect chlorophyll, reduce ROS-induced oxidative damage, and improve crop yield and nutrient efficiency. It also discusses prospects and environmental concerns for safe and sustainable agricultural applications of nanomaterials.

2. Relationship between nanotechnology and photosynthesis

The relationship between nanotechnology and photosynthesis is rooted in the unique ability of nanomaterials to interact directly with the molecular components involved in capturing light and converting it into chemical energy (Bengisu et al., 2026). Photosynthesis is a highly efficient natural process in which chlorophyll and other pigments absorb sunlight, generate excited electrons, and drive biochemical reactions that produce glucose and oxygen (Esgici et al., 2026). Nanotechnology enhances this process by introducing nanoscale materials such as carbon nanotubes, quantum dots, fullerenes, and metal oxide nanoparticles that can act as artificial light antennae, extending the range of wavelengths plants can absorb (Mony et al., 2022). These nanomaterials improve the flow of electrons during the light-dependent reactions, increasing the overall efficiency of energy conversion (Gautam et al., 2025). Nano-fertilizers and nano-carriers further support photosynthesis by providing essential nutrients like nitrogen, magnesium, and iron in a controlled-release form, improving chlorophyll formation, leaf growth, and gas exchange (Bice et al., 2026; Nazeer et al., 2023; Zeng et al., 2024; Kafeel et al., 2022b, 2023).

Another important connection between nanotechnology and photosynthesis is the development of nano-sensors capable of measuring real-time physiological processes within plants. These sensors can monitor chlorophyll fluorescence, stomatal activity, CO_2 uptake, and water transport, giving detailed insights into plant health and photosynthetic performance without causing tissue damage (Lee et al., 2021; Saglam et al., 2019; Urban et al., 2017). This helps in early detection of stress, nutrient deficiency, and disease, allowing for more precise agricultural management. Nanotechnology is also crucial in the emerging field of artificial photosynthesis. Inspired by the natural systems in chloroplasts, nanostructured semiconductors, photocatalysts, and membrane-like assemblies mimic the light reactions that split water and generate clean fuels such as hydrogen (Ozturk et al., 2025; Kafeel et al. 2022b, 2023; Jahan et al., 2022a, 2022b; Lee et al., 2021; Al-Tawaha et al., 2020; Saglam et al., 2019). Other artificial systems use nanomaterials to convert carbon dioxide into energy-rich organic molecules, offering sustainable solutions for renewable energy production and carbon capture (Gautam et al., 2025; Xu et al., 2024; Almomani et al., 2023).

Overall, nanotechnology not only enhances and monitors natural photosynthesis but also provides the foundation for artificial photosynthetic systems that could address global challenges related to energy, food security, and environmental sustainability.

3. Importance of photosynthesis

As the cornerstone of planetary biogeochemical cycles, photosynthesis represents the principal mechanism for energy conversion and carbon sequestration in the biosphere, with its fundamental importance extending across ecological, atmospheric, and climatic domains. This complex photo-biochemical process, whereby phototrophic organisms utilize photon energy to drive the reduction of carbon dioxide into organic compounds, constitutes the primary entry point of energy into Earth's ecosystems (Alam et al., 2025; Knoflacher, 2024). It thereby supports the entirety of the global trophic pyramid, from primary producers to apex consumers. Beyond its trophic indispensability, photosynthesis exerts a dominant influence on atmospheric composition and global climate dynamics (Kaiser et al., 2015). It serves as the largest flux in the biological carbon cycle, acting as a critical sink for atmospheric CO_2, a key greenhouse gas, and mitigating the anthropogenic forcing of climate change (Kaiser et al., 2015). Concurrently, the evolutionary emergence of oxygenic photosynthesis was the pivotal event that transformed a reducing atmosphere into an oxidizing one, enabling the rise of aerobic metabolism and complex multicellular life. Contemporary research continues to elucidate the intricate regulatory networks of photosynthetic machinery, with implications for enhancing crop yields, developing sustainable bioenergy sources, and innovating novel photochemical technologies (Li et al., 2025a, 2025b; Sharma et al., 2024). Furthermore, understanding the acclimation and adaptation of photosynthetic pathways to environmental stressors is paramount for modeling ecosystem responses to global change and predicting future carbon-climate feedbacks. Thus, the study of photosynthesis remains a critical frontier, not only for advancing fundamental knowledge of biological energy transduction but also for informing strategies addressing the intertwined challenges of food security, climate change mitigation, and sustainable energy production.

3.1 Mechanism of photosynthesis

Green plants and certain other organisms convert light energy into chemical energy through a process called photosynthesis. Light energy is gathered and utilized during photosynthesis in green plants to change water, carbon dioxide, and minerals into oxygen- and energy-rich organic molecules (Sekar and Ramasamy, 2015). Thus, photosynthesis is only the process of turning solar energy (light) into chemical energy (Barber, 2007). It is the conversion of CO_2 into carbohydrates. Other organisms, including cyanobacteria, purple bacteria, and green sulfur bacteria, engage in photosynthesis in addition to green plants. These organisms' photosynthesis is comparable to that of green plants (Faisal et al., 2024; Sreeharsha and Venkata, 2024). Photosynthesis produces glucose, which powers numerous cellular functions. This physicochemical process generates O_2 as a byproduct. Photosynthesis is the process by which plants, algae, and certain microorganisms convert sunlight, CO_2, and water into sugars, oxygen, and other compounds (Sreeharsha and Venkata, 2024). There are two types of photosynthetic processes:

1. Oxygenic photosynthesis and
2. Anoxygenic photosynthesis.

They both follow very similar principles, but oxygenic photosynthesis is the most common and is seen in plants, algae, and cyanobacteria. Light energy is used in oxygenic photosynthesis to convert H_2O absorbed by plant roots into CO_2 and generate carbohydrates (Haeder, 2022). In this process, the water is oxidized, or loses electrons, while the CO_2 is reduced, or gains electrons. Carbohydrates are generated along with oxygen. By absorbing the CO_2 that is created during respiration and delivering oxygen back into the air, oxygenic photosynthesis acts as a counterweight to respiration (Li et al., 2025a, 2025b; Sharma et al., 2024; Mony et al., 2022). According to Anoxygenic Photosynthetic Bacteria, the mechanism typically occurs in bacteria such as green sulfur bacteria and phototrophic purple bacteria. However, anoxygenic photosynthesis uses electron donors other than water and does not create oxygen (Knoflacher, 2024).

Though both types of photosynthesis are complex, multistep affairs, the overall process can be neatly summarized as a chemical equation. The oxygenic photosynthesis equation is:

$$6CO_2 + 12H_2O + \text{Light Energy} \rightarrow 6H_{12}O_6 + 6O_2 + 6H_2O$$

Here, light energy is used to mix 12 molecules of H_2O with six molecules of CO_2. A single glucose ($C_6H_{12}O_6$) molecule and six molecules each of O_2 and H_2O are produced as a result. Similarly, a single generalized formula can be used to represent all anoxygenic photosynthesis reactions:

$$CO_2 + 2H_2A + \text{Light energy} \rightarrow [CH_2O] + 2A + H_2O$$

The letter A in the equation is a variable, and H_2A represents the potential electron donor. For example, "A" may represent sulfur (S) in the electron donor hydrogen sulfide (H_2S). Photosynthesis in plants mostly takes place in the leaves. Since carbon dioxide, water, and sunlight are required for photosynthesis, each of these elements must be acquired by or transferred to the leaves. Stomata, which are little pores found in plant leaves, are how carbon dioxide is captured. The stomata also allow for the discharge of oxygen (Striker, 2023). The roots of the plant absorb water, which is then transported to the leaves by vascular plant tissue systems. Chlorophyll, a green pigment found in the chloroplasts of plant cells, absorbs sunlight. Photosynthesis takes place in chloroplasts (Striker, 2023). The various structures found in chloroplasts each have a distinct purpose: Inner and outer membranes, which act as a barrier to protect chloroplast structures. Stroma, a viscous fluid found inside chloroplasts the location where CO_2 is converted to sugar. Flattened membrane sac-like structures known as thylakoids are the location where light energy is converted to chemical energy (Gilmore, 2019). Grana, which are thickly stacked stacks of thylakoid sacs locations where light energy is converted to chemical energy. Chlorophyll, a chloroplast-stored green pigment takes in light energy (Gilmore, 2019).

3.2 Abiotic stress effect on photosynthesis

Photosynthesis is crucial for sustaining life on this planet and is necessary for plant growth and development. Abiotic stresses such as high and low temperatures, and excess or deficit of water limit the crucial plant processes, thus threatening the global food security (Li et al., 2025a, 2025b). However, recent molecular approaches allowed elucidation of the photosynthetic components/compounds and their efficiency under stress conditions. In the present scenario, these approaches are not enough to reduce the yield penalty due to the reduction in photosynthetic efficiency (Mony et al., 2022). Therefore, comprehensive data on plant behavior and stress crosstalk networks assist in understanding the in-depth mechanism of photosynthesis (Fig. 1.1). In recent years, information regarding crosstalk, signaling characterization of candidate genes, and responses to multiple stressors has advanced our knowledge to understand the mechanism of photosynthesis (Kumar, 2025; Sekar and Ramasamy, 2015). Various studies have been conducted on photosynthesis under multiple abiotic stress factors that affect the photosynthetic efficiency of a plant. Crosstalk signaling compounds (plant growth

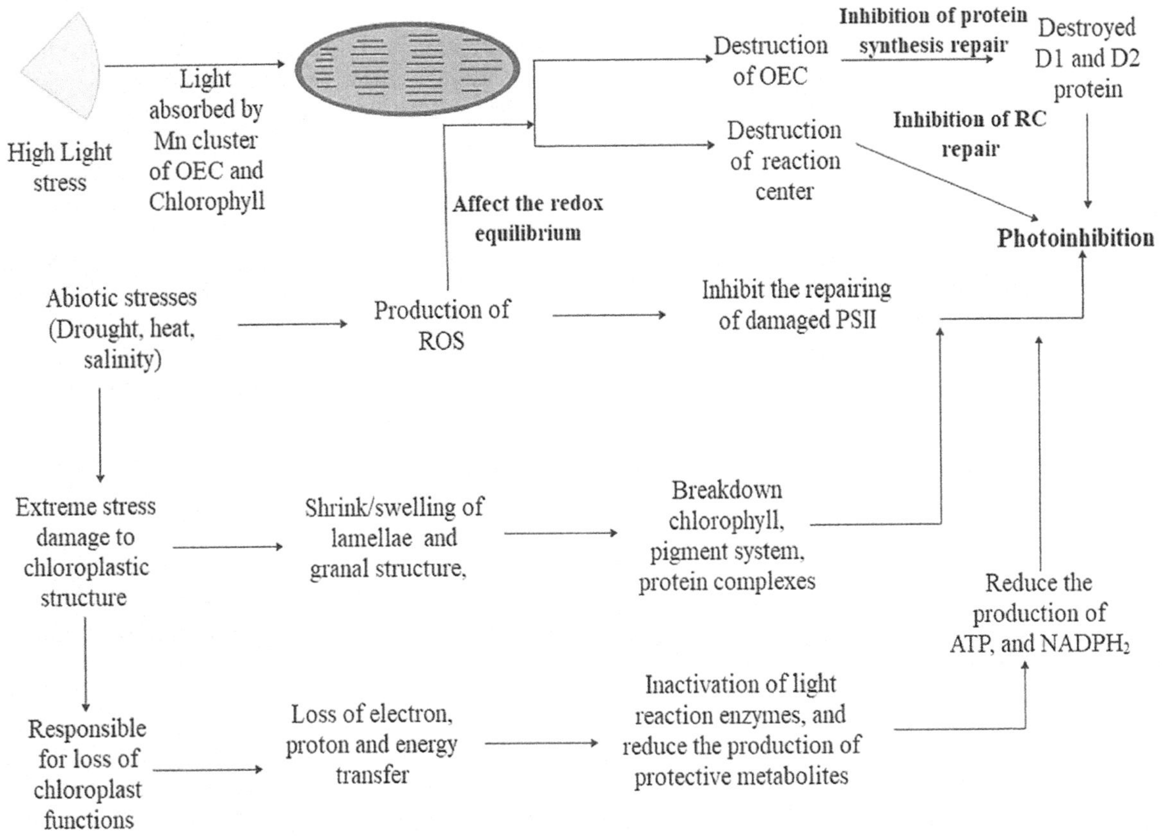

FIGURE 1.1 Impact of various stresses on the process of photosynthesis. Abiotic stresses cause these damages by loss of electron transfer, deactivation of light reaction enzymes and production of reduced ATP, leading to photosynthesis inhibition.*Courtesy: Chauhan, J., Prathibha, M.D., Singh, P., Choyal, P., Mishra, U.N., Saha, D., Kumar, R., Anuragi, H., Pandey, S., Bose, B. and Mehta, B., 2023. Plant photosynthesis under abiotic stresses: damages, adaptive, and signaling mechanisms. Plant Stress, 10, p.100296.*

regulators and micro RNAs) also play an essential role in an in-depth understanding of the photosynthesis mechanism (Zhu et al., 2023; Bora et al., 2022). Based on studies, the mechanism of damage and adaptive response of photosynthesis under multiple stressors to enhance the scientific community's knowledge toward boosting photosynthesis and to accelerate stress tolerance strategies for crop improvement (Kulsoom et al., 2024; Naikoo et al., 2020; Qiao et al., 2024).

4. Impact of nanotechnology on agricultural productivity

Nanotechnology is poised to instigate a paradigm shift in agricultural productivity by introducing novel solutions to persistent biotic and abiotic challenges that constrain global food security (Kafeel et al., 2023; Al-Tawaha et al., 2020; Saglam et al., 2019). The strategic application of engineered nanomaterials (ENMs) and nano-enabled platforms offers unprecedented precision in managing crop growth and protection. This includes the development of nano-formulated agrochemicals, or "smart" fertilizers and pesticides, which utilize stimuli-responsive nanocarriers for the controlled release of active ingredients, thereby enhancing nutrient use efficiency and minimizing deleterious environmental runoff (Palit, 2018). Furthermore, nano sensors enable real-time, in-field monitoring of soil conditions, plant health, and pathogen presence, facilitating data-driven precision agriculture. The unique physicochemical properties of nanomaterials also confer direct bio-efficacy; for instance, nano-scale micronutrients can be engineered for superior foliar or root uptake, while certain nanoparticles themselves exhibit potent antifungal or antibacterial properties, offering alternative modes of action against resistant pathogens (Sreeharsha and Venkata, 2024). However, the translation of these laboratory successes into field-scale applications necessitates a rigorous, interdisciplinary research focus on the long-term fate, transport, and potential ecotoxicological impacts of ENMs within complex agroecosystems (Sreeharsha and Venkata, 2024). Addressing these biosafety and regulatory considerations is paramount to responsibly harnessing nanotechnology for developing sustainable, high-efficiency agricultural practices capable of meeting the demands of a burgeoning global population amidst a changing climate (Rai-Kalal and Jajoo, 2021).

4.1 Effect of nanomaterials on photosynthesis

Nanomaterials exert significant and complex effects on photosynthesis, offering both promising enhancements and potential risks to plant physiological processes. Due to their small size, large surface area, and unique physicochemical properties, nanomaterials can interact directly with cellular structures involved in photosynthesis, particularly chloroplasts and thylakoid membranes (Alam and Faizen, 2024; Alsaleh, 2025). In many cases, engineered nanomaterials such as carbon nanotubes, titanium dioxide nanoparticles, zinc oxide nanoparticles, and quantum dots have been shown to improve photosynthetic efficiency by increasing light absorption, accelerating electron transport, and enhancing chlorophyll synthesis (Elhefnawy and Elsheery, 2023). For example, carbon nanotubes can embed themselves within chloroplasts, functioning as artificial antenna systems that expand the spectrum of usable light, thereby boosting the light-dependent reactions. Metal oxide nanoparticles like TiO_2 can facilitate photoreduction and promote ROS scavenging, reducing oxidative damage and sustaining higher rates of carbon fixation (Striker, 2023). Additionally, nano-fertilizers provide essential nutrients in more bioavailable and controlled-release forms, improving chlorophyll content, stomatal conductance, and overall carbohydrate production (Li et al., 2025a, 2025b).

However, the effects of nanomaterials are strongly concentration-dependent, and excessive exposure can negatively impact photosynthetic machinery. Heavy metal-based nanoparticles such as silver, cadmium sulfide, or lead nanoparticles may generate toxic ROS, disrupt thylakoid membrane integrity, and inhibit key enzymes involved in the Calvin cycle (Amiri et al., 2025). High concentrations of nanoparticles can also reduce chlorophyll content, block stomatal pores, interfere with electron transport chains, and lead to photoinhibition. Moreover, nanoparticle accumulation in leaf tissues may alter chloroplast ultrastructure and impair ATP and NADPH production, ultimately lowering photosynthetic rates (Amiri et al., 2025). The dual nature of nanomaterial impacts highlights the need for careful evaluation of nanoparticle type, size, surface chemistry, dosage, and environmental fate (Sharma et al., 2024). While nanotechnology offers promising avenues for enhancing photosynthesis, improving crop productivity, and mitigating environmental stress through nanosensors and nano-enabled fertilizers, it also requires thorough risk assessment to prevent long-term ecological and physiological harm. Understanding these interactions at molecular and ecosystem levels is essential for the safe and sustainable application of nanomaterials in agriculture and plant biotechnology (Alsaleh, 2025).

4.2 Mechanism of action

The mechanism of action of nanomaterials on photosynthesis involves a series of intricate interactions at cellular, subcellular, and molecular levels that collectively influence the efficiency and regulation of the photosynthetic machinery (Gautam et al., 2025). When nanomaterials enter plant tissues either through foliar uptake, root absorption, or seed priming, they can translocate to chloroplasts, where the primary photosynthetic reactions occur. One important mechanism is the enhancement of light absorption, in which nanomaterials such as carbon nanotubes, quantum dots, and metal oxide nanoparticles act as artificial antenna systems capable of capturing a broader range of wavelengths than natural pigments (Gautam et al., 2025). These materials transfer the absorbed energy to chlorophyll molecules, thereby boosting excitation energy and increasing the rate of electron transport in photosystems I and II. Additionally, metal oxide nanoparticles like TiO_2 and ZnO exhibit photocatalytic properties that support photochemical reactions and facilitate the generation of ATP and NADPH in the light-dependent stage (Gautam et al., 2025). Another critical mechanism involves the modulation of ROS. Nanomaterials at appropriate concentrations can act as antioxidant agents by scavenging excess ROS generated during environmental stress, protecting thylakoid membranes, and maintaining the structural integrity of chloroplasts (Elhefnawy and Elsheery, 2023). Nano-fertilizers further influence photosynthesis by gradually releasing essential nutrients such as nitrogen, magnesium, and iron, which are key components of chlorophyll, photosynthetic enzymes, and electron carriers, thus improving pigment synthesis and CO_2 assimilation (Singh et al., 2025; Kafeel, 2025). However, the mechanism of toxicity arises when nanomaterials accumulate at high concentrations or possess harmful surface chemistries. Toxic nanoparticles such as silver, cadmium-based quantum dots, and lead nanoparticles can induce excessive ROS production, leading to oxidative stress, lipid peroxidation, and damage to chloroplast membranes. This disrupts the organization of grana and stroma thylakoids, impairs the functioning of photosystems, and inhibits enzymes like Rubisco (Xu et al., 2025; Kafeel et al. 2022a, 2022b; Palit, 2018). Some nanoparticles can also bind to sulfhydryl groups in proteins or replace essential metals in chlorophyll biosynthesis pathways, resulting in chlorosis and reduced pigment content. Furthermore, nanoparticles may interfere with stomatal functioning, limiting CO_2 diffusion and restricting Calvin cycle activity (Zhang and Li, 2025). Transport blockage within vascular tissues is another possible mechanism, reducing water supply to photosynthetically active cells (Zhang and Li, 2025). Overall, the mechanism of action of nanomaterials on photosynthesis is highly dose-dependent and influenced by particle size, shape, surface charge, and chemical composition. Understanding these mechanisms provides valuable insights for designing safe, targeted nanomaterials that enhance plant productivity while minimizing physiological and ecological risks (Li et al., 2024).

4.3 Nano fertilizers in optimizing photosynthesis

Photosynthesis, the cornerstone of terrestrial life and agricultural productivity, is an inherently inefficient process, with plants typically converting only a fraction of intercepted solar energy into biomass. This limitation presents a critical bottleneck for global food security, urging the development of innovative agronomic technologies (Sultan et al., 2024; Barrett, 2021). Among the most promising of these are nano fertilizers, which represent a paradigm shift from conventional nutrient delivery systems and offer a sophisticated toolkit for directly optimizing the photosynthetic apparatus. Unlike their bulk counterparts, these nanoengineered particles, typically ranging from 1 to 100 nm, possess unique physico-chemical properties that enable a targeted and controlled release of essential micronutrients and macronutrients directly at the chloroplast level (Zhang and Li, 2025; Li et al., 2024; Elhefnawy and Elsheery, 2023). This precise delivery is crucial for enhancing the light-dependent and light-independent reactions of photosynthesis. For instance, nano-formulations of iron, magnesium, and manganese, key cofactors in chlorophyll synthesis and the photosystem II (PS-II) oxygen-evolving complex, can effectively mitigate nutrient deficiencies that impede electron transport chain efficiency (Sharma et al., 2024; Nayeri et al., 2023; Mony et al., 2022; Rai-Kalal and Jajoo, 2021). By ensuring the structural and functional integrity of photosynthetic pigments and reaction centers, nano fertilizers directly augment light harvesting capacity, reduce photoinhibition, and maximize photochemical yield, as often measured by an increase in chlorophyll fluorescence parameters such as Fv/Fm (Zhang and Li, 2025; Al-Tawaha et al., 2020). Furthermore, certain nanomaterials, including carbon quantum dots and some metal oxide nanoparticles, have demonstrated a unique ability to act as artificial antennas, broadening the spectrum of light available for photosynthesis by converting high-energy wavelengths into usable, lower-energy red and blue light that chlorophylls can absorb more efficiently. This not only boosts the initial energy capture but also supports the Calvin cycle by ensuring a sustained supply of ATP and NADPH (Li et al., 2024; Barrett, 2021). The controlled release mechanism of nano fertilizers also minimizes the luxury consumption of ions that can cause oxidative stress within the chloroplast, thereby protecting the thylakoid membranes from damage and maintaining a favorable redox environment for carbon fixation (Zhang and Li, 2025; Sharma et al., 2024; Nayeri et al., 2023). Consequently, the strategic application of nano fertilizers translates these subcellular enhancements into tangible physiological outcomes: higher net photosynthetic rates, improved stomatal conductance, increased Rubisco enzyme activity, and greater biomass accumulation (Sharma et al., 2024). In conclusion, by moving beyond mere nutrition to active bio-enhancement at the molecular level, nano fertilizers emerge as a transformative technology for overcoming intrinsic photosynthetic inefficiencies, paving the way for a new era of sustainable high-yield agriculture capable of meeting the demands of a growing population under the constraints of a changing climate (Li et al., 2025a, 2025b; Sharma et al., 2024).

5. Future prospects and concerns

While the potential of nano fertilizers to revolutionize agricultural productivity by optimizing photosynthesis is undeniable, their trajectory from laboratory innovation to widespread field application is fraught with both exhilarating prospects and profound concerns that demand a meticulously balanced approach (Kafeel et al., 2023). The prospects are multidimensional, pointing toward an era of smart and precision agro-nanotechnology. The next generation of nano fertilizers will likely be highly tenable, engineered for specific crop genotypes and even responsive to environmental cues such as light intensity or internal plant stress signals, thereby releasing nutrients precisely when photosynthetic demand is highest (Singh et al., 2025; Ahmad and Ahmad, 2024). We can anticipate the development of multi-functional nanoscale platforms that integrate nutrient delivery with photocatalytic properties, pest management, and real-time biosensing, creating an all-in-one solution for crop health monitoring and enhancement (Riseh et al., 2024; Patel et al., 2021). Furthermore, the integration of artificial intelligence and machine learning with nano-biosensor networks could enable closed-loop systems where nanofertilizer application rates and timing are dynamically adjusted based on real-time data on photosynthetic efficiency, ushering in an unprecedented level of resource use efficiency (Supriya et al., 2024). The utilization of biodegradable and biocompatible nanomaterials, such as those derived from chitosan or cellulose, presents a significant avenue for mitigating environmental persistence, aligning the technology with the principles of green chemistry and circular economy (Singh et al., 2025). However, these promising prospects are counterbalanced by significant and legitimate concerns. The foremost issue is the long-term environmental fate and biocompatibility of these engineered particles. The potential for accumulation in soils, leaching into groundwater, and transfer through trophic levels poses unknown risks to ecosystem integrity and food web safety (Xu et al., 2025; Kafeel et al. 2022a, 2022b; Palit, 2018). The impact on soil microbiome diversity and function, which is crucial for nutrient cycling and overall soil health, remains inadequately understood. From a human health perspective, the translocation of nanoparticles within edible portions of crops and their subsequent bioavailability upon consumption requires rigorous toxicological assessment (Amiri et al., 2025). The Trojan horse effect, where nanoparticles facilitate the uptake of other contaminants, is another critical area for investigation (Gautam et al., 2025). Moreover, the economic and social dimensions cannot be ignored; the high cost of research, development, and production could exacerbate global inequalities, making this transformative technology inaccessible to smallholder farmers who need it most, thereby creating a nano-divide (Das et al., 2025a, 2025b). Therefore, the path forward necessitates a robust and interdisciplinary framework involving materials science, plant physiology, ecotoxicology, and social sciences (Qiao et al., 2024; Faraz et al., 2019). A proactive and transparent regulatory regime, coupled with strong public engagement, is imperative to navigate the complex risk-benefit landscape and ensure that the promise of nano fertilizers in optimizing photosynthesis translates into a sustainable and equitable future for global agriculture (Faizan et al., 2023; Li et al., 2025a, 2025b; Sharma et al., 2024).

6. Conclusion

Nano fertilizers show strong potential to optimize photosynthesis and transform modern agriculture. They go beyond nutrient supply by actively enhancing plant physiological performance. Their nanoscale traits enable targeted delivery, controlled release, and stronger interaction with the photosynthetic system. They improve micronutrient availability for chlorophyll synthesis and Photosystem II stability. This strengthens light harvesting and electron transport, improving photosynthetic efficiency indicators. Some nanomaterials can act as artificial antenna systems to expand usable light wavelengths. Improved light reactions boost ATP and NADPH production, supporting the Calvin cycle. This enhances Rubisco activity, carbon fixation, biomass production, and crop yield. However, risks include soil accumulation, food-chain transfer, microbiome disruption, and nanoparticle entry into edible tissues. Future progress requires safe biodegradable designs, precision integration, strong regulation, and equitable access.

References

Ahmad, F., Ahmad, S., 2024. NPs and soil microorganisms interactions in crop management-current status and future prospects. In: Industrial Applications of Soil Microbes, 4, pp. 361–403.

Alam, P., Faizan, M., 2024. Glucose modulates photosynthesis, antioxidant system and morpho-physiology in tomato (Solanum lycopersicum) plants under cadmium stress. Egypt. J. Soil Sci. 64 (3), 845–854.

Alam, P., Yalcin, M., Faizan, M., Albalawi, T., 2025. Response of tomato to silicon dioxide nanoparticles under salinity: impact on photosynthesis, antioxidant enzymes activity, stress biomarkers and osmoregulatory substances. Plant Nano Biol. 13, 100171.

Almomani, F., Abdelbar, A., Ghanimeh, S., 2023. A review of the recent advancement of bioconversion of carbon dioxide to added value products: a state of the art. Sustainability 15 (13), 10438.

Alsaleh, N.B., 2025. Adverse responses of engineered nanomaterials at low and subtoxic exposure levels: current understanding and future perspectives. Nanotoxicology 1–21.

Al-Tawaha, A.R.M.S., Singh, S., Singh, V., Kafeel, U., Naikoo, M.I., Kumari, A., Imran, A., Al-Tawaha, A.R., Qaisi, A.M., Khanum, S., 2020. Improving water use efficiency and nitrogen use efficiency in rice through breeding and genomics approaches. In: Rice Research for Quality Improvement: Genomics and Genetic Engineering: Volume 2: Nutrient Biofortification and Herbicide and Biotic Stress Resistance in Rice. Springer Singapore, Singapore, pp. 307–337.

Amiri, H., Hemmati Hassan Gavyar, P., Farrokhzad, Y., 2025. Nanomaterials for increasing photosynthesis efficiency. In: Handbook of Nanotechnology in Agriculture. Springer Nature Singapore, Singapore, pp. 1–23.

Araus, J.L., Sanchez-Bragado, R., Vicente, R., 2021. Improving crop yield and resilience through optimization of photosynthesis: panacea or pipe dream? J. Exp. Bot. 72 (11), 3936–3955.

Awad, M., Moustafa-Farag, M., Liu, Z., El-Shazoly, R.M., 2022. Combined effect of biochar and salicylic acid in alleviating heavy metal stress, antioxidant enhancement, and Chinese mustard growth in a contaminated soil. J. Soil Sci. Plant Nutr. 22 (4), 4194–4206.

Barber, J., 2007. Biological solar energy. Philos. Trans. R. Soc. A Math. Phys. Eng. Sci. 365 (1853), 1007–1023.

Barrett, C.B., 2021. Overcoming global food security challenges through science and solidarity. Am. J. Agric. Econ. 103 (2), 422–447.

Bengisu, G., Bice, B.T., Atay, U., Alatar, A.A., Faizan, M., Adil, M.F., Faisal, M., 2026. Multiwalled carbon nanotubes-jasmonate interactions enhance biomass production by regulating reactive oxygen species homeostasis, δ-aminolevulinic acid/glutamate-1-semialdehyde metabolism, source–sink balance, and defense responses under chromium stress in soybean. Front. Plant Sci. 17, 1806073.

Bice, B.T., Faizan, M., Rajput, V.D., Alam, P., Albalawi, T., Ahmad, S., 2026. Arsenic-triggered nano-hormonal responses modulate stomatal conductance, primary–secondary metabolites, and nutrient homeostasis in rice (Oryza sativa L.). Nanotechnology 37, 035702.

Bora, K.A., Hashmi, S., Zulfiqar, F., Abideen, Z., Ali, H., Siddiqui, Z.S., Siddique, K.H., 2022. Recent progress in bio-mediated synthesis and applications of engineered nanomaterials for sustainable agriculture. Front. Plant Sci. 13, 999505.

Chauhan, J., Prathibha, M.D., Singh, P., Choyal, P., Mishra, U.N., Saha, D., Kumar, R., Anuragi, H., Pandey, S., Bose, B., Mehta, B., 2023. Plant photosynthesis under abiotic stresses: damages, adaptive, and signaling mechanisms. Plant Stress 10, 100296.

Croce, R., Carmo-Silva, E., Cho, Y.B., Ermakova, M., Harbinson, J., Lawson, T., McCormick, A.J., Niyogi, K. K., Ort, D.R., Patel-Tupper, D., Pesaresi, P., 2024. Perspectives on improving photosynthesis to increase crop yield. Plant Cell 36 (10), 3944–3973.

Das, D., Singh, A., Adari, R., Mudgal, G., Radha, P., 2025. Biocompatible cellulose derivatives: green chemistry and its sustainable applications. In: Functionalized Cellulose Materials: Sustainable Manufacturing and Applications. Springer Nature Switzerland, Cham, pp. 159–190.

Das, H.S., Basak, A., Maity, S., 2025. Materials science and nanotechnology. In: Innovations in Energy Efficient Construction through Sustainable Materials. IGI Global, pp. 175–206.

Denning, G., 2023. Universal Food Security: How to End Hunger while Protecting the Planet. Columbia University Press.

Dey, P., Datta, D., Pattnaik, D., Dash, D., Saha, D., Panda, D., Bhatta, B.B., Parida, S., Mishra, U.N., Chauhan, J., Pandey, H., 2022. Physiological, biochemical, and molecular adaptation mechanisms of photosynthesis and respiration under challenging environments. In: Plant Perspectives to Global Climate Changes. Academic Press, pp. 79–100.

El-Desouki, Z., Xia, H., Abouseif, Y., Cong, M., Zhang, M., Riaz, M., Moustafa-Farag, M., Jiang, C., 2024. Improved chlorophyll fluorescence, photosynthetic rate, and plant growth of Brassica napus L. after co-application of biochar and phosphorus fertilizer in acidic soil. J. Plant Nutr. Soil Sci. 187 (2), 260–273.

Elhefnawy, S.M., Elsheery, N.I., 2023. Use of nanoparticles in improving photosynthesis in crop plants under stress. In: Photosynthesis: From Plants to Nanomaterials.

Nanomaterial-Plant Interactions. Academic Press, pp. 105–135.

Esgici, R., Elicin, A.K., Albalawi, T., Azzam, M.M., Faizan, M., Alam, P., 2026. Nano-biochar regulates soil enzymatic activities and nitric oxide signaling to improve photosynthesis and nutrient balance in Brassica napus under salt stress: implications for precision agriculture. Plant Physiol. Biochem. 235, 111365.

Faisal, M., Faizan, M., Soysal, S., Alatar, A.A., 2024. Synergistic application of melatonin and silicon oxide nanoparticles modulates reactive oxygen species generation and the antioxidant defense system: a strategy for cadmium tolerance in rice. Front. Plant Sci. 15, 1484600.

Faizan, M., Karabulut, F., Alam, P., Yusuf, M., Tonny, S.H., Adil, M.F., Sehar, S., Ahmed, S.M., Hayat, S., 2023. Nanobionics: a sustainable agricultural approach towards understanding plant response to heavy metals, drought, and salt stress. Nanomaterials 13, 974.

Faraz, A., Faizan, M., Sami, F., Siddiqui, H., Pichtel, J., Hayat, S., 2019. Nanoparticles: biosynthesis, translocation and role in plant metabolism. IET Nanobiotechnol. 13 (4), 345–352.

Garcia, A., Gaju, O., Bowerman, A.F., Buck, S.A., Evans, J.R., Furbank, R.T., Gilliham, M., Millar, A.H., Pogson, B.J., Reynolds, M.P., Ruan, Y.L., 2023. Enhancing crop yields through improvements in the efficiency of photosynthesis and respiration. New Phytol. 237 (1), 60–77.

Gautam, K., Singh, H., Sinha, A.K., 2025. Nanotechnology in plant nanobionics: mechanisms, applications, and future perspectives. Adv. Biol. 9 (4), 2400589.

Ghosh, A., Kumar, A., Biswas, G., 2024. Exponential population growth and global food security: challenges and alternatives. In: Bioremediation of Emerging Contaminants from Soils. Elsevier, pp. 1–20.

Gilmore, D., 2019. Biochemistry and Molecular Biology of Plants. Scientific e-Resources.

Haeder, D.P., 2022. Photosynthesis in plants and algae. Anticancer Res. 42 (10), 5035–5041.

Jahan, U., Kafeel, U., Naikoo, M.I., Khan, F.A., 2022a. Nanoparticles and their effects on growth, yield, and crop quality cultivated under polluted soil. In: The Role of Nanoparticles in Plant Nutrition under Soil Pollution: Nanoscience in Nutrient Use Efficiency. Springer International Publishing, Cham, pp. 333–352.

Jahan, U., Kafeel, U., Khan, F.A., Jahan, A., 2022b. Reactive nitrogen species in plants. In: Advances in Plant Nitrogen Metabolism. CRC Press, pp. 203–212.

Jahan, U., Kafeel, U., Naikoo, M.I., Kaifiyan, M., Hasan, M., Khan, F.A., 2023. Trophic transfer, bioaccumulation, and detoxification of lead and zinc via sewage sludge applied soil-barley-aphid-ladybird food chain. Water, Air, Soil Pollut. 234 (8), 508.

Kafeel, U., Jahan, U., Raghib, F., Khan, F.A., 2022a. Global importance and cycling of nanoparticles. In: The Role of Nanoparticles in Plant Nutrition under Soil Pollution: Nanoscience in Nutrient Use Efficiency. Springer International Publishing, Cham, pp. 1–20.

Kafeel, U., Jahan, U., Khan, F.A., 2022b. Response of different wheat cultivars (Triticum aestivum L.) to biotic stress. J. Himalayan Ecol. Sustain. Dev. 17, 1–10.

Kafeel, U., Jahan, U., Khan, F.A., 2023. Role of mineral nutrients in biological nitrogen fixation. In: Sustainable Plant Nutrition. Academic Press, pp. 87–106.

Kafeel, U., 2025. Halophyte biotechnology for environmental restoration. In: Sustainable Utilisation and Bioengineering of Halophytes. Springer Nature Singapore, Singapore, pp. 57–75.

Kaiser, E., Morales, A., Harbinson, J., Kromdijk, J., Heuvelink, E., Marcelis, L.F., 2015. Dynamic photosynthesis in different environmental conditions. J. Exp. Bot. 66 (9), 2415–2426.

Khatri, P., Kumar, P., Shakya, K.S., Kirlas, M.C., Tiwari, K. K., 2024. Understanding the intertwined nature of rising multiple risks in modern agriculture and food system. Environ. Dev. Sustain. 26 (9), 24107–24150.

Knoflacher, M., 2024. Biological energy transformation. In: Relativity of Evolution. Springer Berlin Heidelberg, Berlin, Heidelberg, pp. 327–426.

Kulsoom, U., Naushin, F., Hasan, M., Kafeel, U., Kumar, S., Khan, F.A., 2024. Effects of As, Cd, Hg and Pb mixture on growth and yield of fenugreek (*Trigonella foenum-graecum* L.) in a semi-hydroponic system. Res. J. Agric. Sci. 15 (6), 1318–1326.

Kumar, A., 2025. Engineered nanomaterials and associated threats in the environment risk assessment strategies. In: Biotechnological Interventions in the Removal of Emerging Pollutants. Springer Nature, Singapore, Singapore, pp. 515–537.

Lee, G., Wei, Q., Zhu, Y., 2021. Emerging wearable sensors for plant health monitoring. Adv. Funct. Mater. 31 (52), 2106475.

Li, G., Chen, T., Feng, B., Peng, S., Tao, L., Fu, G., 2021. Respiration, rather than photosynthesis, determines rice yield loss under moderate high-temperature conditions. Front. Plant Sci. 12, 678653.

Li, P., Xia, Y., Song, K., Liu, D., 2024. The impact of nanomaterials on photosynthesis and antioxidant mechanisms in gramineae plants: research progress and future prospects. Plants 13 (7), 984.

Li, C., Du, X., Liu, C., 2025a. Enhancing crop yields to ensure food security by optimizing photosynthesis. J. Genet. Genom. 52 (9), 1082–1095.

Li, W., Luo, Y., Jiang, P., Dong, X., Tang, K., Liang, Z., Shi, Y., 2025b. A sustainable crop protection through integrated technologies: UAV-based detection, real-time pesticide mixing, and adaptive spraying. Sci. Rep. 15 (1), 35748.

Mony, C., Kaur, P., Rookes, J.E., Callahan, D.L., Eswaran, S. V., Yang, W., Manna, P.K., 2022. Nanomaterials for enhancing photosynthesis: interaction with plant photosystems and scope of nanobionics in agriculture. Environ. Sci. Nano 9 (10), 3659–3683.

Naikoo, M.I., Kafeel, U., Naushin, F., Khan, F.A., 2020. Halophytes in India and their role in phytoremediation. In: Handbook of Halophytes: From Molecules to Ecosystems Towards Biosaline Agriculture. Springer International Publishing, Cham, pp. 1–21.

Nayeri, S., Dolatyari, M., Mouladoost, N., Nayeri, S., Zarghami, A., Mirtagioglu, H., Rostami, A., 2023. Ag/ZnO core–shell NPs boost photosynthesis and growth rate in wheat seedlings under simulated full sun spectrum. Sci. Rep. 13 (1), 14385.

Nazeer, A., Ahmad, F., Verma, N., Ahmad, S., 2023. Nanofertilizers in agriculture. In: Targeted Delivery of Nanopesticides and Nanofertilizers in Sustainable Agricultural Farming. Springer Nature Switzerland, Cham, pp. 87–98.

Ozturk, F., Rajput, V.D., Faizan, M., 2025. Sustainable alleviation of cadmium toxicity in soybean through biochar and silicon oxide nanoparticles: insights into photosynthesis and physiological responses. Nanotechnology 36, 425101.

Palit, S., 2018. Recent advances in the application of engineered nanomaterials in the environment industry—a critical overview and a vision for the future. In: Handbook of Nanomaterials for Industrial Applications, pp. 883–893.

Patel, J.K., Patel, A., Bhatia, D., 2021. Introduction to nanomaterials and nanotechnology. In: Emerging Technologies for Nanoparticle Manufacturing. Springer International Publishing, Cham, pp. 3–23.

Popp, J., Pető, K., Nagy, J., 2013. Pesticide productivity and food security. A review. Agron. Sustain. Dev. 33 (1), 243–255.

Qiao, M., Hong, C., Jiao, Y., Hou, S., Gao, H., 2024. Impacts of drought on photosynthesis in major food crops and the related mechanisms of plant responses to drought. Plants 13 (13), 1808.

Rai-Kalal, P., Jajoo, A., 2021. Priming with zinc oxide nanoparticles improve germination and photosynthetic performance in wheat. Plant Physiol. Biochem. 160, 341–351.

Riseh, R.S., Vatankhah, M., Hassanisaadi, M., Tamanadar, E., Skorik, Y., 2024. Nanocomposites in agriculture as pesticides for plant protection: a review. Adv. Nat. Sci. Nanosci. Nanotechnol. 15 (2), 023003.

Saglam, A., Chaerle, L., Van Der Straeten, D., Valcke, R., 2019. Promising monitoring techniques for plant science: thermal and chlorophyll fluorescence imaging. In: Photosynthesis, Productivity and Environmental Stress, pp. 241–266.

Sekar, N., Ramasamy, R.P., 2015. Photosynthetic energy conversion: recent advances and future perspective. Electrochem. Soc. Interface 24 (3), 67.

Sharma, A., Kumar, V., Shahzad, B., Ramakrishnan, M., Singh Sidhu, G.P., Bali, A.S., Handa, N., Kapoor, D., Yadav, P., Khanna, K., Bakshi, P., 2020. Photosynthetic response of plants under different abiotic stresses: a review. J. Plant Growth Regul. 39 (2), 509–531.

Sharma, P., Janaagal, M., Sheoran, A.R., Sharma, C., 2024. Nanoparticle innovations in plant systems: enhancing photosynthesis and nutrient dynamics. Asian J. Adv. Agric. Res. 24 (12), 75–88.

Singh, A., Shah, K.N., Rana, D.K., Singh, V., 2025. Prospective trajectories in nano-enabled sustainable agronomy and ecological resilience. In: Eco-Friendly Nanotechnology: Harnessing Small-Scale Technologies for a Cleaner and Healthier Planet, pp. 199–212.

Sreeharsha, R.V., Venkata, S.M., 2024. Photosynthetic microbes: evolution, classification, and structural physiology. In: Microbial Photosynthesis: From Basic Biology to Artificial Cell Factories and Industrial Applications. Springer Nature Singapore, Singapore, pp. 3–22.

Striker, G.G., 2023. An overview of oxygen transport in plants: diffusion and convection. Plant Biol. 25 (6), 842–847.

Sultan, H., Li, Y., Ahmed, W., Yixue, M., Shah, A., Faizan, M., Ahmad, A., Abbas, M., Nie, L., Khan, M.N., 2024. Biochar and Nano Biochar: enhancing salt resilience in plants and soil while mitigating greenhouse gas emission: a comprehensive review. J. Environ. Manag. 355, 120448.

Supriya, A.B.S., Kumar, P., Tiwari, A., 2024. Dynamic Role of Innovation in Enhancing Productivity and Ensuring Food Security, p. 1.

Urban, L., Aarrouf, J., Bidel, L.P., 2017. Assessing the effects of water deficit on photosynthesis using parameters derived from measurements of leaf gas exchange and of chlorophyll a fluorescence. Front. Plant Sci. 8, 2068.

Wu, H., Bose, J., 2024. Abiotic stress tolerance: adaptations, mechanisms, and new techniques. Crop J. 12 (5), 1271–1273.

Xu, J., Cui, Y., Wang, J., Liu, Z., Zhao, C., Liu, Z., Niu, N., Chen, L., Zhao, X., Fu, Y., 2025. Advances of nano-photo-fertilizers in plant research. Nanoscale 17 (27), 16153–16192.

Xu, L., Tan, X., He, Z.H., Hao, L., Wang, W., Liu, Z.T., Robertson, A.W., Sun, Z., 2024. Emerging green catalytic synthesis of biomolecules from CO_2 and/or nitrogenous small molecules. Matter 7 (1), 59–81.

Yang, S., Ulhassan, Z., Shah, A.M., Khan, A.R., Azhar, W., Hamid, Y., Hussain, S., Sheteiwy, M.S., Salam, A.,

Zhou, W., 2021. Salicylic acid underpins silicon in ameliorating chromium toxicity in rice by modulating antioxidant defense, ion homeostasis and cellular ultrastructure. Plant Physiol. Biochem. 166, 1001–1013.

Zeng, Y., Molnárová, M., Motola, M., 2024. Metallic nanoparticles and photosynthesis organisms: comprehensive review from the ecological perspective. J. Environ. Manag. 358, 120858.

Zhang, H., Li, G., 2025. Microbial strategies for enhancing nickel nanoparticle detoxification in plants to mitigate heavy metal stress. Phyton 94 (5), 1367–1399.

Zhu, J., Cai, Y., Wakisaka, M., Yang, Z., Yin, Y., Fang, W., Xu, Y., Omura, T., Yu, R., Zheng, A.L.T., 2023. Mitigation of oxidative stress damage caused by abiotic stress to improve biomass yield of microalgae: a review. Sci. Total Environ. 896, 165200.

CHAPTER

2

Introduction to nano-harvest: Bridging nanotechnology and photosynthesis innovations

Priyanka Devi[1] *and Monika Sindhu*[2]

[1]Department of Research and Development Cell, Lovely Professional University, Phagwara, Punjab, India; [2]University Centre for Research and Development, Chandigarh University, Mohali, Punjab, India

1. Introduction

Nano-Harvest is a transdisciplinary philosophy proposing the intentional combination of engineered nanomaterials, photosynthetic systems, and artificial photosynthetic systems to increase the performance of biological and hybrid systems driven by sunlight. Instead of addressing nanomaterials as passive additives, Nano-Harvest focuses on engineered nano-bio interfaces that interface optical, electronic, catalytic, and biochemical functionality of nanomaterials to photosynthetic machinery, chloroplasts, thylakoid membranes, reaction centers, and metabolic networks to enhance usable photon capture, induce desirable electron fluxes, and channel reducing equivalents to desired biosynthetic or catalytic pathways (Liu et al., 2024a; Safdar et al., 2022)

The concept is developed on two observations. To begin with, natural photosynthesis is a highly developed but fundamentally restricted process at the field scale by spectral incompatibilities, recombination losses, photorespiration, and metabolic bottlenecks, all of which limit solar-to-biomass conversion (Liu et al., 2024a). Second, engineered nanomaterials exhibit size-dependent absorption and emission, plasmonic field enhancement, high conductivity, and catalytic activity, all of which are tunable properties that can be mapped directly to the physical/chemical processes of light capture, charge separation, and catalytic conversion (Andleeb, 2025; Donahue, 2025; Donahue et al., 2025). Nano-Harvest aims to create systems that maintain biological flexibility by selectively aligning nanomaterial activity with particular photosynthetic bottlenecks and gaining the nanoscale benefits of photonics and catalysis.

1.1 Importance of nanoscale strategy to photosynthesis

Photosynthetic conversion of energy- Photon capture by pigment arrays, ultrafast excitation

Nanotechnology in Photosynthesis
https://doi.org/10.1016/B978-0-443-44376-3.00001-X

transfer to reaction centers, charge separation, eventual CO 2 assimilation, and electron transport- The photosynthetic sequence of steps at nanometer and femto-to picosecond timescales. This natural nanoscale of operation renders photosynthetic systems exceptionally responsive to commensurate nanoscale engineering: slight variations in distance, dielectric environment, or local electromagnetic field can change excitation lifetimes, energy transfer efficiencies, and electron tunneling probabilities. Quantum dots or carbon quantum dots (CQDs) can be designed to absorb wavelengths where the solar spectrum would otherwise not be utilized and reemit at wavelengths that correspond to chlorophyll absorption bands, broadening the usable solar spectrum (Andleeb, 2025; Haider et al., 2024). Plasmonic nanoparticles focus optical energy and can be used to increase the photon density around reaction centers (Donahue, 2025; Donahue et al., 2025). Photogenerated electrons may be conducted through nanowires and carbon nanotubes (CNTs) with low resistance, reducing recombination and enabling faster transfer to downstream acceptors (Safdar et al., 2022). Nanomaterials can be tailored to accumulate at leaf surfaces, cross stomatal or cuticular barriers, or even localize to chloroplast membranes-bringing functionality where it is most useful with reduced off-target interactions because they can be functionalized (e.g., with targeting peptides, DNA, polymer coatings, etc.) (Djanaguiraman et al., 2024) as shown in Fig. 2.1.

1.2 Major objectives and transtheoretical directions

Nano-Harvest has three overlapping objectives:

(i) enhance biological photosynthesis to increase crop productivity, stress resistance, or bioproduct output;
(ii) design biohybrid and artificial photosynthetic reactors that integrate life

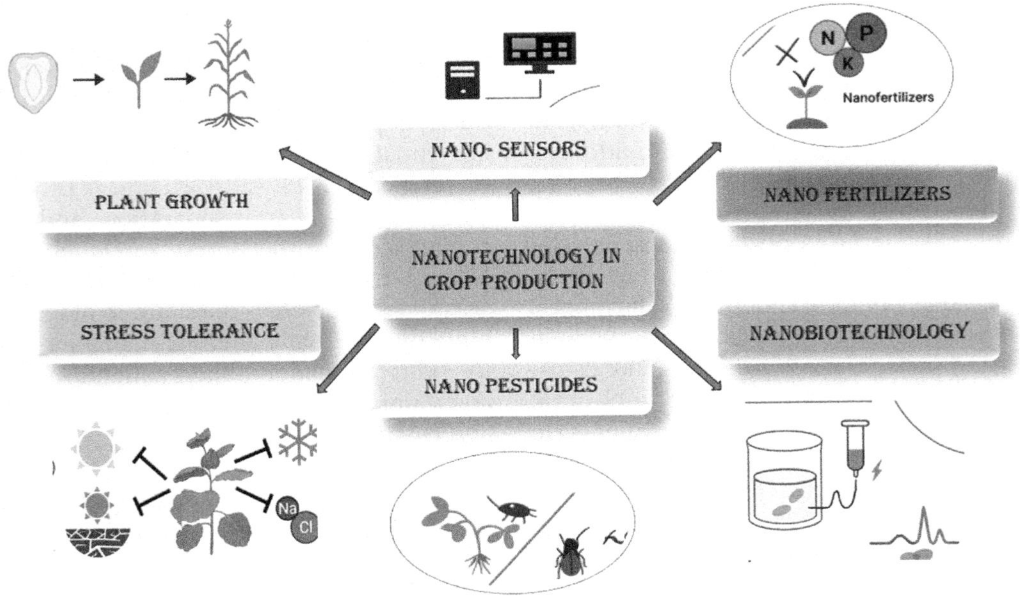

FIGURE 2.1 Role of nanotechnology in crop production.

or biomimetic components with nanomaterial light absorbers and solar fuels or chemical catalysts

(iii) enable precision agronomic interventions, nano-encapsulated nutrient delivery, photonic seed treatments, or foliar photonic sprays, that could minimize input waste and environmental leakage

These objectives can be translatedally subdivided into viable application pathways: (a) in field crops, these nanocatalysts could be applied either foliarly or rootly to enhance chlorophyll content, stomatal activity, and nutrient use efficiency (Andleeb, 2025; Khundi, 2025); (b) in model plants and algae, spectral responsiveness or enhanced electron extraction could be achieved by organelle-targeted assemblies (e.g., DNA-coated CNTs) (Djanaguiraman et al., 2024; Sidhu et al., 2024) (Table 2.1).

2. Anatomy of photosynthesis and light harvesting

Photosynthesis is the mechanism by which photosynthetic organisms absorb solar energy and convert it into chemical energy to facilitate carbon fixation and biomass production. The very essence of this process is that it depends on several orchestrated processes: light absorption by pigment-protein complexes, excitation energy transfer (EET) to reaction centers (RCs), charge separation and the creation of reducing equivalents, electron transport, generation of proton motive force, and further carbon dioxide (CO_2) assimilation through the Calvin Benson Bassham (CBB) cycle (Iwai et al., 2024). Being elegantly effective at the molecular scale, the total solar-to-biomass efficiency of most crops at conventional scales (to whole-plant and field scales) does not scale beyond three to six per cent (perspective suggests this limit is one of the main factors behind yield limits) (Maity and Kleinekathöfer, 2023; Iwai et al., 2024). Knowing these mechanistic layers at the horizon of Nano-Harvest is important, since every step offers opportunities for nano-intervention and bottlenecks.

Photon capture, the first step, is mediated by light-harvesting complexes (LHCs), which contain chlorophylls, carotenoids, and related proteins. They are located in the thylakoid of chloroplasts in plants (and similar systems in algae/cyanobacteria). Excitonic energy in the

TABLE 2.1 Goals, examples, and main risks of nano-harvest.

Nano-harvest goal	Representative nanomaterials/ approach	Example benefit (reported)	Main risk
Broaden spectral capture	Carbon quantum dots, QDs	Increased chlorophyll & stomatal conductance (Andleeb, 2025; Haider et al., 2024)	Dose-dependent phytotoxicity (Wang et al., 2025)
Improve electron extraction	DNA-coated CNTs, conductive nanowires	Enhanced electron capture in chloroplasts (Safdar et al., 2022)	Potential membrane disruption; unknown long-term effects (Djanaguiraman et al., 2024)
Controlled nutrient delivery	Nanofertilizers (polymeric carriers)	Improved nutrient use efficiency (Ayenew et al., 2025; Demeke et al., 2025)	Soil accumulation; regulatory gaps (Sidhu et al., 2024)
Artificial CO_2 conversion	Nanocatalyst–enzyme biohybrids	Lab CO_2 → organics conversion (Mori et al., 2025)	Scalability and catalyst stability (Mori et al., 2025)

form of electron-hole pairs (excitons) generated by photons absorbed by pigments migrates to RCs via EET and is converted to chemical potential via a charge-separation process. There, the electrons pass through a series of redox carriers (such as quinones, iron-sulfur clusters, and plastocyanin) of the photosynthetic electron transport chain (PETC). This leads to the production of ATP, with the redundant equivalents (e.g., NADPH) used to trap CO_2 in organic matter.

2.1 Structure and function of light-harvesting complexes

Light-harvesting complexes are multi-molecule complexes that assemble into the light-independent pigment-binding thylakoid membrane (TM), any of which can be disturbed by photons of light:

✓ The antenna system of photosynthesis is light-harvesting complexes: these are structures that accumulate photons and rapidly and efficiently (subpicosecond to picosecond) channel excitonic energy to the reaction center (RC) and allow quantum efficiencies of almost one under favorable conditions (Fiebig et al., 2023). In their recent study of the structure and function of LH2/LH3 proteins in bacterial membranes, the group of Schlau-Cohen at MIT has shown that a somewhat disordered arrangement of the proteins can enable long-distance energy transfer, indicating that structural flexibility, rather than rigid ordering, can optimize exciton migration (MIT News, 2023).

✓ Particularly, in the closer proximity of proteins, the study revealed times of energy transfer to be about 6 ps compared to about 15 ps in more remote cases, which confirmed that nanoscale distances (less than 3 nm) are crucial to energy transfer efficiency in proteins. The larger antenna complexes (including LHCII) of eukaryotic plants and algae adopt oligomeric geometries (e.g., trimers), the geometry of which affects energy acquisition efficiency and photoprotective reactions. Indicatively, a study on pea (Pisum sativum L.) revealed that the greater the LHCII oligomerization, the higher the electron transport efficiency, and the photosynthetic performance index (PIABS, P Itotal) (Turner, 2023).

✓ Structurally, it has been shown that atomic and multiscale modeling has shown key parameters, such as site energies, excitonic coupling and spectral densities define the ultrafast exciton dynamics; a recent review of the state of modeling of LHCs, by Maity and Kleinekathofer (2023) highlights the importance of accurate models of the protein-pigment scaffold in understanding the way exciton migration can be manipulated by nanoscale insertion or modification.

✓ Photoprotection is also found in LHCs: when the surroundings are bright, the antenna complexes are switched to dissipative states by NPQ mechanisms, hence preventing the photodamage of RCs. Review of literature Maity and colleagues describe the influence of pH, salt concentrations, and the protein PsbS on these transitions (Maity and Kleinekathöfer, 2023). The light-capture and photoprotection regimes are therefore crucial to Nano-Harvest strategies; one does not wish to increase absorption without considering damage mitigation.

2.2 Natural photosynthetic efficiency limitations

Whereas molecular quantum efficiencies can be close to unity, especially under controlled conditions, the overall process (photon collection, conversion to biomass), is much lower.

Key limitations include:

(a) Spectral Limitations and Absorption. Photosynthetic pigments (carotenoids, chlorophyll *a*/b) absorb part of the solar spectrum. A lot of near-infrared (NIR) and ultraviolet (UV) light is either unused or wasted. The structural study of eukaryotic LHC proteins shows that there are intrinsic limits on capture cross-sections (Iwai et al., 2024).

(b) Charge Separation Losses/Exciton Transfer. The exciton migration may be slowed by distance, orientation, and dielectric heterogeneity, or recombination may be enhanced. In point of fact, the MIT work (2023) suggests that the spacing between proteins is a major contributor to transfer time and, consequently, efficiency.

(c) Dissipation under photoprotection. At photon fluxes exceeding the metabolic sink capacity, NPQ dissipates excess energy as heat, minimizing the conversion of useful energy.

(d) Electron Transport Bottlenecks. Charge or back-electron recombination is an energy-loss mechanism.

(e) Metabolic/Biochemical Constraints. Rubisco (ribulose-1,5-bisphosphate carboxylase/oxygenase) is a relatively sluggish enzyme that can catalyze O_2 in place of CO_2, causing photorespiration; this is especially constrained in C3 crops under warmer or higher light conditions (Wu and Huang, 2024). Studies underline that enhancing photosynthesis to boost crop production should be at different levels-starting with pigment activity, all the way to the activity of enzymes (Wu and Huang, 2024).

(f) Supply of CO_2/Stomata/Diffusion Mesophyte Limits. Assimilation rates are also limited by the supply of CO_2 to Rubisco and by diffusive constraints in alveoli.

(g) Effects of Scaling & Canopy Architecture. Most optimisation strategies can be leaf-level efficient without considering canopy, seasonal, and whole-plant trade-offs.

Nano-Harvest:

✓ To the Nano-Harvest agenda, the mechanistic understanding implies that there are three primary loci of application:

✓ The enhancement of photon capture—e.g., nanoparticles, which extend absorption into NIR, or which tune antenna orientation.

✓ Plasma/excitation control—e.g., nanowires or conductive particles to speed up electron recovery, slow recombination.

✓ Metabolic or diffusion support—e.g., targeted nutrient delivery, CO_2-concentrating nano-systems, protective nanomaterials inhibiting ROS or thermal/photodamage.

All these need to work together in synergy, since upstream bottlenecks can be improved, and downstream bottlenecks that are not removed can result in diminishing individual returns or even maladaptive responses (i.e., more absorption without increased assimilation, leading to photodamage). Nano-Harvest thus stresses nanobi design rather than ad hoc nano-particle addition. Lastly, the field should also consider dynamic environments (changes in light, temperature, and water stress) and biological mechanisms (photoprotection, reorganisation of the antenna). PH, salts, and PsbS-mediated NPQ studies of LHC dynamics indicate that the entire antenna system is a responsive structure rather than a static one.

3. Definitions and classification of nanomaterials

Nanotechnology is the management and application of materials at the nanoscale, characterized by dimensions of 1–100 nm in one or more dimensions, where individual physical, chemical, and biological characteristics emerge (Ashfaq et al., 2025). Nanomaterials related to the nano-harvest paradigm in the context of agriculture and photosynthetic systems can be semiconductor quantum dots (QDs), carbon quantum dots (CQDs), carbon nanotubes (CNTs), and graphene derivatives, metal and metal-oxide nanoparticles (e.g., Au, Ag, TiO_2, and CeO_2), and plasmonic nanostructures (Donahue, 2025; Ashfaq et al., 2025). These materials are likely to include inorganic nanomaterials (metal/metal oxides, ceramics), organic nanomaterials (polymer- and lipid-based), and carbon-based nanostructures (CNTs, graphene, and CQDs). Table 2.2 gives a representative classification. The chemistry, size distribution, surface functionalization, solubility, and agglomeration behavior of these materials are highly diverse and can affect their interactions with biological systems and, therefore, their efficacy or hazard (Zhang and Su, 2024; Djanaguiraman et al., 2024), as shown in Fig. 2.2.

3.1 The nanomaterials are in contact with the biological components, specifically the biological cells

In the case of nanoharvesting, it is important to understand how nanomaterials interact with plant systems. The interactions range from nano-material uptake at leaf or root surfaces, transit through tissues (apoplastic and symplastic routes), and organelle (chloroplast) association to ultimate fate and transformation (Djanaguiraman et al., 2024).

A recent literature review notes that, in terrestrial plants, the uptake and translocation of nanoparticles are strongly influenced by particle size, surface charge, coating, plant species, and exposure route (Zhang and Su, 2024):

✓**Routes of uptake and entry:** Nanoparticles can enter plant leaves through stomatal pores, cuticular cracks, or hydathodes; they can be taken up by roots through root hairs, apoplastic bypass, or endocytosis (Djanaguiraman et al., 2024). Translocation within the organism occurs through the xylem (water transport), the phloem (sugar/solute transport), or both, depending on the size and solubility of the nanomaterial (Zhang and Su, 2024).

✓**Internal interactions:** When nanomaterials impregnate the plant cell, they interact with plant cell walls, membranes, and organelles. As an illustration, CNTs have the ability to interact with chloroplast membranes and to regulate electron-transfer pathways (Safdar et al., 2022) as shown in Table 2.2.

TABLE 2.2 Key nanomaterials properties influencing plant interaction.

Property	Effect on plant uptake/fate	Design implication
Size (nm)	Smaller sizes (<20–30 nm) cross cell walls and membranes more readily; larger sizes may be retained or aggregated.	Synthesis and size-fractionation are critical.
Surface charge/coating	Positive charges often enhance interaction with negatively-charged cell walls; coating affects biocompatibility and aggregation.	Use biocompatible coatings and reduce aggregation.
Surface functionalization	Targeting ligands or biomolecules alters localization (e.g., chloroplast targeting)	Custom ligand design
Agglomeration/solubility	Aggregates reduce mobility, change deposition	Stabilizing agents needed
Chemical reactivity/dissolution	Metal ions may leach or catalyze ROS generation	Choose an appropriate stabilized form

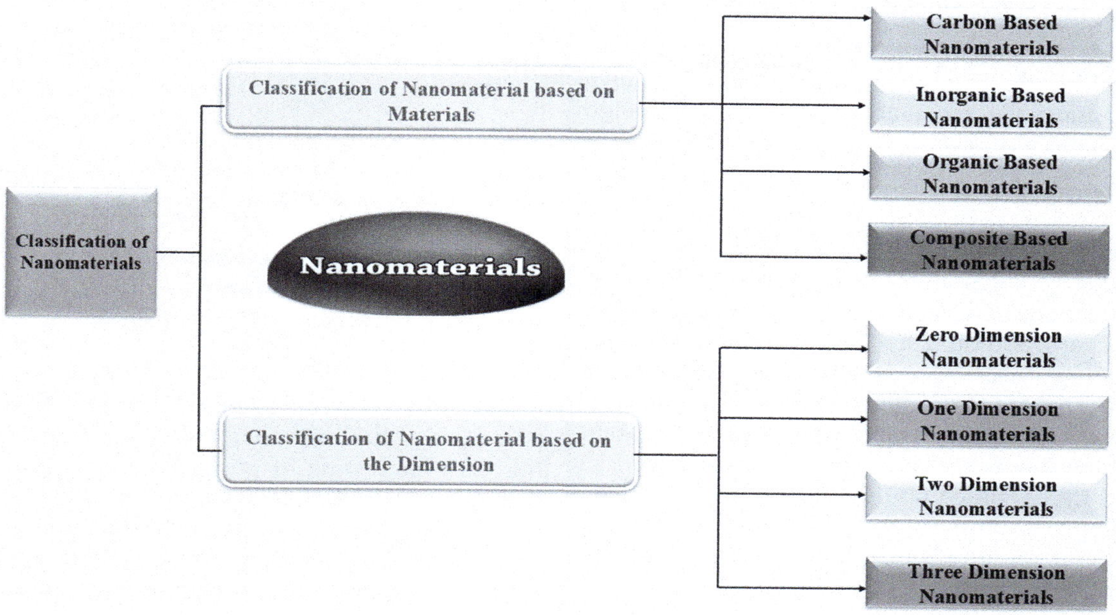

FIGURE 2.2 Different classification of nanomaterial.

3.2 Nano-bio interface in plants mechanisms

The nano-bio interface is the interface where nanomaterials are engineered to interact with biological systems, including physical, chemical, optical, and biological processes. In the case of nano-harvest, three mechanistic pathways are particularly important: optical/photonic coupling, electronic/charge coupling, and catalytic/biochemical coupling.

Optical/photonic coupling: The nanomaterials can alter the behavior of light within plant tissues. for example, plasmonic nanoparticles focus electromagnetic fields and increase the photon density around pigment complexes (Donahue, 2025; Donahue et al., 2025). Quantum dots, or carbon quantum dots, can absorb wavelengths outside the natural pigment range (such as NIR or blue-UV) and reemit photons at wavelengths optimally absorbed by chlorophyll, thereby extending the range of proper sunlight (Andleeb, 2025). This could increase the photon flux to reaction centers, enhancing light-harvesting efficiency.

Electronic/charge coupling: Once photon capture and charge separation have occurred, efficient electron transport is needed. Nanomaterials can serve as electron conductors (e.g., carbon nanotubes or conductive nanowires), which can transfer electrons generated by photon absorption to the PETC more efficiently or to a preengineered acceptor to minimize recombination losses (Safdar et al., 2022). Lab-scale demonstrations have shown that CNTs can be incorporated into chloroplasts to modify electron flow pathways. The main difficulty is ensuring biocompatibility and stable insertion without interfering with normal organelle functions (Gowtham, 2024; Gowtham et al., 2024; Ashfaq et al., 2025).

3.3 Design issues and problems

Effective implementation of nanomaterials in photosynthetic systems cannot merely involve adopting the most advanced. Several design limitations and issues need to be addressed:

✓ **Biocompatibility and toxicity:** It has been observed several times that nanoparticles can cause oxidative stress, alter gene expression, disrupt membranes or organelles, and affect nutrient uptake (Djanaguiraman et al., 2024; Zhang and Su, 2024). It is also imperative to know dose response, long-term effects of exposure, and ecological impacts. As an example, a review has noted that the impacts of nanoparticles on plant physiology are different based on the type of nanoparticles, concentration, route of exposure, and the plant species (Djanaguiraman et al., 2024).

✓ **Stability and fate of the material:** Nanomaterials can aggregate, dissolve into ions, chemically convert in soil or plant tissue, or become trapped in the vacuoles instead of interacting with the photosynthetic apparatus (Zhang and Su, 2024). The main challenge is to ensure the effect is stable enough to achieve the desired outcome while maintaining minimal long-term persistence in the environment.

✓ **Targeting and localization:** Nanomaterials can be required to reach the leaves, chloroplasts, or a particular organelle to affect the photosystems. Passive foliar sprays can lead to either low uptake or nonselective distribution. Nanomaterials that target organelles (with the help of peptides and ligands) have potential but require advanced design and testing.

✓ **Scale and cost:** Scale nano-bio is being carried out on a lab/greenhouse scale. When lifting to field conditions, light, temperature, soil conditions, biotic stress, and deposition routes are varied. In addition, there are additional constraints on the production costs of nanomaterials, the mode of application, and the regulatory burden.

4. Mechanisms and pathways nano-augmented photosynthesis

4.1 Quantum dots and plasmonic nanoparticles are among the alternatives that have been recommended to enhance the light absorption

One of the key processes of the nano-harvest paradigm is the improvement of light absorption and harvesting of photosynthetic systems through nanomaterials, which are artificial antennas, and to complement the natural pigment arrays. In the conventional photosynthetic systems, the absorption spectrums of the native pigments, primarily the chlorophylls *a*/*b*, carotenoids, and other associated antenna pigments, restrict the range of spectral bands used, including those not well used by a photosynthetic system, for example, one of the blue/UV wavelengths or NIR bands (Liu et al., 2024a,b,c). In response to this, scientists have loaded quantum dots (QDs) or carbon quantum dots (CQDs) into plant leaves or roots that absorb photons at underutilized wavelengths and reemit them at wavelengths that trigger chlorophyll absorption reactions, thereby augmenting the spectrum of usable light (Andleeb, 2025). As an illustration, exposure of maize to fluorescent g-C3N4 nanosheets exhibited a 44.6% enhancement in electron transfer and up-regulation of psbA and psaA genes when used at 10 mg/L-this is evidence of the possibility of these artificial light-harvesting antennas within the plant tissues (Chen et al., 2024).

Another pathway is plasmonic nanoparticles (i.e., gold or silver nanorods). By concentrating the electromagnetic field on pigment complexes (through localized surface plasmon resonance), they increase local photon density and excitation rates. Even though much of this research is limited to isolated pigment-nanoparticle assemblies, the characterization of this technology in whole-plant systems is increasing, owing to enhanced biocompatibility and targeted delivery. In one review, it is stressed that plant nanobiotechnology has the potential to enhance photosynthetic activity by simultaneously targeting light capture, electron transfer, and carbon assimilation (Liu et al., 2024a,b,c; Andleeb, 2025).

4.2 Carbon quantum dots have the potential to be used as internal light-harvesting antennas

Functionalization and specificity: The nano-bio interface can be modified with surface modification (e.g., PEGylation, ligand conjugation, peptide or antibody targeting, etc.): redistribution, organelle targeting, reduced phytotoxicity, and increased functionality (Ashfaq et al., 2025). Specifically, carbon-based nanocarriers are being targeted to be used in the delivery of micronutrients or biomolecules on-demand (Ashfaq et al., 2025). As an example, carbon nanotubes (CNTs) and conductive nanowires have been introduced into the isolated chloroplasts or plant systems to act as a direct electron conduit to minimize recombination and increase throughput (Safdar et al., 2022). In a groundbreaking work, DNA-surfaced CNTs have entered the chloroplast envelope of Arabidopsis thaliana and enhanced the rate of electron transfer and expanded the light response. Although previous studies were on the lab scale, recent reviews show an increase in the translation of crop species (Liu et al., 2024a,b,c). The design factors to consider when making improvements in electron transfer are: the alignment and orientation of nanowires in case of the chloroplast thylakoids, the chemical compatibility of the conductive materials with the aqueous and ionic environment found in the chloroplast, and the need to maintain membrane potential and proton gradient. Also, it should be biocompatible and should not be damaged by photothermal or photochemical effects of metallic materials. According to a recent review, nanomaterials should be assessed based on their influence on the formation of ROS and antioxidant defense (Li et al., 2024).

4.3 Nanointerventions to improve carbon fixation and energy conversion

Although improved light capture and electron transfer increase the upstream potential of photosynthesis, the final performance measure would be the fixation of carbon (CO_2 assimilation) and the downstream conversion of reducing equivalents to sugars, oil, or other biomass/bioproducts. Nanofertilizers and nanocarriers that transport the micronutrients (FE, Zn, Mo) stimulate the chlorophyll production, enzyme cofactors in the Calvin–Benson cycle, and overall assimilation rates (Liu et al., 2024b). As an example, a review concluded that nano-fertilizers improved photosynthetic efficiency, chlorophyll content, biomass, and nutrient uptake in comparison to conventional fertilizers, but the mechanisms by which this is observed remain under study (Liu et al., 2024a, b,c).

Nanocomposites (e.g., chitosan-selenium NPs) enhanced the gas exchange and chlorophyll fluorescence data of leaves under salinity stress, which indicated enhanced net photosynthetic rate and stress tolerance (Saeedi et al., 2024). These studies reveal that the application of nanomaterials can be used to alter the photosynthetic apparatus to use more electrons to convert into carbon assimilation instead of waste or dissipation. Further, although not currently field-ready, hybrid or artificial photosynthesis systems suggest possibilities in the future: nanomaterial light-harvesting material + biological redox enzymes + CO_2-fixing catalysts bring initial success in the field of solar fuels (Mori et al., 2025). In the case of plants, the corresponding would be to couple a higher light capture and electron transfer to metabolic modules that are more effective in converting reducing power to storage compounds or gases.

4.4 Nano-harvest systems integration problems and concerns

Photosynthetic nano-augmentation has a number of caveats and challenges, which need to be overcome to develop the concepts to scale to the field level and to robust systems of agricultural biomass and bioenergy:

✓ **Downstream bottlenecks:** It is of little use to improve light capture and electron transfer when CO_2 assimilation, metabolic conversion, or sink strength (storage capacity) is not improved. Recent review articles focus on the need to see the system as a whole, orchestrating the moderation of a single node can potentially produce a small or even adverse payoff when other nodes are unable to follow suit (Liu et al., 2024a,b,c).

✓ **Environmental and biological variability:** Field conditions cause variability in light, temperature, drought, nutrient conditions, microbiome of the soil, and plant phenology-conditions that can alter nanoparticle uptake, distribution, and effectiveness. The phylogosphere and root microbiome change in the study with ZnO-tea plant (Chen et al., 2024) indicates the importance of ecological interaction and its potential to affect the results.

✓ **Disposition of the material, transformation, and persistence:** Nanomaterials used on plants can accumulate in the soil or on the leaf surfaces, dissolve to form an ionic form, or be trapped in vacuoles-making them less functional. As stated in reviews, the factors of fate and effect depend on NP size, surface coating, and environmental chemistry (Li et al., 2024; Zhang and Su, 2024).

✓ **Dosage and phytotoxicity:** Although, in many instances, low or moderate dosage levels of nanomaterials are beneficial, higher doses can cause oxidative stress, membrane damage, or interfere with nutrient binding (Li et al., 2024). Therefore, it is essential to conduct dose-response and safety profiling.

✓ **Regulatory, cost, and scalability:** On a translational perspective cost of high-quality nanomaterials, their scale production, application techniques (foliar sprays, seed coating, and soil application), regulatory approval of use by agriculture, and potential environmental release are of major concern. There is a need for early life-cycle analysis and techno-economic analysis shown as Table 2.3.

5. Nanotechnology has been used in improving the efficiency of photosynthesis

Nanotechnology has become a revolutionary technology in plant science, which can be used to optimize photosynthesis by providing technology to capture light, convert it into energy, deliver nutrients, and resilience to stress. The concept of Nano-Harvest proposes the use of nanoparticles (NPs) and nanoscale materials to complement natural photosynthetic machineries, to connect physical nanomaterials with the biological complex of light-harvesting. This combination enables more precise regulation of photosynthetic pathways, resulting in higher carbon assimilation and yields (Torney et al., 2023).

Alongside the development and characterization of nanomaterials, researchers can now synthesize nanoparticles that can simulate or aid chloroplast processes, expand the absorption spectrum, or transfer electrons across the thylakoid membrane (Siddiqui et al., 2024). These

TABLE 2.3 Recent nano-augmented photosynthetic interventions.

System	Nanomaterial/Approach	Observed benefit	Key References
Tea plants with ZnO NPs	ZnO nanoparticles sprayed on leaves	Increased chlorophyll, upregulated photosynthesis genes, improved shoot development, and phyllosphere microbiome (Chen et al., 2024)	Chen et al. (2024)
Wheat with Fe_2O_3 NPs	Iron-oxide nanoparticles applied to roots	Increased chlorophyll and biomass under controlled conditions (Li et al., 2024)	Li et al. (2024)
Citrus seedlings under salinity	Chitosan–selenium nanocomposite seed/foliar treat	Improved gas exchange, chlorophyll fluorescence, and photosynthetic performance (Saeedi et al., 2024)	Saeedi et al. (2024)
Maize with g-C_3N_4 nanosheets	Fluorescent nanosheets applied to roots	Up to 44.6% increase in electron transfer rate, 51.8% increase in chlorophyll content, up-regulation of psbA/psaA genes	Hu et al. (2023)

innovations will have a paradigm shift with the traditional biochemical or genetic methods, where an engineering dimension will be added in the optimization of photosynthesis.

5.1 Nanomaterials toward improved light capture

Photosynthetic exploitation of solar energy is limited by the limited spectral absorption range of chlorophyll pigments (including 400,700 nm). Nanotechnology offers several solutions to prolong or enhance light absorption:

(a) Plasmonic Nanoparticles
Surface plasmon resonance (SPR) Nanoparticles are surface oscillations of electrons that increase local electromagnetic fields, especially metallic particles, such as gold (Au), silver (Ag), and aluminum (Al). These nanoparticles increase photon uptake and energy flow to the chloroplasts when added to plant tissues or used as foliar sprays (Zhou et al., 2024). For example, spinach treated with AuNPs showed a 30% increase in chlorophyll fluorescence and a higher quantum yield because AuNPs increased local light intensity (Singh et al., 2024).

(b) Upconversion Nanoparticles (UCNPs)
There are various methods of synthesizing Upconversion Nanoparticles (UCNPs). UCNPs transform near-infrared (NIR) light into visible wavelengths that apply to the chlorophyll pigments. Nanoparticles, including $NaYF_4:Yb^{3+}$, Er^{3+}, have been used on leaves to facilitate NIR-driven photosynthesis, thereby expanding the usable solar spectrum (Zhang et al., 2023a,b). The method has the potential to improve total photon use efficiency by 1015% when sunlight is available, with NIR accounting for close to half of the overall irradiance.

(c) Nanomaterials based on Carbon.
Carbon nanotubes (CNTs) and graphene quantum dots (GQDs) exhibit unique photothermal and electron-transport properties. Their integration within chloroplasts maximizes charge separation and minimizes recombination losses (Tripathi et al., 2024). Experimental systems have shown that single-walled CNTs enhance chloroplast ATP production by 49%, demonstrating the ability of nanostructured carbon to affect photophosphorylation directly (Kwak et al., 2023).

5.2 Nanotechnology of efficient electron transport and separation of charges

The speed of photosynthesis, in large part, relies on the rapid transfer of photoexcited electrons via photosystem II (PSII) and photosystem I (PSI). It is possible to design nanomaterials that do not hinder electron flow, minimize recombination, and act as redox mediators:

(a) Nanoparticles of Semiconductor.
Artificial electron sinks or conduits are semiconducting materials like TiO_2, ZnO, and CdS nanoparticles, which enhance the separation of charge carriers. In isolated chloroplasts in which TiO_2 nanoparticles were incorporated, the rate of electron transport and oxygen evolution was significantly increased (Das et al., 2023). TiO_2 will serve as an electron mediator to suppress back-electron transfer, hence maintaining high photochemical efficiency.

(b) Quantum Dots (QDs)
Quantum dots (e.g., CdSe, PbS, $CuInS_2$) have size-tunable band gaps, allowing them to be accurately tuned to match chlorophyll absorption bands. This helps in the efficient transfer of energy to the reaction center (RC). The artificial antenna systems can also be used with QDs, which take in a wide range of light and release it at photosynthetically active wavelengths (Li et al., 2023).

(c) Metal Oxide Nanostructures.
Fe_2O_3 and MnO_2 metal oxides in nanoparticles can be used as redox catalysts during electron transport chains in photosynthesis. For example, MnO_2 nanoclusters can mimic the natural oxygen-evolving complex (OEC) of PSII and promote water-splitting reactions (Yadav and Sharma, 2023). The damaged PSII units under stress can eventually be replaced or reinforced by such biomimetic systems.

5.3 Nanocarriers with a targeted delivery and CO_2 enrichment

In addition to light harvesting and electron carriers, nanotechnology aids the delivery of nutrients, CO_2, and biomolecules to the photosynthetic organelles with ease, as shown in Table 2.4:

TABLE 2.4 Nanomaterials and their roles in enhancing photosynthesis.

Nanomaterial type	Mechanism of Action	Photosynthetic target	Reported outcome	References
Gold nanoparticles (AuNPs)	Plasmonic field enhancement	Light absorption	↑ Quantum yield by 30%	Singh et al. (2024)
TiO_2 nanoparticles	Electron mediation	PSII–PSI chain	↑ Electron transport rate	Das et al. (2023)
Upconversion nanoparticles ($NaYF_4$:Yb,Er)	NIR → visible light conversion	Light harvesting	+15% light-use efficiency	Zhang et al., 2023a,b
Carbon nanotubes (SWCNTs)	Charge transport facilitation	Thylakoid membranes	+49% ATP synthesis	Kwak et al. (2023)
Cerium oxide nanoparticles (CeO_2)	ROS scavenging	Chloroplast redox balance	Improved drought tolerance	Kumar et al., 2024a,b
MOF-based nanocarriers	CO_2 enrichment	Rubisco activity	↓ Photorespiration	Ramasamy et al. (2023)

(a) Nanocarriers Nutrient and Hormone Delivery.
Nanoparticles such as silica and chitosan are typically used as nanocarriers for micronutrients (Fe, Zn, Mg) or growth regulators (cytokinins and gibberellins). They are optimally released in chloroplasts because they are controlled, thereby improving chlorophyll production and enzyme activity (Patra et al., 2024).

(b) Carbon dioxide Nanocapture Systems.
To enhance local CO_2 enrichment near Rubisco enzyme locations, nanoengineered CO_2 delivery systems have been developed using mesoporous materials, such as MOFs (metal-organic frameworks) (Ramasamy et al., 2023). This method will decrease photorespiration and increase the effectiveness of carbon fixation, especially in C_3 plants like wheat and rice.

(c) Nanoenzymes (Artificial Catalysts)
Nanozymes are nanomaterials that mimic Rubisco or superoxide dismutase (SOD) functions by being engineered to behave like enzymes. An example is cerium oxide nanoparticles (CeO_2 NPs), which can switch between Ce^{3+} and Ce^{4+} oxidation states and, therefore, eliminate reactive oxygen species (ROS) and balance redox homeostasis in response to high-light stress (Chen et al., 2023). This maintains photosynthetic activity during environmental changes.

5.4 Stress tolerance amelioration through nanotechnology

Photosynthetic efficiency is decreased during abiotic stresses like heat, drought, and toxicity to heavy metals. Nanotechnology would provide measures to alleviate stress and provide a stable photosynthetic performance:

(a) Antioxidant Nanomaterials.
Oxidative damage in chloroplasts is caused by the accumulation of ROS in the presence of stress. Oxidative stress may be alleviated by incorporating antioxidant nanoparticles, such as cerium oxide (CeO_2), zinc oxide (ZnO), and selenium nanoparticles (SeNPs) (Bose et al., 2023). For example, the foliar application of CeO_2 NPs increased the membrane stability index and chlorophyll retention in maize under drought stress by 22% (Kumar et al., 2024a,b).

(b) Stress Protection on Light and Heat.
AuNPs and SiO_2 nanoparticles are reported to increase heat tolerance by stabilizing thylakoid membranes and modulating NPQ (Rahman et al., 2023). On the same note, TiO_2 nanocoatings on leaves act as light filters, minimizing photoinhibition during extreme sunlight (Ali et al., 2024).

(c) Heavy Metal Detoxification.
Nanoparticles such as Fe_3O_4 and biochar-based nanocomposites capture heavy metals, preventing their translocation into chloroplasts and preserving the integrity of photosystems (Adeleye et al., 2023).

5.5 Difficulties and safety hurdles

Although there have been many positive results, there are various challenges to incorporating nanomaterials into photosynthetic systems. They are nanoparticle biocompatibility, cytotoxicity, bioaccumulation, and environmental persistence (Usman et al., 2024). Nanoparticles are absorbed and transported based on their size, charge, and coating, and when they accumulate uncontrolled, they will interfere with the chloroplast ultrastructure. Nanotechnology and photosynthesis integration is on the frontier of agricultural innovations. Improved future studies should be done on bio-compatible, degradable, and precision-targeted nanomaterials. The prospective areas include machine-learning-based nanomaterial design, in vivo imaging of nanobio interactions, and nanosensors to monitor real-time parameters of photosynthesis (Patra et al., 2024).

6. Nano-bio interface and chloroplast engineering

The most crucial aspect of Nano-Harvest is the active interface between nanomaterials and biological systems, where synthetic structures communicate with living chloroplasts, thylakoid membranes, and stromal enzymes. The interaction between nanoparticles (NPs), quantum dots (QDs), and nanocarriers with photosynthetic machinery determines the electron flow, ion balance, and regulation of genes (Sarma et al., 2024). This interface involves the ability to combine knowledge on plant cell biology, surface chemistry, and biophysics to understand. The plant cells, unlike mammalian cells, have a rigid cell wall, plasmodesmata, and several internal membranes. Nanoengineering of chloroplasts must therefore be able to circumvent these barriers safely and efficiently to succeed. Recently, bio-compatible, charge-tunable, and stimulus-responsive nanomaterials have been proposed (which can penetrate cell walls and target chloroplasts) without cytotoxicity (Mahapatra and Gupta, 2024).

6.1 Pathways to plant cells and chloroplasts of nanomaterials

There are two primary routes of entry into chloroplasts by nanoparticles:

Apoplastic and symplastic movement by stomata, cuticle ruptures, or root uptake, and then xylem or phloem movement. Trafficking that makes intracellular translocation possible (through endocytosis or plasmodesmata) (Zhu et al., 2023). Permeability is dependent on size, surface charge, and coating. Usually, NPs less than 50 nm can pass through cell wall pores, whereas positively charged or amphiphilic coatings enhance adhesion to negatively charged cell surfaces (Kang et al., 2024). For example, cationic chitosan nanoparticles enter mesophyll cells and accumulate around the chloroplasts, whereas neutral PEG-coated particles remain in the apoplast (Wang and Chen, 2024). Chloroplast targeting within is based on redox gradient and electrostatic affinity. Experiments with fluorescently labeled silica nanoparticles have shown selective concentration in chloroplasts in 24 h of foliar exposure (Ali et al., 2024).

6.2 Physicochemical interactions of the nano-bio interface

The molecular interactions comprise several nano-biological interactions at the nano-bio interface:

✓ Electrostatic (between NP surface charge and chloroplast envelope lipids).
✓ Hydrogen bonding and van der Waals (with membrane proteins or pigments).
✓ Redox and catalytic reactions (in the case of NPs being involved in electron transfer).

Photo system II (PSII) and electron transport chain Photosystem II (PSII) can expose temporary electron bridges with metal oxide nanoparticles, including TiO_2, ZnO, and Fe_2O_3, to facilitate charge transfer (Das et al., 2023). In the meantime, nanostructures based on carbon (graphene, CNTs) serve as conductive scaffolds that can be linked to thylakoid membranes, thereby reducing charge recombination (Tripathi et al., 2024; Huang et al., 2024).

6.3 Nanotechnology in engineering chloroplasts

Engineering of chloroplasts intends to alter chloroplast genomes, protein expression, or membrane activities to optimize photosynthetic. Nanotechnology can help in this using several strategies:

(a) Vehicle-based Gene Delivery.
In conventional chloroplast transformation, particle bombardment or Agrobacterium-mediated transfer is used, which is often

inefficient or species-restrictive. DNA, RNA, or CRISPR components can be transported into chloroplasts using nanocarriers, including carbon nanotubes (CNTs), mesoporous silica nanoparticles (MSNs), and lipid-polymer hybrids with no mechanical damage (Demirer et al., 2023). Gene delivery using CNTs resulted in a 5-fold improvement in transient GFP expression in chloroplasts of Nicotiana benthamiana compared to conventional delivery methods (Kwak et al., 2023).

(b) Regulation by Nanozyme.
Nanozymes: Nano-sized particles with the characteristics of enzymatic catalysts are under development as a substitute or complement to the most important chloroplast enzymes. For example, CeO2 nanoparticles that serve as SOD and catalase analogs help maintain redox homeostasis during photo-oxidative stress (Chen et al., 2023). In the meantime, Fe3O4 NPs can also mimic peroxidase activity, aiding ROS detoxification and preserving PSI integrity (Li et al., 2024).

(c) Nanostructured Artificial Thylakoids.
Scientists are working on hybrid thylakoid membranes, which are biological lipid-protein matrices and conductive nanomaterials (e.g., graphene oxide sheets or metallic nanowires) (Qiao et al., 2023). These hybrids enhance charge separation and can be prepared in vitro or in living chloroplasts to improve photosynthetic efficiency.

(d) Nanoscale Repair Systems
PSII has a damaged D1/D2 protein that can be targeted using self-assembled peptide-nanoparticle complexes, and damaged reaction centers can be repaired in situ (Shao and Lin, 2023). These systems recreate the effects of natural turnover processes but at an accelerated rate, prolonging the chloroplast lifespan during periods of stress.

6.4 Nano-bio modulation of chloroplast signaling and metabolism

In addition to structural incorporation, nanoparticles can modify cellular signaling in photosynthesis. For example, ZnO NPs regulate calcium-dependent signaling pathways that control stomatal behavior and CO_2 assimilation (Liu et al., 2024a,b,c). Equally, SiO_2 nanocarriers loaded with salicylic acid (SA) elicit retrograde signaling between chloroplasts and nuclei, leading to stress tolerance (Patra et al., 2024). The concept of nano-enabled modulation is also applicable to metabolite regulation; carbon nanotubes enhance NADPH regeneration and ATP/ADP ratios by influencing the kinetics of stromal enzymes (Mahapatra and Gupta, 2024). These relations establish an integrated layer of control between nanomaterials and biochemical feedback.

6.5 Visualization and analytical nano-bio interface

The primary focus of unraveling nano-bio interactions is advanced imaging and spectroscopy. Confocal laser scanning microscopy (CLSM), transmission electron microscopy (TEM), and atomic force microscopy (AFM) can be used to obtain localization and structural information, and X-ray photoelectron spectroscopy (XPS) and Raman mapping can be used to detect surface chemical states (Raina et al., 2023). Exciton transfer and redox dynamics in nanomodified chloroplasts are monitored using new in situ methods, such as time-resolved fluorescence spectroscopy and ultrafast transient absorption (TA) imaging (Fiebig et al., 2023). A combination of these tools facilitates the rational designing of nanomaterials having optimized interface compatibility.

6.6 Safety, stability and bioethical concerns

Chloroplast engineering using nanotechnology promises a lot of potential, but there are a number of safety issues. The uncontrolled accumulation of nanoparticles can alter chloroplast ultrastructure, enzymatic functions, or induce oxidative stress (Usman et al., 2024). The nanoparticles that enter the edible tissues increase bioaccumulation and food-chain concerns. Researchers support the principles of safe-by-design, namely the creation of biodegradable nanomaterials (e.g., Mg, Si, and Fe oxides) and the reduction of persistence in the soil-plant system (Duhan et al., 2024). Also, the regulatory and ethical frameworks should be made to be transparent, sharing of data and acceptable to the masses. Future Nano-Harvest efforts will be expected to be drawn to programmable nano-bio assemblies with sensing, delivery, and self-repair capabilities. Advances in synthetic biology would allow bio-templated nanofabrication, in which the chloroplasts themselves produce useful nanomaterials (Sarma et al., 2024). The combination of AI-directed design and molecular dynamics simulations will enable predictive simulation of nanobio interactions at the atomic scale, eliminating the risks of insecurity and inefficiency prior to experimentation. Nanotechnology In the long term, nanotechnology can be used to engineer chloroplasts to sustain bio-hybrid energy devices-to join agriculture, renewable energy, and carbon-neutral technologies as shown in Table 2.5.

7. Bioethical and environmental nano-harvesting

The fast introduction of nanotechnology in agricultural and photosynthesis improvement is subject to deep environmental and ethical concerns. Although Nano-Harvest technologies have high potential for enhancing crop yields, light absorption, and improved resource use, their usage introduces materials and mechanisms, the ecological performance and long-term safety of which are not well comprehended (Zhang and Su, 2024). Introduction of engineered nanomaterials (ENMs) to bioecosystems,

TABLE 2.5 Representative nanomaterials strategies for chloroplast engineering.

Nanomaterial/System	Mechanism	Chloroplast target	Reported effect	References
CNT-DNA complexes	Gene delivery	Chloroplast genome	↑ Transformation efficiency (5×)	Demirer et al. (2023)
CeO_2 nanoparticles	SOD/CAT nanozyme	Thylakoid stroma	↓ ROS accumulation	Chen et al. (2023)
Fe_3O_4 nanoparticles	Peroxidase mimic	PSI region	Stabilized electron transport	Li et al. (2024)
Graphene–lipid hybrids	Artificial thylakoid	PSII/ETC	↑ Charge separation	Qiao et al. (2023)
Peptide–Au nanoclusters	Repair complex	D1/D2 proteins	Enhanced photostability	Shao and Lin (2023)
SiO_2–SA nanocarriers	Retrograde signaling	Chloroplast–nucleus	↑ Stress tolerance	Patra et al. (2024)

particularly soils, waters, and air require a comprehensive review that incorporates toxicology, environmental chemistry, conservation of biodiversity, and ethical management (Ashfaq et al., 2025).

7.1 Nanomaterials environmental fate and transformation

The NP released, whether via foliar sprays, seed coatings, or soil amendments, may undergo various physicochemical changes, including oxidation, dissolution, aggregation, and interactions with natural organic matter (Zhang and Su, 2024). The changes dictate their survival and movement through soil, plant, and water substrates. For example, metal oxide nanoparticles such as TiO_2 or ZnO can partially dissolve, releasing metal ions that participate in redox reactions that influence soil chemistry (Djanaguiraman et al., 2024). Equally, organic contaminants can be adsorbed onto carbon-based nanomaterials, thereby altering the bioavailability of pollutants (Safdar et al., 2022; Gowtham, 2024; Gowtham et al., 2024).

7.2 Bioaccumulation and ecotoxicological effects

Nanoparticles can bioaccumulate in plant tissues, animals, and aquatic organisms (Djanaguiraman et al., 2024). As soon as they are internalized, they can be transported across trophic levels into the food chain. For example, Ag and CeO_2 nanoparticles were detected in edible plant components such as lettuce and rice grains after being sprayed onto foliage (Zhang and Su, 2024). With prolonged use, plants may experience oxidative stress, membrane damage, and dysfunctional photosynthetic enzymes, whereas aquatic organisms may exhibit gill and liver dysfunction after ingesting nanoparticles (Gowtham, 2024; Gowtham e al., 2024). The potential of the trophic transfer of nanoparticles is determined by their bioavailability and persistence. Recent microcosm research demonstrated that TiO_2 nanoparticles can be carried by soil microbes to nematodes and, consequently, to their predators, thereby increasing ecological risks (Donahue, 2025; Donahue et al., 2025; Ashfaq et al., 2025). This ambiguity underscores the need for detailed ecotoxicity profiling of biomarkers using omics, real-time nanosensors, and multi-species bioassays.

7.3 Soil and microbiological ecosystem effects

The microbial communities in the soil are important in the cycling of nutrients, the fixation of nitrogen, and the decomposition of organic matter. Nano-Harvest formulations of nanomaterials have the potential to alter microbial diversity and enzyme activity (Chowdhury et al., 2024). For example, CeO_2 nanoparticles can prevent colonization of legume roots by Rhizobium, whereas CNTs can disrupt fungal hyphal networks and reduce soil aggregation (Safdar et al., 2022). On the other hand, at low concentrations, some nanomaterials can also promote positive microbial growth by altering redox or enzyme functions. It has been demonstrated that TiO_2 nanoparticles enhanced the level of dehydrogenase and phosphatase in neutral soils and indicated a dose-related response (Ashfaq et al., 2025).

7.4 Ethical and social aspects

In addition to environmental safety, ethical factors influence people's trust and acceptance of Nano-Harvest technologies. There is an ethical debate that covers five key dimensions:

✓ **Biosafety and Precaution**—The invisibility of nanotechnology to the senses of a human being makes it difficult to perceive risk. Precautionary testing and open communication regarding uncertainties are required for ethical deployment (Donahue, 2025; Donahue et al., 2025).

✓ **Equity and Access**—Developed nanotechnologies can be available to high-income agricultural sectors mainly, which worsens technological disparities.

✓ **Intellectual Property (IP) and Data Transparency Patents**—Patents on nanomaterials or bio-nano hybrids can be a barrier to open innovation. Safety profiles need to be stored in shared data repositories (Ashfaq et al., 2025).

✓ **Regulatory Governance**—It is essential to harmonize nanoparticle safety standards on an international level. Current frameworks (OECD, REACH, and EPA) are being revised, and plant-integrated nanomaterials have yet to have a specific protocol (Zhang and Su, 2024).

7.5 Frameworks of sustainable nano-harvest development

To make deployment ethical and environmentally sound, several cohesive frameworks have been suggested:

a. Safe-by-Design (SbD):
Unintended toxicity is reduced by incorporating safety measures at the very early stages of material synthesis, controlling size, surface chemistry, and dissolution properties (Ashfaq et al., 2025). For example, biodegradable polymers on metal nanoparticles reduce ion leaching and persistence in the environment.

b. Circular Nano-Economy:
Closed-loop systems for recycling nanomaterials and eliminating environmental leakage are beneficial for promoting sustainability (Chowdhury et al., 2024). For example, nutrient-delivery nanocarriers can be recycled through irrigation runoff.

c. Green Nanotechnology:
Nanoparticles made with plant extracts or biopolymers (phytonanotechnology) eliminate the use of hazardous reagents, consistent with the principles of green chemistry (Safdar et al., 2022).

d. Ethical Governance and Engagement of Stakeholders:
Legitimacy is enhanced by the involvement of the population in the policymaking process. Responsible innovation can be achieved by open discussions between scientists, regulators, farmers, and communities (Donahue, 2025; Donahue et al., 2025).

8. Conclusion

Combining nanotechnology and photosynthesis in the concept of Nano-Harvest represents a revolutionary shift in how science addresses energy capture, crop yields, and environmental sustainability. As mentioned in this chapter, nanomaterials provide unprecedented control over light absorption, electron transport, and carbon fixation in photosynthetic systems. Nanoscience has transformed plant biology by offering improved ways to enhance chloroplast efficiency and stress resistance, as well as to develop artificial photosynthetic devices, due to its role in the new age of sustainable innovation (Singh et al., 2023).

8.1 The convergence of nanotechnology and photosynthesis

Initially, Nano-Harvest fills the gap between the physical finesse of nanotechnology and the biological flair of photosynthesis. The conventional processes of photosynthesis, although efficient in nature, are also constrained by evolutionary factors, including photoinhibition, poor use of light, and inefficiency in water-use (Kumar et al., 2024a,b). The solution to such biological limitations lies in nanomaterials, particularly quantum dots, carbon nanotubes, and plasmonic nanoparticles, which can enhance photon capture and accelerate charge separation and electron flow (Zhang et al., 2023a,b).

Recent developments have demonstrated that nanoparticles can serve as artificial antenna assemblies, improving the collection of unused light spectra, such as near-infrared or ultraviolet, and increasing the overall photosynthetic rate (Liu et al., 2023). On the same note, natural enzymatic processes are being modeled with the aid of nanostructured catalysts, allowing semi-synthetic photosystems that can generate hydrogen or other solar fuels (Hu et al., 2023).

8.2 Implications on sustainable agriculture

Sustainable agriculture is one of the areas where Nano-Harvest technologies are most likely to be used. Increased crop yields, more efficient nutrient use, and greater stress resistance can be achieved through nanoparticle-mediated improvements in photosynthesis under the pressure of climate change (Patra et al., 2022). As an example, nanoparticles of titanium dioxide (TiO2) and zinc oxide (ZnO) have been mentioned to enhance chlorophyll content, CO_2-absorption, and antioxidant enzymes in maize, wheat, and rice (Jiang et al., 2023). These improvements not only help in crop yield, but also make them resistant to drought, heat, or salinity stressful factors (Kaur et al., 2024). Nevertheless, nanotechnology in agriculture should be given careful consideration. There is also a limited number of long-term studies of the fate of nanoparticles, their accumulation, and ecotoxicity (Lee et al., 2024). It will be essential to follow the principles of green nanotechnology, that is, use of biodegradable, biogenic, or plant-derived nanomaterials to ensure ecological harmony (Das et al., 2023).

Consent for publicaton

All authors confirm that they have reviewed the final version of this manuscript and consent to this publication.

Conflict of interest

The authors declare that there is no conflict of interest regarding the publication of this chapter.

Acknowledgments

The authors acknowledge the support of all the contributors involved in this Work.

References

Adeleye, A.S., Lin, S., Keller, A.A., Zhang, W., 2023. Engineered nanomaterials for environmental remediation: plant-based uptake and detoxification. Environ. Sci. Technol. 57 (8), 3552–3563.

Ali, R., Khan, S., Mehmood, A., 2024. TiO_2 nanocoatings for enhanced photoprotection in crop plants. Plant Physiol. Rep. 29 (2), 122–131.

Andleeb, N., 2025. Carbon Quantum Dots as Versatile Nanomaterials for Improving Soil Health and Plant Performance. Frontiers in Nanobiotechnology.

Ashfaq, M., Gupta, G., Verma, N., 2025. Carbon-based nanocarriers for plant growth promotion: fuelling when needed. Nanoscale 17, 616–634. https://doi.org/10.1039/D4NR03268C.

Ayenew, B.M., Satheesh, N., Zegeye, Z.B., Kassie, D.A., 2025. A review on the production of nanofertilizers and its application in agriculture. Heliyon 11 (1).

Bose, S., Patel, R., Nair, P., 2023. Nanoparticle-mediated antioxidant defense in higher plants. Plant Sci. Front. 12, 1085–1098.

Chen, X., Zhao, Y., Li, J., 2023. Nanozyme-assisted modulation of plant redox systems under oxidative stress. Nano Lett. 23 (7), 2985–2993.

Chen, H., Song, Y., Wang, Y., et al., 2024. ZnO nanoparticles: improving photosynthesis, shoot development, and phyllosphere microbiome composition in tea plants. J. Nanobiotechnol. 22, 389. https://doi.org/10.1186/s12951-024-02667-2.

Das, P., Sharma, M., Roy, D., 2023. Green synthesis of biocompatible nanoparticles for sustainable agriculture. J. Nanobiotechnol. 21 (5), 112–130.

Demeke, E.D., Benti, N.E., Terefe, M.G., Anbessa, T.T., Mengistu, W.M., Mekonnen, Y.S., 2025. A comprehensive review on nano-fertilizers: preparation, development, utilization, and prospects for sustainable agriculture in Ethiopia. Nanoscale Adv. 7 (8), 2131–2144.

Demirer, G.S., Zong, Y., Landry, M.P., 2023. Nanocarrier-mediated delivery of biomolecules into plant chloroplasts. Nat. Nanotechnol. 18, 611–622.

Djanaguiraman, M., Anbazhagan, V., Dhankher, O.P., Vara Prasad, P.V., 2024. Uptake, translocation, toxicity, and impact of nanoparticles on plant physiological processes. Plants 13 (22), 3137. https://doi.org/10.3390/plants13223137.

Donahue, E., 2025. Ethical dimensions of plasmonic nanotechnology: toward responsible innovation. Commun. Chem. 8 (5), 421–432.

Donahue, E., Malina, T., Smith, E., Pšenčík, J., Sprague-Klein, E.A., 2025. Photophysics of plasmonically enhanced self-assembled artificial light-harvesting nanoantennas. Commun. Chem. 8 (1), 263.

Duhan, J.S., Kumar, R., Kaur, N., 2024. Nanotechnology in sustainable agriculture: safety, toxicity, and regulatory aspects. Environ. Nanotechnol. 11 (1), 115–129.

Fiebig, O.C., Harris, D., Wang, D., Hoffmann, M.P., Schlau-Cohen, G.S., 2023. Ultrafast dynamics of photosynthetic light harvesting: strategies for acclimation across organisms. Annual Rev. Phys. Chem. 74 (1), 493–520.

Gowtham, K., 2024. Nanoceria-mediated reactive oxygen regulation and ecological safety. Plant Physiol. Rep. 29 (3), 455–472.

Gowtham, H.G., Shilpa, N., Singh, S.B., Aiyaz, M., Abhilash, M.R., Nataraj, K., Amruthesh, K.N., Ansari, M.A., Alomary, M.N., Murali, M., 2024. Toxicological effects of nanoparticles in plants: mechanisms involved at morphological, physiological, biochemical and molecular levels. Plant Physiol. Biochem. 210, 108604.

Haider, Z., Yang, C., Ahmad, I., Zia, S., Javaid, M.H., Rehman, M., Yasin, M.U., Ali, B., Nana, C., Gan, Y., 2024. Bioderived carbon quantum dots boost maize growth and photosynthesis by augmenting UV spectrum absorption and carbon assimilation regulatory genes. J. Environ. Chem. Eng. 12 (6), 114396.

Hu, J., Wang, T., Zhou, X., 2023. Biohybrid nanostructures for enhanced solar-to-chemical conversion. Nat. Nanotechnol. 18 (9), 712–726.

Huang, Y., Jiang, T., Li, P., 2024. Biocorona formation on plant-interacting nanoparticles: implications for bioavailability and toxicity. Environ. Sci. Nano 11 (1), 57–71.

Iwai, M., Patel-Tupper, D., Niyogi, K.K., 2024. Structural diversity in eukaryotic photosynthetic light harvesting. Annu. Rev. Plant Biol. 75.

Jiang, Y., Chen, Q., Luo, M., 2023. Nanoparticles in improving photosynthetic efficiency and crop productivity. Plant Sci. Today 10 (1), 45–59.

Kang, D., Park, E., Lee, S., 2024. Size-dependent uptake and transport of engineered nanoparticles in higher plants. Environ. Toxicol. Chem. 43 (2), 155–167.

Kaur, G., Singh, A., Bhatt, P., 2024. Smart nanosensors for real-time plant health monitoring. Sensor. Actuator. B Chem. 415, 134917.

Khundi, Q., 2025. Nanofertilizers for sustainable African agriculture: a global review. Nanomaterials.

Kumar, V., Mehta, R., Yadav, D., 2024a. Nano-enabled fertilizers for sustainable crop production: a review. J. Agricult. Nanotechnol. 12 (3), 201–218.

Kumar, V., Sharma, M., Singh, A., 2024b. CeO_2 nanoparticles improve drought resilience and photosynthetic stability in maize. Plant Stress 7, 100175.

Kwak, S.Y., Wong, M.H., Strano, M.S., 2023. Carbon nanotube–chloroplast interactions for enhanced energy conversion. Nat. Nanotechnol. 18, 451–463.

Lee, J., Park, H., Kim, S., 2024. Ethical and safety perspectives in agricultural nanotechnology. NanoImpact 32, 100485.

Li, X., Wu, F., Chen, Z., 2023. Quantum dots as artificial antennas in photosynthetic enhancement. ACS Appl. Nano Mater. 6 (2), 1799–1809.

Li, C., Zhou, P., Chen, X., 2024. AI-assisted nanomaterial design for enhanced photosynthesis. Artif. Intell. Life Sci. 5, 100135.

Liu, H., Zhang, W., Xu, L., 2023. Molecular insights into nanoparticle-chloroplast interactions. Plant Physiol. Rep. 28 (4), 411–428.

Liu, J., Zhang, D., Bao, L., Mao, T., Zhao, L., Liu, C., Huang, G., Zhai, Y., 2024a. Plant nanobiotechnology: a new strategy to enhance crop photosynthesis. Environ. Sci. Nano 11, 4655–4665. https://doi.org/10.1039/D4EN00520A.

Liu, X., Nadeem, M., Rui, Y., 2024b. Effects of nanofertilizers on the mechanism of photosynthetic efficiency in plants: a review. Phyton – Int. J. Experim. Botany 93 (12), 3197–3216. https://doi.org/10.32604/phyton.2024.059281.

Liu, Z., Zheng, P., Huang, Y., 2024c. ZnO nanoparticles modulate calcium signaling and stomatal behavior in Arabidopsis. Plant Sci. Front. 15, 1139–1152.

Mahapatra, A., Gupta, S., 2024. Nanotechnology and chloroplast metabolism: emerging molecular perspectives. Trends Plant Sci. 29 (3), 215–229.

Maity, S., Kleinekathöfer, U., 2023. Recent progress in atomistic modelling of light-harvesting complexes: a mini review. Photosynth. Res. 156, 147–162.

MIT News. (July 3, 2023). Chemists Discover Why Photosynthetic Light-Harvesting Is So Efficient. MIT News.

Mori, S., Hashimoto, R., Hisatomi, T., Domen, K., Saito, S., 2025. Artificial photosynthesis directed toward organic synthesis. Nature Commun. 16 (1), 1797.

Patra, S., Das, A., Sen, P., 2022. Nanotechnology for climate-smart agriculture: challenges and opportunities. Agricult. Nanotechnol. 9 (3), 122–140.

Patra, R., Gupta, D., Rai, M., 2024. Smart nanocarriers for nutrient and growth regulator delivery in crops. Agronom. Adv. 15 (4), 221–234.

Qiao, Y., Sun, M., Li, B., 2023. Hybrid thylakoid membranes incorporating graphene oxide for enhanced photoconversion. Adv. Funct. Mater. 33 (45), 2307861.

Rahman, M.M., Ali, S., Azam, M., 2023. Gold nanoparticle-induced thermotolerance and photoprotection in wheat. Environ. Exp. Bot. 208, 105102.

Raina, S., Joshi, D., Mehta, N., 2023. Nanoparticle tracking and imaging in plant cells: recent advances. Microsc. Res. Tech. 86 (11), 1450–1462.

Ramasamy, M., Sahoo, P.K., Tripathi, R., 2023. Metal–organic frameworks for CO_2 enrichment in plant micro-environments. J. CO_2 Util. 72, 102222.

Saeedi, R., Seyedi, A., Esmaeilizadeh, M., et al., 2024. Improving the performance of the photosynthetic apparatus of *Citrus sinensis* with the use of chitosan-selenium nanocomposite (CS + Se NPs) under salinity stress. BMC Plant Biol. 24, 745. https://doi.org/10.1186/s12870-024-05462-1.

Safdar, M., Kim, W., Park, S., Gwon, Y., Kim, Y.O., Kim, J., 2022. Engineering plants with carbon nanotubes: a sustainable agriculture approach. J. Nanobiotechnol. 20 (1), 275.

Sarma, B., Rao, G., Das, D., 2024. The nano-bio interface in plant systems: mechanisms and design strategies. Plant Nanobiotechnol. Rev. 12 (1), 45–68.

Shao, L., Lin, Z., 2023. Peptide-stabilized gold nanoclusters for in situ repair of PSII damage. ACS Appl. Bio Mater. 6 (8), 3021–3031.

Siddiqui, M.H., Al-Whaibi, M.H., Ali, H.M., 2024. Nanotechnology in photosynthesis: prospects and challenges. Front. Plant Sci. 15, 1147852.

Singh, P., Verma, A., Dasgupta, R., 2023. Nano-assisted photosynthesis: emerging frontiers and applications. Nanosci. Plant Biol. 4 (2), 55–74.

Sidhu, A.K., Sharma, M., Bhickchand Agrawal, S., Pradip Bhavsar, Samota, M.K., 2024. Nanomaterial strategies for enhancing plant resilience in the face of temperature stress. CABI Agricult. Biosci. 5 (1), 60.

Singh, R., Patel, M., Gupta, N., 2024. Plasmonic nanostructures for augmenting light capture in plants. Nanoscale Adv. 6 (5), 2850–2863.

Torney, F., Rees, T., Choi, J., 2023. Engineering photosynthesis with nanotechnology: a review. Trends Biotechnol. 41 (9), 1035–1051.

Tripathi, D.K., Mishra, R.K., Chauhan, D.K., 2024. Graphene and carbon nanostructures for enhanced plant productivity. Nanomater. 14 (2), 339–354.

Turner, A.S., 2023. The role of light-harvesting complex II organization in the efficiency of light-dependent reactions in the photosynthetic apparatus of *Pisum sativum* L. Plants 14 (12), 1846.

Usman, M., Sohail, A., Ahmad, M., 2024. Biosafety considerations in agricultural nanotechnology. Environ. Res. 238, 117601.

Wang, C., Chen, H., 2024. Surface chemistry-dependent nanoparticle uptake in mesophyll tissues. Plant Physiol. Biochem. 205, 107–119.

Wang, Q., Lv, K., Song, J., Li, M., Ouyang, X., Liu, C., Gong, S., Wang, J., Li, J., Zhang, Z., 2025. The dose-dependent effect of carbon quantum dots as a photosynthesis enhancer on soybean plant growth. Nanomaterials 15 (20), 1603.

Wu, J.Y., Huang, Z., 2024. Improving photosynthesis efficiency in potato: a review of genetic and agronomic approaches. J. Energy Biosci. 15 (6), 337–348.

Yadav, R., Sharma, K., 2023. Mimicking photosystem II with MnO_2 nanoparticles: a step toward artificial photosynthesis. J. Photochem. Photobiol. B Biol. 245, 112556.

Zhang, H., Su, W., 2024. Classification, uptake, translocation, and detection methods of nanoparticles in crop plants: a review. Environ. Sci. Nano 11, 1847–1870.

Zhang, H., Li, Y., Sun, Q., 2023a. Plasmonic nanostructures for light-harvesting enhancement in plants. Nano Lett. 23 (7), 3124–3138.

Zhang, L., Yu, P., Zhou, Q., 2023b. Upconversion nanoparticles for light-harvesting extension in crop photosynthesis. Nano Energy 108, 107614.

Zhou, L., Chen, J., Gao, M., 2024. Surface plasmonic nanoparticles in biohybrid light-harvesting systems. Adv. Funct. Mater. 34 (5), 2401528.

Zhu, S., Han, L., Peng, X., 2023. Mechanisms of nanoparticle entry into plant cells: a review. Environ. Nanosci. 10 (9), 851–869.

CHAPTER

3

Nanoscale transformation within plant cell

Ghafoor Unnisa[1], *S. Maqbool Ahmed*[1], *Ahmad Faraz*[2], *Nayab Ishrat*[3], *Vishnu D. Rajput*[4] *and Mohammad Faizan*[1]

[1]Botany Section, School of Sciences, Maulana Azad National Urdu University, Hyderabad, Telangana, India; [2]Department of Botany, Shibli National College, Azamgarh, Uttar Pradesh, India; [3]Department of Biotechnology, Era University, Lucknow, Uttar Pradesh, India; [4]Academy of Biology and Biotechnology, Southern Federal University, Rostov-On-Don, Russia

1. Introduction

Plant absorbs water, nutrients, and gases which are essential for its growth and development. Water, absorbed through root hairs via osmosis, serves as both a solvent and a medium for nutrient transport (Ahanger et al., 2017; Ludwiczak et al., 2021). Root hairs significantly enhance water absorption and soil adhesion in species such as *Arabidopsis thaliana* (Choi and Cho, 2019). Plants acquire macronutrients such as nitrogen, phosphorus, and potassium, and micronutrients including iron, zinc, and copper, through passive and active uptake mechanisms (Kumar et al., 2021). Passive diffusion allows ion entry along concentration gradients (Canarini et al., 2019; Muratore et al., 2021) while active transport depends on energy-driven carrier proteins such as K^+ transporters and channels OsGORK, OsAKT1, OsHAK1, OsHAK5, and OsHAK21 (Lian et al., 2020). After absorption, the xylem and phloem facilitate the long-distance transport of water, minerals, and photosynthates throughout the plant (Lian et al., 2020). Transpiration through stomata generates tension that drives upward water and nutrient movement (Brewer et al., 2022; Wheeler and Stroock, 2008), while gas exchange enables CO_2 uptake for photosynthesis and O_2 release (Driesen et al., 2020).

Understanding nanomaterial (NM) uptake, translocation, and transformation is crucial for evaluating their biological impacts. NMs can be introduced via foliar, root, or seed-based applications (Wang et al., 2022c; Zhao et al., 2020), with nanoparticle (NP) size, charge, and surface chemistry determining entry and mobility. Cell walls and membranes act as selective barriers, while transformations such as ion release and reactive oxygen species (ROS) generation particularly by ZnO nanoparticles can influence plant physiology and metabolism (Khan et al., 2016;

Nanotechnology in Photosynthesis
https://doi.org/10.1016/B978-0-443-44376-3.00003-3

Molnár et al., 2020; Wagner et al., 2016). This chapter explores the mechanisms governing NP uptake, translocation, transformation, and their consequent effects on plant metabolic and physiological processes.

2. Translocation of nanomaterials in plants

The translocation of nanomaterials (NMs) within plants is governed by structural, physicochemical, and electrostatic barriers presented by the cell wall and plasma membrane. These cellular interfaces regulate the movement, accumulation, and transformation of nanoparticles (NPs), making their characterization critical for designing effective nanomaterial-based agricultural applications.

2.1 Pore size and cell wall structure

The plant cell wall, composed primarily of cellulose, hemicelluloses, lignin, and pectins, provides mechanical strength and selectivity that restricts NM entry (Zeng et al., 2017; Shomer et al., 2003). The pore size of this cross-linked polymeric network determines the permeability of nanoparticles, varying among plant species and tissues (Carpita et al., 1979; Lew et al., 2018). In wheat (*Triticum aestivum*), pore radii range from 1.5–3 nm (Chesson et al., 1997), while maize (*Zea mays*) stover cell walls exhibit wider pores of 10–1000 nm (Chundawat et al., 2011). Pea (*Pisum sativum*) epidermal cells display pore diameters of approximately 5.5 ± 1.4 nm in elongating walls and 13.4 ± 4.3 nm in non-elongating walls (Fujino and Itoh, 1998). Experimental observations confirm that NPs smaller than 13 nm can traverse cell walls through pores, whereas larger particles (40–60 nm) penetrate via intercellular spaces (Wu and Li, 2022). Transmission electron microscopy has revealed the presence of 11 nm particles on maize leaves (Hu et al., 2020) and 10 nm Zr/CeOx NPs within leaf tissues (Schwabe et al., 2015). Conversely, Fe_2O_3 nanoparticles exceeding 18 nm exhibited limited root-to-shoot transport in maize (Li et al., 2016), demonstrating the strong size-dependence of translocation efficiency.

2.2 Interactions between nanomaterials and the cell wall

Beyond size exclusion, chemical and electrostatic interactions influence NM mobility. Cell walls are dynamic structures that can undergo remodeling in response to nanoparticle exposure (Rajput et al., 2026). For instance, ZnO NPs adhered to root cell walls of *Brassica napus*, indicating ion–polymer binding (Molnár et al., 2020). Exposure to Y_2O_3 NPs altered tobacco (*Nicotiana tabacum*) cell walls, increasing thickness up to 12-fold while raising pectin content and reducing hemicelluloses (Chen et al., 2021). Similarly, Ag NPs induced reactive oxygen species (ROS) generation, facilitating Ag^+ binding to hydroxyl-cellulose groups and disrupting wall hydrogen bonds (de Paiva Pinheiro et al., 2021). TiO_2 nanoparticles modified polysaccharide composition in *Pisum sativum* (Fan et al., 2014) and *Solanum lycopersicum* leaves, affecting xyloglucan and galacturonic acid levels (Liné et al., 2021). Furthermore, nano zero-valent iron (nZVI) at 0.5 g/L promoted root elongation in *Arabidopsis thaliana* through ROS-mediated wall loosening and H_2O_2 release (Kim et al., 2014). These findings suggest that nanomaterials can actively remodel wall architecture, influencing both transport and physiological responses.

2.3 Electrostatic properties and nanoparticle charge

The electrical potential of the cell wall, typically ranging from −50 to −110 mV (Shomer et al., 2003), plays a significant role in NP binding and translocation. Negatively charged cell walls tend to retain positively charged nanoparticles, reducing their mobility (Bao et al., 2016; Juárez-Maldonado et al., 2019). In *Arabidopsis thaliana*, negatively charged Au NPs were internalized by root cells, whereas positively charged Au NPs were excluded (Avellan et al., 2017). In *Brassica chinensis*, positively charged Se NPs showed greater accumulation in shoots, suggesting charge-dependent mobility (Wang et al., 2022b). Likewise, positively charged ZnO NPs exhibited improved leaf surface adsorption and uniform distribution compared to negatively charged counterparts (Zhu et al., 2021).

2.4 Membrane transport and internalization

Following wall penetration, nanoparticles must cross the plasma membrane, typically 5–10 nm thick, to enter the cytoplasm (Wu and Li, 2022). Uptake mechanisms include endocytosis, transport through protein channels, or lipid exchange processes (Perozo et al., 2002; Wong et al., 2016; Lv et al., 2021). The plasma membrane potential (≈ -120 mV) further regulates electrostatic interactions and ion-mediated transport (Wu et al., 2013). Nanoparticles with high zeta potential can penetrate protoplasts effectively (Lew et al., 2018). For example, 5 nm Au NPs readily crossed plasma membranes regardless of charge (Milewska-Hendel et al., 2019). Once internalized, NPs can reach organelles such as chloroplasts or mitochondria (Liu et al., 2021; Zhou et al., 2021), potentially influencing metabolic and signaling pathways.

3. Transport mechanisms of nanoparticles into plants

The translocation of nanoparticles (NPs) in plants occurs through two principal mechanisms: **assisted delivery** and **passive delivery**. Assisted delivery employs external physical forces to facilitate NP entry into plant tissues, overcoming biological barriers such as cell walls and membranes (Wu and Li, 2022; Rossi et al., 2019). The **biolistic (gene gun) method** uses compressed gas or helium to propel nanoparticles into plant tissues (Rajkumari, 2020) demonstrated that silver nanoparticles (Ag NPs) functioned as efficient gene carriers in *Nicotiana tabacum* L., outperforming traditional gold micro carriers. **Sonoporation**, employing ultrasonic waves, transiently permeabilizes membranes to enhance NP uptake. **Magnetofection** utilizes magnetic nanoparticles guided by external magnetic fields for targeted tissue delivery (Molnár et al., 2020). Their magnetic characteristics can be studied through bulk techniques such as vibrating sample magnetometry (VSM), superconducting quantum interference device (SQUID), or X-ray magnetic circular dichroism (XMCD), as well as nanoscale analyses using magnetic force microscopy (MFM) and single-particle magnetometry (Perozo et al., 2002). Other assisted delivery methods include **electroporation**, where electrical pulses induce transient pores in membranes (Lv et al., 2021), and **micro injection**, enabling precise NP introduction into single cells (Wong et al., 2016). Such methods are predominantly used in vitro for instance, in plant embryos, callus tissues, or protoplast systems-under highly controlled conditions (Zhou et al., 2021). Despite their precision, large-scale agricultural applications remain limited due to high energy requirements, technical complexity, and inconsistent field reproducibility (Juárez-Maldonado, 2022; Rossi et al., 2019). Environmental fluctuations and resource constraints further hinder scalability (Wu and Li, 2022).

In contrast, **passive delivery** exploits the plant's intrinsic physiological processes, allowing nanoparticles to enter without external intervention. The two major routes are **root uptake** and **foliar uptake** (Wang et al., 2022a; Lian et al., 2020). During root uptake, NPs enter with water and dissolved solutes through apoplastic or symplastic pathways (Rastogi et al., 2019). In foliar application, NPs penetrate through the stomatal pores or diffuse across the cuticle (Lv et al., 2019; Lian et al., 2020) observed that *Zea mays* L. exhibited higher NP assimilation efficiency via foliar exposure than through roots. Once internalized, NPs are distributed through the **xylem** and **phloem**, facilitating systemic transport to aerial and subterranean tissues (Wang et al., 2022b). Passive routes are environmentally compatible and cost-effective but constrained by factors such as pore size, NP charge, aggregation, and surface reactivity (Wu and Li, 2022).

4. Plant metabolism: Effects of nanomaterials

Nanomaterials (NMs) can profoundly affect plant metabolic networks, influencing physiological, biochemical, and molecular pathways. NPs may serve as nutrient sources, growth regulators, or stress inducers depending on their type, concentration, and reactivity (Hossain et al., 2015; Nair and Chung, 2014). Metal-based nanoparticles such as CuO, ZnO, and TiO_2 have been shown to alter cellular redox balance by modulating reactive oxygen species (ROS) generation (Liu and Lal, 2015). Excessive ROS accumulation disrupts electron transport in chloroplasts and mitochondria, impairing photosynthesis and respiration (Foyer and Shigeoka, 2011; Kalaji et al., 2014).

Nanoparticles can also modify **plant hormone homeostasis**, influencing growth and development. For instance, CuO NPs reduced indole-3-acetic acid (IAA) and abscisic acid (ABA) levels in cotton (*Gossypium* spp.), while Ag NPs increased cytokinin levels in pepper, collectively disturbing hormonal balance and metabolic regulation (O'Brien and Benková, 2013). Transcriptomic analysis revealed that Y_2O_3 NPs exposure upregulated genes associated with cell wall metabolism and remodeling (Chen et al., 2021). Moreover, NPs can affect secondary metabolism through ROS-mediated signaling, leading to changes in the synthesis of flavonoids, alkaloids, and phenolic compounds (Marslin et al., 2017). The **dosage-dependent effects** of nanoparticles are critical for determining physiological outcomes. Even moderate levels of CeO_2 NPs (25–100 mg L^{-1}) significantly altered root and leaf secondary metabolism in *Phaseolus vulgaris* L. (Salehi et al., 2020). At optimal concentrations, certain NPs enhance photosynthetic performance and productivity, as observed with ZnO and SiO_2 NPs improving chlorophyll content and yield (Wang et al., 2022a). However, excessive exposure may lead to metabolic inhibition, oxidative stress, and structural damage. Thus, the balance between beneficial and phytotoxic thresholds is essential for sustainable NM application in agriculture.

5. Future transport recognition sites: Plant stems

Recent studies highlight the **plant stem** as a promising yet underexplored route for nanoparticle delivery. Compared with foliar or root applications, stem-mediated transport offers several structural and physiological advantages. The stem's vascular architecture comprising the **xylem** and **phloem** possesses larger conduits that facilitate the movement of micron and nanoscale particles with minimal obstruction or surface deposition (Sembada et al., 2024). This architecture supports **long-distance, unidirectional transport**, aligning naturally with the upward translocation of water, nutrients, and signaling molecules from roots to shoots. Stem tissues also provide greater **mechanical stability** and lower permeability variability compared to roots or leaves, reducing risks of leakage and external loss. Direct stem introduction ensures rapid systemic distribution and minimal environmental exposure, making it particularly suitable for applications such as nanofertilizer delivery, systemic pest management, and transgenic nanoparticle-based enhancement strategies. Sembada et al. (2024) successfully demonstrated the integration of 110 nm SiO_2 nanoparticles via stem-cutting in tomato (*Solanum lycopersicum*), achieving stable nanoparticle distribution without impairing plant viability. Future advancements in stem-mediated nano delivery will require optimization of particle **size, charge, and surface functionalization** to enhance vascular compatibility and bioavailability. Moreover, field-scale validation, safety assessment, and regulatory frameworks must be established to ensure environmentally responsible deployment of nanotechnology in crop systems.

6. Conclusions

Nanoparticle transport within plants represents a rapidly evolving domain with vast potential to revolutionize agriculture, plant physiology, and biotechnology. Owing to their nanoscale dimensions, high surface-to-volume ratios, and adjustable physicochemical properties, nanoparticles serve as efficient carriers for nutrients, genetic materials, and agrochemicals, enabling enhanced precision and controlled delivery. However, realizing their full potential requires balancing delivery efficiency with critical considerations of nanoparticle toxicity, bioaccumulation, persistence, and broader ecological impact. Plant-based nanoparticle delivery operates primarily through **passive** and **assisted** mechanisms, each offering distinct advantages based on the intended application. Passive uptake via roots or leaves provides a sustainable and straightforward route, though it is constrained by inherent biological barriers such as cell walls and cuticular structures that limit nanoparticle penetration. Conversely, **stem-mediated delivery** has emerged as a promising alternative, leveraging the wider and more continuous vascular channels of the xylem and phloem to facilitate efficient systemic transport and minimize surface deposition.

Acknowledgement

The research was supported by the Strategic Academic Leadership Program of the Southern Federal University ("Priority 2030").

References

Ahanger, M.A., Tomar, N.S., Tittal, M., Argal, S., Agarwal, R.M., 2017. Plant growth under water/salt stress: ROS production; antioxidants and significance of added potassium under such conditions. Physiol. Mol. Biol. Plants 23 (4), 731–744. https://doi.org/10.1007/s12298-017-0462-7.

Avellan, A., Schwab, F., Masion, A., Chaurand, P., Borschneck, D., Vidal, V., Rose, J., Santaella, C., Levard, C., 2017. Nanoparticle uptake in plants: gold nanomaterial localized in roots of *Arabidopsis thaliana* by X-ray computed nanotomography and hyperspectral imaging. Environ. Sci. Technol. 51 (15), 8682–8691. https://doi.org/10.1021/acs.est.7b01133.

Bao, W., Wang, J., Wang, Q., O'Hare, D., Wan, Y., 2016. Layered double hydroxide nanotransporter for molecule delivery to intact plant cells. Sci. Rep. 6. https://doi.org/10.1038/srep26738.

Brewer, K., Clulow, A., Sibanda, M., Gokool, S., Odindi, J., Mutanga, O., Naiken, V., Chimonyo, V.G.P., Mabhaudhi, T., 2022. Estimation of maize foliar temperature and stomatal conductance as indicators of water stress based on optical and thermal imagery acquired using an unmanned aerial vehicle (UAV) platform. Drones 6 (7). https://doi.org/10.3390/drones6070169.

Canarini, A., Kaiser, C., Merchant, A., Richter, A., Wanek, W., 2019. Root exudation of primary metabolites: mechanisms and their roles in plant responses to environmental stimuli. Front. Plant Sci. 10, 157. https://doi.org/10.3389/fpls.2019.00157.

Carpita, N., Sabularse, D., Montezinos, D., Delmer, D.P., 1979. Determination of the pore size of cell walls of living plant cells. Science 205 (4411), 1144–1147. https://doi.org/10.1126/science.205.4411.1144.

Chen, F., Wang, C., Yue, L., Zhu, L., Tang, J., Yu, X., Cao, X., Schröder, P., Wang, Z., 2021. Cell walls are remodeled to alleviate nY2O3 cytotoxicity by elaborate regulation of de novo synthesis and vesicular transport. ACS Nano 15 (8), 13166–13177. https://doi.org/10.1021/acsnano.1c02715.

Chesson, A., Gardner, P.T., Wood, T.J., 1997. Cell wall porosity and available surface area of wheat straw and wheat gram fractions. J. Sci. Food Agric. 75 (3), 289–295. https://doi.org/10.1002/(SICI)1097-0010(199711)75:3<289::AID-JSFA879>3.0.CO;2-R.

Choi, H.S., Cho, H.T., 2019. Root hairs enhance *Arabidopsis* seedling survival upon soil disruption. Sci. Rep. 9, 11181. https://doi.org/10.1038/s41598-019-47733-0.

Chundawat, S.P.S., Donohoe, B.S., Da Costa Sousa, L., Elder, T., Agarwal, U.P., Lu, F., Ralph, J., Himmel, M. E., Balan, V., Dale, B.E., 2011. Multi-scale visualization and characterization of lignocellulosic plant cell wall deconstruction during thermochemical pretreatment. Energy Environ. Sci. 4 (3), 973–984. https://doi.org/10.1039/c0ee00574f.

de Paiva Pinheiro, S.K., Miguel, T.B.A.R., Chaves, M.de M., Barros, F.C.de F., Farias, C.P., Moura, T.A.de, Ferreira, O. P., Rocha Paschoal, A., Gomes Souza Filho, A., Miguel, E. de C., Wu, H., 2021. Silver nanoparticles (AgNPs) internalization and passage through the Lactuca sativa (Asteraceae) outer cell wall. Funct. Plant Biol. 48 (11), 1113–1123. https://doi.org/10.1071/fp21161.

Driesen, E., Ende, W.Van den, De Proft, M., Saeys, W., 2020. Influence of environmental factors light, CO_2, temperature, and relative humidity on stomatal opening and development: a review. Agronomy 10 (12), 1975. https://doi.org/10.3390/agronomy10121975.

Fan, R., Huang, Y.C., Grusak, M.A., Huang, C.P., Sherrier, D. J., 2014. Effects of nano-TiO2 on the agronomically-relevant Rhizobium-legume symbiosis. Sci. Total Environ. 466–467, 503–512. https://doi.org/10.1016/j.scitotenv.2013.07.032.

Foyer, C.H., Shigeoka, S., 2011. Understanding oxidative stress and antioxidant functions to enhance photosynthesis. Plant Physiol. 155 (1), 93–100. https://doi.org/10.1104/pp.110.166181.

Fujino, T., Itoh, T., 1998. Changes in pectin structure during epidermal cell elongation in pea (*Pisum sativum*) and its implications for cell wall architecture. Plant Cell Physiol. 39 (12), 1315–1323. https://doi.org/10.1093/oxfordjournals.pcp.a029336.

Hossain, Z., Mustafa, G., Komatsu, S., 2015. Plant responses to nanoparticle stress. Int. J. Mol. Sci. 16 (11), 26644–26653. https://doi.org/10.3390/ijms161125980.

Hu, P., An, J., Faulkner, M.M., Wu, H., Li, Z., Tian, X., Giraldo, J.P., 2020. Nanoparticle charge and size control foliar delivery efficiency to plant cells and organelles. ACS Nano 14 (7), 7970–7986. https://doi.org/10.1021/acsnano.9b09178.

Juárez-Maldonado, A., Ortega-Ortíz, H., Morales-Díaz, A.B., González-Morales, S., Morelos-Moreno, Á., Cabrera-De la Fuente, M., Sandoval-Rangel, A., Cadenas-Pliego, G., Benavides-Mendoza, A., 2019. Nanoparticles and nanomaterials as plant biostimulants. Int. J. Mol. Sci. 20 (1), 162. https://doi.org/10.3390/ijms20010162.

Juárez-Maldonado, A., 2022. Impact of nanomaterials on plants: what other implications do they have? Biocell 46, 651–654. https://doi.org/10.32604/biocell.2022.017350.

Kalaji, H.M., Schansker, G., Ladle, R.J., Goltsev, V., Bosa, K., Allakhverdiev, S.I., Brestic, M., Bussotti, F., Calatayud, A., Dąbrowski, P., Elsheery, N.I., Ferroni, L., Guidi, L., Hogewoning, S.W., Jajoo, A., Misra, A.N., Nebauer, S.G., Pancaldi, S., Penella, C., Poli, D., Pollastrini, M., Romanowska-Duda, Z.B., Rutkowska, B., Serôdio, J., Suresh, K., Szulc, W., Tambussi, E., Yanniccari, M., Zivcak, M., 2014. Frequently asked questions about in vivo chlorophyll fluorescence: practical issues. Photosynth. Res. 122 (2), 121–158. https://doi.org/10.1007/s11120-014-0024-6.

Khan, M.F., Ansari, A.H., Hameedullah, M., Ahmad, E., Husain, F.M., Zia, Q., Baig, U., Zaheer, M.R., Alam, M. M., Khan, A.M., Alothman, Z.A., Ahmad, I., Ashraf, G. M., Aliev, G., 2016. Sol-gel synthesis of thorn-like ZnO nanoparticles endorsing mechanical stirring effect and their antimicrobial activities: potential role as nano-antibiotics. Sci. Rep. 6. https://doi.org/10.1038/srep27689.

Kim, J.H., Lee, Y., Kim, E.J., Gu, S., Sohn, E.J., Seo, Y.S., An, H.J., Chang, Y.S., 2014. Exposure of iron nanoparticles to *Arabidopsis thaliana* enhances root elongation by triggering cell wall loosening. Environ. Sci. Technol. 48 (6), 3477–3485. https://doi.org/10.1021/es4043462.

Kumar, S., Kumar, S., Mohapatra, T., 2021. Interaction between macro-and micro-nutrients in plants. Front. Plant Sci. 12, 665583. https://doi.org/10.3389/fpls.2021.665583.

Lew, T.T.S., Wong, M.H., Kwak, S.Y., Sinclair, R., Koman, V. B., Strano, M.S., 2018. Rational design principles for the transport and subcellular distribution of nanomaterials into plant protoplasts. Small 14 (44). https://doi.org/10.1002/smll.201802086.

Li, J., Hu, J., Ma, C., Wang, Y., Wu, C., Huang, J., Xing, B., 2016. Uptake, translocation and physiological effects of magnetic iron oxide (γ-Fe2O3) nanoparticles in corn (*Zea mays* L.). Chemosphere 159, 326–334. https://doi.org/10.1016/j.chemosphere.2016.05.083.

Lian, J., Zhao, L., Wu, J., Xiong, H., Bao, Y., Zeb, A., Tang, J., Liu, W., 2020. Foliar spray of TiO2 nanoparticles prevails over root application in reducing Cd accumulation and mitigating Cd-induced phytotoxicity in maize (*Zea mays* L.). Chemosphere 239, 124794. https://doi.org/10.1016/j.chemosphere.2019.124794.

Liné, C., Reyes-Herrera, J., Bakshi, M., Wazne, M., Costa, V., Roujol, D., Jamet, E., Castillo-Michel, H., Flahaut, E., Larue, C., 2021. Fourier transform infrared spectroscopy contribution to disentangle nanomaterial (DWCNT, TiO 2) impacts on tomato plants. Environ. Sci. Nano 8 (10), 2920–2931. https://doi.org/10.1039/d1en00455g.

Liu, R., Lal, R., 2015. Potentials of engineered nanoparticles as fertilizers for increasing agronomic productions. Sci. Total Environ. 514, 131–139. https://doi.org/10.1016/j.scitotenv.2015.01.104.

Liu, J., Li, G., Chen, L., Gu, J., Wu, H., Li, Z., 2021. Cerium oxide nanoparticles improve cotton salt tolerance by enabling better ability to maintain cytosolic K+/Na+ ratio. J. Nanobiotechnol. 19 (1). https://doi.org/10.1186/s12951-021-00892-7.

Ludwiczak, A., Osiak, M., Cárdenas-Pérez, S., Lubińska-Mielińska, S., Piernik, A., 2021. Osmotic stress or ionic composition: which affects the early growth of crop species more? Agronomy 11 (3), 435. https://doi.org/10.3390/agronomy11030435.

Lv, J., Christie, P., Zhang, S., 2019. Uptake, translocation, and transformation of metal-based nanoparticles in plants: recent advances and methodological challenges. Environ. Sci. Nano 6 (1), 41–59. https://doi.org/10.1039/C8EN00645H.

Lv, Z., Sun, H., Du, W., Li, R., Mao, H., Kopittke, P.M., 2021. Interaction of different-sized ZnO nanoparticles with maize (*Zea mays*): accumulation, biotransformation and phytotoxicity. Sci. Total Environ. 796, 148927. https://doi.org/10.1016/j.scitotenv.2021.148927.

Marslin, G., Sheeba, C.J., Franklin, G., 2017. Nanoparticles alter secondary metabolism in plants via ROS burst. Front. Plant Sci. 8. https://doi.org/10.3389/fpls.2017.00832.

Milewska-Hendel, A., Zubko, M., Stróż, D., Kurczyńska, E. U., 2019. Effect of nanoparticles surface charge on the *Arabidopsis thaliana* (L.) roots development and their movement into the root cells and protoplasts. Int. J. Mol. Sci. 20 (7), 1650. https://doi.org/10.3390/ijms20071650.

Molnár, Á., Rónavári, A., Bélteky, P., Szőllősi, R., Valyon, E., Oláh, D., Rázga, Z., Ördög, A., Kónya, Z., Kolbert, Z., 2020. ZnO nanoparticles induce cell wall remodeling and modify ROS/RNS signalling in roots of Brassica seedlings. Ecotoxicol. Environ. Saf. 206, 111158. https://doi.org/10.1016/j.ecoenv.2020.111158.

Muratore, C., Espen, L., Prinsi, B., 2021. Nitrogen uptake in plants: the plasma membrane root transport systems from a physiological and proteomic perspective. Plants 10 (4), 681. https://doi.org/10.3390/plants10040681.

Nair, P.M.G., Chung, I.M., 2014. Impact of copper oxide nanoparticles exposure on *Arabidopsis thaliana* growth, root system development, root lignificaion, and molecular level changes. Environ. Sci. Pollut. Control Ser. 21 (22), 12709–12722. https://doi.org/10.1007/s11356-014-3210-3.

O'Brien, J.A., Benková, E., 2013. Cytokinin cross-talking during biotic and abiotic stress responses. Front. Plant Sci. 4. https://doi.org/10.3389/fpls.2013.00451.

Perozo, E., Cortes, D.M., Sompornpisut, P., Kloda, A., Martinac, B., 2002. Open channel structure of MscL and the gating mechanism of mechanosensitive channels. Nature 418 (6901), 942–948. https://doi.org/10.1038/nature00992.

Rajkumari, N., 2020. Silver Nanoparticles for Biolitics Based Gene Delivery in Plants. Department of Plant Biotechnology, College of Agriculture, Vellayani.

Rajput, P., Mondal, G., Dutta, P., Upadhyay, S.K., Alliluev, I., Bhattacharyya, P., Rajput, V.D., 2026. Nanotoxicomics: understanding the molecular toxicity of nanoparticles and microplastics in plants. Environ. Nanotechnol. Monitor. Manage., 101159 https://doi.org/10.1016/j.enmm.2026.101159.

Rastogi, A., Zivcak, M., Tripathi, D.K., Yadav, S., Kalaji, H. M., 2019. Phytotoxic effect of silver nanoparticles in *Triticum aestivum*: improper regulation of photosystem I activity as the reason for oxidative damage in the chloroplast. Photosynthetica 57 (1), 209–216. https://doi.org/10.32615/ps.2019.019.

Rossi, L., Fedenia, L.N., Sharifan, H., Ma, X., Lombardini, L., 2019. Effects of foliar application of zinc sulfate and zinc nanoparticles in coffee (*Coffea arabica* L.) plants. Plant Physiol. Biochem. 135, 160–166. https://doi.org/10.1016/j.plaphy.2018.12.005.

Salehi, H., Miras-Moreno, B., Rad, A.C., Pii, Y., Mimmo, T., Cesco, S., Lucini, L., 2020. Relatively low dosages of CeO_2 nanoparticles in the solid medium induce adjustments in the secondary metabolism and ionomic balance of bean (*Phaseolus vulgaris* L.) roots and leaves. J. Agric. Food Chem. 68 (1), 67–76. https://doi.org/10.1021/acs.jafc.9b05107.

Schwabe, F., Simon, T., Schulin, R., Rotzetter, A., Stark, W., Albrecht, von Q., Nowack, B., 2015. Dissolved cerium contributes to uptake of Ce in the presence of differently sized CeO_2 -nanoparticles by three crop plants. Metallomics 7 (3), 466–477. https://doi.org/10.1039/c4mt00343h.

Sembada, A.A., Fukuhara, T., Suzuki, T., Lenggoro, I.W., 2024. Stem cutting: a novel introduction site for transporting water-insoluble particles into tomato (*Solanum lycopersicum*) seedlings. Plant Physiol. Biochem. 206. https://doi.org/10.1016/j.plaphy.2023.108297.

Shomer, I., Novacky, A.J., Pike, S.M., Yermiyahu, U., Kinraide, T.B., 2003. Electrical potentials of plant cell walls in response to the ionic environment. Plant Physiol. 133 (1), 411–422. https://doi.org/10.1104/pp.103.024539.

Wagner, G., Korenkov, V., Judy, J.D., Bertsch, P.M., 2016. Nanoparticles composed of Zn and ZnO inhibit *Peronospora tabacina* spore germination *in vitro* and *P. tabacina* infectivity on tobacco leaves. Nanomaterials 6 (3). https://doi.org/10.3390/nano6030050.

Wang, C., et al., 2022a. Carbon dots improve nitrogen bioavailability to promote the growth and nutritional quality of soybeans under drought stress. ACS Nano 16, 12415–12424. https://doi.org/10.1021/acsnano.2c03591.

Wang, Y., et al., 2022b. Therapeutic delivery of nanoscale sulfur to suppress disease in tomatoes: in vitro imaging and orthogonal mechanistic investigation. ACS Nano 16, 11204–11217. https://doi.org/10.1021/acsnano.2c04073.

Wang, C., Liu, X., Chen, F., Yue, L., Cao, X., Li, J., Cheng, B., Wang, Z., Xing, B., 2022c. Selenium content and nutritional quality of *Brassica chinensis* L enhanced by selenium engineered nanomaterials: the role of surface charge. Environ. Pollut. 308. https://doi.org/10.1016/j.envpol.2022.119582.

Wheeler, T.D., Stroock, A.D., 2008. The transpiration of water at negative pressures in a synthetic tree. Nature 455 (7210), 208–212. https://doi.org/10.1038/nature07226.

Wong, M.H., Misra, R.P., Giraldo, J.P., Kwak, S.Y., Son, Y., Landry, M.P., Swan, J.W., Blankschtein, D., Strano, M.S., 2016. Lipid exchange envelope penetration (LEEP) of nanoparticles for plant engineering: a universal localization mechanism. Nano Lett. 16 (2), 1161–1172. https://doi.org/10.1021/acs.nanolett.5b04467.

Wu, H., Li, Z., 2022. Nano-enabled agriculture: how do nanoparticles cross barriers in plants? Plant Commun. 3 (6), 100346. https://doi.org/10.1016/j.xplc.2022.100346.

Wu, H., Shabala, L., Barry, K., Zhou, M., Shabala, S., 2013. Ability of leaf mesophyll to retain potassium correlates with salinity tolerance in wheat and barley. Physiol. Plantarum 149 (4), 515–527. https://doi.org/10.1111/ppl.12056.

Zeng, Y., Himmel, M.E., Ding, S.Y., 2017. Visualizing chemical functionality in plant cell walls. Biotechnol. Biofuels 10 (1). https://doi.org/10.1186/s13068-017-0953-3.

Zhao, L., Lu, L., Wang, A., Zhang, H., Huang, M., Wu, H., Xing, B., Wang, Z., Ji, R., 2020. Nano-biotechnology in agriculture: use of nanomaterials to promote plant growth and stress tolerance. J. Agric. Food Chem. 68 (7), 1935–1947. https://doi.org/10.1021/acs.jafc.9b06615.

Zhou, H., Wu, H., Zhang, F., Su, Y., Guan, W., Xie, Y., Giraldo, J.P., Shen, W., 2021. Molecular basis of cerium oxide nanoparticle enhancement of rice salt tolerance and yield. Environ. Sci. Nano 8 (11), 3294–3311. https://doi.org/10.1039/d1en00390a.

Zhu, J., Wang, J., Zhan, X., Li, A., White, J.C., Gardea-Torresdey, J.L., Xing, B., 2021. Role of charge and size in the translocation and distribution of zinc oxide particles in wheat cells. ACS Sustain. Chem. Eng. 9 (34), 11556–11564. https://doi.org/10.1021/acssuschemeng.1c04080.

CHAPTER

4

Translocation and bioaccumulation of nanoparticles in plant tissues

Haider Sultan[1], Jingdong Chen[1], Amjad Mehsud[2], Aqsa Bibi[3], Heping Wan[1] and Changli Zeng[1]

[1]Hubei Engineering Research Center for Protection and Utilization of Special Biological Resources in the Hanjiang River Basin, College of Life Sciences, Jianghan University, Wuhan, China; [2]Key Laboratory of Agro-Forestry Environmental Processes and Ecological Regulation of Hainan Province, School of Environmental Science and Engineering, Hainan University, Haikou, China; [3]School of Chinese Medicine, Medical Department, Jianghan University, Wuhan, China

1. Introduction

Over the past 2 decades, the rapid expansion of nanotechnology has led to an unprecedented increase in the production, application, and environmental dissemination of engineered nanoparticles (ENPs). These nanoscale materials, typically ranging from 1–100 nm, possess unique physicochemical properties including increased surface area-to-volume ratio, quantum effects, and surface reactivity that differentiate them fundamentally from their bulk counterparts. While these characteristics enable remarkable applications in medicine, catalysis, energy systems, and agriculture, they also introduce new biological interactions and ecological behaviors that are not easily predicted from conventional toxicological knowledge (Oliveira et al., 2023). As plants represent the base of terrestrial food webs and serve as primary biological interfaces between soil, water, and the atmosphere, understanding how nanoparticles interact with plant systems has become an essential component of modern environmental nanoscience. Plants are sessile organisms that constantly interact with complex environmental matrices containing both natural and anthropogenic particles (Zijun Wang et al., 2023). With the growing use of nano-enabled fertilizers, nano-pesticides, and soil amendments, deliberate addition of nanoparticles to agricultural fields is increasing. Simultaneously, indirect exposure pathways such as atmospheric deposition, wastewater irrigation, biosolids application, industrial runoff, and degradation of nano-containing products release nanoparticles into agroecosystems. Consequently, plant tissues have become both intentional targets and unintended sinks for nanoparticles, making them critical subjects for

Nanotechnology in Photosynthesis
https://doi.org/10.1016/B978-0-443-44376-3.00005-7

evaluating nanoparticle fate, behavior, and ecological safety (Ghassemi-Golezani and Rahimzadeh, 2024; Hao Chen et al., 2024). The interaction between plants and nanoparticles is governed by a series of physicochemical, anatomical, and physiological events (Sultan et al., 2025). Nanoparticles must first encounter plant surfaces, roots, leaves, or seeds before breaching biological barriers such as the root epidermis, the cuticular membrane, or seed coat (Khan et al., 2025; Nauman Khan et al., 2024). Once internalized, nanoparticles can negotiate through cellular structures, cross the plasma membrane, and engage in intracellular trafficking processes involving the cytoplasm, vacuoles, endosomes, and organelles (Wang et al., 2012). These processes are strongly influenced by particle attributes such as size, shape, crystallinity, surface charge, coating, agglomeration state, and dissolution potential (Khan et al., 2024). Likewise, plant characteristics including cell wall porosity, vascular anatomy, transpiration rate, and developmental stage play decisive roles in determining the efficiency of nanoparticle uptake and mobility (Djanaguiraman et al., 2024). One of the most scientifically intriguing aspects of nanoparticle–plant interactions is the ability of nanoparticles to enter the vascular system and undergo long-distance translocation (Oliveira et al., 2023). Unlike larger particles, nanoparticles can move through the xylem driven by transpiration flow or be redistributed through the phloem in accordance with source–sink dynamics (Wang et al., 2012). This capability allows nanoparticles to reach aerial tissues, reproductive organs, and storage structures, resulting in bioaccumulation in fruits, grains, seeds, and leaves (Cifuentes et al., 2010). Such accumulation raises important questions regarding food safety, trophic transfer, and potential impacts on human and animal health. At the same time, controlled translocation of beneficial nanoparticles can enhance nutrient use efficiency, stimulate growth, or deliver agrochemicals more effectively, underscoring the dual nature of nanoparticle plant interactions as both opportunity and risk. Mechanistically, the behavior of nanoparticles in plant tissues is highly complex due to dynamic transformations that occur within biological environments (Xiong et al., 2021). Nanoparticles may undergo dissolution, oxidation–reduction reactions, sulfidation, or interact with biomolecules such as proteins, polysaccharides, organic acids, and phenolic compounds. These transformations not only alter the chemical identity and mobility of nanoparticles but also influence toxicity profiles and biological responses (Zhang et al., 2020). Plants themselves respond to nanoparticle exposure through changes in gene expression, metabolic pathways, antioxidant defense systems, and hormone signaling networks. The interplay of these responses determines whether nanoparticles will be sequestered (e.g., into vacuoles), degraded, or transported further into the vascular system (Wei et al., 2024). The scientific importance of studying translocation and bioaccumulation of nanoparticles extends beyond plant physiology into broader environmental and societal dimensions (Munir et al., 2023). From an ecological standpoint, accumulation of nanoparticles in plant tissues can facilitate entry into higher trophic levels, potentially altering soil–plant–herbivore–predator interactions. From an agricultural perspective, controlled delivery of engineered nanomaterials has potential to improve crop nutrition, stress tolerance, and productivity, but only if their movement and accumulation patterns are well understood. From a human-health perspective, assessing the potential presence of nanoparticles in edible plant parts is essential for establishing regulatory guidelines, exposure limits, and product safety standards. Despite significant progress, many scientific gaps remain. The size thresholds that govern nanoparticle passage across cell walls and membranes are not fully understood. The relative contributions of apoplastic versus symplastic pathways remain debated for different materials. The conditions under which nanoparticles preferentially accumulate in roots versus shoots differ widely among species and NP types (Gao et al., 2023). Furthermore, the long-term ecological consequences of chronic low-dose nanoparticle exposure in agricultural soils are still largely unknown (Abbas et al., 2020). These unresolved questions emphasize the need for an integrated, multidisciplinary approach that combines plant physiology, material science, soil chemistry, imaging technologies, and molecular biology. This chapter addresses these knowledge gaps by systematically examining the mechanisms of nanoparticle entry, translocation, and accumulation within plant tissues. It synthesizes current scientific understanding of the pathways that nanoparticles use to breach plant barriers, the intracellular processes that govern their movement, and the physiological and biochemical factors that shape their distribution across organs. By integrating mechanistic insights with environmental and agricultural implications, the chapter aims to provide a comprehensive foundation for evaluating the role of nanoparticles in plant systems and guiding the future development of safe and effective nanotechnologies.

2. Pathways of nanoparticle entry into plants

Nanoparticles can enter plants through two main pathways: root uptake from the soil and foliar uptake via aerial parts such as leaves. These entry routes allow nanoparticles to interact with plant tissues, potentially leading to their absorption, translocation, and accumulation within the plant system. The efficiency and mechanisms of uptake depend on various factors, including the physicochemical properties of the nanoparticles and the specific barriers presented by plant structures. As summarized in Table 4.1, nanoparticle entry into plants occurs primarily through root uptake and foliar pathways, with each route governed by distinct anatomical barriers, transport drivers, and particle size constraints. Understanding these pathways is essential for applications in nanoagriculture, environmental remediation associated with nanoparticle exposure in plants.

2.1 Root uptake

Roots are the primary interface between plants and the soil environment, making them the most significant entry point for nanoparticles (NPs) entering terrestrial ecosystems (Afzal et al., 2022). Root uptake of nanoparticles is a multifactorial process controlled by soil chemistry, nanoparticle physicochemical properties, root structural characteristics, and dynamic interactions occurring at the rhizosphere level (Gao et al., 2023; Wei et al., 2021). Unlike conventional ions or molecules, nanoparticles must first navigate through physical barriers, selective cell wall pores, and physiological transport mechanisms, making the root uptake process highly size- and property-dependent (Xiong et al., 2021). Understanding these mechanistic pathways is essential for predicting nanoparticle fate in planta, assessing ecological risks, and designing nano-enabled agricultural inputs.

2.1.1 Interaction of nanoparticles with the rhizosphere

Before reaching the root surface, nanoparticles undergo significant transformations within the rhizosphere, a biologically active zone influenced by root exudates, microorganisms, and soil minerals. Root exudates comprising organic acids, amino acids, polysaccharides, phenolic compounds, and enzymes which can alter nanoparticle stability, aggregation, dissolution, or surface chemistry. For example: Organic acids such as citrate and malate can chelate metal ions, increasing

TABLE 4.1 Comparison of nanoparticle entry pathways in plants.

Entry pathway	Primary barrier(s)	Drivers of entry	Typical NP size range	Examples of NP types commonly entering this route
Root uptake	Cell wall, endodermis, casparian strip	Water flow, root exudate interactions	<50 nm for efficient internalization	Metal oxides (ZnO, CuO), carbon dots, Fe_3O_4
Foliar uptake (stomatal)	Stomatal pore size, guard cell membrane	Transpiration pull, capillary forces	<200 nm	Ag NPs, TiO_2, SiO_2
Foliar uptake (cuticular)	Cuticle waxes, cutin matrix	Humidity-induced swelling, surfactants	<5–20 nm	Carbon dots, ultra-small metal nanoclusters
Seed uptake	Seed coat (testa), micropyle	Imbibition-driven water flow	<50 nm	ZnO, Fe-based nanoparticles, QDs

dissolution of metal oxide nanoparticles (e.g., ZnO, CuO) (Zhang et al., 2020). Polysaccharides and mucilage can stabilize or destabilize nanoparticles by promoting steric hindrance or bridging flocculation. Microbial biofilms may act as sinks for nanoparticles or mediate redox reactions that change particle valence states (Abbas et al., 2020). These transformations define the "biologically available" nanoparticle fraction that ultimately interacts with the root epidermis (Khan et al., 2024). Thus, root uptake is not determined solely by original NP properties but also by their condition at the moment of root contact. Fig. 4.1 illustrates how rhizosphere processes, including nanoparticle aggregation, dissolution, and corona formation, modify nanoparticle bioavailability and influence subsequent root uptake.

2.1.2 *Penetration of the root surface*

The first physical barrier encountered by nanoparticles is the root epidermis. For water and dissolved ions, entry is relatively straightforward, but nanoparticles must navigate through two broad pathways:

(a) *Apoplastic Pathway:* Movement through Cell Walls and Intercellular Spaces: The apoplastic route involves NP movement through the porous cell wall matrix composed of cellulose, hemicellulose, and pectins. The cell wall pore size typically ranges from 5–20 nm, creating a size-dependent filter that allows only sufficiently small nanoparticles or nanoparticle aggregates to pass through (Gao et al., 2023; Kumar et al., 2025; Schwab et al., 2016). Key aspects include:

- **Size exclusion:** Nanoparticles larger than approximately 20–30 nm generally cannot penetrate intact cell walls unless structural modifications occur.
- **Surface charge effects:** Negatively charged cell wall pectins readily bind positively charged nanoparticles, potentially immobilizing them.
- **Chemical interactions:** Redox-active nanoparticles (e.g., Ag NPs) may undergo surface reactions with wall components, altering mobility. In some cases, environmental stress or nanoparticle-induced cell wall remodeling may temporarily increase pore size, enhancing permeability.

(b) ***Symplastic Pathway: Entry through living cells and plasmodesmata.***
Nanoparticles that cross the plasma membrane enter the symplast, the interconnected network of plant cell cytoplasm linked through plasmodesmata (Lv et al., 2019). Entry into the symplast requires:

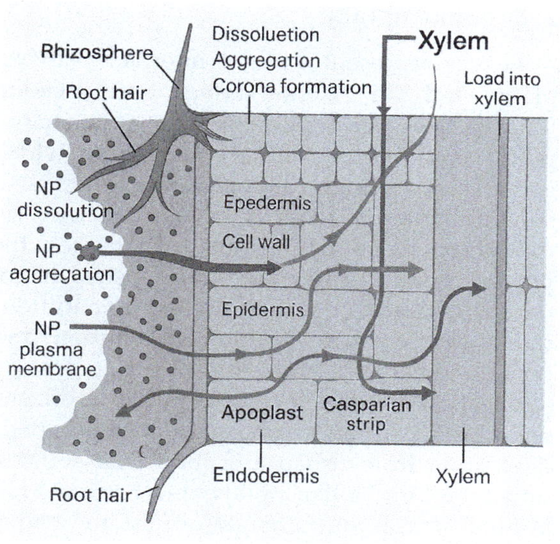

FIGURE 4.1 Transformation and uptake of nanoparticles in the rhizosphere, including aggregation, dissolution, corona formation, and loading into the xylem.

- Endocytosis, where vesicles form around nanoparticles and internalize them into the cytoplasm. This is currently the most widely supported mechanism for uptake of metal and carbon nanomaterials.
- Transporter-mediated interactions, in which nanoparticles or dissolved ionic species interact with nutrient transport channels (e.g., Fe, Zn, Ca transporters).
- Direct membrane diffusion, possible only for ultra-small nanoparticles (<5 nm), such as carbon quantum dots or metal clusters.
  Once inside the symplast, nanoparticles may move intercellularly through plasmodesmata, although size constraints (~3–5 nm) make this pathway selective and material-dependent.

2.1.3 *The role of the endodermis and casparian strip*

After passing through epidermal and cortical layers, nanoparticles encounter the endodermis, a critical regulatory layer that controls solute movement into the vascular system (Hernandez-Viezcas et al., 2013). The Casparian strip; a band of lignin and suberin encircling endodermal cells-blocks apoplastic flow, forcing most nanoparticles to enter the symplast to reach the xylem (Schwab et al., 2016).

Key implications include:

- **Barrier function:** Large or aggregated nanoparticles cannot bypass the Casparian strip unless it is structurally compromised.
- **Selective permeability:** Only those NPs that successfully internalize into endodermal cells via endocytosis or dissolution (releasing ions) can progress toward the stele.
- **Developmental variation:** Young roots, lateral root emergence sites, and root tips may have less developed Casparian strips, providing "leakage points" for enhanced NP entry.

Thus, the endodermis acts as a biological checkpoint, controlling whether nanoparticles accumulate in roots or gain access to aerial tissues (Zhao et al., 2012). As shown in Fig. 4.2,

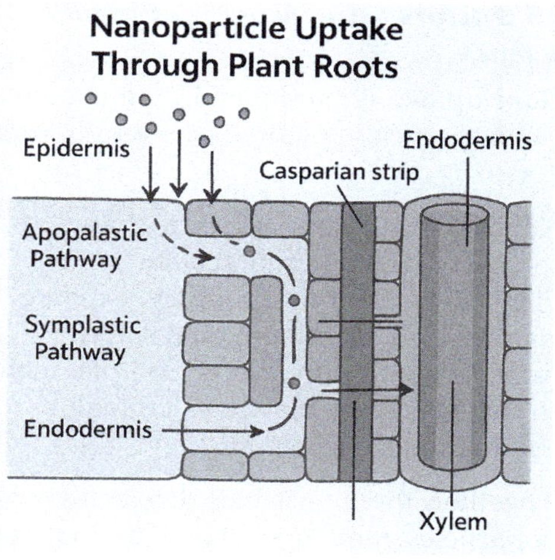

FIGURE 4.2 Apoplastic and symplastic pathways governing nanoparticle transport across root tissues and entry into the xylem.

nanoparticles entering through the root epidermis may follow either apoplastic or symplastic pathways, with the Casparian strip acting as a critical checkpoint regulating access to the vascular system.

2.1.4 *Entry into the xylem and onward transport*

When nanoparticles reach the stele, they may enter xylem vessels and move upward via transpiration-driven flow (de la Rosa et al., 2021). This process marks the transition from root uptake to long-distance translocation. The likelihood of xylem entry depends on:

- Nanoparticle size (small particles enter more readily)
- Stability in the xylem sap (aggregation can lead to entrapment)
- Surface chemistry and interactions with xylem cell walls

Dissolution of metal-based nanoparticles into ionic species (e.g., Zn^{2+}, Cu^{2+}) significantly enhances their mobility because ions are readily transported systemically (Kumar et al., 2025; Zhang et al., 2020) (P.).

2.1.5 *Factors influencing root uptake efficiency*

Root uptake is modulated by a combination of nanoparticle, plant, and environmental factors:

- Nanoparticle properties: Size, charge, coating, solubility, agglomeration state
- Plant traits: Root morphology, exudate composition, developmental stage
- Environmental conditions: Soil pH, ionic strength, organic matter, microbial communities

Together, these variables determine whether nanoparticles remain confined to root tissues, accumulate in specific cell types, or progress toward long-distance translocation (Upadhyay et al., 2025; Afzal et al., 2022).

2.2 Foliar uptake

Leaves represent a major interface between plants and the external atmosphere, making foliar surfaces an important pathway for nanoparticle (NP) entry. Unlike root uptake, which is mediated through soil interactions and rhizosphere dynamics, foliar uptake is governed by the structural complexity of leaf surfaces, the physicochemical properties of deposited nanoparticles, and the environmental conditions influencing particle adhesion, dissolution, and penetration (Hong et al., 2021). With increasing application of nano-enabled foliar fertilizers, nanopesticides, and atmospheric nanoparticle deposition from urban and industrial sources, understanding foliar uptake mechanisms has become crucial for both agricultural innovation and environmental safety assessment. Foliar uptake of nanoparticles involves a sequence of steps beginning with nanoparticle deposition on the leaf surface, followed by penetration through physical or biological openings, and finally internalization into leaf tissues (Xueran Wang et al., 2023). These processes are influenced by factors such as leaf wettability, cuticle thickness, stomatal morphology, humidity, and nanoparticle surface chemistry.

2.2.1 *Deposition and interaction with the leaf surface*

The leaf surface is covered by a multilayered structure known as the cuticle, composed of cutin, waxes, and embedded polysaccharides. The cuticle functions as a hydrophobic barrier that limits water loss and prevents the entry of pathogens and particulate matter (Schwab et al., 2016). Nanoparticles that land on the leaf surface by foliar spraying, aerosol deposition, or atmospheric transport must first adhere to this cuticular layer.

Key interactions at this stage include:

- Surface adhesion: Determined by leaf surface hydrophobicity, presence of trichomes, epicuticular wax crystallinity, and particle–surface electrostatic forces.

- Aggregation or dispersion: Nanoparticles may aggregate under low humidity, reducing penetration potential, whereas high humidity can facilitate dispersion and spreading.
- Dissolution: Metal oxide nanoparticles may partially dissolve in leaf surface moisture, producing ionic species that can penetrate more easily than intact particles.

These early interactions govern whether nanoparticles remain on the leaf surface or proceed toward deeper tissue penetration (Schwab et al., 2016).

2.2.2 Penetration through stomatal openings

Stomata are the most widely recognized entry route for foliar nanoparticle uptake. These microscopic pores, bordered by guard cells, regulate gas exchange and transpiration. Their dimensions, typically 200–500 nm in width when open, can allow the passage of nanoscale materials under suitable conditions (Avellan et al., 2021).

Mechanisms of stomatal penetration include:

(a) Direct entry through open stomata
Nanoparticles may physically enter stomatal pores when the stomata are open, particularly for spherical particles and carbon-based nanomaterials with diameters below ~200 nm. The transpiration stream can pull nanoparticles inward, facilitating entry into the substomatal cavity and eventually leaf mesophyll tissues (Khan et al., 2022).

(b) Capillary-driven transport
Under high humidity or spray application conditions, water films form over stomatal openings. Nanoparticles suspended in these droplets may be drawn into stomatal pores via capillary forces, enhancing uptake, especially for hydrophilic nanoparticles.

(c) Stomatal endocytosis and guard cell interactions
Studies indicate that certain nanoparticles can interact with guard cell membranes, triggering endocytosis-like processes. Though less understood than mechanical pore entry, this mechanism may be relevant for ultra-small nanoparticles (<50 nm) (Avellan et al., 2021). Stomatal uptake efficiency depends strongly on stomatal density, size, opening state, and environmental cues such as light, CO_2 concentration, and plant water status (Shen et al., 2020). Fig. 4.3 further illustrates the mechanistic diversity of foliar nanoparticle uptake, highlighting the roles of stomatal openings, cuticular pores and cracks, and epidermal imperfections in facilitating particle entry.

2.2.3 Penetration through the cuticular layer

Although the cuticle is traditionally considered impermeable to particles, numerous studies show that nanoparticles can penetrate through this layer under specific conditions (Kumar et al., 2025). Several pathways contribute to cuticular penetration:

(a) Cuticular pores and nanochannels
The cuticle contains aqueous pores with estimated diameters between 0.5 and 4 nm small enough to restrict most particles but potentially allowing passage of:

- Ultra-small nanoparticles
- Metal ions derived from NP dissolution
- Carbon quantum dots

Even when particles exceed pore size, plasticization of the cuticle under high humidity or chemical modification (by surfactants or organic solvents) can increase permeability.

(b) Diffusion through cuticular cracks and imperfections

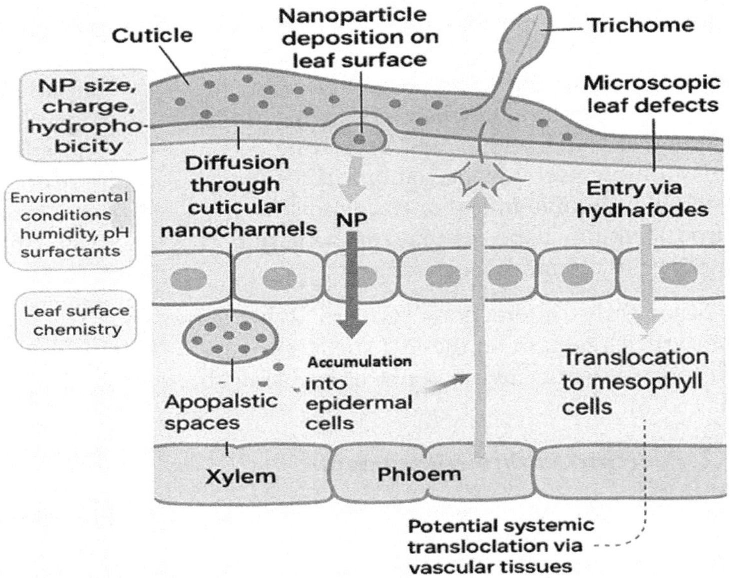

FIGURE 4.3 Schematic representation of foliar nanoparticle uptake through cuticular diffusion, stomatal openings, hydathodes, and subsequent translocation via vascular tissues.

Mechanical stress, environmental damage, herbivory, and leaf age can produce microcracks in the cuticle. Nanoparticles may exploit these defects to gain entry into epidermal tissues.

(c) Cuticular penetration enhanced by surfactants
Many foliar agricultural formulations contain surfactants that reduce surface tension, improve spreading, and increase cuticular solubility. These additives can significantly increase nanoparticle penetration depth (Hong et al., 2021).

2.2.4 *Entry through specialized leaf structures*

In addition to stomata and the cuticle, several specialized structures facilitate nanoparticle entry:

(a) ***Hydathodes:*** Located at leaf margins or tips, hydathodes excrete guttation droplets and maintain a direct connection to the vascular system (Larue et al., 2014). Because hydathode pores remain open, nanoparticles present in guttation fluid or deposited externally may penetrate easily and move into xylem tissues.

(b) ***Trichomes (Leaf Hairs):*** Trichomes can capture, trap, or even absorb nanoparticles due to their porous cell walls. Glandular trichomes may also accumulate particles from atmospheric deposition (Schwab et al., 2016).

(c) ***Lenticular Openings and Epidermal Defects:*** Less common in leaves than stems, these structures can nonetheless permit localized entry of nanoparticles. These alternative pathways may be particularly relevant in species with dense trichome coverage or naturally leaky epidermal regions.

2.2.5 *Intracellular internalization and movement within leaf tissues*

Once nanoparticles cross the cuticle or stomata, they come into contact with epidermal or mesophyll cells. Internalization can occur through:

- Endocytosis: The predominant mechanism for metal and carbon nanoparticles.
- Membrane penetration: Possible for ultra-small or highly hydrophobic nanoparticles.
- Transporter interactions: Ionic species released from dissolving nanoparticles may utilize nutrient transport proteins.

After internalization, nanoparticles may localize in:

- Cytoplasm
- Vacuoles (detoxification compartments)
- Chloroplasts (particularly carbon-based nanoparticles)
- Intercellular spaces

Movement through mesophyll tissues depends on particle size, surface chemistry, and interactions with cell walls. As illustrated in Fig. 4.4, nanoparticles deposited on leaf surfaces can enter plant tissues through multiple foliar pathways, including diffusion across the cuticle, stomatal penetration, and entry via specialized structures such as trichomes and hydathodes.

2.2.6 Factors influencing foliar uptake efficiency

Multiple factors govern the rate and extent of nanoparticle penetration through leaves. The efficiency of foliar nanoparticle uptake is governed by a combination of nanoparticle properties, plant anatomical traits, and environmental conditions, as summarized in Fig. 4.5 The extent of foliar penetration depends on nanoparticle characteristics (size, shape, charge, hydrophobicity, dissolution behavior, and formulation additives), plant traits (cuticle thickness, stomatal patterning, trichome coverage, leaf age), and environmental conditions (humidity, light, temperature, presence of surfactants) (Kohay et al., 2024; Xueran Wang et al., 2023). These elements together determine whether particles remain at the leaf surface, penetrate superficially, or move into internal

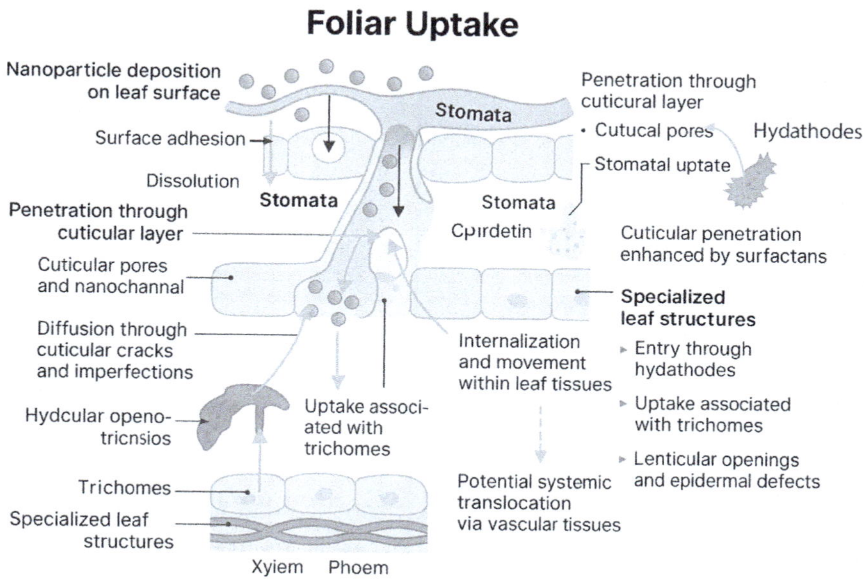

FIGURE 4.4 Multiple foliar entry routes for nanoparticles, including stomatal uptake, cuticular pores and cracks, trichomes, and hydathodes.

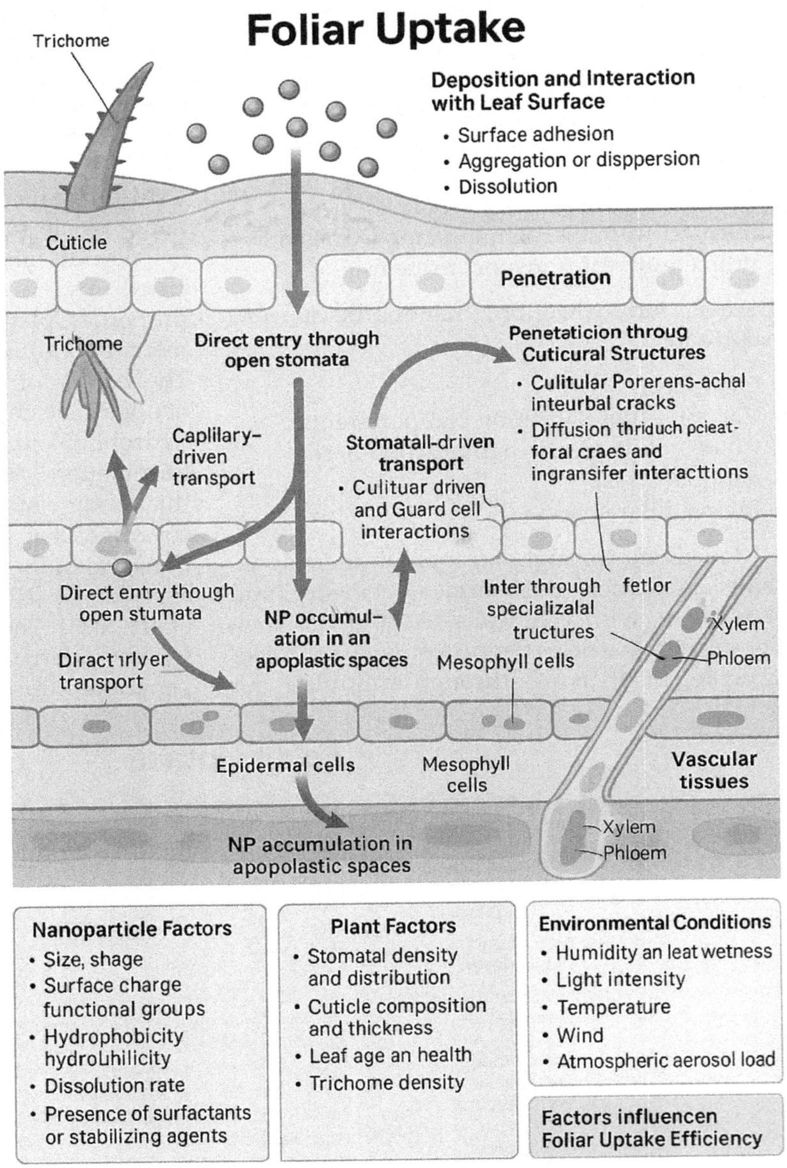

FIGURE 4.5 Nanoparticle, plant, and environmental factors governing foliar uptake efficiency.

tissues and the vasculature. These combined factors determine whether nanoparticles remain superficial, penetrate epidermal tissues, or reach deeper mesophyll and vascular systems.

2.3 Seed uptake

Seeds represent a unique and increasingly important point of nanoparticle (NP) entry into plant systems. Unlike roots and leaves, which encounter nanoparticles throughout the plant's life cycle, seeds interact with nanoparticles primarily during early developmental stages imbibition, germination, and seedling establishment (Acharya et al., 2020). These stages involve extensive physiological changes, including rehydration, enzyme activation, cell expansion, and mobilization of stored reserves. Because seed coats hydrate, swell, and undergo structural loosening during germination, they create temporally dynamic windows in which nanoparticles can penetrate internal tissues (Mahakham et al., 2017). Moreover, the growing popularity of seed priming with nanoscale materials has intensified scientific interest in understanding the mechanisms of NP uptake through seeds and their subsequent impact on plant developmental biology (Yang et al., 2024). Seed priming with nanoparticles has gained attention due to its potential to enhance germination and early growth. During imbibition, seeds absorb water along with dissolved or suspended nanoparticles, enabling NPs to enter embryonic tissues (Kasote et al., 2019). Although this route is less studied, evidence suggests that nanoparticles can remain within growing tissues and influence early seedling development. Seed uptake encompasses the adhesion, penetration, internal diffusion, and systemic movement of nanoparticles introduced through intentional seed treatments or environmental exposure. The process depends on seed coat morphology, porosity, integument chemistry, and the physicochemical properties of the nanoparticles. The plant cell wall presents the first intracellular barrier for nanoparticles. With a typical size exclusion limit of 5–20 nm, only nanoparticles smaller than this threshold can passively diffuse through the pores. Larger nanoparticles may cause localized cell wall loosening or become trapped on the surface. Interactions with cellulose, pectin, lignin, and other wall components influence whether nanoparticles accumulate, transform, or move further into the cell (do Espirito Santo Pereira et al., 2021).

2.3.1 *Interaction of nanoparticles with the seed surface*

The first stage of seed uptake involves the interaction between nanoparticles and the dry or hydrated seed coat. The seed coat (testa) is composed of multiple layers epidermis, sclerenchyma, parenchyma each contributing to barrier functions, mechanical protection, and water regulation (Khodakovskaya et al., 2011). Key factors influencing NP interaction with the seed surface include:

- **Surface charge:** Many seed coats possess negatively charged polysaccharides (e.g., pectins), which can attract positively charged nanoparticles or repel negatively charged ones.
- **Hydrophobicity:** Waxy or lignified seed coats limit wettability and reduce NP adhesion unless surfactants or priming solutions alter surface tension.
- **Seed coat ornamentation:** Microstructures such as ridges, pits, or trichomes can trap nanoparticles, increasing residence time and potential penetration.

In dry conditions, nanoparticles often remain adhered to surface irregularities. Upon hydration, swelling of the seed coat and dissolution of mucilage layers can mobilize nanoparticles, facilitating deeper penetration (Mahakham et al., 2017).

2.3.2 *Penetration pathways through seed coats*

Nanoparticles can enter seeds through multiple structural features of the seed coat. One major entry point is the micropyle, a natural opening that allows water uptake during imbibition (Terskikh et al., 2005). Because it provides a direct passage to internal tissues, the micropyle readily admits nanoparticles suspended in water. Capillary forces can pull nanoparticles into this channel, while rapid hydration currents during the earliest stages of germination may carry them deeper into embryonic tissues (Geremew et al., 2025). As a result, micropyle-mediated uptake is most pronounced in the first hours of water absorption. Another route involves seed coat cracks and hydration-induced microfissures (Koizumi et al., 2008). As seeds swell during imbibition, the outer coat-particularly in species with hardened, lignified structures may develop temporary cracks. These openings permit the passage of larger nanoparticles that would otherwise be excluded by intact barriers. Because these fissures appear only during specific hydration phases and close once tissues rehydrate, uptake efficiency can vary significantly depending on when the nanoparticles are introduced (Ameen et al., 2021). Seeds may also take up nanoparticles through natural pores and hilum-associated structures. Some species possess small openings, intercellular spaces, or loosely organized tissues around the hilum that act as passive diffusion channels for hydrated nanoparticles (Avellan et al., 2021). These features provide an additional, species-dependent entry route. Finally, chemical softening of seed coats by priming agents can enhance nanoparticle penetration (Shah et al., 2021). Priming treatments involving osmotic solutions, electrolytes, or biologically active compounds can plasticize or soften the polymers within the seed coat, increasing water permeability and facilitating nanoparticle movement. Agents such as hydrogen peroxide or organic acids may alter the seed coat's chemistry, making it more receptive to nanoparticle diffusion.

2.3.3 *Internal movement of nanoparticles during germination*

Once nanoparticles cross the seed coat, they interact with internal tissues, including the endosperm or perisperm and the embryo itself (Shelar et al., 2021). As germination progresses and hydration resumes, metabolic activity generates internal gradients that help transport nanoparticles throughout the seed (Terskikh et al., 2005). Diffusion and capillary flow in hydrating tissues play a major role, as rehydrated cell walls become more permeable and water channels reopen, allowing nanoparticles to move through both apoplastic and symplastic pathways (Linlin Chen et al., 2024). Nanoparticles may also engage in interactions with storage molecules. Proteins, lipids, and carbohydrates in the seed can bind or adsorb nanoparticles, influencing their mobility and potentially altering their availability for biochemical reactions. Additionally, uptake by embryonic cells can occur through mechanisms such as endocytosis, sequestration within vacuoles, or movement through plasmodesmata, particularly for very small particles under 5 nm (Zhai et al., 2014). These early interactions can shape metabolic activities that ultimately influence seedling vigor and development.

2.3.4 *Systemic movement into emerging seedlings*

As seeds transition into active growth and the radicle emerges, nanoparticles already present in embryonic tissues can become integrated into the developing seedling (Koizumi et al., 2008). This systemic movement occurs as the xylem resumes water transport, the phloem establishes functional connections between cotyledons and the shoot, and stored nutrients are mobilized and redistributed (Sun and Yuan, 2022; Terskikh et al., 2005).

During these processes, nanoparticles or their dissolved ionic forms may be transported along with nutrient flows. Nanoparticles situated in root or shoot primordia may become embedded in new tissues, meaning their influence whether beneficial or harmful can persist throughout the plant's life cycle. This mechanism explains the long-term effects often observed with nanopriming treatments.

2.3.5 *Factors affecting seed uptake efficiency*

The efficiency with which seeds absorb nanoparticles is determined by characteristics of the particles themselves, the seeds, and the environmental conditions during treatment (Yang et al., 2024). Nanoparticle factors such as size, surface charge, solubility, and functionalization strongly influence uptake; for example, ultra-small nanoparticles under 10 nm penetrate more easily, while surface coatings may either enhance or hinder entry. Seed factors include species-specific variations in coat thickness, porosity, chemical composition, dormancy level, and the presence of mucilage, all of which shape how readily nanoparticles can pass through the integument (Dong et al., 2025; Qu et al., 2015; Sun and Yuan, 2022).

Environmental and treatment variables also play a critical role. Conditions such as temperature, water availability, duration of soaking or priming, presence of surfactants or osmotic agents, and the pH or ionic strength of the priming solution can significantly alter seed permeability. Together, these factors determine whether nanoparticles merely remain on the outer surface, reach internal embryonic tissues, or become widely distributed throughout the developing seedling.

3. Intracellular and intercellular movement of nanoparticles

Once nanoparticles (NPs) penetrate external plant barriers such as the root epidermis, leaf cuticle, or seed coat, they encounter a second level of biological filtering: the internal cellular architecture (Shen et al., 2020). Intracellular and intercellular movement governs whether nanoparticles remain confined to initial entry tissues, become sequestered, or disperse systemically through vascular networks. This process involves complex interactions with the cell wall, plasma membrane, cytoskeletal components, endomembrane systems, and intercellular channels such as plasmodesmata (Singh et al., 2021). Following entry into plant tissues, the intracellular and intercellular movement of nanoparticles is regulated by interactions with the cell wall, plasma membrane, and plasmodesmata, as depicted in Fig. 4.6. The internal mobility of nanoparticles depends heavily on their physicochemical traits size, charge, hydrophobicity, shape but also on plant-specific anatomy, developmental stage, and physiological conditions. As outlined in Table 4.2, the internal movement of nanoparticles within plants is regulated by multiple biological barriers, including the cell wall, plasma membrane, plasmodesmata, and vascular tissues, each imposing distinct size and transport constraints.

3.1 Interactions with the cell wall

The plant cell wall is the first internal barrier that nanoparticles encounter after entering plant tissues. Structurally distinct from the plasma membrane, the cell wall forms a robust, semi-permeable network composed primarily of cellulose microfibrils, hemicellulose, pectins, lignin, and various structural proteins (Zhang et al., 2019). Its primary roles include providing

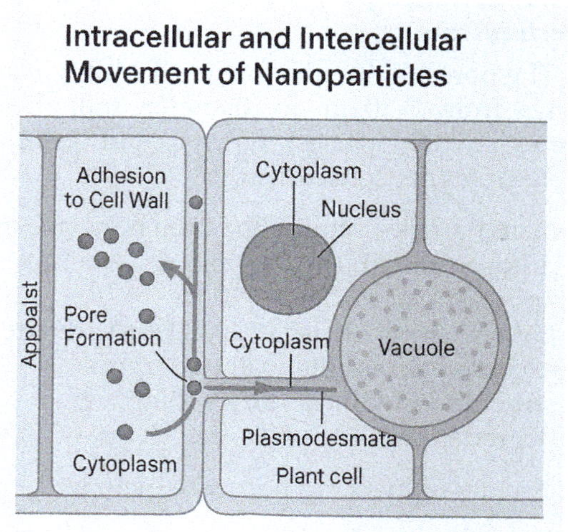

FIGURE 4.6 Intracellular and intercellular movement of nanoparticles through cell walls, cytoplasm, plasmodesmata, and vacuoles.

TABLE 4.2 Biological barriers that regulate nanoparticle movement inside plants.

Plant barrier	Structural features	Barrier function	Nanoparticle behaviors at this barrier	Possible outcomes
Cell wall	Cellulose + pectin matrix	Size-exclusion, charge selectivity	Large NPs trapped; small NPs diffuse; positive NPs bind to pectins	Immobilization, aggregation, limited entry
Plasma membrane	Lipid bilayer with proteins	Selective transport, endocytosis	NP–protein interactions, vesicle formation	Endocytic uptake, membrane penetration
Plasmodesmata	Cytoplasmic nanochannels	Symplastic transport	Ultra-small NPs (<5 nm) may pass; larger NPs excluded	Limited intercellular movement
Xylem vessels	Hollow lignified tubes	Upward bulk flow	Aggregation, deposition on vessel walls	Systemic shoot transport
Phloem sieve tubes	Living tubes + companion cells	Source–sink transport	Ionic derivatives move easily; NPs require loading	Redistribution to reproductive tissues

mechanical support, mediating water and solute movement, and regulating pathogen resistance. For nanoparticles, however, the cell wall acts as a size- and charge-dependent molecular sieve.

3.1.1 *Cell wall pore size and size-exclusion mechanisms*

The pores within the plant cell wall typically range from 5–20 nm in diameter, though this varies across species, tissues, and developmental stages. Consequently:

- Nanoparticles smaller than the pore size may diffuse through the wall matrix, entering the apoplast.
- Larger nanoparticles (>20–30 nm) are often excluded unless the wall undergoes structural modification, swelling, or degradation.

Factors influencing pore size include:

- Growth stage: Rapidly expanding cells have more flexible walls.
- Stress conditions: Abiotic stress or NP exposure can alter wall porosity.
- Cell type: Parenchyma cells typically allow more permeability than sclerenchyma or collenchyma.

This structural selectivity determines whether nanoparticles continue toward the plasma membrane or remain immobilized in the apoplast.

3.1.2 *Electrostatic and chemical interactions*

The polysaccharides forming the cell wall carry abundant carboxyl and hydroxyl groups, imparting a predominantly negative charge. As a result:

- Positively charged nanoparticles (e.g., amine-functionalized NPs, some metal oxides) bind strongly to pectins and hemicelluloses, often becoming immobilized.
- Negatively charged nanoparticles may experience electrostatic repulsion, improving their mobility through the apoplast.
- Metal nanoparticles (Ag, Cu, Zn) may undergo oxidation or partial dissolution due to interactions with phenolic compounds or reactive oxygen species.

These binding interactions not only influence mobility but can trigger biochemical responses such as cell wall stiffening or remodeling.

3.1.3 *Cell wall remodeling in response to nanoparticles*

Exposure to nanoparticles can induce structural and chemical changes in the cell wall, including:

- Upregulation of peroxidases leading to reinforcement of lignin or phenolic cross-linking
- Loosening of cell wall networks through alteration of expansin activity
- Formation of callose deposits as part of defense responses

These modifications can alter subsequent nanoparticle movement, either by restricting mobility (defensive strengthening) or enhancing permeability (stress-induced loosening).

3.1.4 *Fate of nanoparticles in the apoplast*

Nanoparticles trapped in the apoplast may:

- Accumulate locally, forming clusters
- Undergo chemical transformations or dissolution
- Interfere with water and ion transport
- Participate in redox reactions that generate ROS

For nanoparticles capable of crossing the plasma membrane, the cell wall represents only a transient barrier. For others, it becomes a site of long-term immobilization or detoxification.

3.2 Crossing the plasma membrane

The plasma membrane represents a highly selective biological interface that determines whether nanoparticles gain entry into the symplast, the interconnected network of living cell cytoplasm. Unlike the porous cell wall, the plasma membrane is a dynamic phospholipid bilayer with embedded proteins that regulate nutrient uptake, signaling, and cellular homeostasis. Crossing this barrier requires specific physical or biochemical mechanisms.

3.2.1 *Endocytosis: the primary uptake mechanism*

Endocytosis is widely regarded as the dominant pathway for intracellular entry of nanoparticles.

In plant cells, endocytosis occurs mainly through, clathrin-mediated endocytosis (CME), clathrin-independent endocytosis (CIE) and caveolae-like domains (in some species) (Bandmann and Homann, 2012), which further processed in different steps involved:

- Nanoparticles bind to the cell membrane, often via electrostatic or ligand-receptor interactions.
- Membrane invagination forms a vesicle around the nanoparticle.
- The vesicle detaches and enters the cytoplasm.
- Nanoparticles are trafficked via endosomes to vacuoles or other organelles.

Similarly there are many factors enhancing endocytosis such as nanoparticle size <100 nm, moderate positive surface charge, presence of protein coronas or plant-derived biomolecules and surface functionalization that mimics nutrient molecules. Endocytosis explains why nanoparticles often accumulate in vesicles, endosomes, and vacuoles within plant cells.

3.2.2 *Direct penetration and passive diffusion*

Although less common, some nanoparticles especially ultra-small NPs (<5 nm)—can penetrate the plasma membrane directly through:

- **Lipid bilayer insertion:** Hydrophobic particles integrate into the membrane.

- **Transient nanopores:** Formed by mechanical pressure, osmotic imbalance, or membrane destabilization.
- **Passive diffusion:** Applicable to extremely small and neutral particles (e.g., carbon dots, metal clusters).

These pathways enable faster entry but are highly size-dependent and rarely observed for larger nanoparticles.

3.2.3 *Transporter-mediated or ion-mimic pathways*

Nanoparticles and their dissolved ionic forms may enter cells by exploiting nutrient transport proteins that normally facilitate the movement of essential elements. Metal ions released from nanoparticles such as ZnO, CuO, or Fe_3O_4 can be taken up through metal transporters belonging to the ZIP, NRAMP, or IRT families. In some cases, hydrated nanoclusters or very small carbon nanodots may pass through aquaporins, which typically regulate water movement. Nanoparticles carrying functional groups resembling phosphate or sulfate may also utilize the corresponding nutrient transporters (Upadhyay et al., 2025). This mechanism is especially important for nanoparticles capable of partial dissolution, as their ionic products can behave like natural nutrients and therefore move through established cellular uptake systems.

3.2.4 *Post-entry intracellular localization*

After nanoparticles cross the plasma membrane, they can settle in several intracellular compartments, each influencing their behavior and biological impact. Some nanoparticles remain dispersed within the cytoplasm, while others are transported into vacuoles, where they are sequestered as part of the cell's detoxification mechanisms (Adil et al., 2024; Zhai et al., 2014). Vesicular trafficking may direct certain nanoparticles to organelles such as the endoplasmic reticulum or Golgi apparatus. Carbon-based nanoparticles, in particular, have been observed in chloroplasts and mitochondria, where they may interact with metabolic processes (Dang et al., 2024). In rare instances, especially when the particles are extremely small, nanoparticles may even pass through nuclear pores and enter the nucleus. The specific sites where nanoparticles accumulate determine their effects on cellular function, potentially disrupting photosynthesis, respiration, enzymatic activity, or signal transduction pathways.

3.2.5 *Intercellular movement via plasmodesmata*

Nanoparticles that enter the symplast can move between cells through plasmodesmata, the microscopic channels that link adjacent plant cells (Hong et al., 2021). These structures impose a size restriction known as the size exclusion limit (SEL), typically around 3–5 nm, meaning that only ultra-small nanoparticles or their dissociated ionic forms can diffuse through them. Under certain developmental stages or stress conditions, the SEL may widen slightly, but larger nanoparticles are generally excluded due to mechanical constraints. As a result, particles smaller than 5 nm or nanoparticles that degrade into ionic species have the highest potential for symplastic transport. Once past the cell wall, nanoparticles must cross the plasma membrane to enter the symplast. Mechanisms for membrane crossing include endocytosis, transporter-assisted uptake, or passive diffusion for very small particles such as carbon dots or quantum dots. Endocytosis is the most common mechanism for metal-based NPs, involving vesicle formation around the particle. After internalization, NPs may localize in vacuoles, cytoplasm, or organelles depending on their chemical nature (Singh et al., 2024).

4. Translocation within plant tissues

Once nanoparticles (NPs) cross external barriers and intracellular interfaces, their capacity for long-distance movement determines how widely they distribute throughout the plant body. Translocation-movement through the xylem and phloem allows nanoparticles to reach stems, leaves, flowers, fruits, and seeds, influencing physiological processes and contributing to bioaccumulation in edible tissues. Understanding these pathways is critical for evaluating food-chain safety, optimizing nano-enabled fertilizers and pesticides, and predicting nanoparticle fate in agroecosystems.

Nanoparticle translocation involves two primary vascular systems:

1. Xylem—transporting water and dissolved minerals upward from roots to shoots.
2. Phloem—distributing photosynthates and signals bidirectionally among source and sink organs.

The movement of nanoparticles within these vascular pathways depends on particle size, agglomeration behavior, surface chemistry, solubility, and interactions with the vascular cell walls and sap components (Wang et al., 2012).

4.1 Xylem transport

The xylem serves as the main pathway for transporting nanoparticles from the roots to above-ground plant parts. This upward movement is powered by transpiration pull, root pressure, and capillary forces, which together generate a strong negative pressure capable of lifting water along with suspended nanoparticles through the vascular system.

4.1.1 Entry of nanoparticles into the xylem

Before nanoparticles can enter the xylem, they must overcome the barrier formed by the endodermis and its Casparian strip. They may accomplish this through symplastic transport into endodermal cells followed by release into the xylem parenchyma (Feng et al., 2021), by moving through immature or unsuberized regions of the endodermis, or by entering through lateral root emergence zones where the casparian strip is naturally interrupted. Once nanoparticles reach the stele, they encounter xylem vessel elements dead, hollow, lignified cells specialized for efficient water conduction.

4.1.2 Physical and chemical behavior of nanoparticles in xylem sap

Inside the xylem, nanoparticles are suspended in a dilute sap composed mainly of water, inorganic ions, small organic molecules, and occasional proteins. Their behavior in this medium depends on several factors. Particle size and aggregation state influence how long they remain suspended, with smaller nanoparticles remaining mobile while larger aggregates may settle or adhere to vessel walls. Surface charge also plays a role: neutral or negatively charged particles disperse more effectively, whereas positively charged particles tend to bind to negatively charged lignin or pectin. Solubility influences transport as well, since metal oxide nanoparticles may partially dissolve, allowing ions such as Zn^{2+}, Cu^{2+}, or Fe^{3+} to move more readily than intact particles. Additionally, interactions with organic ligands such as phenolics or organic acids can stabilize or destabilize nanoparticles (Yasin et al., 2024; Zia-ur-Rehman et al., 2023). These combined properties determine whether nanoparticles remain mobile, aggregate, undergo chemical changes, or become immobilized.

4.1.3 Mechanisms of xylem transport

Once inside the xylem vessels, nanoparticles are carried upward primarily through bulk flow generated by transpiration. Capillary action in narrow tracheids contributes to their movement, and in certain species, transcellular transfer through xylem parenchyma cells may occur (Feng et al., 2021). Nanoparticles can also move between adjacent vessels through pit membranes, although this process is size-restricted. Particles smaller than roughly 20–50 nm can pass through these porous structures, enabling circulation between older and newly formed vessels.

4.1.4 *Fate of nanoparticles during xylem transport*

As nanoparticles travel through the xylem, several outcomes are possible. They may reach stems and leaves, where they accumulate in vascular or mesophyll tissues, or they may adhere to vessel walls, forming localized accumulation zones. In some cases, aggregated nanoparticles could partially obstruct xylem conduits, potentially impairing water movement. During transport, nanoparticles may also undergo chemical transformations, such as dissolution or redox reactions. Additionally, transfer from the xylem into the phloem can occur via xylem–phloem parenchyma interfaces (Feng et al., 2021). Ultimately, the mobility of nanoparticles within the xylem strongly influences how widely they are distributed throughout the plant's aerial organs.

4.2 Phloem transport

While the xylem supports the upward movement of water and suspended nanoparticles, the phloem provides a bidirectional pathway for transporting photosynthates, hormones, signaling molecules, and, in certain circumstances nanoparticles. Because the phloem links mature leaves (source tissues) to roots, young shoots, flowers, fruits, and seeds (sink tissues), phloem-based nanoparticle movement has important implications for plant development, reproductive structures, and food safety (Pinna and Husted, 2025).

4.2.1 *Entry of nanoparticles into the phloem*

Nanoparticles can access the phloem through several routes. Following foliar application, particles that penetrate the leaf cuticle and mesophyll may move into phloem tissues (Avellan et al., 2021). They can also cross bundle sheath cells and enter companion cells, or transfer from the xylem through specialized xylem-phloem transfer cells and vascular parenchyma. In some cases, surface-bound sugars on nanoparticles may facilitate their phloem loading by promoting interactions with sugar transport pathways (Wang et al., 2012). Because phloem loading involves living cells, nanoparticles must cross plasma membranes, usually via endocytosis or by mimicking ions when dissolved forms are present.

4.2.2 *Behavior of nanoparticles in phloem sap*

Once inside the phloem, nanoparticles encounter a dense, nutrient-rich solution primarily composed of sucrose, other sugars, amino acids, hormones, RNAs, and proteins. This environment differs markedly from the dilute xylem sap. The high osmotic pressure and viscosity can slow the movement of larger nanoparticle aggregates, while the abundance of biomolecules promotes the formation of a biocorona on particle surfaces. Smaller nanoparticles, generally below 10–20 nm, move most effectively through phloem channels, whereas dissolved ionic species readily enter and travel with nutrient streams. Because phloem elements are living cells, their membranes and plasmodesmatal connections impose strict selective barriers that restrict the passage of larger particles (Singh et al., 2021).

4.2.3 *Mechanisms of phloem transport*

Phloem transport operates according to the pressure-flow hypothesis. High turgor pressure in source tissues drives solutes toward sink tissues with lower pressure. Nanoparticles can be carried along by this bulk flow, move symplastically through plasmodesmata linking companion cells and sieve elements, or undergo sequential unloading and reloading during extended transport. However, the small size exclusion limit of plasmodesmata typically around 3–5 nm means that larger nanoparticles cannot readily move intact unless the cell walls loosen or plasmodesmata dilate under stress.

4.2.4 *Consequences of phloem transport*

Movement through the phloem can result in nanoparticle accumulation in sink tissues such as roots or in developing reproductive structures, including fruits and seeds, raising considerations for food safety. Nanoparticles may also reach meristems, where they can influence growth regulation and hormonal signaling. Dissolved ions derived from nanoparticles can be redistributed throughout the plant in ways that resemble nutrient transport. Consequently, phloem mobility is particularly important for nanoparticles introduced through foliar application or those capable of dissolving and integrating into the plant's nutrient transport pathways (Kohay et al., 2024).

4.3 Redistribution between organs

Redistribution of nanoparticles between plant organs is a dynamic, highly regulated process that begins once nanoparticles have entered the vascular system. Their movement throughout the plant is shaped by developmental stage, source sink relationships, transpiration patterns, and any physicochemical changes that occur during transport (Cifuentes et al., 2010). This redistribution ultimately determines where nanoparticles accumulate, whether in vegetative tissues such as roots, stems, and leaves; in metabolically active regions like meristems; or in reproductive organs such as flowers, fruits, and seeds. Understanding these patterns is critical for assessing long-term physiological effects, potential food-chain transmission, and the fate of nanoparticles within the plant body. Importantly, redistribution involves selective loading, unloading, and storage of nanoparticles or their ionic forms rather than simple passive movement.

4.3.1 *Root-to-shoot redistribution via the xylem*

After uptake by roots, nanoparticles are transported upward through the xylem, and their redistribution between roots and shoots depends on several factors. Transpiration strength plays a major role: high transpiration rates common under conditions such as bright light, low humidity, or elevated temperatures enhance nanoparticle movement to leaves and reduce retention in roots, whereas low transpiration favors root accumulation. Particle size also affects redistribution, as nanoparticles smaller than roughly 30–50 nm remain suspended and mobile in the xylem sap, while larger particles or aggregates tend to adhere to vessel walls, block pit membranes, or remain trapped in the stele (Schwab et al., 2016; Zhao et al., 2012). Vascular connectivity influences partitioning as well, with younger leaves that are actively integrating into the xylem network typically receiving larger amounts of nanoparticles. Additionally, transformations during transport, such as partial dissolution of metal oxide nanoparticles into ions, can increase mobility and allow dissolved species to follow nutrient-like pathways into developing tissues.

4.3.2 *Shoot-to-root redistribution via the phloem*

The phloem serves as the main route for downward redistribution, particularly for nanoparticles absorbed through foliage. Source-sink gradients drive movement, with photosynthetically, active leaves acting as sources and roots, meristems, and developing organs functioning as sinks. Phloem mobility is strongly restricted by particle size: nanoparticles under 10–20 nm or dissolved ions can load into sieve elements more efficiently, while particles larger than the plasmodesmatal size exclusion limit of approximately 3–5 nm generally enter only through mechanisms such as endocytosis. Redistribution to root tissues is determined by the extent of phloem loading, unloading at sink areas, and retention within phloem parenchyma, all of which are influenced by the nanoparticles' physicochemical properties (Avellan et al., 2021).

4.3.3 *Redistribution to reproductive organs*

Reproductive structures including flowers, fruits, and seeds act as strong metabolic sinks, making redistribution to these organs particularly important for evaluating food safety, inheritance, and potential reproductive effects (Radziemska et al., 2021). Nanoparticles may accumulate in floral tissues such as sepals, petals, anthers, and stigmatic surfaces through phloem connections active during flower development (Sharma et al., 2021). In some cases, nanoparticles have been detected in pollen, raising questions about pollinator exposure and generational transfer. Developing fruits also attract nanoparticles due to their high carbohydrate demand during filling, leading to accumulation in the peel, flesh, or vascular bundles and to transformations from interactions with sugars and organic acids. Seeds are a major endpoint of phloem-delivered solutes, and nanoparticles may enter cotyledons, embed in embryonic tissues, or remain concentrated in seed coats, with significant implications for edible seed safety, nano-fortification, and potential vertical transmission.

4.3.4 *Redistribution to meristems and growing points*

Meristems, such as shoot and root apical meristems, function as strong sinks due to their rapid cell division and intense metabolic activity (Mehmood et al., 2023). Nanoparticle accumulation in these regions can influence plant growth, alter hormone signaling pathways, or modify gene expression through stress-related or nutrient-related responses. Because meristems regulate the formation of new tissues, even small amounts of nanoparticles in these zones can have disproportionately large physiological effects.

4.3.5 *Sequestration, storage, and transformation during redistribution*

During movement through the plant, nanoparticles may undergo sequestration, storage, or chemical transformation. Vacuolar sequestration is a common cellular strategy to reduce toxicity, limiting further systemic movement and often leading to accumulation in roots or older leaves (Rahimzadeh and Ghassemi-Golezani, 2022). Nanoparticles can also bind strongly to cell walls especially lignified or pectin-rich regions reducing their mobility but promoting long-term retention (Ning et al., 2025). Chemical transformations such as dissolution, oxidation–reduction reactions, sulfidation, or biocorona formation can either increase mobility (as with dissolution into ions) or decrease it (as with aggregation). Under certain developmental or stress conditions, sequestered nanoparticles may be remobilized into the vascular system, such as during leaf senescence or nutrient reallocation.

4.3.6 *Factors governing redistribution patterns*

Redistribution patterns are shaped by interactions among nanoparticle properties, plant physiological factors, and environmental conditions. Key nanoparticle traits such as size, shape, surface charge, solubility, dissolution rate, and surface functionalization affect mobility and organ targeting (Kumar et al., 2025). Plant-related factors, including transpiration rate, source sink dynamics, developmental stage, vascular architecture, and stress conditions like drought or salinity, also strongly influence redistribution. Environmental variables such as temperature, humidity, soil nutrient status, and light intensity further modify how nanoparticles are partitioned among organs (Upadhyay et al., 2025). Together, these elements determine whether nanoparticles accumulate primarily in roots, shoot tips, aging leaves, fruits, or seeds.

5. Bioaccumulation in different plant organs

Bioaccumulation describes the long-term buildup of nanoparticles or their ionic derivatives within plant tissues after initial uptake through roots, leaves, or seeds. Once inside the plant, nanoparticles may become immobilized, stored, chemically transformed, or redistributed depending on physiological conditions and environmental influences (Singh et al., 2024). Whereas translocation focuses on movement, bioaccumulation emphasizes retention, tissue-specific deposition, and final localization (Linlin Chen et al., 2024). Understanding how different organs accumulate nanoparticles is essential for evaluating toxicity risks, nutrient-enhancement strategies, environmental impacts, and food safety. Nanoparticles can accumulate due to physical trapping in apoplastic spaces, adhesion to cell walls and membranes, sequestration in vacuoles, adsorption to biomolecules, precipitation or aggregation, and chemical transformations such as dissolution or oxidation (Isayenkov and Maathuis, 2019). The extent and pattern of accumulation depend on nanoparticle size, solubility, surface chemistry, transport pathway, organ structure, and developmental stage.

5.1 Root accumulation

Roots frequently show the highest levels of nanoparticle accumulation because they are in direct and continuous contact with soil or hydroponic environments. Root-associated bioaccumulation reflects both the initial uptake and the limited mobility of many nanoparticles, particularly those that do not readily move through the xylem.

5.1.1 Mechanisms of root accumulation

Root accumulation occurs through several mechanisms. The first is cell wall binding, where nanoparticles attach to pectins, hemicelluloses, and lignin, causing immobilization and surface aggregation while reducing movement into the symplast (Ning et al., 2025). Positively charged nanoparticles show especially strong affinity for the negatively charged components of root cell walls. Apoplastic trapping is also common; limited pore size, restricted movement near the Casparian strip, and interactions with extracellular enzymes cause deposition in the epidermis and cortex. Once inside the symplast, nanoparticles may undergo vacuolar sequestration, where vacuoles store or detoxify them, often serving as long-term reservoirs (Djanaguiraman et al., 2024). Finally, chemical transformations such as dissolution, redox reactions, and complexation with organic acids modify nanoparticle behavior, influencing mobility and persistence in root tissues.

5.1.2 Spatial distribution in root tissues

Root bioaccumulation typically follows recognizable spatial trends. The epidermis and root hairs exhibit the highest adsorption due to direct environmental exposure. The cortex shows moderate accumulation, particularly in its intercellular spaces, while the endodermis often acts as a selective barrier that restricts further movement (Zhao et al., 2012). The stele generally contains lower nanoparticle levels unless the particles or their ionic forms successfully cross the endodermal boundary. Thus, roots function both as a protective barrier and as a storage site for nanoparticles.

5.1.3 *Implications of root accumulation*

Accumulation in roots can influence water and nutrient uptake, root elongation and branching, and interactions with symbiotic organisms such as rhizobia and mycorrhizal fungi. It can also modify rhizosphere chemistry. Depending on the nanoparticle type and concentration, root bioaccumulation can either cause toxicity or provide beneficial nutrient reserves.

5.2 Stem and leaf accumulation

Nanoparticles that move beyond the root system or are absorbed directly through the leaves can accumulate in stems and leaves. The degree of accumulation is shaped by transpiration rates, vascular connectivity, leaf anatomy, and nanoparticle characteristics.

5.2.1 *Mechanisms of stem accumulation*

In stems, accumulation often occurs along vascular bundles. Nanoparticles traveling in the xylem tend to deposit on vessel walls, tracheids, or pit membranes, especially when they are large or aggregated. Stems' high lignin content encourages strong binding of positively charged nanoparticles, reducing mobility (Ameen et al., 2021). Ray parenchyma cells may also store dissolved ions released through nanoparticle dissolution, aiding in radial movement and retention (Feng et al., 2021).

5.2.2 *Mechanisms of leaf accumulation*

Leaves show complex accumulation patterns due to their diverse tissues and physiological activities. Transpiration-driven flow pulls nanoparticles into leaves, where they deposit in xylem endings, mesophyll tissues, and intercellular spaces. Foliar-applied nanoparticles may enter through stomata, accumulating in stomatal chambers, guard cells, or substomatal cavities (Eichert et al., 2008). Certain nanoparticles, such as carbon dots or TiO_2 particles, can reach chloroplasts, where they may become localized in the stroma or thylakoid membranes, potentially enhancing or disrupting photosynthesis. Hydrophobic nanoparticles or adhesive formulations also bind strongly to the leaf cuticle.

5.2.3 *Physiological consequences of stem and leaf accumulation*

The presence of nanoparticles in stems and leaves can influence photosynthesis, transpiration, stomatal behavior, reactive oxygen species generation, hormonal signaling, and leaf senescence. Depending on their material and concentration, some nanoparticles may improve photosynthetic capacity or stress tolerance, whereas others can trigger oxidative stress or toxicity.

5.3 Accumulation in reproductive tissues

Nanoparticle accumulation in flowers, fruits, and seeds is of critical importance because it determines potential entry into the food chain, transmission to the next generation, and impacts on reproductive biology.

5.3.1 *Accumulation in flowers*

Nanoparticles reach flowers primarily through phloem transport or direct foliar exposure during flowering. Within floral organs, nanoparticles may accumulate in anthers, potentially affecting pollen development; in pollen grains, raising concerns about pollinator exposure and generational transfer; and in ovules, where they may directly influence seed formation. Accumulation in petals and sepals reflects systemic movement rather than functional effects.

5.3.2 *Accumulation in fruits*

Fruits act as strong sinks during development, receiving substantial phloem flow.

Nanoparticles may accumulate in the skin, flesh, or vascular bundles, depending on the transport pathway. Fruit chemistry-including acidity, sugar concentration, and metabolic enzyme activity—can further modify nanoparticles through dissolution, binding, or transformation. Because fruits are consumed, understanding nanoparticle behavior in edible tissues is especially important.

5.3.3 *Accumulation in seeds*

Seeds represent the ultimate deposition site in reproductive tissues. Nanoparticles may reach seeds through maternal phloem delivery, entry via the micropyle during fruit development, or direct penetration of the seed coat when exposed externally. Accumulation commonly occurs in the seed coat due to strong binding, but nanoparticles may also localize in cotyledons or, more rarely, in the embryonic axis. These patterns have significant implications for vertical transmission, seed safety, germination behavior, and ecological impacts.

5.3.4 *Biotransformation in reproductive tissues*

Reproductive tissues have high metabolic activity, which promotes chemical transformations of nanoparticles. These may include dissolution (such as ZnO dissolving into Zn^{2+} in acidic fruit environments), protein corona formation, oxidation, and chelation by organic acids. Such transformations may alter retention, toxicity, and bioavailability.

6. Factors influencing uptake, translocation, and bioaccumulation

The behavior and ultimate fate of nanoparticles in plant systems arise from a complex interaction among nanoparticle properties, plant physiology, and environmental conditions. These combined influences determine how efficiently nanoparticles are taken up, how far they travel through vascular networks, and where they ultimately accumulate within plant organs. Understanding these variables is essential for predicting the outcomes of engineered nanomaterial use in agriculture and for developing safe, effective nano-enabled technologies.

6.1 Nanoparticle properties

The physicochemical characteristics of nanoparticles are among the most influential factors in determining their interactions with plants as presented in Table 4.3. These properties shape nanoparticle solubility, mobility, uptake pathways, and potential toxicity.

6.1.1 *Size and size distribution*

Size is one of the most decisive traits affecting plant–nanoparticle interactions. Smaller nanoparticles, generally under 20–30 nm, more readily penetrate cell walls, move through plasmodesmata, and travel through both xylem and phloem. Larger particles, typically above 40–50 nm, often remain confined to entry sites such as the root epidermis or stomatal openings, or may deposit within vascular tissues. Nanoparticles with a broad size distribution exhibit mixed behavior: smaller fractions may move systemically while larger ones remain localized. Particle size also influences aggregation tendency, surface area-to-volume ratio, and dissolution rate.

6.1.2 *Shape and aspect ratio*

Nanoparticle geometry plays a significant role in determining mobility and barrier interactions. Spherical nanoparticles tend to diffuse more readily through pores in cell walls, while rod-shaped or tubular forms—such as carbon nanotubes—may align with plant tissue structures but can be hindered by steric restrictions. Plate-like nanoparticles often adhere strongly to cell surfaces or stack within cell wall matrices.

TABLE 4.3 Nanoparticle key properties influencing plant uptake, translocation, and bioaccumulation.

Nanoparticle property	Description	Influence on plant interaction
Size	Diameter of particles (1–100 nm)	Smaller NPs penetrate cell walls more easily, travel farther in vascular tissues, and show higher systemic mobility.
Shape	Spherical, rod-like, tubular, plate-like	Spherical NPs move efficiently; high aspect-ratio NPs may get trapped or cause mechanical disruption.
Surface charge (zeta potential)	Positive, negative, or neutral surface potential	Positive charge → strong binding to cell walls; neutral/negative → enhanced mobility in apoplast and vascular tissues.
Surface coating/ Functionalization	Organic polymers, surfactants, biomolecules	Improves dispersion, reduces aggregation, alters toxicity, and may mimic biological ligands to enhance uptake.
Solubility/ Dissolution	Rate at which NPs release ions	High dissolution increases ionic transport via nutrient channels; insoluble NPs accumulate as particulates.
Aggregation state	Stability of NPs in soil/ water	Aggregated NPs have reduced mobility; stable dispersions improve uptake and translocation.

High aspect ratio materials may cause more physical disruption to cells.

6.1.3 *Surface charge (zeta potential)*

Surface charge strongly influences nanoparticle adhesion and movement. Positively charged nanoparticles bind tightly to negatively charged plant cell walls and cuticles, promoting surface adsorption but reducing internalization. Negatively charged nanoparticles experience fewer electrostatic barriers and move more freely through apoplastic pathways. Neutral nanoparticles tend to be the most mobile in both xylem and phloem. Charge also affects suspension stability, with highly charged particles dispersing more readily and avoiding aggregation.

6.1.4 *Surface coating and functionalization*

Surface modifications can dramatically alter nanoparticle behavior in plant environments. Polymer coatings such as PEG or PVP improve dispersibility and reduce toxicity. Biological functionalization with amino acids, sugars, or peptides may enable nanoparticles to mimic natural plant metabolites and increase uptake. Surfactants in foliar formulations enhance cuticle penetration and stomatal entry. Conversely, cationic coatings increase membrane interactions but often reduce long-distance mobility. Surface chemistry also shapes how plant enzymes and biomolecules form a "biocorona" around nanoparticles.

6.1.5 *Solubility and dissolution behavior*

Dissolution is particularly important for metal-based nanoparticles. Materials such as ZnO, CuO, and Ag can partially or fully dissolve, releasing ions that translocate efficiently through nutrient transport pathways. Dissolved ions typically show much higher phloem mobility than intact particles. Insoluble nanoparticles, such as TiO_2 or SiO_2, remain in particulate form and tend to accumulate in apoplastic spaces.

6.1.6 *Aggregation and stability*

Nanoparticles often aggregate in environmental media, which affects their entry, transport, and deposition. Aggregated nanoparticles have reduced uptake efficiency and are more likely to settle on plant surfaces or become trapped in vascular tissues. In contrast, stable, well-dispersed suspensions promote higher uptake and mobility.

6.2 Plant species and physiology

Plant structural and physiological characteristics significantly affect nanoparticle uptake and distribution as shown in Table 4.4.

6.2.1 Root system architecture

Root traits such as root hair density, total surface area, the presence of lateral roots, and root cap integrity influence nanoparticle exposure and absorption. Species with extensive root systems, such as many cereals, may accumulate more nanoparticles simply due to greater soil contact.

6.2.2 Leaf anatomy and stomatal traits

Foliar uptake depends on stomatal density and size, cuticle thickness, trichome abundance, and leaf wettability. Plants with high stomatal density typically allow more nanoparticle entry through stomatal openings, particularly under humid conditions that hydrate stomatal chambers.

6.2.3 Vascular architecture

The organization of xylem and phloem strongly shapes systemic movement. Broad xylem vessels facilitate long-distance upward transport, while highly active phloem networks enhance redistribution to growing tissues. Variations in vascular connectivity can result in preferential nanoparticle accumulation in specific organs.

6.2.4 Developmental stage

Young and developing tissues are generally more permeable and metabolically active. New leaves have high transpiration rates, meristems act as strong sinks, and the immature endodermis in young roots allows increased penetration. Developmental timing therefore affects both uptake and redistribution.

6.2.5 Metabolic activity and stress response

Stress conditions such as salinity, drought, or pathogen infection can modify uptake pathways. Increased transpiration under drought may enhance nanoparticle entry, while biotic stress can alter cell wall porosity or trigger hormonal shifts that change stomatal behavior.

TABLE 4.4 Plant traits influencing nanoparticles uptake and transport.

Plant trait	Specific characteristics	Impact on nanoparticle behavior
Root architecture	Root hair density, root branching, total root surface area	Larger and more branched root systems increase nanoparticle contact and uptake; root hairs enhance nanoparticle binding and absorption.
Leaf anatomy	Stomatal density, cuticle thickness, presence of trichomes	High stomatal density facilitates foliar nanoparticle uptake; thick cuticles restrict penetration; trichomes can trap nanoparticles on leaf surfaces.
Vascular system anatomy	Xylem vessel diameter, phloem loading pathways	Wider xylem vessels promote faster upward transport of nanoparticles; phloem loading mechanisms influence nanoparticle redistribution within the plant.
Developmental stage	Seedling versus mature tissues	Young and actively growing tissues are more permeable and act as strong sinks for nanoparticle accumulation.
Physiological status	Transpiration rate, hormone levels, stress responses	High transpiration rates enhance xylem-mediated nanoparticle transport; stress conditions can alter membrane permeability and increase nanoparticle entry.

Stress often increases nanoparticle internalization by altering physiochemical barriers.

6.2.6 *Species-specific nanoparticle sensitivity*

Different plant species exhibit unique sensitivities and uptake behaviors. Legumes exude organic acids that can alter nanoparticle dissolution, while Brassica species naturally accumulate more metals due to hyperaccumulation traits. Aquatic plants absorb nanoparticles readily due to thin epidermal layers. Thus, plant identity plays a major role in determining nanoparticle fate.

6.3 Environmental conditions

Environmental factors shape nanoparticle behavior before and during plant exposure, influencing mobility, transformation, and bioavailability.

6.3.1 *Soil properties*

Soil pH, texture, and organic matter strongly affect nanoparticle availability. Acidic soils enhance dissolution of metal nanoparticles, while organic-rich soils bind nanoparticles and limit mobility. Clay-rich soils promote aggregation and hinder uptake, whereas sandy soils enhance leaching and root exposure.

6.3.2 *Water chemistry*

Ionic strength, water hardness, and dissolved organic carbon influence nanoparticle stability. High ionic strength promotes aggregation, reducing nanoparticle uptake and mobility.

6.3.3 *Environmental transformations*

Nanoparticles may undergo abiotic transformations that alter their behavior. These include oxidation–reduction reactions in anaerobic soils, sulfidation of metals like silver in sulfur-rich conditions, photochemical modification on leaf surfaces, and coating with humic substances in soil. Such transformations can increase or decrease mobility and toxicity.

6.3.4 *Climatic conditions*

Climate influences plant–nanoparticle interactions through humidity, temperature, and light intensity. High humidity improves foliar uptake by hydrating stomata and swelling the cuticle, while high temperatures increase transpiration rates. Light intensity affects stomatal aperture and xylem flow, and rainfall can redistribute foliar-applied nanoparticles or promote runoff.

6.3.5 *Application method and exposure route*

The manner in which plants are exposed to nanoparticles, whether through soil application, foliar spraying, seed priming, contaminated irrigation water, or atmospheric deposition strongly affects uptake patterns. Each route interacts differently with environmental constraints, shaping both bioavailability and bioaccumulation outcomes.

7. Physiological and biochemical effects on plants

Nanoparticles can induce both beneficial and adverse effects depending on concentration, exposure duration, and material type. Positive effects include enhanced nutrient uptake, biomass accumulation, and stress tolerance. However, high NP concentrations may generate reactive oxygen species (ROS), leading to oxidative stress, membrane damage, enzyme inhibition, and impaired photosynthesis. Plants respond by activating antioxidant systems such as superoxide dismutase, catalase, and glutathione pathways. Changes in nutrient distribution or hormone signaling may also occur.

8. Methods for detecting and tracking nanoparticles

Understanding nanoparticle uptake, translocation, and bioaccumulation in plant tissues requires advanced analytical tools capable of resolving nanoscale structures, identifying elemental composition, and capturing dynamic interactions within complex biological matrices. Because plant tissues contain diverse organic and mineral components, no single technique can fully characterize nanoparticle behavior. Instead, researchers rely on a suite of imaging, spectroscopic, chemical, and isotopic methods to distinguish intact nanoparticles from dissolved ions and to map their distribution, morphology, and transformations inside plant systems.

8.1 Electron microscopy techniques

8.1.1 *Transmission electron microscopy (TEM)*

TEM is a foundational method for visualizing nanoparticles at sub-nanometer resolution. It allows direct imaging of particle size, shape, and aggregation and enables confirmation of nanoparticle localization within intracellular structures such as chloroplasts, vacuoles, and mitochondria. Ultrathin tissue sections are used to determine whether nanoparticles have penetrated cell walls and plasma membranes.

8.1.2 *Scanning electron microscopy (SEM)*

SEM provides high-resolution images of external plant surfaces, including stomata, cuticles, trichomes, and root epidermal layers. It is also valuable for detecting nanoparticle aggregates on seed coats or leaf surfaces. When combined with energy-dispersive X-ray spectroscopy (EDS), SEM-EDS offers spatially resolved elemental identification of nanoparticles in situ.

8.2 Spectroscopic and elemental analysis techniques

8.2.1 *Inductively coupled plasma mass spectrometry (ICP-MS)*

ICP-MS is widely used for quantitative elemental analysis after tissue digestion. Its strengths include high sensitivity, multi-element detection, and robust quantification of metal-based nanoparticles. However, it cannot distinguish between dissolved ions and intact particles without additional fractionation techniques.

8.2.2 *Single-particle ICP-MS (spICP-MS)*

spICP-MS detects individual nanoparticles, enabling measurement of particle size distributions, differentiation between ionic and particulate forms, and quantification of nanoparticle number concentrations in plant extracts.

8.2.3 *X-ray absorption spectroscopy (XAS)*

XAS provides detailed information on oxidation states and chemical speciation. It is particularly useful for monitoring transformations such as dissolution, oxidation, and sulfidation in hydrated plant tissues without destructive sample preparation.

8.2.4 *Raman and FTIR spectroscopy*

Raman spectroscopy is effective for carbon-based nanomaterials, while FTIR identifies changes in plant biochemical composition resulting from nanoparticle exposure. Both methods can map nanoparticle distribution and characterize interactions between nanoparticles and biomolecules.

8.3 Fluorescent or isotopically labeled nanoparticles

8.3.1 *Fluorescent nanoparticles*

Quantum dots, carbon dots, and fluorescent dye–labeled nanoparticles allow real-time tracking of nanoparticle movement in living tissues. These tools offer high sensitivity and the ability to monitor dynamic processes, although some materials may suffer from photobleaching or ion release.

8.3.2 *Isotopic labeling*

Stable or radioactive isotopes provide precise tracers for quantifying nanoparticle dissolution and movement. Isotopic methods can differentiate natural background metals from nanoparticle-derived metals and are ideal for long-term tracking studies.

8.4 Synchrotron-based imaging and tomography

8.4.1 *Synchrotron X-ray fluorescence (XRF) microscopy*

XRF microscopy generates high-resolution elemental maps, allowing visualization of metal-based nanoparticles within vascular bundles or other tissues at scales ranging from micrometers to nanometers.

8.4.2 *X-ray computed nanotomography*

Nanotomography offers three-dimensional, non-destructive imaging of internal plant structures and nanoparticle distributions, making it suitable for intact tissues such as roots, leaves, and seeds.

8.5 Other complementary techniques

Techniques such as dynamic light scattering (DLS), zeta potential analysis, laser ablation ICP-MS (LA-ICP-MS), and confocal laser scanning microscopy (CLSM) provide additional insights into nanoparticle size, stability, elemental distribution, and fluorescent localization.

8.6 Importance of multi-method approaches

Because nanoparticles undergo constant transformations including dissolution, aggregation, and biocorona formation—no single analytical technique can capture all aspects of their behavior. Combining structural, chemical, and functional data across multiple platforms is essential to accurately track nanoparticles, distinguish particle versus ion effects, and understand their interactions within plant tissues.

9. Environmental and agricultural implications

The incorporation of nanotechnology into agriculture brings both opportunities and challenges. The movement and accumulation of nanoparticles in plant tissues affect plant productivity, soil ecosystems, food chain dynamics, and ultimately human health. A thorough understanding of nanoparticle behavior is therefore necessary to optimize benefits while minimizing environmental risks.

9.1 Positive agricultural implications

9.1.1 *Nano-fertilizers*

Nanoparticles can improve nutrient delivery by enhancing uptake efficiency, enabling controlled release, and reducing fertilizer loss and soil pollution. Metal oxide nanoparticles such as ZnO and Fe_2O_3, along with nano-chelated nutrients, support improved micronutrient assimilation.

9.1.2 *Nano-pesticides*

Nanoformulated pesticides enable targeted delivery, improved efficacy, reduced chemical usage, and better adhesion to leaf surfaces. Examples include nano-copper fungicides, nano-sulfur formulations, and nano-encapsulated insecticides.

9.1.3 *Growth promotion and stress tolerance*

Certain nanoparticles promote photosynthesis, root growth, antioxidant activity, and resistance to abiotic stresses like drought, heat, and salinity. Carbon-based nanoparticles and SiO_2 nanomaterials are particularly noted for enhancing stress resilience.

9.2 Environmental and health risks

Despite the growing interest in nanotechnology for agricultural applications, increasing evidence suggests that the environmental and health risks associated with nanoparticle use cannot be overlooked. Owing to their small size, high reactivity, and persistence, nanoparticles may interact with biological systems in unintended ways, leading to accumulation, ecological disruption, and potential toxicity. Once introduced into agroecosystems, nanoparticles can migrate through soil-plant systems, enter food chains, and influence both plant physiology and non-target organisms. These concerns have prompted a shift from performance-oriented research toward comprehensive risk evaluation, emphasizing the need to assess nanoparticle fate, bioavailability, and long-term impacts under realistic field conditions. Understanding these risks is essential for ensuring that the benefits of nanotechnology in agriculture are realized without compromising environmental sustainability or food safety. he majors ecological and food safety risks associated with plant–nanoparticle interactions are summarized in Table 4.5, highlighting key exposure pathways, underlying mechanisms, and potential consequences. These risks underscore the importance of cautious nanopaterial design and responsible agricultural deployment, as discussed below.

9.2.1 *Accumulation in edible tissues*

When nanoparticles accumulate in edible plant organs, they raise concerns regarding human exposure, long-term toxicity, and dietary safety. Thorough risk assessment is required before widespread agricultural adoption.

9.2.2 *Impact on soil microbiomes*

Nanoparticles entering soil can disrupt microbial communities, including nitrogen-fixing bacteria and mycorrhizal fungi, which play crucial roles in nutrient cycling. Silver and copper nanoparticles, with strong antimicrobial activities, pose particular risks.

9.2.3 *Trophic transfer*

Nanoparticles accumulated in plants may transfer to insects, grazing animals, and humans, contributing to bioaccumulation and biomagnification across food webs.

9.2.4 *Alteration of plant physiology*

Excessive nanoparticle accumulation may impair photosynthesis, generate reactive oxygen species, disrupt membranes, and interfere with hormonal pathways, resulting in reduced growth or reproductive success.

9.3 Regulatory and safety considerations

9.3.1 *Need for standardized testing*

Regulatory frameworks must address nanoparticle size, coating, solubility, transformation products, species-specific responses, and long-term ecological impacts. Standardized international testing protocols are urgently needed.

9.3.2 *Safe-by-design nanomaterials*

Developing nanoparticles that degrade harmlessly, exhibit minimal phytotoxicity, and avoid accumulation in edible tissues is key to environmentally responsible nanotechnology.

TABLE 4.5 Potential ecological and food safety risks associated with plant–nanoparticle interactions.

Risk category	Description	Mechanism or cause	Potential consequences
Trophic transfer	Movement of NPs to herbivores or humans	NP accumulation in edible tissues	Bioaccumulation in food chain, chronic exposure
Soil microbiome disruption	Negative effects on beneficial microbes	Antimicrobial properties of Ag, Cu, ZnO NPs	Reduced nutrient cycling, impaired N fixation
Phytotoxicity	Damage to plant cells and metabolism	ROS generation, membrane disruption	Reduced growth, altered physiology
Environmental persistence	Long-term NP retention in soil	Low degradation rate of many NPs	Accumulation over seasons, ecological imbalance
Human dietary exposure	Consumption of NP-containing crops	Translocation to seeds, fruits, leaves	Unknown long-term health effects

9.3.3 *Responsible field applications*

Best practices include optimized dosing, controlled-release formulations, monitoring for leaching, and integrating nanoparticle fate data into agricultural policy guidelines.

9.4 Balancing benefits and risks

Although nanotechnology holds significant potential to improve crop production, enhance nutrient use, and increase stress tolerance, its unintended ecological consequences must be carefully considered. Achieving a sustainable balance requires comprehensive toxicity testing, long-term environmental studies, and the design of safe, biodegradable nanomaterials aligned with sustainable agricultural principles. Ultimately, the responsible use of nanoparticles depends on a deep understanding of their interactions with plants, soils, and ecosystems at multiple scales.

10. Conclusion

Nanoparticles navigate intricate pathways as they enter, move within, and accumulate throughout plant systems, with each stage shaped by nanoparticle characteristics, plant anatomical barriers, and environmental conditions. After entering through roots, leaves, or seed coats, they encounter structures such as the cell wall and plasma membrane before accessing the vascular network, where the xylem and phloem facilitate systemic transport to organs including roots, stems, leaves, flowers, and seeds. Through these interactions, plants act both as selective biological filters and as reservoirs for engineered nanomaterials. Although nano-enabled fertilizers, pesticides, and growth enhancers offer clear benefits for nutrient efficiency and crop productivity, the potential for nanoparticle persistence, transformation, and accumulation in edible tissues highlights the need for comprehensive risk evaluation. Gaining a deeper understanding of nanoparticle fate within plants is therefore crucial for ensuring the safe, sustainable, and responsible application of nanotechnology in modern agriculture.

AI disclosure

During the preparation of this work the author(s) used Chat GPT in order to improve English writing. After using this tool/service, the author(s) reviewed and edited the content as needed and take(s) full responsibility for the content of the publication.

References

Abbas, Q., Yousaf, B., Amina, Ali, M.U., Munir, M.A.M., El-Naggar, A., Rinklebe, J., Naushad, M., 2020. Transformation pathways and fate of engineered nanoparticles (ENPs) in distinct interactive environmental compartments: a review. Environ. Int. 138. https://doi.org/10.1016/j.envint.2020.105646.

Acharya, P., Jayaprakasha, G.K., Crosby, K.M., Jifon, J.L., Patil, B.S., 2020. Nanoparticle-mediated seed priming improves germination, growth, yield, and quality of watermelons (*Citrullus lanatus*) at multi-locations in Texas. Sci. Rep. 10 (1). https://doi.org/10.1038/s41598-020-61696-7.

Adil, M.F., Sehar, S., Ma, Z., Tahira, K., Askri, S.M.H., El-Sheikh, M.A., Ahmad, A., Zhou, F., Zhao, P., Shamsi, I. H., 2024. Insights into the alleviation of cadmium toxicity in rice by nano-zinc and Serendipita indica: modulation of stress-responsive gene expression and antioxidant defense system activation. Environ. Pollut. 350. https://doi.org/10.1016/j.envpol.2024.123952.

Afzal, S., Aftab, T., Singh, N.K., 2022. Impact of zinc oxide and iron oxide nanoparticles on uptake, translocation, and physiological effects in *Oryza sativa* L. J. Plant Growth Regul. 41 (4), 1445–1461. https://doi.org/10.1007/s00344-021-10388-1.

Ameen, F., Alsamhary, K., Alabdullatif, J.A., Alnadhari, S., 2021. A review on metal-based nanoparticles and their toxicity to beneficial soil bacteria and fungi. Ecotoxicol. Environ. Saf. 213. https://doi.org/10.1016/j.ecoenv.2021.112027.

Avellan, A., Yun, J., Morais, B.P., Clement, E.T., Rodrigues, S.M., Lowry, G.V., 2021. Critical review: role of inorganic nanoparticle properties on their foliar uptake and in planta translocation. Environ. Sci. Technol. 55 (20), 13417–13431. https://doi.org/10.1021/acs.est.1c00178.

Bandmann, V., Homann, U., 2012. Clathrin-independent endocytosis contributes to uptake of glucose into BY-2 protoplasts. Plant J. 70 (4), 578–584. https://doi.org/10.1111/j.1365-313X.2011.04892.x.

Chen, H., Song, Y., Wang, Y., Wang, H., Ding, Z., Fan, K., 2024. Zno nanoparticles: improving photosynthesis, shoot development, and phyllosphere microbiome composition in tea plants. J. Nanobiotechnol. 22 (1). https://doi.org/10.1186/s12951-024-02667-2.

Chen, L., Zhu, L., Cheng, H., Xu, W., Li, G., Zhang, Y., Gu, J., Chen, L., Xie, Z., Li, Z., Wu, H., 2024. Negatively charged carbon dots employed symplastic and apoplastic pathways to enable better plant delivery than positively charged carbon dots. ACS Nano 18 (34), 23154–23167. https://doi.org/10.1021/acsnano.4c05362.

Cifuentes, Z., Custardoy, L., de la Fuente, J.M., Marquina, C., Ibarra, M.R., Rubiales, D., Pérez-de-Luque, A., 2010. Absorption and translocation to the aerial part of magnetic carbon-coated nanoparticles through the root of different crop plants. J. Nanobiotechnol. 8. https://doi.org/10.1186/1477-3155-8-26Spain.

Dang, K., Wang, Y., Tian, H., Bai, J., Cheng, X., Guo, L., Zhang, Q., Geng, Y., Shao, X., 2024. Impact of ZnO NPs on photosynthesis in rice leaves plants grown in saline-sodic soil. Sci. Rep. 14 (1). https://doi.org/10.1038/s41598-024-66935-9.

Djanaguiraman, M., Anbazhagan, V., Dhankher, O.P., Prasad, P.V.V., 2024. Uptake, translocation, toxicity, and impact of nanoparticles on plant physiological processes. Plants 13 (22). https://doi.org/10.3390/plants13223137.

Dong, Q., Huang, T., Zhou, C., Wan, X., He, X., Miao, P., Cheng, H., Wang, X., Yu, H., Hu, M., Dong, L., Qu, B., Wu, Y., Li, J., Pan, C., 2025. Nano-priming with selenium nanoparticles reprograms seed germination, antioxidant defense, and phenylpropanoid metabolism to enhance Fusarium graminearum resistance in maize seedlings. J. Adv. Res. https://doi.org/10.1016/j.jare.2025.05.057.

Eichert, T., Kurtz, A., Steiner, U., Goldbach, H.E., 2008. Size exclusion limits and lateral heterogeneity of the stomatal foliar uptake pathway for aqueous solutes and water-suspended nanoparticles. Physiol. Plantarum 134 (1), 151–160. https://doi.org/10.1111/j.1399-3054.2008.01135.x.

do Espirito Santo Pereira, A., Oliveira, H.C., Fernandes Fraceto, L., Santaella, C., 2021. Nanotechnology potential in seed priming for sustainable agriculture. Nanomaterials 11 (2), 267. https://doi.org/10.3390/nano11020267.

Feng, X., Hussain, T., Guo, K., An, P., Liu, X., 2021. Physiological, morphological and anatomical responses of Hibiscus moscheutos to non-uniform salinity stress. Environ. Exp. Bot. 182, 104301. https://doi.org/10.1016/j.envexpbot.2020.104301.

Gao, M., Chang, J., Wang, Z., Zhang, H., Wang, T., 2023. Advances in transport and toxicity of nanoparticles in plants. J. Nanobiotechnol. 21 (1). https://doi.org/10.1186/s12951-023-01830-5.

Geremew, A., Stovall, L., Woldesenbet, S., Ma, X., Carson, L., 2025. Nanopriming with zinc oxide: a novel approach to enhance germination and antioxidant systems in amaranth. Front. Plant Sci. 16. https://doi.org/10.3389/fpls.2025.1599192.

Ghassemi-Golezani, K., Rahimzadeh, S., 2024. Biochar-based nanoparticles mitigated arsenic toxicity and improved physiological performance of basil via enhancing cation exchange capacity and ferric chelate reductase activity. Chemosphere 362, 142623. https://doi.org/10.1016/j.chemosphere.2024.142623.

Hernandez-Viezcas, J.A., Castillo-Michel, H., Andrews, J.C., Cotte, M., Rico, C., Peralta-Videa, J.R., Ge, Y., Priester, J. H., Holden, P.A., Gardea-Torresdey, J.L., 2013. In situ synchrotron X-ray fluorescence mapping and speciation of CeO_2 and ZnO nanoparticles in soil cultivated soybean (Glycine max). ACS Nano 7 (2), 1415–1423. https://doi.org/10.1021/nn305196q.

Hong, J., Wang, C., Wagner, D.C., Gardea-Torresdey, J.L., He, F., Rico, C.M., 2021. Foliar application of nanoparticles: mechanisms of absorption, transfer, and multiple impacts. Environ. Sci. Nano 8 (5), 1196–1210. https://doi.org/10.1039/d0en01129k.

Isayenkov, S.V., Maathuis, F.J.M., 2019. Plant salinity stress: many unanswered questions remain. Front. Plant Sci. 10. https://doi.org/10.3389/fpls.2019.00080.

Kasote, D.M., Lee, J.H.J., Jayaprakasha, G.K., Patil, B.S., 2019. Seed priming with iron oxide nanoparticles modulate antioxidant potential and defense-linked hormones in watermelon seedlings. ACS Sustain. Chem. Eng. 7 (5), 5142–5151. https://doi.org/10.1021/acssuschemeng.8b06013.

Khan, A.R., Salam, A., Li, G., Iqbal, B., Ulhassan, Z., Liu, Q., Azhar, W., Liaquat, F., Shah, I.H., Ul Hassan, S.S., Du, D., 2024. Nanoparticles and their crosstalk with stress mitigators: a novel approach towards abiotic stress tolerance in agricultural systems. Crop J. 12 (5), 1280–1298. https://doi.org/10.1016/j.cj.2024.09.010.

Khan, I., Awan, S.A., Rizwan, M., Hassan, Z.U., Akram, M. A., Tariq, R., Brestic, M., Xie, W., 2022. Nanoparticle's uptake and translocation mechanisms in plants via seed priming, foliar treatment, and root exposure: a review. Environ. Sci. Pollut. Control Ser. 29 (60), 89823–89833. https://doi.org/10.1007/s11356-022-23945-2.

Khan, M.N., Li, Y., Mu, Y., Sultan, H., Baloch, A., Din, I., Fu, C., Li, J., Khan, Z., Kumar, S., Wu, H., Grillo, R., Nie, L., 2025. Recent advances in nano-enabled plant salt tolerance: methods of application, risk assessment, opportunities and future prospects. J. Integr. Agric. 24 (5), 1611–1630. https://doi.org/10.1016/j.jia.2024.05.028.

Khodakovskaya, M.V., de Silva, K., Nedosekin, D.A., Dervishi, E., Biris, A.S., Shashkov, E.V., Galanzha, E.I., Zharov, V.P., 2011. Complex genetic, photothermal, and photoacoustic analysis of nanoparticle-plant interactions. Proc. Natl. Acad. Sci. 108 (3), 1028–1033. https://doi.org/10.1073/pnas.1008856108.

Kohay, H., Wielinski, J., Reiser, J., Perkins, L.A., Ristroph, K., Giraldo, J.P., Lowry, G.V., 2024. Nanocarrier foliar uptake pathways affect delivery of active agents and plant physiological response. Environ. Sci. Nano 12 (1), 660–674. https://doi.org/10.1039/d4en00547c.

Koizumi, M., Kikuchi, K., Isobe, S., Ishida, N., Naito, S., Kano, H., 2008. Role of seed coat in imbibing soybean seeds observed by micro-magnetic resonance imaging. Ann. Bot. 102 (3), 343–352. https://doi.org/10.1093/aob/mcn095.

Kumar, V., Naik, I.S., Das, B., Singh, A., Nayak, P., Mohapatra, C., Debnath, D., Tripathy, M., Behera, K., Masika, F.B., Manohar, S., Chung, S.M., Kherawat, B.S., Hemalatha, M., Kesawat, M.S., 2025. Nanoparticles in plant system: a comprehensive review on their role in diverse stress management and phytohormone signaling. Plant Stress 18. https://doi.org/10.1016/j.stress.2025.101000.

Larue, C., Castillo-Michel, H., Sobanska, S., Cécillon, L., Bureau, S., Barthès, V., Ouerdane, L., Carrière, M., Sarret, G., 2014. Foliar exposure of the crop *Lactuca sativa* to silver nanoparticles: evidence for internalization and changes in Ag speciation. J. Hazard Mater. 264, 98–106. https://doi.org/10.1016/j.jhazmat.2013.10.053.

Lv, J., Christie, P., Zhang, S., 2019. Uptake, translocation, and transformation of metal-based nanoparticles in plants: recent advances and methodological challenges. Environ. Sci. Nano 6 (1), 41–59. https://doi.org/10.1039/c8en00645h.

Mahakham, W., Sarmah, A.K., Maensiri, S., Theerakulpisut, P., 2017. Nanopriming technology for enhancing germination and starch metabolism of aged rice seeds using phytosynthesized silver nanoparticles. Sci. Rep. 7 (1). https://doi.org/10.1038/s41598-017-08669-5.

Mehmood, S., Ou, W., Ahmed, W., Bundschuh, J., Rizwan, M., Mahmood, M., Sultan, H., Alatalo, J.M., Elnahal, A.S.M., Liu, W., Li, W., 2023. ZnO nanoparticles mediated by Azadirachta indica as nano fertilizer: improvement in physiological and biochemical indices of Zea mays grown in Cr-contaminated soil. Environ. Pollut. 339. https://doi.org/10.1016/j.envpol.2023.122755.

Munir, N., Gulzar, W., Abideen, Z., Hasanuzzaman, M., El-Keblawy, A., Zhao, F., 2023. Plant–nanoparticle interactions: transcriptomic and proteomic insights. Agronomy 13 (8), 2112. https://doi.org/10.3390/agronomy13082112.

Nauman Khan, M., Fu, C., Liu, X., Li, Y., Yan, J., Lin, Y., Li, J., Khan, Z., Nie, L., Wu, H., 2024. Nanopriming with selenium doped carbon dots improved rapeseed germination and seedling salt tolerance. Crop J. 12 (5), 1333–1343. https://doi.org/10.1016/j.cj.2024.03.007.

Ning, J., Hou, R., Wei, C., Wu, H., Fu, G., Han, Z., Zhang, B., Zhang, Z., Tang, W.W., 2025. A pesticide delivery system for pH/enzyme response: pectin-coated O-CMCS nanoparticles significantly enhance control of *Cyperus rotundus* L. J. Agric. Food Chem. 73 (46), 29520–29537. https://doi.org/10.1021/acs.jafc.5c12087.

Oliveira, H.C., Seabra, A.B., Kondak, S., Adedokun, O.P., Kolbert, Z., 2023. Multilevel approach to plant–nanomaterial relationships: from cells to living ecosystems. J. Exp. Bot. 74 (12), 3406–3424. https://doi.org/10.1093/jxb/erad107.

Pinna, A., Husted, S., 2025. Translocation of foliar-applied nanoparticles: a critical review from a plant science perspective. Physiol. Plantarum 177 (5). https://doi.org/10.1111/ppl.70476.

Qu, Y., Ma, Q., Deng, J., Shen, W., Zhang, X., He, Z., Nostrand, J.D.V., Zhou, J., Zhou, J., 2015. Responses of microbial communities to single-walled carbon nanotubes in phenol wastewater treatment systems. Environ. Sci. Technol. 49 (7), 4627–4635. https://doi.org/10.1021/es5053045.

Radziemska, M., Gusiatin, Z.M., Cydzik-Kwiatkowska, A., Cerdà, A., Pecina, V., Bes, A., Datta, R., Majewski, G., Mazur, Z., Dzięcioł, J., Danish, S., Brtnický, M., 2021. Insight into metal immobilization and microbial community structure in soil from a steel disposal dump phytostabilized with composted, pyrolyzed or gasified wastes. Chemosphere 272, 129576. https://doi.org/10.1016/j.chemosphere.2021.129576.

Rahimzadeh, S., Ghassemi-Golezani, K., 2022. Biochar-based nutritional nanocomposites altered nutrient uptake and vacuolar H+-Pump activities of dill under salinity. J. Soil Sci. Plant Nutr. 22 (3), 3568–3581. https://doi.org/10.1007/s42729-022-00910-z.

de la Rosa, G., Vázquez-Núñez, E., Molina-Guerrero, C., Serafín-Muñoz, A.H., Vera-Reyes, I., 2021. Interactions of nanomaterials and plants at the cellular level: current knowledge and relevant gaps. Nanotechnol. Environ. Eng. 6 (1). https://doi.org/10.1007/s41204-020-00100-1.

Schwab, F., Zhai, G., Kern, M., Turner, A., Schnoor, J.L., Wiesner, M.R., 2016. Barriers, pathways and processes for uptake, translocation and accumulation of nanomaterials in plants—critical review. Nanotoxicology 10 (3), 257–278. https://doi.org/10.3109/17435390.2015.1048326.

Shah, T., Latif, S., Saeed, F., Ali, I., Ullah, S., Alsahli, A.A., Jan, S., Ahmad, P., 2021. Seed priming with titanium dioxide nanoparticles enhances seed vigor, leaf water status, and antioxidant enzyme activities in maize (*Zea mays* L.) under salinity stress. J. King Saud Univ. Sci. 33 (1), 101207. https://doi.org/10.1016/j.jksus.2020.10.004.

Sharma, D., Afzal, S., Singh, N.K., 2021. Nanopriming with phytosynthesized zinc oxide nanoparticles for promoting germination and starch metabolism in rice seeds. J. Biotechnol. 336, 64–75. https://doi.org/10.1016/j.jbiotec.2021.06.014.

Shelar, A., Singh, A.V., Maharjan, R.S., Laux, P., Luch, A., Gemmati, D., Tisato, V., Singh, S.P., Santilli, M.F., Shelar, A., Chaskar, M., Patil, R., 2021. Sustainable agriculture through multidisciplinary seed nanopriming: prospects of opportunities and challenges. Cells 10 (9). https://doi.org/10.3390/cells10092428.

Shen, Y., Borgatta, J., Ma, C., Elmer, W., Hamers, R.J., White, J.C., 2020. Copper nanomaterial morphology and composition control foliar transfer through the cuticle and mediate resistance to root fungal disease in tomato (*Solanum lycopersicum*). J. Agric. Food Chem. 68 (41), 11327–11338. https://doi.org/10.1021/acs.jafc.0c04546.

Singh, A., Sharma, A., Singh, O., Rajput, V.D., Movsesyan, H., Minkina, T., Alexiou, A., Papadakis, M., Singh, R.K., Singh, S., Sousa, J.R., El-Ramady, H.R., Zulfiqar, F., Kumar, R., Al-Ghamdi, A. A., Ghazaryan, K., 2024. In-depth exploration of nanoparticles for enhanced nutrient use efficiency and abiotic stresses management: present insights and future horizons. Plant Stress 14. https://doi.org/10.1016/j.stress.2024.100576.

Singh, A., Tiwari, S., Pandey, J., Lata, C., Singh, I.K., 2021. Role of nanoparticles in crop improvement and abiotic stress management. J. Biotechnol. 337, 57–70. https://doi.org/10.1016/j.jbiotec.2021.06.022.

Sultan, H., Abbas, H.M.M., Faizan, M., Emamverdian, A., Shah, A., Bahadur, S., Li, Y., Khan, M.N., Nie, L., 2025. Residual effects of biochar and nano-modified biochar on growth and physiology under saline environment in two different genotype of *Oryza sativa* L. J. Environ. Manag. 373, 123847. https://doi.org/10.1016/j.jenvman.2024.123847.

Sun, L., Yuan, Z., 2022. Seed morphology of soybean. Adv. Bot. Res. 102, 349–375. https://doi.org/10.1016/bs.abr.2022.03.004.

Terskikh, V.V., Feurtado, J.A., Ren, C., Abrams, S.R., Kermode, A.R., 2005. Water uptake and oil distribution during imbibition of seeds of western white pine (Pinus monticola Dougl. ex D. Don) monitored in vivo using magnetic resonance imaging. Planta 221 (1), 17–27. https://doi.org/10.1007/s00425-004-1426-z.

Upadhyay, S.K., Kumar, P., Rajput, V.D., Mandzhieva, S.S., Minkina, T., 2025. Mechanistic insights in to biocompatible-metal nanoparticles in modulating soil enzyme and its convergence in soil health improvement. Plant Nano Biol. 13. https://doi.org/10.1016/j.plana.2025.100186.

Wang, X., Xie, H., Wang, P., Yin, H., 2023. Nanoparticles in plants: uptake, transport and physiological activity in leaf and root. Materials 16 (8), 3097. https://doi.org/10.3390/ma16083097.

Wang, Z., Xie, X., Zhao, J., Liu, X., Feng, W., White, J.C., Xing, B., 2012. Xylem- and phloem-based transport of CuO nanoparticles in maize (*Zea mays* L.). Environ. Sci. Technol. 46 (8), 4434–4441. https://doi.org/10.1021/es204212z.

Wang, Z., Wang, S., Ma, T., Liang, Y., Huo, Z., Yang, F., 2023. Synthesis of zinc oxide nanoparticles and their applications in enhancing plant stress resistance: a review. Agronomy 13 (12), 3060. https://doi.org/10.3390/agronomy13123060.

Wei, X., Cao, P., Wang, G., Liu, Y., Song, J., Han, J., 2021. CuO, ZnO, and γ-Fe2O3 nanoparticles modified the underground biomass and rhizosphere microbial community of Salvia miltiorrhiza (Bge.) after 165-day exposure. Ecotoxicol. Environ. Saf. 217, 112232. https://doi.org/10.1016/j.ecoenv.2021.112232.

Wei, L., Liu, J., Jiang, G., 2024. Nanoparticle-specific transformations dictate nanoparticle effects associated with plants and implications for nanotechnology use in agriculture. Nat. Commun. 15 (1). https://doi.org/10.1038/s41467-024-51741-8.

Xiong, T., Zhang, S., Kang, Z., Zhang, T., Li, S., 2021. Dose-dependent physiological and transcriptomic responses of lettuce (*Lactuca sativa* L.) to copper oxide nanoparticles—insights into the phytotoxicity mechanisms. Int. J. Mol. Sci. 22 (7), 3688. https://doi.org/10.3390/ijms22073688.

Yang, Le, Zhang, L., Qi, Z., Wei, J., Zhao, X., Zheng, Z., Chen, B., Xu, Z., 2024. Nanopriming boost seed vigor: deeper insights into the effect mechanism. Plant Physiol. Biochem. 214, 108895. https://doi.org/10.1016/j.plaphy.2024.108895.

Yasin, M.U., Hannan, F., Munir, R., Muhammad, S., Iqbal, M., Yasin, I., Khan, M.S.S., Kanwal, F., Chunyan, Y., Fan, X., Gan, Y., 2024. Interactive mode of biochar-based silicon and iron nanoparticles mitigated Cd-toxicity in maize. Sci. Total Environ. 912. https://doi.org/10.1016/j.scitotenv.2023.169288.

Zhai, G., Walters, K.S., Peate, D.W., Alvarez, P.J.J., Schnoor, J.L., 2014. Transport of gold nanoparticles through plasmodesmata and precipitation of gold ions in woody poplar. Environ. Sci. Technol. Lett. 1 (2), 146–151. https://doi.org/10.1021/ez400202b.

Zhang, P., Guo, Z., Zhang, Z., Fu, H., White, J.C., Lynch, I., 2020. Nanomaterial transformation in the soil–plant system: implications for food safety and application in agriculture. Small 16 (21). https://doi.org/10.1002/smll.202000705.

Zhang, Q., Li, M., Guo, C., Jia, Z., Wan, G., Wang, S., Min, D., 2019. Fe3O4 nanoparticles loaded on lignin nanoparticles applied as a peroxidase mimic for the sensitively colorimetric detection of H2O2. Nanomaterials 9 (2), 210. https://doi.org/10.3390/nano9020210.

Zhao, L., Peralta-Videa, J.R., Ren, M., Varela-Ramirez, A., Li, C., Hernandez-Viezcas, J.A., Aguilera, R.J., Gardea-Torresdey, J.L., 2012. Transport of Zn in a sandy loam soil treated with ZnO NPs and uptake by corn plants: electron microprobe and confocal microscopy studies. Chem. Eng. J. 184, 1–8. https://doi.org/10.1016/j.cej.2012.01.041.

Zia-ur-Rehman, M., Anayatullah, S., Irfan, E., Hussain, S. M., Rizwan, M., Sohail, M.I., Jafir, M., Ahmad, T., Usman, M., Alharby, H.F., 2023. Nanoparticles assisted regulation of oxidative stress and antioxidant enzyme system in plants under salt stress: a review. Chemosphere 314. https://doi.org/10.1016/j.chemosphere.2022.137649.

CHAPTER

5

Nanomaterials for improving photosynthetic efficiency: current innovations

Aina Inam[1], *Asma Ahmad*[1], *Maham Sheikh*[1], *Sumera Javad*[1], *Sumera Iqbal*[1] *and Ozair Chaudhry*[2]

[1]Department of Botany, Lahore College for Women University, Lahore, Pakistan; [2]DECAC, Community Growth & Economic Dev. Department, The Regional Municipality Durham, Durham, ON, Canada

1. Introduction

Global food security is increasingly at risk due to population growth, climate change, and unsustainable agricultural practices. With the world population projected to approach 10 billion by 2050, it is essential to achieve sustainable growth in food production (Dillard, 2019). Climate change exacerbates these challenges by intensifying droughts, accelerating soil degradation, and increasing the frequency of extreme weather events, all of which hinder agricultural productivity (Ghosh et al., 2022).

The United Nations estimates that the global population will continue to rise, reaching approximately 10.3 billion by the mid-2080s before gradually declining to about 10.2 billion by the end of the century. To support this growing population, many countries are adopting various techniques and strategies to enhance cereal production (Qaseem and Wu, 2021). Additionally, a range of scientific technologies is being employed to increase grain yields on existing agricultural land in the coming decades. These efforts aim to ensure food security while also improving the efficiency of water and fertilizer use to maintain a healthy environment (Csikós and Tóth, 2023).

After billions of years of evolution, photosynthesis has been refined rather than maximized. The complexity of this process enables plants to adapt to various environments by utilizing optimal combinations of photosynthetic traits, which are essential for their reproduction and survival, fundamental needs for any living organism (Hohmann-Marriott and Blankenship, 2011).

The complexity of photosynthesis has led to uncertainty about the possibility of modifying the process to improve crop yields. However, recent bioengineering studies have demonstrated that enhancing photosynthetic electron

Nanotechnology in Photosynthesis
https://doi.org/10.1016/B978-0-443-44376-3.00004-5

transport rates, bypassing part of the photorespiratory pathway, and increasing catalytic flux through the Calvin-Benson-Bassham (CBB) cycle can significantly boost photosynthetic rates and crop biomass (Yoon et al., 2020).

1.1 Limitations of natural photosynthetic efficiency in plants

The natural photosynthetic efficiency in plants is significantly limited, usually achieving only 1%–4% conversion of total incident solar energy into biomass under field conditions (Burgess and Cardoso, 2023). In contrast, theoretical upper limits range from 4.6%–12%, depending on the light spectrum and underlying assumptions. These limitations stem from a combination of biochemical, biophysical, physiological, and environmental factors that collectively result in more than 95% of available sunlight being wasted (Falkowski et al., 2017).

1.2 Natural limitations of photosynthesis

Photosynthesis in plants is inherently inefficient, leading to notable energy losses. In C_3 plants, such as wheat, rice, and soybeans, the photorespiratory pathway is a key contributor to this inefficiency. During photorespiration, the enzyme Rubisco reacts with oxygen instead of carbon dioxide, resulting in a loss of fixed carbon as CO_2 and decreased photosynthetic efficiency (Caemmerer et al., 2014).

RuBisCO, the most abundant enzyme on Earth, performs two main functions: carboxylation, which facilitates productive carbon dioxide (CO_2) fixation, and oxygenation, which leads to photorespiration. However, RuBisCO has a poor specificity factor ($S \approx 80$–100 in C_3 plants), which results in a tendency to bind oxygen (O_2) wastefully under typical environmental conditions (Pradhan et al., 2022). In temperate C_3 crops such as wheat and rice, photorespiration consumes 20%–40% of the carbon fixed during photosynthesis each day (Ishaq et al., 2025). This process recycles glycine through mitochondria and peroxisomes, at a net cost of 3.5-5 ATP and 2 NADPH for each molecule of CO_2 that escapes, resulting in an energy loss equivalent to 25%–30%. This loss is exacerbated when temperatures exceed 25°C or when CO_2 levels drop below 200 ppm (Khan et al., 2025).

The regeneration phase of the Calvin-Benson-Bassham (CBB) cycle tends to lag the light reactions. Enzymes such as sedoheptulose-1,7-bisphosphatase (SBPase) and fructose-1,6-bisphosphatase often operate below saturation levels, limiting the maximum carboxylation velocity even in the presence of sufficient ATP and NADPH (Raines, 2022). Furthermore, triose phosphate utilization becomes a limiting factor when the demand for sugars exceeds the supply, typically exporting only about 50% of the fixed carbon as sugars.

1.3 Environmental stress factors

Environmental stressors, such as drought, extreme temperatures, and nutrient deficiencies, exacerbate the inefficiencies in photosynthesis. For example, water stress can lead to stomatal closure, reducing CO_2 availability for photosynthesis and resulting in decreased PE (Flexas et al., 2016).

Drought and heat stress significantly impacts the light-dependent reactions of photosynthesis, which are crucial for photosynthetic efficiency. These reactions take place in the thylakoid membranes of chloroplasts, where light energy is converted into ATP and NADPH through photosystems II (PS II) and I (PS I) (Tripathy and Oelmüller, 2014). Since these processes rely heavily on water, drought can severely disrupt them, ultimately reducing plant productivity.

PS II is particularly vulnerable to drought conditions, losing both stability and effectiveness when water availability is low. This impairment leads to a decrease in the efficiency of light energy conversion. The D1 protein of PS II, essential for the electron transport chain, is susceptible to damage from reactive oxygen species (ROS) generated during drought. This damage hampers water splitting, results in reduced oxygen production and electron transfer, and diminishes photochemical efficiency. Consequently, this increases the risk of photoinhibition and photodamage (Qiao et al., 2024).

Drought and heat stressors impact photosynthesis significantly, particularly harming light reactions. PSI is less affected than PS II, but reduced electron flow leads to an imbalance in the electron transport chain, causing excessive ROS generation and damage to the photosynthetic machinery (Liu et al., 2018). Drought inhibits linear electron flow by damaging components and increasing resistance in the proton gradient, which raises energy of excitation. If not dissipated, this energy can cause photodamage to PSI and PS II, decreasing productivity (Qiao et al., 2024).

The dark reactions, particularly the Calvin cycle, also suffer under drought. This cycle, which uses ATP and NADPH from light reactions to fix carbon, experiences reduced enzyme activity and carbon fixation due to limited CO_2 uptake. Key enzyme Rubisco is sensitive to drought, with its activity inhibited and its synthesis impaired due to oxidative stress (Parry et al., 2002). Other enzymes like PRK and GAPDH also show decreased activity. Drought disrupts the export of triose phosphates, leading to their accumulation and further inhibiting the Calvin cycle. This results in lower sugar availability and shows negative impacts on growth processes, ultimately compromising the plant's ability to adapt to drought (Paul and Pellny, 2003).

1.4 Need for innovative strategies to enhance photosynthesis

Photosynthesis is crucial for life on Earth, fueling plant growth and supporting food chains, yet its efficiency in converting solar energy is often below that is, 1% in crops. This low efficiency offers a significant opportunity for improvement, particularly given the rising global population and increasing agricultural demand due to climate change (Croce et al., 2024).

While advancements have been made in other aspects of crop yield, photosynthesis remains largely underexplored. The complexity of this process, involving many genes and biochemical pathways, makes it challenging to optimize. Unlike traits such as disease resistance, which can be targeted through traditional breeding, enhancing photosynthesis requires advanced molecular understanding and genetic manipulation (Singha et al., 2022). Recently, improved tools and knowledge have made this new frontier for boosting crop productivity by using nanotechnology (Singh et al., 2024).

1.5 Climate change and increasing population

The impact of climate change on both natural and managed systems is a major concern, particularly regarding how increasing temperatures affect the production of major crops. To meet the rising demand for food for both humans and livestock, which is expected to potentially double between 2010 and 2050 (van Dijk et al., 2021). Therefore, enhancing major crop yields is crucial (Smith et al., 2013). This improvement can be achieved by increasing yield resilience in suboptimal conditions (Rivero et al., 2022).

Historically, increases in crop yields have been driven using Green Revolution cultivars, which required more intensive irrigation, fertilization, and mechanization, along with continued selective breeding (Crespo-Herrera et al., 2018). These strategies resulted in a doubling of wheat yields and a tripling of crop yields, accomplished with only a 30% increase in cultivated land area (Garcia et al., 2023). To avoid expanding agricultural land and increasing fertilizer use, it is essential to find new ways to enhance crop yields. Improving yield potential by using nanotechnology is one such opportunity.

1.6 Emerging role of nanotechnology in plant sciences

Nanotechnology is an exciting and rapidly advancing field of science that is transforming research across multiple domains. The breakthroughs in nanotechnology are set to drive innovative applications in biotechnology and agriculture. In areas such as electronics, energy, medicine, and life sciences, nanotechnology is leading the way in significant research advancements (Ranjha et al., 2022). Plant nano biotechnology is an emerging field that combines nanotechnology and plant biology to enhance plant capabilities beyond their natural limits. By integrating engineered nanoparticles (1–100 nm) into plant tissues, scientists aim to improve photosynthesis efficiency, optimize nutrient uptake, and bolster resistance to environmental stressors like drought and salinity (Sharma et al., 2024; Francis et al., 2024). These innovations address critical global issues such as food security and climate change. For instance, using nanomaterials like carbon nanotubes can enhance light absorption and energy conversion, leading to increased plant productivity (Li et al., 2021).

Nanotechnology in plant science integrates nanomaterial particles sized (1–100 nm) into plant tissues to enhance their natural capabilities. These nanomaterials interact with plant biological processes at molecular and cellular levels, improving physiological functions or introducing new ones (Butnariu and Butu, 2019). Due to their high surface area and reactivity, nanoparticles can easily enter plant cells, bind to biomolecules, and influence key biochemical pathways (Cai et al., 2020).

1.6.1 *Nanofertillizers*

Fertilizers play a crucial role in promoting plant growth and development. However, a significant portion of the applied fertilizers becomes unavailable to plants due to various factors, such as leaching, photolysis, hydrolysis, and decomposition. Therefore, it is essential to minimize nutrient losses during fertilization and enhance crop yields by utilizing innovative approaches involving nanotechnology and nanomaterials. Nanofertilizers, or nano-encapsulated nutrients, exhibit properties that can benefit crops by delivering nutrients on-demand. They also enable controlled release of chemical fertilizers, which helps regulate plant growth and improve efficacy (Derosa et al., 2010; Nair et al., 2010).

Nanofertilizers represent a revolutionary approach to nutrient management, addressing key drawbacks of traditional fertilizers such as nutrient leaching, low bioavailability, and environmental contamination, which can lead to nutrient losses and degradation (Chhipa, 2017). By incorporating nanotechnology, nanofertilizers encapsulate essential nutrients in nanoscale carriers for controlled and sustained release. This targeted delivery ensures longer nutrient availability to plants, minimizes fertilizer use, reduces economic costs, and decreases environmental impacts, particularly in nutrient-deficient soils (Nehra et al., 2024).

The small size and high surface area of nanoparticles enhance interaction with plant roots and soil particles, improving nutrient bioavailability. For instance, zinc oxide nanoparticles can boost zinc uptake, which is crucial for plant growth, especially in zinc-deficient areas.

Additionally, the controlled release of nutrients reduces the frequency of applications and minimizes nutrient loss through runoff or leaching (Guo et al., 2018).

Nanofertilizers also improve nitrogen and phosphorus uptake in plants, leading to better growth while reducing reliance on bulk fertilizers and the environmental impact of excess nitrogen and phosphorus (Fatima et al., 2021; El-Saadony et al., 2021).

1.6.2 *Chelated nanoparticles*

In high pH soils, the availability of micronutrients like iron is often limited due to the formation of insoluble compounds, which hampers plant growth. Metal nanoparticles functionalized with chelating agents, such as EDTA, provide a solution by preventing nutrient precipitation and ensuring a steady supply of micronutrients to plants (Dealba-Montero et al., 2017). Chelating agents bind to metal ions, keeping them soluble and accessible to plant roots.

Iron oxide nanoparticles, when combined with chelating agents like EDTA, enhance iron availability, promoting better plant uptake and preventing insoluble compound formation (Jung et al., 2022). Chelated nanoparticles are particularly effective in soils with nutrient immobilization issues, improving overall nutrient availability and aiding plant absorption (Huhtinen et al., 2005). Studies have shown that iron oxide nanoparticles functionalized with EDTA significantly increase iron bioavailability in calcareous soils and cause ultimate increase in the efficiency of photosynthesis.

1.6.3 *Nanoparticles as soil remediators*

Certain nanoparticles are gaining recognition for their ability to improve nutrient uptake and remediate contaminated soils. Heavy metal pollution, often from industrial activities and the excessive use of agrochemicals, significantly impacts soil quality and plant growth. Metals like cadmium, lead, arsenic, and mercury can accumulate in soil, harming plants and posing risks to the food chain.

Nanoparticles such as titanium dioxide (TiO_2) and carbon-based materials (e.g., graphene oxide, activated carbon) help mitigate these risks by adsorbing and immobilizing toxic metals, thus reducing their bioavailability (Sarkar et al., 2019). TiO_2 nanoparticles are particularly effective at binding heavy metals like arsenic and lead, preventing their leaching and protecting plants from toxicity (Wu et al., 2021). Additionally, TiO_2's photocatalytic properties can break down organic contaminants, further improving soil quality (Solymos et al., 2024).

Carbon-based nanomaterials also show promise in adsorbing heavy metals. Their high surface area and functional groups can form stable complexes with metal ions, reducing their mobility. For instance, graphene oxide nanoparticles have effectively removed lead and cadmium from soil, enhancing nutrient uptake in plants (Li et al., 2021).

Utilizing nanoparticles for soil remediation is especially vital in areas with degraded soil quality due to chemical fertilizers and industrial waste. By immobilizing heavy metals, nanoparticles enhance plant health and productivity, ensuring food safety and sustainability (Shipley et al., 2011). This approach also supports the circular economy by restoring polluted lands for agricultural use.

2. Overview of photosynthetic mechanisms

Photosynthesis is a vital biological process through which green plants, algae, and specific bacteria harness light energy from the sun, transforming it into chemical energy. This transformation occurs primarily in the chloroplasts of plant cells and pigments like chlorophyll capture sunlight. Energy is then used to convert carbon dioxide from the air and water from the soil into organic compounds, notably glucose, which serves as a primary energy source for the organism (Björn and Govindjee, 2008; Stirbet et al., 2020). As a remarkable byproduct of this transformation, oxygen is released into the atmosphere, enriching the air we breathe. This intricate process not only fuels the growth and development of plants but also underpins nearly all life on Earth, providing essential energy and organic matter that sustain ecosystems across the globe (Johnson, 2016).

2.1 Historical perspective

- **1771: Joseph Priestley**, an English chemist and clergyman, made a groundbreaking discovery by revealing that plants have the remarkable ability to produce oxygen, a vital gas for life (Gest, 2000; Govindjee and Krogmann, 2004).
- **1845: Julius Robert Mayer**, a prominent German physician and physicist, proposes a revolutionary idea that plants not only absorb sunlight but also convert light energy into chemical energy through photosynthesis, fundamentally changing our understanding of energy transformation in nature (Björn and Govindjee, 2008).
- **1931: Cornelius van Niel**, an influential microbiologist, presents a key insight that challenges previous beliefs by suggesting that the oxygen released during photosynthesis actually comes from water molecules rather than carbon dioxide, paving the way for further research in the field (Govindjee et al., 2017).
- **1954–1961: Melvin Calvin**, an American chemist, conducts extensive research that leads to the elucidation of the dark reaction pathway, known as the Calvin Cycle. His work unveils the intricate processes by which plants convert carbon dioxide into glucose, highlighting the complexity and efficiency of photosynthesis (Schreier and Hibberd, 2019).

The global significance of our ecosystem is profound, as it plays a vital role in several critical processes. Every year, around 105–130 Pg of carbon are fixed through primary production, a measure of the amount of organic carbon created by photosynthetic organisms (He et al., 2021). Additionally, these ecosystems contribute approximately 70% of the oxygen present in our atmosphere, which is essential for the survival of almost all life on Earth (Mouw et al., 2019). Furthermore, the ability of these systems to sequester carbon is crucial for regulating our climate, helping to mitigate the impacts of climate change, and maintaining a balanced environment (Friedlingstein et al., 2025).

2.2 Chloroplast structure and functional organization

2.2.1 *Chloroplast ultrastructure*

- **Outer Membrane**: This membrane is permeable to small molecules, allowing for the free exchange of substances between the chloroplast and its surrounding environment (Shimoni et al., 2005; Jarvis and López-Juez, 2013).
- **Inner Membrane:** Characterized by its selective permeability, the inner membrane utilizes specific transporters to regulate the passage of ions and larger molecules, ensuring the chloroplast maintains its essential internal conditions (Bölter and Soll, 2016).
- **Stroma**: The stroma is a gel-like fluid that fills the interior of the chloroplast, serving as the site for the dark reactions (Calvin cycle). It contains a rich mixture of key components, including enzymes necessary for biochemical reactions, as well as DNA and ribosomes, supporting the synthesis of proteins within the chloroplast (Linka and Weber, 2010).
- **Thylakoids:** These membrane-bound sacs are the sites of the light-dependent reactions of

photosynthesis, organized into stacks known as grana that increase surface area for capturing light energy (Kirchhoff, 2014).

2.2.2 *Photosynthetic membranes and compartmentalization*

- **Granal Thylakoids**: These are the stacked thylakoid membranes that house an abundance of Photosystem II, playing a critical role in harnessing light energy to drive the initial steps of photosynthesis (Dekker and Boekema, 2005).
- **Stromal Thylakoids**: Unstacked thylakoid membranes that interconnect the grana; they are rich in Photosystem I and facilitate the transfer of energy during the light reactions (Hertle et al., 2013).
- **Lumen**: The lumen is the internal space of the thylakoids, functioning as a reservoir for proton accumulation, which is essential for generating a proton gradient used in ATP synthesis during photosynthesis.

2.3 Light-dependent reactions

2.3.1 *Photophysical principles*

- **Photon Absorption:** Chlorophyll, the primary pigment in photosynthesis, along with various accessory pigments such as carotenoids, plays an essential role in capturing sunlight. These pigments absorb specific wavelengths of light energy (primarily in the blue and red regions of the spectrum) and convert that energy into chemical forms that can initiate the process of photosynthesis (Croce and van Amerongen, 2014).
- **Excitation Energy Transfer:** Once chlorophyll absorbs light, the energy excites electrons within the chlorophyll molecules (Mirkovic et al., 2017). This excitation energy is transferred through a series of pigments in a process known as resonance energy transfer, efficiently funneling energy to the reaction center located in each photosystem (Şener et al., 2011).
- **Redox Potentials:** The movement of electrons creates redox potentials, which drive the establishment of an energy gradient across the thylakoid membrane. This gradient is vital for generating ATP, the energy currency of the cell (Semenov and Tikhonov, 2023).

2.3.2 *Photosystem II (PSII)—water-oxidizing complex*

2.3.2.1 Structure and function

- **Reaction Center:** The core of PSII features a P680 chlorophyll dimer, which is primarily responsible for absorbing light and converting this energy into high-energy electrons to be transferred through the electron transport chain (Nelson and Junge, 2015).
- **Oxygen-Evolving Complex (OEC):** This complex includes a unique cluster of four manganese ions (Mn_4CaO_5) that plays a crucial role in the photolysis of water. This reaction is fundamental to photosynthesis, as it splits water molecules, releasing oxygen as a byproduct while providing electrons to replace those lost by P680 (Chen et al., 2020).
- **Photolysis of Water:** The overall reaction of photolysis can be summarized as $2H_2O \rightarrow O_2 + 4H^+ + 4e^-$. This reaction not only generates molecular oxygen but also produces protons that contribute to the proton gradient essential for ATP synthesis (Govindjee et al., 2017).

2.3.3 *Electron transport chain*

2.3.3.1 Components and sequence

- **Pheophytin:** Acting as the primary electron acceptor, this molecule captures the energized electrons from P680, initiating the flow of electrons through the electron transport chain (Klimov, 2003).

- **Plastoquinone (PQ):** This small, mobile electron carrier transports electrons from PSII to the cytochrome b_6f complex (Liu and Lu, 2016). It plays a critical role in transmitting energy and contributing to the creation of the proton gradient across the thylakoid membrane (Havaux, 2020).
- **Cytochrome b_6f Complex:** A multi-subunit complex that facilitates both electron transfer and proton pumping. It captures electrons from plastoquinone and uses the energy from this transfer to actively transport protons into the thylakoid lumen (Tikhonov, 2024).
- **Plastocyanin:** A copper-containing protein that functions as a mobile electron carrier, transferring electrons from the cytochrome b_6f complex to Photosystem I, thereby linking the two photosystems (Hope, 2000).
- **Proton Gradient Formation:**
- **Q-cycle Mechanism:** This mechanism within the cytochrome b_6f complex is critical for the establishment of a proton gradient, which is essential for driving ATP synthesis. It involves multiple steps of electron transfer and proton translocation (Osyczka et al., 2005).

2.3.4 *Photosystem I (PSI)—NADPH production*

2.3.4.1 Structure and function

- **Reaction Center:** The reaction center of PSI consists of a P700 chlorophyll dimer that absorbs light energy, further energizing electrons for transfer to the subsequent components of the electron transport chain (Cherepanov et al., 2023).
- **Final Electron Acceptors:** Electrons released in PSI are ultimately transferred to iron-sulfur (Fe-S) clusters, which hold these high-energy electrons until they are passed on to $NADP^+$(Tikhonov, 2023).
- **Ferredoxin-NADP + Reductase (FNR):** This enzyme catalyzes the final reduction step in the light reactions, converting NADP + into NADPH. NADPH serves as a crucial reducing agent for the biochemical reactions of the Calvin cycle (Spaans et al., 2015).

2.4 Light-independent reactions (Calvin Cycle)

The Calvin Cycle, also known as the light-independent reactions or the carbon fixation cycle, is the biochemical pathway that converts atmospheric carbon dioxide (CO_2) into organic sugars. This process utilizes the energy carriers ATP and NADPH, which are produced during the light-dependent reactions. Unlike the light-dependent phase, the Calvin Cycle occurs in the stroma of chloroplasts and does not require light directly; however, it is indirectly regulated by light-driven processes. The cycle is named after Melvin Calvin, who clarified its steps using radioactive carbon tracing in the mid-20th century.

1. **Carboxylation**: In this initial phase, carbon dioxide (CO_2) is fixed onto ribulose bisphosphate (RuBP) by the enzyme RuBisCO, leading to the production of an unstable six-carbon compound that quickly splits into two molecules of 3-phosphoglycerate (3PGA) (Tcherkez, 2016). The reaction is represented as RuBP (5C) + $CO_2 \rightarrow 2 \times$ 3-phosphoglycerate (3PGA). RuBisCO catalyzes this reaction, transforming inorganic carbon into a stable organic molecule.
2. **Reduction:** The second phase involves using ATP and NADPH generated during the light-dependent reactions to convert 3PGA into glyceraldehyde-3-phosphate (G3P). This three-carbon sugar is essential for the synthesis of glucose and other carbohydrates (Evans, 2013).
3. **Regeneration:** The final phase of the Calvin cycle focuses on regenerating RuBP from

G3P. This regeneration involves a series of enzymatic reactions that require additional ATP, enabling the cycle to continue for further CO_2 fixation (Ali et al., 2020).

2.4.1 *Regulation and integration*

2.4.1.1 Light-mediated regulation of redox signaling

The thioredoxin system plays a pivotal role in regulating metabolic pathways in response to light conditions. Activation of this system allows for the modulation of enzyme activities, particularly those involved in the Calvin cycle (Gulyás et al., 2023).

2.4.1.2 Ferredoxin-thioredoxin reductase

This enzyme is critical for connecting the light reactions to the activation of key enzymes necessary for carbon fixation. Enzymes such as RuBisCO activase and others are targets of this regulation (Nikkanen and Rintamäki, 2019).

2.4.1.3 Allosteric regulation metabolite feedback

Different metabolites, including 3-phosphog lycerate (PGA) and inorganic phosphate (Pi), provide feedback to regulate enzymatic activities, ensuring that the Calvin cycle operates efficiently according to the plant's energetic and carbon needs (Hartman et al., 2023).

2.4.1.4 pH and Mg^{2+} effects

Environmental changes, such as a rise in stromal pH and the influx of Mg^{2+} ions, enhance the activity of several enzymes in the Calvin cycle, coordinating metabolic processes with light availability.

2.4.1.5 Enzyme modulation

The activity of RuBisCO activase is influenced by the ratio of ADP to ATP, linking the plant's energy state to its capacity for carbon fixation (Scales et al., 2014).

2.5 Factors affecting photosynthetic rate (light intensity, CO_2 concentration, temperature, water availability, nutrient status)

The photosynthetic rate is influenced by several key environmental factors, including light intensity, carbon dioxide (CO_2) concentration, temperature, water availability, and nutrient status. Each of these factors has an optimal range; beyond this range, the photosynthetic rate declines due to limitations or stress. These factors interact dynamically, often resulting in one factor becoming the limiting factor.

2.5.1 *Light intensity*

Higher light intensity increases the photosynthetic rate up to a saturation point, where additional increases may instead be limited by enzyme activity or CO_2 availability (Pappert et al., 2025). Too much light can lead to photoinhibition, damaging photosystem II. Red and blue wavelengths are the most effective in driving photosynthesis, while green light contributes less (Tang et al., 2022; Boretti, 2025).

2.5.2 *CO_2 concentration*

Photosynthesis accelerates as CO_2 levels rise because CO_2 fuels the Calvin cycle. However, saturation occurs at current atmospheric levels or higher in controlled settings. Low CO_2 levels limit carbon fixation, whereas elevated levels in greenhouses can enhance crop yields, such as in tomatoes. Stomatal closure under stress can also further restrict CO_2 uptake (Wang et al., 2009).

2.5.3 *Temperature*

Enzymatic activity in both light-dependent and Calvin cycle reactions typically peaks at temperatures between 25°C and 35°C for most plants, with moderate increases in temperature raising the photosynthetic rate (Yamasaki et al., 2002). Extreme heat (above 40°C) can denature enzymes, slowing processes down, while cold temperatures (below 10°C) reduce diffusion rates and enzymatic activity. Optimal temperatures can vary by plant type; for instance, C4 plants are generally more tolerant of higher temperatures (Mathur et al., 2014; Bunce, 2008).

2.5.4 *Water availability*

Water stress leads to stomatal closure to prevent water loss, which limits CO_2 intake and, consequently, photosynthesis (Song et al., 2020). It also impedes light reactions by reducing the electron supply from photolysis. Drought conditions can trigger a buildup of reactive oxygen species, which can damage photosynthetic machinery (Sachdev et al., 2021).

2.5.5 *Nutrient status*

Nutrients such as nitrogen are essential for chlorophyll synthesis and enzyme production, enhancing the photosynthetic rate when present in ample quantities (Evans and Clarke, 2019). Deficiencies, such as low nitrogen or magnesium levels, can reduce pigment levels and the activity of Rubisco, ultimately limiting photosynthesis (Ye et al., 2019). Additionally, phosphorus is vital for ATP formation in light reactions.

2.6 Photosynthetic efficiency and its measurement parameters (Pn, Fv/Fm, ETR, etc.)

Photosynthetic efficiency measures how effectively plants convert light energy into chemical energy through photosynthesis. The key parameters for this measurement include the net photosynthesis rate (Pn), the maximum quantum yield of photosystem II (Fv/Fm), and the electron transport rate (ETR). These parameters are commonly assessed using chlorophyll fluorescence or gas exchange methods (Walker et al., 2024; Nowak et al., 2024).

2.6.1 *Net photosynthesis (Pn)*

Pn quantifies the overall rate of CO_2 assimilation minus respiration, typically measured using gas exchange analyzers such as LI-COR systems. Under optimal conditions, values for crops usually range from 10–40 $\mu mol\ m^{-2}\ s^{-1}$, reflecting whole-leaf efficiency. This measurement combines the activities of light reactions and the Calvin cycle but requires controlled environmental conditions (Yu et al., 2020).

2.6.2 *Fv/Fm ratio*

The Fv/Fm ratio indicates the maximum photochemical efficiency of photosystem II (PSII) in dark-adapted leaves. It is calculated using the formula (Fm–F0)/Fm, where Fm is the maximal fluorescence and F0 is the minimal fluorescence (Nowak et al., 2024). Healthy leaves typically show values close to 0.8, while values dropping below 0.6 may indicate photoinhibition or damage. This ratio is measured non-destructively with PAM fluorometers (Dziwulska-Hunek et al., 2025).

2.6.3 *Electron transport rate (ETR)*

ETR estimates the flow of electrons in PSII, derived from fluorescence parameters using equation $Y(II) \times PAR \times 0.5 \times 0.84$, where Y(II) represents effective quantum yield. Typical ETR values range from 50–200 $\mu mol\ e^{-}\ m^{-2}\ s^{-1}$ under full sunlight, linking light harvesting to carbon fixation. ETR can reveal limitations in electron acceptors or downstream processes (Dziwulska-Hunek et al., 2025).

3. Definition and scope of nanomaterials in agriculture

3.1 Nanomaterials in agriculture

Nanomaterials in agriculture refer to engineered materials that fall within the nanoscale range of 1–100 nm (Chowdhury et al., 2023). These materials leverage their unique physical and chemical properties, such as enhanced surface area and increased reactivity, to significantly boost crop production and promote sustainability in farming practices (Munir et al., 2024). Their applications are broad, spanning from optimization of agricultural inputs to effective management strategies for post-harvest processes, thus tackling critical challenges like nutrient inefficiency, pest resistance, and environmental degradation (Upadhyay and Bhargava, 2025).

Nanomaterials encompass a variety of nanoscale particles designed explicitly for agricultural applications. Key types include nanofertilizers, nanopesticides, and nanosensors. These materials can take forms such as carbon-based nanostructures (e.g., carbon nanotubes), metal oxides (e.g., nano-zinc oxide), and biopolymers (Zhang et al., 2024). The design of these materials aims to improve the bioavailability of nutrients, thereby facilitating plant uptake and minimizing the unintended environmental consequences associated with traditional agricultural inputs, such as chemical runoff (Tripathi et al., 2023).

3.1.1 *Applications*

As described earlier, these innovative particles can be used as fertilizers, pesticides or sensors (Sales et al., 2024; Mgadi et al., 2024; Tripathi et al., 2023).

3.1.2 *Soil and water benefits*

Nanomaterials play a crucial role in environmental remediation efforts. They can effectively adsorb heavy metals from contaminated soils, making them safer for agricultural use (Yu et al., 2021). In addition, nanomaterials have proven beneficial in the treatment of wastewater used for irrigation, promoting cleaner and more sustainable water sources (Kumar et al., 2026). Furthermore, these materials can enhance the resilience of plants against abiotic stressors, such as drought and salinity, using gene delivery vectors that improve plant adaptations (Khalid et al., 2022).

3.1.3 *Emerging scope*

The future of nanomaterials in agriculture is promising, especially with the integration of advanced technologies like artificial intelligence (AI) and the Internet of Things (IoT). These integrations facilitate the creation of smart agricultural systems capable of automated nutrient dosing and monitoring, helping to achieve sustainable yields even amidst the challenges posed by climate change. Continuous research is essential to ensure the safety of nanomaterials in agricultural applications, focusing on assessing and mitigating potential toxicity risks to human health and the environment. This ongoing exploration aims to harness the full potential of nanotechnology while safeguarding ecological integrity and agricultural sustainability (Meng et al., 2025; Mgadi et al., 2024; Zhang et al., 2024).

3.2 Classification of nanomaterials relevant to plant systems

Nanomaterials that are relevant to plant systems can be classified by composition, structure, and origin into three main categories: organic, inorganic, and carbon-based (Rizwan et al., 2021). Typically sized between 1 and 100 nm, these materials enable unique interactions, such as enhanced uptake through cell walls or membranes (Singh et al., 2020). The classification of nanomaterials determines their roles in promoting plant growth, enhancing stress tolerance, delivering nutrients, and aiding in phytoremediation (Ali et al., 2021). The effects of these nanomaterials can vary based on factors such as particle size, concentration, surface charge, and the specific plant species involved (Juárez-Maldonado et al., 2019; Nair et al., 2010).

3.2.1 *Metal oxide nanoparticles*

Inorganic nanomaterials, including nano-zinc oxide (ZnO), nano-titanium dioxide (TiO_2), nano-iron oxide (Fe_3O_4), and nano-silver (Ag), release ions that can have antimicrobial effects or catalyze enzyme activity within plants (Nisar et al., 2019). At low doses, these nanoparticles can boost photosynthesis and root elongation, but at higher levels, they may induce oxidative stress. They are commonly used in nanofertilizers and nanopesticides (Fatima et al., 2021).

3.2.2 *Carbon-based nanoparticles*

Carbon-based materials, such as fullerenes, carbon nanotubes (single- and multi-walled), graphene, and nanodiamonds, feature a high surface area suitable for gene delivery or water retention in plant tissues (Yazdani et al., 2024). Carbon nanotubes can penetrate cell walls, promoting seed germination and stress resistance, while graphene enhances conductivity for electron transport in chloroplasts. Their biocompatibility makes them suitable for non-toxic applications (Chandrasekaran et al., 2019; Farmand et al., 2022).

3.2.3 *Polymeric nanoparticles*

Biodegradable organic nanomaterials like chitosan, alginate, and polyethylene glycol (PEG)-based nanoparticles can encapsulate agrochemicals for controlled release (Vemula and Reddy, 2023). These materials protect active ingredients from degradation, ensure targeted delivery to roots or leaves, and promote nutrient uptake without leaving harmful residues. They are ideal for developing sustainable pesticide and fertilizer systems (An et al., 2022).

3.2.4 *Quantum dots*

Semiconductor nanocrystals, such as cadmium selenide (CdSe) and zinc sulfide (ZnS), exhibit tunable fluorescence and can serve as imaging probes or sensors for real-time monitoring of nutrient uptake and pathogen detection in plants (Oliveira et al., 2023). Their size-dependent emission can track the translocation of nanomaterials from roots to shoots. However, potential toxicity arising from heavy metals limits their widespread use (Zhang et al., 2024).

3.3 Mechanisms of nanoparticle uptake and translocation in plants

Nanoparticles (NPs) play a critical role in plant interactions and can enter through two primary pathways: the roots and leaves. These tiny particles can navigate barriers such as cell walls through size-dependent mechanisms, allowing them to translocate via the plant's vascular tissues (Wang et al., 2023). The efficiency of NP uptake is influenced by several factors, including the physical characteristics of the particles themselves, such as their size, which is most effective when under 50 nm, favoring symplastic pathways, and their surface charge, which can impact how well they adhere to plant tissues (An et al., 2022). Furthermore, the method of application, whether through the roots, foliar sprays, or seed priming, also significantly affects the extent of uptake. The subsequent movement of NPs within the plant can lead to varied physiological responses, from promoting growth and vitality to causing toxicity (Khan et al., 2022; Djanaguiraman et al., 2024).

3.3.1 *Root uptake mechanisms*

Nanoparticles access the roots through two primary mechanisms: drainage from foliar sprays or direct exposure from soil. Once they reach the rhizodermal region, they can cross the protective barriers through either apoplastic routes, navigating the spaces between cell walls, or symplastic routes, moving through the interconnected cytoplasm of plant cells via plasmodesmata (Wang et al., 2023). Smaller nanoparticles, those less than 50 nm in size, usually enter roots effectively through endocytosis or specialized water channels known as aquaporins (Khan et al., 2022; Bárzana et al., 2022). In contrast, larger particles, measuring between 50 and 200 nm, predominantly utilize the apoplastic pathway. Positively charged nanoparticles have a propensity to bind to the negatively charged components of cell walls, which enhances their adhesion but can complicate their penetration (Forest and Pourchez, 2017). Notably, when these particles dissolve into ions, they can promote their uptake through transporters embedded in the root tissue (Lesniak et al., 2013).

3.3.2 *Foliar uptake mechanisms*

When it comes to leaf absorption, NPs infiltrate mainly through hydrophilic cuticles, stomata, or trichomes, the small hair-like structures on leaf surfaces. The uptake of particles less than 1 μm is primarily facilitated through diffusion and endocytosis (Hong et al., 2021). Interestingly, the shape of nanoparticles significantly influences their entry: rod-shaped NPs make more effective contact with leaf surfaces relative to spherical ones, improving their chances of successful penetration (Bhati et al., 2026). Hydrophilic pathways can restrict the uptake of larger NPs, whereas stomatal pores, measuring between 20 and 50 μm, create a gateway for larger particles to enter the leaf interior (Wang et al., 2023; Avellan et al., 2022).

3.3.3 *Translocation pathways*

After successful uptake, nanoparticles travel through the plant. From the roots, they ascend through the xylem, following the upward flow of water driven by passive transpiration into the shoots. In contrast, NPs that enter through the leaves move downward through the phloem, utilizing a process known as symplastic loading (Vijay et al., 2025). The phloem's sieve tubes can accommodate small particles, although the mobility of these particles can be affected by transformations such as aggregation or dissolution (Mullendore et al., 2010; Shukla et al., 2016).

3.3.4 *Influencing factors*

The uptake of nanoparticles is not uniform; it's influenced by multiple environmental factors. Different plant species exhibit varying responses to NP exposure, and soil pH levels can alter the chemical environment, affecting how NPs behave (Ahmed et al., 2021). Additionally, the plant's microbiome the community of microbes residing in the rhizosphere, can play a vital role in enhancing NP dissolution and uptake (Zhang et al., 2024). The nanoparticles' size, electric charge (zeta potential), and concentration critically shape their fate within the plant system: low concentrations often stimulate growth, while elevated levels can lead to the production of reactive oxygen species (ROS), inducing oxidative stress and potential harm to the plant (Su et al., 2019).

4. Mechanistic insights: how nanomaterials influence photosynthesis

In plants, photosynthesis drives growth and biomass production by turning sunlight into chemical energy and changing atmospheric CO_2 into organic compounds. Its efficiency not only decides how productive a plant can be but also affects how well it handles stresses like drought, heat, and salinity. Traditional ways to improve photosynthesis and stress tolerance often don't meet the growing demands of food production, especially with changing climate. Recent studies show that nanotechnology may offer a new way to address these challenges by acting directly at molecular and cellular levels (Li et al., 2021; Akhavan and Gonçalves, 2021; Tripathi et al., 2017).

Nanomaterials, including bio-based, metal oxide, and carbon-based types, have special properties because of their tiny size, high surface reactivity, and adjustable surfaces. These properties allow them to interact with plant tissues and organelles like chloroplasts in ways normal agrochemicals cannot. For example, some nanoparticles can boost chlorophyll levels or change leaf light absorption. This helps plants use light more efficiently and improves photosynthesis (Tripathi et al., 2017; Wang et al., 2020; Singh et al., 2021).

Nanomaterials can also affect enzyme activity and gene expression. For instance, zinc oxide (ZnO) nanoparticles can increase Rubisco, the key enzyme for CO_2 fixation, and turn on genes involved in light harvesting. These changes often improve photosystem II performance and carbon assimilation under both normal and stressful conditions (Li et al., 2021; Faizan et al., 2018).

Nanoparticles may also protect plants by managing reactive oxygen species (ROS) and stabilizing photosynthetic structures. Stress often causes too much ROS, which damages chloroplasts and electron transport, reducing photosynthesis. Certain nanomaterials support plant antioxidants, helping keep thylakoid membranes and photosystems stable (Tripathi et al., 2017; Akhavan and Gonçalves, 2021). The effects depend on dosage. Low doses help nutrient uptake, light capture, enzyme activity, and stress tolerance, while high doses may be toxic. Overall, the result depends on balancing helpful effects with possible stress.

In short, nanotechnology boosts photosynthesis by improving light use, regulating key enzymes, and protecting photosynthetic structures. Understanding these effects is key to designing nanomaterial-based strategies that increase crop productivity (Li et al., 2021; Wang et al., 2020).

4.1 Nanomaterial interaction with chloroplasts

Chloroplast is important for photosynthesis and global food production. But improving chloroplast function is challenging because of technical limits like low transformation efficiency, species barriers, and lack of tools to monitor in real time (Tripathi et al., 2017; Akhavan and Gonçalves, 2021; Zhang et al., 2019).

Nanotechnology offers a solution by allowing targeted delivery of molecules into chloroplasts. Specially designed nanoparticles can cross cell walls and chloroplast membranes to deliver DNA, RNA, proteins, or small molecules. This can work across many species and improve photosynthetic enzyme expression, carbon fixation, and crop yield (Li et al., 2021; Tripathi et al., 2017; Akhavan and Gonçalves, 2021). Beyond delivering genes, nanoparticles can directly improve chloroplast function. They may increase light absorption, stabilize carbon-fixing enzymes, and improve electron transport. Under stress like drought, salinity, or high light, nanoparticles can reduce ROS in chloroplasts and protect photosystem structures (Li et al., 2021).

Using peptides from Rubisco helps nanoparticles reach chloroplasts more efficiently. Quantum dots with β-cyclodextrin can carry molecules that control ROS levels in chloroplasts without affecting other cell parts (Tripathi et al., 2017; Akhavan and Gonçalves, 2021; Zhang et al., 2019).

Nanoparticles enter plants through roots or leave and travel via xylem and phloem to reach chloroplast-rich tissues. Size, charge, and surface chemistry influence uptake and mobility. Low doses improve growth, germination, root length, and photosynthesis, while high doses may cause oxidative stress or DNA damage. Biodegradable, bio-based nanoparticles are safer, and soil factors like pH and organic matter also affect their stability (Li et al., 2021; Wang et al., 2020). In short, nanoparticles can improve chloroplast function, photosynthesis, and plant resilience. Using precise delivery with safe nanoparticles opens new possibilities for advanced and sustainable agriculture (Akhavan and Gonçalves, 2021; Zhang et al., 2019; Tripathi et al., 2017). Fig. 5.1.

4.2 Role in enhancing chlorophyll biosynthesis

Bio-based nanoparticles made from natural polymers like cellulose, chitosan, and lignin are biodegradable and safe for the environment. They work at cell and molecular levels,

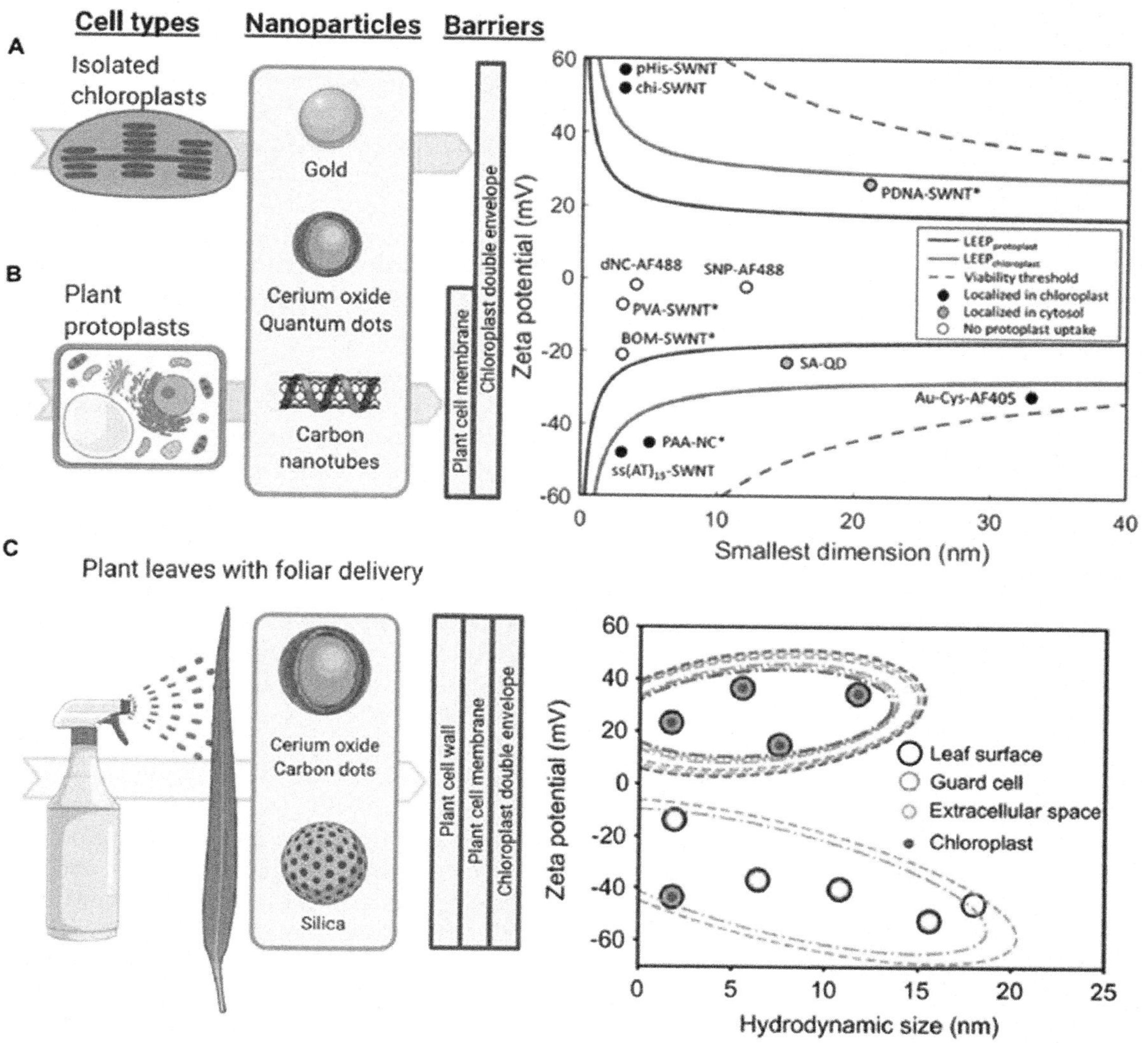

FIGURE 5.1 Understanding and modeling nanoparticle-chloroplast interactions (Newkirk et al., 2021). *Newkirk, G.M., de Allende, P., Jinkerson, R.E., Giraldo, J.P., 2021. Nanotechnology approaches for chloroplast biotechnology advancements. Front. Plant Sci., 12. https://doi.org/10.3389/fpls.2021.691295.*

influencing plant growth and development (Malode et al., 2025; Li et al., 2021).

These nanoparticles regulate hormones like auxins, cytokinins, and gibberellins, which control roots, shoots, and reproduction. Their small size helps deliver nutrients efficiently, boosting chlorophyll production, enzyme activity, and energy metabolism. This supports growth and strengthens photosynthetic tissues (Wang et al., 2020). Bio-based nanoparticles also help plants handle stress like drought, salinity, or nutrient shortages by stabilizing cells and reducing oxidative damage. Compared to synthetic nanoparticles, they are safer and leave less environmental residue (Li et al., 2021).

Zinc oxide nanoparticles (ZnO-NPs) are effective in enhancing photosynthesis. They improve enzyme activity, chlorophyll content, and nutrient balance, which increases carbon fixation and biomass. ZnO-NPs also support shoot growth and positive plant–microbe interactions, boosting productivity (Adil et al., 2022).

4.3 Improvement of light absorption and energy transfer

Nanoparticles improve photosynthesis by enhancing light capture and energy flow. They interact with chloroplasts, stabilize thylakoid membranes, increase light-harvesting complexes, and promote chlorophyll production. Some, like carbon dots, can act as artificial electron donors, improve electron flow and reducing energy loss. Nanoparticles also boost Rubisco activity for better carbon fixation. By managing ROS, they prevent light damage and may activate genes for light harvesting, pigments, and electron transport. Root or leaf uptake ensures chloroplast-rich tissues get the nanoparticles.

ZnO-NPs and TiO_2-NPs enhance photochemical efficiency, stabilize photosystems, improve electron transport, and increase nutrient availability. Fluorescent carbon dots convert UV light into usable wavelengths, reducing photoinhibition and increasing ATP and NADPH. Overall, nanoparticles support photosynthesis, growth, biomass, and stress tolerance, making them a practical tool for modern agriculture. NPs may augment the photosynthetic efficiency through

(1) expanding the absorption wavelength range,
(2) enhancing Hill reaction,
(3) harvesting more electrons,
(4) accelerating electron transfer,
(5) scavenging ROS (Li et al., 2021; Tripathi et al., 2017; Singh et al., 2024; Wang et al., 2020; Dang et al., 2024; Li et al., 2023).

4.4 Regulation of stomatal conductance and CO_2 assimilation

Stresses like drought, salinity, and cold can potentially reduce growth, photosynthesis, and yield. Nanoparticles can help plants cope by improving photosynthesis, stomatal function, antioxidants, and metabolism. Silicon nanoparticles (SiO_2-NPs) help cotton under salt and cold stress. They improve stomata size, density, and opening, allowing more CO_2 into leaves. SiO_2-NPs also boost chlorophyll and maintain photochemical efficiency, electron transport, and Rubisco activity, improving carbon fixation. A dose of $100\ mg\ L^{-1}$ works well under stress (Liang et al., 2023; Raza et al., 2025).

Zinc oxide nanoparticles (ZnO-NPs) improve drought tolerance in coriander. Foliar spray of 100 ppm ZnO-NPs increases photosynthesis, stomatal conductance, and transpiration. They raise chlorophyll *b*, sugars, phenolics, and flavonoids, protecting cells and improving water use. ZnO-NPs adjust stomata size and lower ABA, keeping stomata open and supporting carbon fixation (Ahmed et al., 2023). Effectiveness depends on nanoparticle size, charge, composition, and application. NPs deliver nutrients and active molecules, regulate stress genes, and improve antioxidants and osmolytes.

Optimizing stomata, chlorophyll, and PS II helps plants maintain energy conversion, photosynthesis, and growth under stress. In short, SiO_2-NPs and ZnO-NPs improve stress tolerance by enhancing stomata, photosynthesis, and antioxidant defenses (Faizan et al., 2021; Alam et al., 2025).

4.5 Impact on rubisco activity and carbon fixation

Carbon dots (CDs) are tiny, water-soluble nanoparticles with low toxicity and unique light-absorbing properties. They help plants grow by improving photosynthesis, enzyme activity, and carbon metabolism. CDs absorb UV and blue light and re-emit it as light chloroplasts can use. This gives Rubisco more energy for CO_2 fixation. They also increase enzyme and sugar transporter expression, improving the Calvin cycle and moving carbohydrates from leaves to roots, stems, and grains. This helps faster growth and better grain filling. CDs support stomata under stress by reducing ABA-induced closure, keeping CO_2 available. Their antioxidant action protects chloroplasts, stabilizes photosystems, and maintains Rubisco under heat, drought, or salinity (Li et al., 2020, 2021; Guirguis et al., 2023). Proper dosage is key: low doses help growth and carbon fixation, while too much may block light or cause mild toxicity. Soil, water, and temperature also affect results. CDs can carry nutrients or growth regulators to chloroplasts, making them useful for precision agriculture. In short, CDs improve Rubisco, carbon fixation, and sugar transport, boosting growth, yield, and stress resistance (Raveena et al., 2026).

4.6 Reduction of reactive oxygen species (ROS)

Heat and drought increase ROS in chloroplasts, mitochondria, and peroxisomes. Too much ROS damages proteins, pigments, and membranes, reducing photosynthesis and causing photoinhibition. Plants use antioxidants like superoxide dismutase, catalase, ascorbate, and glutathione to control ROS. Severe stress can overwhelm them. Nanoparticles help by managing ROS production and boosting antioxidants. This protects chloroplasts, keeps PS II stable, and reduces photoinhibition. ROS also act as stress signals, so they shouldn't be fully removed. Nanoparticles balance ROS, keeping enough for signaling but not enough to harm cells. Challenges remain, like nanoparticle buildup in edible parts, environmental safety, and cost. Using safe, eco-friendly nanoparticles is critical. In short, nanoparticles reduce ROS damage, protect photosynthesis, and help plants handle heat and drought, supporting growth and productivity (Yanga et al., 2017; Molnár et al., 2020).

5. Types of nanomaterials used to enhance photosynthesis

Nanotechnology is transforming plant science by providing innovative strategies to improve photosynthesis, the core process through which plants convert light energy into chemical energy. Photosynthesis is highly sensitive to environmental conditions, nutrient availability, and various stresses. Conventional approaches, such as fertilizers or selective breeding, often fall short because nutrients may not be efficiently absorbed, and environmental factors can limit growth. Nanomaterials, in contrast, act at the cellular and tissue level, reaching chloroplasts and other organelles directly, thereby enhancing light capture, energy transfer, and the synthesis of key biomolecules.

Nanomaterials, typically ranging from 1–100 nm, possess unique surface properties and a high surface-to-volume ratio, which allow them to interact intimately with plant cells and organelles. These interactions improve light absorption, pigment production, and electron transport, while scavenging reactive oxygen species (ROS) and supporting nutrient uptake. Such properties enable plants to maintain growth and photosynthetic efficiency even under challenging conditions, including drought, high salinity, or nutrient deficiencies.

Overall, nanomaterials enhance photosynthesis, growth, and stress resilience. The principal categories include metal and metal oxide nanoparticles, carbon-based nanomaterials, and polymeric or biogenic nanomaterials, each with specific mechanisms and advantages. Fig. 5.2.

5.1 Metal and metal oxide nanoparticles

5.1.1 *Titanium dioxide (TiO_2) nanoparticles—light harvesting and ROS scavenging*

Titanium dioxide nanoparticles (TiO_2 NPs) are extensively studied for their photocatalytic activity and ability to interact directly with leaf tissues and chloroplasts. TiO_2 NPs improve light absorption, electron transport, and chlorophyll content, collectively boosting photosynthesis, plant growth, and biomass accumulation. Their small size and large surface area facilitate penetration into leaves, allowing direct access to photosynthetic machinery. These nanoparticles enhance light capture by modulating light-harvesting complexes (LHCs). For instance, in *Arabidopsis thaliana*, TiO_2 NPs increased LHCII gene expression and protein abundance, enhancing the absorption of blue (400–500 nm) and red (650–680 nm) light. This promotes efficient energy transfer to PSII and PSI, accelerating electron flow and oxygen evolution. TiO_2 NPs also support electron transport chain efficiency. Studies in spinach chloroplasts revealed elevated electron flow, ATP production, and photophosphorylation compared to untreated plants. Acting as photocatalysts, they facilitate water splitting, supplying electrons for NADPH generation, which is essential for carbon fixation in the Calvin cycle (Ze et al., 2011). Environmental conditions, particularly light intensity, influence their performance. In radish, TiO_2 NPs maintained higher PSII efficiency and electron transport under low-light conditions, suggesting their potential to enhance adaptability to diverse environments. Additionally, TiO_2 NPs improve nutrient uptake, especially magnesium and nitrogen, which are critical for chlorophyll synthesis and Rubisco activity, further enhancing photosynthetic efficiency (Tan et al., 2018; Vatankhah et al., 2023).

FIGURE 5.2 **Different modes of application of nanoparticulate formulations in plants.** Methods of NP application on plants.*Made by BioRenders.com.*

Key mechanisms of TiO_2 NPs:

- Upregulation of LHCII, enhancing light harvesting.
- Improved electron transport and ATP/NADPH production.
- Increased chlorophyll biosynthesis via better nutrient uptake.
- Enhanced adaptability to varying light conditions.

5.1.2 *Zinc oxide (ZnO) nanoparticles—enzyme activation and chlorophyll synthesis*

Zinc oxide nanoparticles (ZnO NPs) are biocompatible, nutritionally important, and highly reactive due to their nanoscale size. Zinc is a cofactor for key photosynthetic and stress-related enzymes, including carbonic anhydrase, Rubisco, and superoxide dismutase (SOD). Compared to bulk zinc, ZnO NPs penetrate plant tissues more effectively, improving nutrient assimilation and physiological performance (Hamada et al., 2024). ZnO NPs activate photosynthetic enzymes, enhancing carbonic anhydrase and Rubisco activity, thereby promoting CO_2 fixation in the Calvin cycle. They also stabilize δ-aminolevulinic acid dehydratase (ALAD), which increases chlorophyll *a* and *b* content and optimizes the chlorophyll a/b ratio, improving light capture. These nanoparticles strengthen antioxidant defenses, raising activities of SOD, catalase, and peroxidase, which protects chloroplasts under drought, salinity, and other stresses. ZnO NPs also improve nutrient uptake and ionic balance, supporting enzymatic and metabolic processes. Low-to-moderate doses (10–100 ppm) enhance photosynthetic efficiency, whereas higher doses may induce oxidative stress. Interestingly, ZnO NPs can indirectly protect photosynthesis by reducing pest damage. Green-synthesized ZnO NPs increased chlorophyll content and defense metabolites in tomato plants exposed to *Tuta absoluta*. Gene expression studies confirm that ZnO NPs upregulate genes associated with chlorophyll biosynthesis, Rubisco, and antioxidant defense, enhancing stress adaptability (Raghib et al., 2025; Mardani-Talaee et al., 2025).

Summary of ZnO NP effects:

- Activation of photosynthetic enzymes
- Enhanced chlorophyll content and light harvesting
- Strengthened antioxidant defense against ROS
- Improved nutrient uptake and ionic balance
- Reduced pest damage, preserving photosynthetic tissues
- Upregulation of photosynthesis and stress-related genes

5.1.3 Iron oxide (Fe_2O_3) nanoparticles—chloroplast development and electron transport

Iron oxide nanoparticles (Fe_2O_3 NPs) supply bioavailable iron, essential for chlorophyll biosynthesis, photosystem assembly, and electron transport chain function. Unlike conventional iron fertilizers, Fe_2O_3 NPs penetrate tissues efficiently and are readily assimilated by plant cells. Fe_2O_3 NPs promote chloroplast development by enhancing granum stacking, organizing thylakoids, and increasing chloroplast density, which improves light capture and energy conversion. Iron supports redox proteins such as ferredoxin and cytochromes, accelerating electron flow from PSII to PSI, boosting ATP and NADPH production, and maintaining photosynthetic efficiency under moderate stress. They also enhance pigment content by activating iron-dependent enzymes in the porphyrin pathway, increasing chlorophyll a, chlorophyll *b*, and carotenoids while optimizing the chlorophyll a/b ratio. Fe_2O_3 NPs strengthen antioxidant defenses through Fe–SOD activity, stabilizing membranes and protecting photosynthetic proteins from ROS damage. Enhanced iron nutrition further promotes uptake of magnesium and manganese, supporting coordinated metabolism and resilience. Moderate doses (50–200 mg/L) improve photosynthesis, whereas excessive levels may cause iron toxicity and oxidative stress. Co-application with TiO_2 or ZnO NPs can produce synergistic effects, further improving chloroplast structure, photosynthesis, and stress tolerance. At the molecular level, Fe_2O_3 NPs upregulate genes involved in chlorophyll biosynthesis, ferredoxin, and cytochrome proteins (Feng et al., 2022; Jampílek and Kráľová, 2019).

5.1.4 *Silver (Ag) nanoparticles—electron transport efficiency*

Silver nanoparticles (Ag NPs) enhance photosynthesis beyond their antimicrobial effects. Their nanoscale size enables efficient penetration into plant tissues, improving chloroplast function and electron transport. Ag NPs accelerate electron flow from PSII to PSI, increasing ATP and NADPH production and supporting Calvin cycle activity. This enhances photosynthetic efficiency under mild stress, such as drought or salinity. They stabilize chlorophyll a, chlorophyll *b*, and carotenoids, protecting pigments from ROS and improving light capture. Chloroplast ultrastructure also benefits, with improved thylakoid organization, granum stacking, and increased chloroplast density, leading to higher photosynthetic rates and biomass accumulation. Ag NPs enhance antioxidant enzyme activities (SOD, CAT, POD), maintaining redox balance. Ag NPs support nutrient uptake (Mg, Fe, Mn), aiding chlorophyll biosynthesis, electron transport, and energy metabolism. Low-to-moderate concentrations (10–100 mg/L) are effective, while higher doses may be phytotoxic. Co-application with TiO_2 or ZnO further improves photosynthesis and stress tolerance. Ag NPs also modulate gene expression of photosystem proteins, chlorophyll biosynthesis enzymes, and antioxidant enzymes (Liu et al., 2021; Gupta et al., 2018).

5.2 Carbon-based nanomaterials

5.2.1 *Carbon nanotubes (CNTs)—water and CO_2 transport*

Carbon nanotubes (CNTs), including single-walled (SWCNTs) and multi-walled (MWCNTs), enhance water and CO_2 transport due to their tubular structure, high aspect ratio, and large surface area. CNTs penetrate cell walls and integrate into vascular tissues, improving root water absorption, stomatal conductance, and CO_2 assimilation, collectively boosting photosynthesis and growth (Khodakovskaya et al., 2009; Safdar et al., 2022). They improve root hydraulic properties by modifying membrane permeability and improving aquaporins, facilitating water transport under salinity or drought stress. In rice, CNTs increased root hair density, xylem cell size, chlorophyll content, and photosynthetic performance. Low-to-moderate doses enhance plant performance, while high concentrations may induce nanotoxicity (Raveena et al., 2026).

5.2.2 *Graphene oxide (GO) and carbon quantum dots (CQDs)—light utilization and photochemical activity*

Graphene oxide (GO) and carbon quantum dots (CQDs) improve photosynthesis by enhancing light absorption, electron transport, and photochemical efficiency. GO increases PSII efficiency, strengthens connectivity between PSII units, and elevates chlorophyll content, resulting in higher photosynthetic rates and biomass accumulation. GO also facilitates nutrient uptake (N, Fe, Cu), supporting enzymatic and photochemical processes. CQDs act as nano–light converters, transforming UV light into visible wavelengths usable by chlorophyll. In rice seedlings, CQDs increased chlorophyll content, Rubisco activity, and growth traits, broadening light utilization. Both GO and CQDs reduce photoinhibition and ROS formation, protecting chloroplasts under stress (Guirguis et al., 2023).

5.3 Polymeric and biogenic nanomaterials

5.3.1 *Slow-release nutrient nanocarriers*

Polymeric and biogenic slow-release nanocarriers improve nutrient availability by releasing nutrients gradually in sync with plant demand. Made from chitosan, alginate, or hydrogels, they protect macro- and micronutrients from leaching and degradation, supporting chlorophyll biosynthesis, electron transport, and Calvin cycle activity. These carriers enhance nutrient uptake, chlorophyll content, stomatal conductance, and photosynthetic efficiency in crops such as rice and wheat, even at reduced application rates. Micronutrient-enriched nanocarriers stabilize pigments, improve electron transport proteins, and activate antioxidant enzymes. They also promote beneficial soil microbes, enhancing nutrient mineralization and water transport, indirectly supporting photosynthesis (Banerjee et al., 2023).

5.3.2 *Green synthesized nanomaterials*

Green synthesized nanomaterials, derived from plant extracts, bacteria, fungi, or biopolymers, offer an eco-friendly alternative to conventional nanoparticles. Biological molecules act as reducing and capping agents, enhancing nanoparticle stability and biocompatibility (Iravani, 2011).

These nanoparticles improve nutrient availability, chlorophyll stability, electron transport, and antioxidant activity while regulating stomatal behavior and water relations, ensuring efficient CO_2 diffusion. Biodegradable and minimally toxic, green nanomaterials support soil microbial health and offer a sustainable strategy to enhance photosynthesis, growth, and resilience under environmental stress.

5.4 Summary of key studies showing improved photosynthetic performance using nanomaterials

Recent research clearly explains that nanomaterials can truly enhance photosynthetic performance in plants by improving chlorophyll biosynthesis, gas exchange, electron transport efficiency, and carbon assimilation. These physiological improvements frequently translate into enhanced growth and yield under both optimal and stress conditions, highlighting the practical relevance of nanotechnology in plant production systems. Zinc oxide nanoparticles (ZnO NPs) are among the most widely studied nanomaterials in this context. In tea plants (*Camellia sinensis*), ZnO NP application increased chlorophyll content, Rubisco activity, and the expression of photosynthesis-related genes, leading to higher net photosynthetic rates and improved shoot growth compared with untreated controls. ZnO NP treatment also effect the phyllosphere microbial community, suggesting that improved photosynthesis may partly arise from enhanced plant–microbe interactions (Chen et al., 2024).

In cereal crops, field-based studies further confirm these benefits. Foliar application of ZnO NPs in rice (*Oryza sativa* L.) enhanced leaf area index, net photosynthetic rate, and overall photosynthetic potential. These effects delayed leaf senescence, sustained carbon assimilation during grain filling, and resulted in yield increases of 1.5%–6.4% (Kumar et al., 2025). The extended duration of active photosynthesis under field conditions highlights the suitability of ZnO NPs for practical agricultural use.

Carbon-based nanomaterials have shown strong positive effects on photosynthesis, particularly under abiotic stress. In tomato (*Solanum lycopersicum*) exposed to salinity, nano-biochar application improved chlorophyll content, net photosynthetic rate, and antioxidant enzyme activity while enhancing plant water status. These findings indicate that carbon nanomaterials can indirectly support photosynthetic efficiency by mitigating stress-induced physiological limitations (Shahzadi et al., 2025).

Similarly, green-synthesized ZnO NPs improved photosynthetic performance in chilli (*Capsicum annuum* L.) under saline conditions by enhancing chlorophyll synthesis, promoting shoot and root growth, and reducing oxidative stress markers such as hydrogen peroxide and malondialdehyde. These responses demonstrate the role of ZnO NPs in maintaining photosynthetic efficiency through improved physiological resilience (Adnan et al., 2024).

Iron oxide nanoparticles (Fe_2O_3 NPs) have also been reported to enhance photosynthesis. In rice callus cultures, biogenic Fe_2O_3 NPs increased antioxidant activity and stabilized physiological processes, suggesting improved photosynthetic capacity through better redox balance and metabolic regulation (Ullah et al., 2024).

Advanced carbon nanomaterials such as graphene oxide further broaden these effects. In *Lavandula angustifolia*, graphene oxide treatment increased chlorophyll accumulation, enhanced light absorption, and stimulated growth, which was associated with higher essential oil production (Asadi et al., 2025).

Overall, these studies demonstrate that nanomaterials including ZnO, Fe_2O_3, nano-biochar, and graphene oxide enhance photosynthetic performance by improving pigment content, gas exchange, electron transport, and protection of the photosynthetic apparatus under stress conditions. Evidence from both controlled experiments and field trials confirms their potential to enhance crop productivity and resilience.

5.5 Comparative analysis across plant types (C_3, C_4, CAM plants)

Plants utilize distinct photosynthetic pathways—C_3, C_4, and CAM—that differ in carbon fixation strategies, water-use efficiency, and stress tolerance. These inherent differences strongly influence how plants respond to nanomaterial exposure.

C_3 plants fix CO_2 directly through the Calvin–Benson cycle in mesophyll cells, a process that is efficient under moderate environmental conditions but highly sensitive to photorespiration, particularly during heat or drought stress (Smith et al., 2020). In contrast, C_4 plants possess a CO_2-concentrating mechanism that spatially separates initial carbon fixation from the Calvin cycle, reducing photorespiration and improving water-use efficiency. CAM plants employ a temporal separation strategy, fixing CO_2 at night and releasing it during the day, which enables exceptional water conservation under arid conditions but often limits growth rate (Gilman et al., 2023).

These physiological distinctions result in different responses to nanomaterials. Comparative studies have shown that C_4 plants generally exhibit greater tolerance to nanoparticle exposure than C_3 species. Under identical treatments with metallic and carbon-based nanomaterials, C_3 plants such as rice and soybean displayed pronounced reductions in chlorophyll content and photosystem II efficiency, whereas C_4 species such as maize and amaranth experienced only minor declines. Higher nanoparticle accumulation in C_3 plants contributed to greater disruption of chloroplast structure and photosynthetic performance (Bai et al., 2021).

This differential sensitivity reflects broader physiological principles. The CO_2-concentrating mechanism in C_4 plants buffers photosynthesis against disruptions in carbon availability, reducing vulnerability to nanomaterial-induced stress. In contrast, C_3 plants are more susceptible because small disturbances can intensify photorespiration and suppress net photosynthesis (Li et al., 2021).

Leaf anatomy and nanoparticle uptake pathways further influence these responses. Nanoparticles typically enter leaves through stomata or cuticular pores and may translocate to chloroplasts. Species with thicker cuticles and more rigid vascular tissues—common in many C_4 and CAM plants—often exhibit reduced nanoparticle penetration and fewer disruptions to electron transport processes compared with thinner-leafed C_3 plants (An et al., 2022).

CAM plants remain relatively understudied in nanotechnology research. Their nocturnal stomatal opening and internal CO_2 storage may alter the timing and extent of nanoparticle uptake, potentially leading to unique interaction patterns depending on application timing and environmental conditions (Mercado-Díaz de León et al., 2024). Nanomaterials also influence antioxidant defenses and gene expression. Because C_3, C_4, and CAM plants differ in baseline reactive oxygen species management and stress signaling, nanoparticle exposure can trigger plant-type-specific changes in antioxidant enzyme activity and regulation of photosynthesis-related genes. Understanding these pathway-specific responses is essential for optimizing nanomaterial application strategies tailored to plant type and environmental context.

5.6 Highlights from controlled environment and field trials

Research on nanomaterial-mediated enhancement of photosynthesis has progressed from laboratory-scale experiments to controlled environment studies and field trials, providing both mechanistic understanding and real-world validation. Controlled environment studies allow precise regulation of light, temperature, water, nutrients, and nanomaterial exposure, enabling detailed analysis of photosynthetic responses.

Under controlled conditions, ZnO nanoparticles consistently improved chlorophyll content, net photosynthetic rate, stomatal conductance, and CO_2 assimilation across multiple crops. In hydroponically grown rice, ZnO NP treatments enhanced photosynthetic efficiency more effectively than conventional zinc fertilizers, likely due to improved micronutrient bioavailability (Singh et al., 2024). Similar results were observed in tomato, where biogenic ZnO NPs increased photosystem II efficiency and electron transport rates (Faizan et al., 2018). Combined nanomaterial applications have also shown synergistic effects. In greenhouse-grown *Brassica napus*, joint application of ZnO and TiO_2 nanoparticles increased pigment concentration, improved chlorophyll fluorescence indices, and enhanced antioxidant enzyme activity, protecting the photosynthetic machinery under high light conditions.

Carbon-based nanomaterials such as nanobiochar improved photosynthetic pigment stability, CO_2 assimilation, and biomass accumulation in lettuce, largely through improved root function and rhizosphere interactions. Carbon quantum dots have further demonstrated the ability to enhance light utilization and electron transfer efficiency in controlled microalgal systems (Wang et al., 2025).

Field trials confirm that these benefits translate to agricultural environments. In rice, foliar application of ZnO NPs increased chlorophyll content and photosynthetic rates, leading to higher grain yields compared with bulk zinc fertilizers (Kumar et al., 2025). Nano-enabled fertilizer formulations in wheat improved photosynthetic efficiency, biomass accumulation, and grain quality under variable field conditions. In maize, green-synthesized Fe_2O_3 nanoparticles enhanced chlorophyll stability and drought tolerance, supporting sustained CO_2 assimilation during water-limited periods (Manzoor et al., 2021; Yousaf et al., 2024).

While responses are often stronger under controlled conditions, field studies demonstrate that nanomaterial benefits are robust but influenced by soil properties, microbial interactions, application timing, and delivery method. Foliar applications at key developmental stages generally provide the most consistent improvements in photosynthetic performance.

6. Integrative role of nanomaterials in stress-induced photosynthetic decline

Photosynthesis is a highly stress-sensitive physiological process in plants. Abiotic stresses such as drought, salt, and heavy metals significantly affect photosynthesis by disturbing water relations, ion balance, chloroplast ultrastructure, enzyme functions, and redox states. These stress-induced disturbances ultimately lead to reduced carbon fixation, biomass, and crop productivity. In the past few years, nanomaterials (NMs) have been recognized as novel tools to counter stress-induced reductions in photosynthesis because of their unique physicochemical properties, such as high surface area, reactivity, and ability to interact at the cellular and subcellular levels.

Nanomaterials play a role of integration in the protection of photosynthesis by enhancing stress tolerance, protecting chloroplasts, and strengthening antioxidant systems. This section will highlight the role of nanomaterials in the alleviation of drought, salinity, and heavy metal stress by protecting photosynthesis, with a focus on chloroplast protection and the management of oxidative stress.

6.1 Mitigation of drought, salinity, and heavy metal stress via photosynthetic protection

6.1.1 Drought stress and photosynthetic impairment

Drought stress inhibits photosynthesis mainly due to stomatal closure, reduced CO2 diffusion, chlorophyll loss, and inhibition of photosystem II (PSII). Chronic drought stress causes an overproduction of ROS, which causes oxidative damage to the thylakoid membranes and photosynthetic proteins, such as the D1 protein of PSII (FLEXAS et al., 2004; Chaves et al., 2009).

Nanomaterials including silicon dioxide (SiO_2), zinc oxide (ZnO), titanium dioxide (TiO_2), and carbon nanomaterials have been shown to reduce drought-induced losses in photosynthesis. Nanomaterials increase water use efficiency by optimizing water uptake by roots, modulating stomatal conductance, and minimizing transpiration. In addition, they promote maintenance of chlorophyll and photosynthetic efficiency by protecting photosynthetic systems from oxidative stress and maintaining electron transport function (Dimkpa et al., 2017; Tripathi et al., 2017).

Silicon nanomaterials, especially, strengthen cell walls and cuticular layers, hence lowering water loss and enhancing leaf water potential. The enhanced water status allows the photosynthetic apparatus to function more effectively under conditions of reduced water availability.

6.1.2 Salinity stress and ionic toxicity

Photosynthesis is negatively affected by salinity stress primarily because of the presence of Na+ and Cl-ions, which produce ionic toxicity when they build up, osmotic stress, and nutrient imbalance. The accumulation of excess sodium disrupts potassium uptake, alters enzyme function, and disrupts chloroplast membrane integrity, ultimately resulting in decreased chlorophyll synthesis and disrupted transport of photosynthetic electrons (Munns and Tester, 2008).

Nanomaterials reduce salinity stress-induced damage to photosynthesis by improving ionic balance and strengthening antioxidant activities. Certain metal nanoparticles such as ZnO, Fe3O4, and TiO2 appear to stimulate chlorophyll production and enhance photo-pigment stability under salinity conditions and increase the rate of photosynthesis. Furthermore, nanomaterials also decrease Na + ion uptake while increasing K+ ion retention, thus supporting metabolic pathways needed for carbon fixation (Dimkpa et al., 2017; Nair et al., 2010).

Nanomaterials also aid in regulating the accumulation of osmolytes (e.g., proline, soluble sugars), thereby maintaining cellular turgor and preventing the denaturation of photosynthetic enzymes due to sodium. As a consequence of this enhanced efficiency, many plants that are treated with nanomaterial exhibit an increase in photosynthetic efficiency and enhanced growth when exposed to salinity stress.

6.1.3 Heavy metal stress and photosynthetic disruption

Heavy metals (e.g., cadmium (Cd), lead (Pb), chromium (Cr), and mercury (Hg)) are extremely toxic to photosynthesis due to their interference with the biosynthesis of chlorophyll, damage to photosynthetic membranes, and inhibition of the activity of key enzymes required for the assimilation of carbon. Heavy metals cause an excessive production of reactive oxygen species (ROS), leading to the oxidative damage of lipids and the breakdown of chloroplast ultrastructure (Shahid et al., 2017).

Nanomaterials mitigate the decline in photosynthesis from heavy metal exposure by immobilizing, chelating, and decreasing the amount of toxic metals that translocate to the aerial parts of the plant. Nanomaterials such as biochar, iron oxide, and silica nanoparticles can bind to heavy metals located within the rhizosphere and thus limit their bioavailability and uptake by plants (El-Saadony et al., 2021). Additionally, nanomaterials enhance the activities of antioxidant enzymes and protect the plant's photosynthetic pigments and proteins from oxidative damage. By providing a double benefit—detoxifying metal ions and reducing oxidative stress—nanomaterials lead to increased chlorophyll content, enhanced photosynthetic efficiency, and greater overall performance of plants under the stress-induced by heavy metals (Rizwan et al., 2019).

6.2 Restoration of chloroplast integrity and antioxidant defense

6.2.1 Protection and restoration of chloroplast structure

The main cause of abiotic stress is the injury done to the chloroplasts that results from the impact of environmental factors on their thylakoid membranes. This results in an unorganized thylakoid membrane, reduced grana stacking and swelling of chloroplasts. The result of such changes affects the ability of a plant to effectively harvest light and transport electrons, which ultimately leads to decreased photosynthesis (Baker, 2008).

Nanomaterials maintain the ultrastructure of chloroplasts by stabilizing the chloroplast membrane lipids and proteins. The use of transmission electron microscopy has enabled the confirmation that plants treated with nanomaterials exhibit a well-organized thylakoid membrane and intact grana while under stress when compared to untreated stressed plants. Nanomaterials made up of titanium dioxide and silicon are particularly effective because they protect chloroplast integrity by decreasing the amount of peroxidation of the chloroplast membrane lipids and preventing the loss of chlorophyll due to stress (Tripathi et al., 2017; Rastogi et al., 2017).

The recovery of chloroplast structure appears to correlate positively with improvements in the photochemical efficiency (Fv/Fm) of the photosystem, increased electron transport and continued carbon assimilation, illustrating the central role that nanomaterials play in protecting the photosynthetic process.

6.2.2 Regulation of oxidative stress and antioxidant defense

Drought, salinity, and heavy metal exposure can be serious sources of oxidative stress. Excess ROS can lead to oxidative damage to photosynthesis by damaging important components such as pigments, proteins and nucleic acids, among others. The maintenance of redox homeostasis through the use of both enzymatic (e.g., superoxide dismutase [SOD], catalase, ascorbate peroxidase) and non-enzymatic antioxidants is critical for plant survival (Gill and Tuteja, 2010).

Nanomaterials can be utilized to support oxidative stress through both direct and indirect antioxidant actions. Specific nano-sized materials, such as cerium oxide (CeO_2), are known to have enzyme mimicking properties and therefore can help to remove excess ROS through a cyclic oxidation–reduction process. Some nano-sized materials also help to promote the expression/activity of endogenous antioxidant enzymes, thus minimizing the negative effects of oxidative stress on photosynthesis (Das et al., 2017).

Due to the ability of nanomaterials to maintain redox homeostasis, photoinhibition can be prevented, PSII reaction centers can be protected, and efficient repair of photosynthetic-related proteins can occur through the increased antioxidant capacity afforded by nanomaterials, which is a significant factor contributing to the restoration of photosynthesis under stress conditions.

6.3 Integrative perspective

Nanomaterials have multiple means of protecting photosynthesis; these include an interaction of many physiological and biochemical processes as well as structural responses in plants. Nanomaterials enhance various aspects of plant physiology by improving water relation (i.e., transpiration), enhancing ionic balance with the plant system(s), and by aiding metal detoxification (i.e., assisting plants to detoxify metal ions that may pose a toxicity risk). In addition, they stabilize chloroplast membranes and provide antioxidant defense to plants during periods of stress. Therefore, through this combined approach, nanomaterials can be used as multi-functional agriculture practices in the agricultural sector, especially when crops are grown in an environment prone to stress.

7. Nanomaterial localization and tracking in leaf tissues

Being aware of how plants use, transfer, and store nanomaterials throughout their cells is essential to understanding how these materials impact the plant's physiological processes.

Methods such as Electron Transmission Microscopy (TEM) & Energy Dispersive X-Ray Spectroscopy (EDX), Confocal Laser Scanning Microscopy (CLSM) with fluorescently tagged nanoparticles, and X-Ray Fluorescence (XRF) mapping have been used extensively for tracking the localization of nanomaterials. These methods enable us to visualize where nanomaterials are located: for example, in cell walls, cytoplasmic areas, chloroplast regions, and/or vascular systems, allowing for evidence of nanomaterials interacting with and possibly influencing photosynthesis (Schwab et al., 2016).

The location of nanomaterials within chloroplasts has great importance as it relates to their hypothesized effects on thylakoid membrane stabilization, improvement of transport of electrons, and regulation of antioxidant production during stressful periods.

7.1 Assessment of photosynthetic performance

Combination of methods such as chlorophyll fluorescence; gas exchange; light and electron microscopy; and spectroscopic techniques provides an overall perspective on the effects of nanomaterials on photosynthetic metabolism under abiotic stress factors. An integrated approach will allow for better understanding of how to optimize dosages of nanomaterials, and ultimately the effects on plant physiology and on potential ecological consequences.

8. Conclusion

Nanomaterials have emerged as powerful tools for enhancing photosynthetic efficiency by directly and indirectly modulating key physiological, biochemical, and molecular processes in plants. Current innovations demonstrate that engineered nanomaterials like metal and metal oxide nanoparticles, carbon-based nanostructures, and nano-enabled formulations, can improve light harvesting, can optimize energy transfer within photosystems, enhance chloroplast ultrastructure, regulate stomatal conductance and carbon fixation pathways. By influencing electron transport chains, Rubisco activity, pigment stability, and antioxidant defense systems, nanomaterials contribute to improved photosynthetic performance under both optimal and stress conditions. Beyond direct photosynthetic enhancement, nanomaterials play a significant role in mitigating abiotic stresses such as salinity, drought, heavy metal toxicity, and temperature extremes, which otherwise limit photosynthetic capacity. Nano-enabled delivery of nutrients and signaling molecules ensures improved bioavailability, controlled release, and targeted action, thereby maintaining redox homeostasis and protecting photosynthetic machinery from oxidative damage. These multifunctional properties position nanomaterials as promising agents for sustaining crop productivity in the face of climate variability. In conclusion, nanomaterials offer transformative potential for improving photosynthetic efficiency and crop resilience, representing a frontier in sustainable agriculture. Strategic integration of nanotechnology with plant physiology, molecular biology, and agronomic practices will be essential to harness these innovations responsibly and effectively for global food security.

AI disclosure

During the preparation of this work the author(s) used chatgpt, Gemini in order to For Literature review. After using this tool/service, the author(s) reviewed and edited the content as needed and take(s) full responsibility for the content of the publication.

References

Adil, M., Bashir, S., Bashir, S., Aslam, Z., Ahmad, N., Younas, T., Asghar, R.M.A., Alkahtani, J., Dwiningsih, Y., Elshikh, M.S., 2022. Zinc oxide nanoparticles improved chlorophyll contents, physical parameters, and wheat yield under salt stress. Front. Plant Sci. 13. https://doi.org/10.3389/fpls.2022.932861.

Adnan, M., Mahmood, F., Zhao, Z., Khaliq, H., Usman, M., Muhammad, T., Ashraf, G.A., 2024. Effect of the foliar application of biogenic-ZnO nanoparticles on physio-chemical analysis of chilli (*Capsicum annum* L.) in a salt stress environment. Environ. Sci. Adv. 4 (2), 306–317. https://doi.org/10.1039/D4VA00167B.

Ahmed, B., Rizvi, A., Ali, K., Lee, J., Zaidi, A., Khan, M.S., Musarrat, J., 2021. Nanoparticles in the soil–plant system: a review. Environ. Chem. Lett. 19 (2), 1545–1609. https://doi.org/10.1007/s10311-020-01138-y.

Ahmed, S., Khan, M.T., Abbasi, A., Haq, I.U., Hina, A., Mohiuddin, M., Tariq, M.A.U.R., Zaheer Afzal, M., Zaman, Q., Ng, A.W.M., Li, Y., 2023. Characterizing stomatal attributes and photosynthetic induction in relation to biochemical changes in *Coriandrum sativum* L. by foliar-applied zinc oxide nanoparticles under drought conditions. Front. Plant Sci. 13. https://doi.org/10.3389/fpls.2022.1079283.

Akhavan, A., Gonçalves, P., 2021. Managing the trade-off between groundwater resources and large-scale agriculture: the case of pistachio production in Iran. Syst. Dyn. Rev. 37 (2–3), 155–196. https://doi.org/10.1002/sdr.1689.

Alam, P., Yalcin, M., Faizan, M., Albalawi, T., 2025. Response of tomato to silicon dioxide nanoparticles under salinity: impact on photosynthesis, antioxidant enzymes activity, stress biomarkers and osmoregulatory substances. Plant Nano Biol. 13. https://doi.org/10.1016/j.plana.2025.100171.

Ali, I., Amiri, S., Ullah, N., Younas, M., Rezakazemi, M., Park, J., 2020. Electrocatalytic CO_2 fixation by regenerating reduced cofactor NADH during Calvin Cycle using glassy carbon electrode. PLoS One 15 (9), e0239340. https://doi.org/10.1371/journal.pone.0239340.

Ali, S., Mehmood, A., Khan, N., Tan, B., 2021. Uptake, translocation, and consequences of nanomaterials on plant growth and stress adaptation. J. Nanomater. 2021, 1–17. https://doi.org/10.1155/2021/6677616.

An, C., Sun, C., Li, N., Huang, B., Jiang, J., Shen, Y., Wang, C., Xiang, Z., Cui, B., Wang, C., Li, X., Zhan, S., Gao, F., Zeng, Z., Cui, H., Wang, Y., 2022. Nanomaterials and nanotechnology for the delivery of agrochemicals: strategies towards sustainable agriculture. J. Nanobiotechnol. 20 (1). https://doi.org/10.1186/s12951-021-01214-7.

Asadi, M., Nasiri, Y., Rasouli, F., Karim, K., Morshedloo, M. R., Ercisli, S., Skrovankova, S., Mlcek, J., 2025. Application of graphene oxide nanoparticles for improvement of growth parameters, photosynthetic pigments, and essential oil quality and yield of Lavandula angustifolia Mill. under green manure incorporation. BMC Plant Biol. 25 (1). https://doi.org/10.1186/s12870-025-07364-2.

Avellan, A., Rodrigues, S.M., Morais, B.P., Therrien, B., Zhang, Y., Rodrigues, S., Lowry, G.V., 2022. Biological Barriers, Processes, and Transformations at the Soil–Plant– Atmosphere Interfaces Driving the Uptake, Translocation, and Bioavailability of Inorganic Nanoparticles to Plants Inorganic Nanopesticides and Nanofertilizers: A View from the Mechanisms of Action to Field Applications. Springer International Publishing, pp. 123–152. https://doi.org/10.1007/978-3-030-94155-0_4.

Bai, T., Zhang, P., Guo, Z., Chetwynd, A.J., Zhang, M., Adeel, M., Li, M., Guo, K., Gao, R., Li, J., Yi, H., Rui, Y., 2021. Different physiological responses of C3 and C4 plants to nanomaterials. Environ. Sci. Pollut. Control Ser. 28 (20), 25542–25551. https://doi.org/10.1007/s11356-021-12507-7.

Baker, N.R., 2008. Chlorophyll fluorescence: a probe of photosynthesis in vivo. Annu. Rev. Plant Biol. 59, 89–113. https://doi.org/10.1146/annurev.arplant.59.032607.092759.

Banerjee, S., Roy, P., Nandi, S., Roy, S., 2023. Advanced biotechnological strategies towards the development of crops with enhanced micronutrient content. Plant Growth Regul. 100 (2), 355–371. https://doi.org/10.1007/s10725-023-00968-4.

Bárzana, G., Garcia-Gomez, P., Carvajal, M., 2022. Nanomaterials in plant systems: smart advances related to water uptake and transport involving aquaporins. Plant Nano Biol. 1, 100005. https://doi.org/10.1016/j.plana.2022.100005.

Bhati, V.S., Singh, P.K., Parmar, Y.J., Rokins, P.G., 2026. Foliar uptake and translocation of nanoparticles in the plants: a review. J. Plant Nutr. 49 (2), 246–283. https://doi.org/10.1080/01904167.2025.2538638.

Björn, L.O., Govindjee, 2008. The evolution of photosynthesis and its environmental impact. In: Photobiology: The Science of Life and Light, second ed., pp. 255–287. https://doi.org/10.1007/978-0-387-72655-7_12.

Bölter, B., Soll, J., 2016. Once upon a time—chloroplast protein import research from infancy to future challenges. Mol. Plant 9 (6), 798–812. https://doi.org/10.1016/j.molp.2016.04.014.

Boretti, A., 2025. Evaluating water use efficiency and CO_2 absorption in plants under rising atmospheric carbon dioxide levels. J. Atmos. Sol. Terr. Phys. 266, 106409. https://doi.org/10.1016/j.jastp.2024.106409.

Bunce, 2008. Acclimation of photosynthesis to temperature in *Arabidopsis thaliana* and *Brassica oleracea*. Photosynthetica 46 (4), 517–524. https://doi.org/10.1007/s11099-008-0088-7.

Burgess, A.J., Cardoso, A.A., 2023. Throwing shade: limitations to photosynthesis at high planting densities and how to overcome them. Plant Physiol. 191 (2), 825–827. https://doi.org/10.1093/plphys/kiac567.

Butnariu, M., Butu, A., 2019. Plant nanobionics: application of nanobiosensors in plant biology. Nanotechnol. Life Sci. 337–376. https://doi.org/10.1007/978-3-030-16379-2_12.

Caemmerer, S.von, Ghannoum, O., Pengelly, J.J.L., Cousins, A.B., 2014. Carbon isotope discrimination as a tool to explore C4 photosynthesis. J. Exp. Bot. 65 (13), 3459–3470. https://doi.org/10.1093/jxb/eru127.

Cai, L., Cai, L., Jia, H., Liu, C., Wang, D., Sun, X., 2020. Foliar exposure of Fe3O4 nanoparticles on Nicotiana benthamiana: evidence for nanoparticles uptake, plant growth promoter and defense response elicitor against plant virus. J. Hazard Mater. 393. https://doi.org/10.1016/j.jhazmat.2020.122415.

Chandrasekaran, R., Rajiv, P., Abd-Elsalam, K.A., 2019. Carbon nanotubes: plant gene delivery and genome editing. In: Carbon Nanomaterials for Agri-Food and Environmental Applications. Elsevier, pp. 279–296. https://doi.org/10.1016/B978-0-12-819786-8.00014-1.

Chaves, M.M., Flexas, J., Pinheiro, C., 2009. Photosynthesis under drought and salt stress: regulation mechanisms from whole plant to cell. Ann. Bot. 103 (4), 551–560. https://doi.org/10.1093/aob/mcn125.

Chen, Y., Yao, R., Li, Y., Xu, B., Chen, C., Zhang, C., 2020. Mimicking the Mn4CaO5-Cluster in Photosystem II Microbial Photosynthesis. Springer Singapore, pp. 263–284. https://doi.org/10.1007/978-981-15-3110-1_13.

Chen, H., Song, Y., Wang, Y., Wang, H., Ding, Z., Fan, K., 2024. Zno nanoparticles: improving photosynthesis, shoot development, and phyllosphere microbiome composition in tea plants. J. Nanobiotechnol. 22 (1). https://doi.org/10.1186/s12951-024-02667-2.

Cherepanov, D.A., Milanovsky, G.E., Nadtochenko, V.A., Semenov, A.Y., 2023. Energy barrier of photoinduced charge separation in the reaction centers of photosystems I and II. Fed. Russ. J. Phys. Chem. B 17 (3), 594–599. https://doi.org/10.1134/S1990793123030193.

Chhipa, H., 2017. Nanofertilizers and nanopesticides for agriculture. Environ. Chem. Lett. 15 (1), 15–22. https://doi.org/10.1007/s10311-016-0600-4.

Chowdhury, M., Kushwah, A., Satpute, A.N., Singh, S.K., Patil, A.K., 2023. A comprehensive review on potential application of nanomaterials in the field of agricultural engineering. J. Biosyst. Eng. 48 (4), 457–477. https://doi.org/10.1007/s42853-023-00204-x.

Crespo-Herrera, L.A., Crossa, J., Huerta-Espino, J., Vargas, M., Mondal, S., Velu, G., Payne, T.S., Braun, H., Singh, R.P., 2018. Genetic gains for grain yield in CIMMYT's semi-arid wheat yield trials grown in suboptimal environments. Crop Sci. 58 (5), 1890–1898. https://doi.org/10.2135/cropsci2018.01.0017.

Croce, R., Carmo-Silva, E., Cho, Y.B., Ermakova, M., Harbinson, J., Lawson, T., McCormick, A.J., Niyogi, K. K., Ort, D.R., Patel-Tupper, D., Pesaresi, P., Raines, C., Weber, A.P.M., Zhu, X.G., 2024. Perspectives on improving photosynthesis to increase crop yield. Plant Cell 36 (10), 3944–3973. https://doi.org/10.1093/plcell/koae132.

Croce, R., van Amerongen, H., 2014. Natural strategies for photosynthetic light harvesting. Nat. Chem. Biol. 10 (7), 492–501. https://doi.org/10.1038/nchembio.1555.

Csikós, N., Tóth, G., 2023. Concepts of agricultural marginal lands and their utilisation: a review. Agric. Syst. 204, 103560. https://doi.org/10.1016/j.agsy.2022.103560.

Dang, K., Wang, Y., Tian, H., Bai, J., Cheng, X., Guo, L., Zhang, Q., Geng, Y., Shao, X., 2024. Impact of ZnO NPs on photosynthesis in rice leaves plants grown in saline-sodic soil. Sci. Rep. 14 (1). https://doi.org/10.1038/s41598-024-66935-9.

Das, S., Debnath, N., Pradhan, S., Goswami, A., 2017. Enhancement of photon absorption in the light-harvesting complex of isolated chloroplast in the presence of plasmonic gold nanosol—a nanobionic approach towards photosynthesis and plant primary growth augmentation. Gold Bull. 50 (3), 247–257.

Dealba-Montero, Guajardo-Pacheco, J., Morales-Sánchez, E., Araujo-Martínez, R., Becerra, L., Martínez-Castañón, G.-A., Ruiz, F., Compeán Jasso, M.E., 2017. Antimicrobial properties of copper nanoparticles and amino acid chelated copper nanoparticles produced by using a soya extract. Bioinorgan. Chem. Appl. 2017. https://doi.org/10.1155/2017/1064918.

Dekker, J.P., Boekema, E.J., 2005. Supramolecular organization of thylakoid membrane proteins in green plants. Biochim. Biophys. Acta Bioenerg. 1706 (1–2), 12–39. https://doi.org/10.1016/j.bbabio.2004.09.009.

Derosa, M.C., Monreal, C., Morris, S., Walsh, R., Sultan, Y., 2010. Nanotechnology in fertilizers. Nat. Nanotechnol. 5 (2). https://doi.org/10.1038/nnano.2010.2.

van Dijk, M., Morley, T., Rau, M.L., Saghai, Y., 2021. A meta-analysis of projected global food demand and population at risk of hunger for the period 2010–2050. Nat. Food 2 (7), 494–501. https://doi.org/10.1038/s43016-021-00322-9.

Dillard, H.R., 2019. Global food and nutrition security: from challenges to solutions: report of the international congress of plant pathology 2018, Boston, USA, 29th July—3rd August 2018 with the title plant health in a global economy. Food Secur. 11 (1), 249–252. https://doi.org/10.1007/s12571-019-00893-3.

Dimkpa, C.O., Bindraban, P.S., Fugice, J., Agyin-Birikorang, S., Singh, U., Hellums, D., 2017. Composite micronutrient nanoparticles and salts decrease drought stress in soybean. Agron. Sustain. Dev. 37 (1). https://doi.org/10.1007/s13593-016-0412-8.

Djanaguiraman, M., Anbazhagan, V., Dhankher, O.P., Prasad, P.V.V., 2024. Uptake, translocation, toxicity, and impact of nanoparticles on plant physiological processes. Plants 13 (22). https://doi.org/10.3390/plants13223137.

Dziwulska-Hunek, A., Myśliwa-Kurdziel, B., Matwijczuk, A., Szymanek, M., 2025. A case study in photosynthetic parameters of perennial plants growing in natural conditions. BMC Plant Biol. 25 (1). https://doi.org/10.1186/s12870-025-07133-1.

El-Saadony, M.T., Desoky, E.-S.M., Saad, A.M., Eid, R.S.M., Selem, E., Elrys, A.S., 2021. Biological silicon nanoparticles improve *Phaseolus vulgaris* L. yield and minimize its contaminant contents on a heavy metals-contaminated saline soil. J. Environ. Sci. (China) 106, 1–14. https://doi.org/10.1016/j.jes.2021.01.012.

Evans, J.R., 2013. Improving photosynthesis. Plant Physiol. 162 (4), 1780–1793. https://doi.org/10.1104/pp.113.219006.

Evans, J.R., Clarke, V.C., 2019. The nitrogen cost of photosynthesis. J. Exp. Bot. 70 (1), 7–15. https://doi.org/10.1093/jxb/ery366.

Faizan, M., Faraz, A., Yusuf, M., Khan, S.T., Hayat, S., 2018. Zinc oxide nanoparticle-mediated changes in photosynthetic efficiency and antioxidant system of tomato plants. Photosynthetica 56 (2), 678–686. https://doi.org/10.1007/s11099-017-0717-0.

Faizan, M., Bhat, J.A., Chen, C., Alyemeni, M.N., Wijaya, L., Ahmad, P., Yu, F., 2021. Zinc oxide nanoparticles (ZnO-NPs) induce salt tolerance by improving the antioxidant system and photosynthetic machinery in tomato. Plant Physiol. Biochem. 161, 122–130. https://doi.org/10.1016/j.plaphy.2021.02.002.

Falkowski, P.G., Lin, H., Gorbunov, M.Y., 2017. What limits photosynthetic energy conversion efficiency in nature? Lessons from the oceans. Phil. Trans. R. Soc. B. Biol. Sci. 372 (1730). https://doi.org/10.1098/rstb.2016.0376.

Farmand, M., Jahanpeyma, F., Gholaminejad, A., Azimzadeh, M., Malaei, F., Shoaie, N., 2022. Carbon nanostructures: a comprehensive review of potential applications and toxic effects. 3 Biotech 12 (8). https://doi.org/10.1007/s13205-022-03175-6.

Fatima, F., Hashim, A., Anees, S., 2021. Efficacy of nanoparticles as nanofertilizer production: a review. Environ. Sci.

Pollut. Control Ser. 28 (2), 1292–1303. https://doi.org/10.1007/s11356-020-11218-9.

Feng, Y., Kreslavski, V.D., Shmarev, A.N., Ivanov, A.A., Zharmukhamedov, S.K., Kosobryukhov, A., Yu, M., Allakhverdiev, S.I., Shabala, S., 2022. Effects of iron oxide nanoparticles (Fe3O4) on growth, photosynthesis, antioxidant activity and distribution of mineral elements in wheat (*Triticum aestivum*) plants. Plants 11 (14). https://doi.org/10.3390/plants11141894.

Flexas, J., Díaz-Espejo, A., Conesa, M.A., Coopman, R.E., Douthe, C., Gago, J., Gallé, A., Galmés, J., Medrano, H., Ribas-Carbo, M., Tomàs, M., Niinemets, Ü., 2016. Mesophyll conductance to CO_2 and Rubisco as targets for improving intrinsic water use efficiency in C_3 plants. Plant Cell Environ. 39 (5), 965–982. https://doi.org/10.1111/pce.12622.

Flexas, J., Bota, J., Cifre, J., Mariano Escalona, J., Galmés, J., Gulías, J., Lefi, E.-K., Florinda Martínez-Cañellas, S., Moreno, M.T., Ribas-Carbó, M., Riera, D., Sampol, B., Medrano, H., 2004. Understanding down-regulation of photosynthesis under water stress: future prospects and searching for physiological tools for irrigation management. Ann. Appl. Biol. 144 (3), 273–283. https://doi.org/10.1111/j.1744-7348.2004.tb00343.x.

Forest, V., Pourchez, J., 2017. Preferential binding of positive nanoparticles on cell membranes is due to electrostatic interactions: a too simplistic explanation that does not take into account the nanoparticle protein corona. Mater. Sci. Eng. C 70 (Pt 1), 889–896. https://doi.org/10.1016/j.msec.2016.09.016.

Francis, D.V., Abdalla, A.K., Mahakham, W., Sarmah, A.K., Ahmed, Z.F.R., 2024. Interaction of plants and metal nanoparticles: exploring its molecular mechanisms for sustainable agriculture and crop improvement. Environ. Int. 190. https://doi.org/10.1016/j.envint.2024.108859.

Friedlingstein, P., O'Sullivan, M., Jones, M.W., Andrew, R. M., Hauck, J., Landschützer, P., Le Quéré, C., Li, H., Luijkx, I.T., Olsen, A., Peters, G.P., Peters, W., Pongratz, J., Schwingshackl, C., Sitch, S., Canadell, J.G., Ciais, P., Jackson, R.B., Alin, S.R., Arneth, A., Arora, V., Bates, N.R., Becker, M., Bellouin, N., Berghoff, C.F., Bittig, H.C., Bopp, L., Cadule, P., Campbell, K., Chamberlain, M.A., Chandra, N., Chevallier, F., Chini, L.P., Colligan, T., Decayeux, J., Djeutchouang, L. M., Dou, X., Rojas, C.D., Enyo, K., Evans, W., Fay, A.R., Feely, R.A., Ford, D.J., Foster, A., Gasser, T., Gehlen, M., Gkritzalis, T., Grassi, G., Gregor, L., Gruber, N., Gürses, Ö., Harris, I., Hefner, M., Heinke, J., Hurtt, G.C., Iida, Y., Ilyina, T., Jacobson, A. R., Jain, A.K., Jarníková, T., Jersild, A., Jiang, F., Jin, Z., Kato, E., Keeling, R.F., Goldewijk, K.K., Knauer, J., Korsbakken, J.I., Lan, X., Lauvset, S.K., Lefèvre, N., Liu, Z., Liu, J., Ma, L., Maksyutov, S., Marland, G., Mayot, N., McGuire, P.C., Metzl, N., Monacci, N.M., Morgan, E.J., Nakaoka, S.-I., Neill, C., Niwa, Y., Nützel, T., Olivier, L., Ono, T., Palmer, P.I., Pierrot, D., Qin, Z., Resplandy, L., Roobaert, A., Rosan, T.M., Rödenbeck, C., Schwinger, J., Smallman, T.L., Smith, S. M., Sospedra-Alfonso, R., Steinhoff, T., Sun, Q., Sutton, A.J., Séférian, R., Takao, S., Tatebe, H., Tian, H., Tilbrook, B., Torres, O., Tourigny, E., Tsujino, H., Tubiello, F., Werf, G.van der, Wanninkhof, R., Wang, X., Yang, D., Yang, X., Yu, Z., Yuan, W., Yue, X., Zaehle, S., Zeng, N., Zeng, J., 2025. Global carbon budget 2024. Earth Syst. Sci. Data 17 (3), 965–1039. https://doi.org/10.5194/essd-17-965-2025.

Garcia, A., Gaju, O., Bowerman, A.F., Buck, S.A., Evans, J.R., Furbank, R.T., Gilliham, M., Millar, A.H., Pogson, B.J., Reynolds, M.P., Ruan, Y.L., Taylor, N.L., Tyerman, S. D., Atkin, O.K., 2023. Enhancing crop yields through improvements in the efficiency of photosynthesis and respiration. New Phytol. 237 (1), 60–77. https://doi.org/10.1111/nph.18545.

Gest, H., 2000. Bicentenary homage to Dr Jan Ingen-Housz, MD (1730–1799), pioneer of photosynthesis research. Photosynth. Res. 63 (2), 183–190. https://doi.org/10.1023/a:1006460024843.

Ghosh, R.K., Rommel, J., Otto, I.M., 2022. Food Security, Agricultural Productivity, and the Environment: Economic, Sustainability, and Policy Perspectives Frontiers Research Topics. Frontiers Media SA. https://doi.org/10.3389/978-2-88974-992-8.

Gill, S.S., Tuteja, N., 2010. Reactive oxygen species and antioxidant machinery in abiotic stress tolerance in crop plants. Plant Physiol. Biochem. 48 (12), 909–930. https://doi.org/10.1016/j.plaphy.2010.08.016.

Gilman, I.S., Andrew Smith, J.C., Holtum, J.A.M., Sage, R.F., Silvera, K., Winter, K., Edwards, E.J., 2023. The CAM lineages of planet Earth. Ann. Bot. 132 (4), 627–654. https://doi.org/10.1093/aob/mcad135.

Govindjee, Krogmann, D., 2004. Discoveries in oxygenic photosynthesis (1727–2003): a perspective. Photosynth. Res. 80 (1–3), 15–57. https://doi.org/10.1023/b:pres.0000030443.63979.e6.

Govindjee, Shevela, D., Björn, L.O., 2017. Evolution of the Z-scheme of photosynthesis: a perspective. Photosynth. Res. 133 (1–3), 5–15. https://doi.org/10.1007/s11120-016-0333-z.

Guirguis, A., Yang, W., Conlan, X.A., Kong, L., Cahill, D.M., Wang, Y., 2023. Boosting plant photosynthesis with carbon dots: a critical review of performance and prospects. Small 19 (43). https://doi.org/10.1002/smll.202300671.

Gulyás, Z., Székely, A., Kulman, K., Kocsy, G., 2023. Light-dependent regulatory interactions between the redox system and miRNAs and their biochemical and physiological effects in plants. Int. J. Mol. Sci. 24 (9), 8323. https://doi.org/10.3390/ijms24098323.

Guo, H., White, J.C., Wang, Z., Xing, B., 2018. Nano-enabled fertilizers to control the release and use efficiency of nutrients. Curr. Opin. Environ. Sci. Health 6, 77–83. https://doi.org/10.1016/j.coesh.2018.07.009.

Gupta, S.D., Agarwal, A., Pradhan, S., 2018. Phytostimulatory effect of silver nanoparticles (AgNPs) on rice seedling growth: an insight from antioxidative enzyme activities and gene expression patterns. Ecotoxicol. Environ. Saf. 161, 624–633. https://doi.org/10.1016/j.ecoenv.2018.06.023.

Hamada, A.M., Radi, A.A., Al-Kahtany, F.A., Farghaly, F.A., 2024. A review: zinc oxide nanoparticles: advantages and disadvantages. J. Plant Nutr. 47 (4), 656–679. https://doi.org/10.1080/01904167.2023.2280127.

Hartman, M.D., Rojas, B.E., Iglesias, A.A., Figueroa, C.M., 2023. The involvement of allosteric effectors and post-translational modifications in the control of plant central carbon metabolism. Plant J. 114 (5), 1037–1058. https://doi.org/10.1111/tpj.16215.

Havaux, M., 2020. Plastoquinone in and beyond photosynthesis. Trends Plant Sci. 25 (12), 1252–1265. https://doi.org/10.1016/j.tplants.2020.06.011.

He, Q., Ju, W., Dai, S., He, W., Song, L., Wang, S., Li, X., Mao, G., 2021. Drought risk of global terrestrial gross primary productivity over the last 40 years detected by a remote sensing-driven process model. J. Geophys. Res. Biogeosci. 126 (6). https://doi.org/10.1029/2020jg005944.

Hertle, A.P., Blunder, T., Wunder, T., Pesaresi, P., Pribil, M., Armbruster, U., Leister, D., 2013. PGRL1 is the elusive ferredoxin-plastoquinone reductase in photosynthetic cyclic electron flow. Mol. Cell 49 (3), 511–523. https://doi.org/10.1016/j.molcel.2012.11.030.

Hohmann-Marriott, M.F., Blankenship, R.E., 2011. Evolution of photosynthesis. Annu. Rev. Plant Biol. 62, 515–548. https://doi.org/10.1146/annurev-arplant-042110-103811.

Hong, J., Wang, C., Wagner, D.C., Gardea-Torresdey, J.L., He, F., Rico, C.M., 2021. Foliar application of nanoparticles: mechanisms of absorption, transfer, and multiple impacts. Environ. Sci. Nano 8 (5), 1196–1210. https://doi.org/10.1039/d0en01129k.

Hope, A.B., 2000. Electron transfers amongst cytochrome f, plastocyanin and photosystem I: kinetics and mechanisms. Biochim. Biophys. Acta Bioenerg. 1456 (1), 5–26. https://doi.org/10.1016/s0005-2728(99)00101-2.

Huhtinen, P., Kivelä, M., Kuronen, O., Hagren, V., Takalo, H., Tenhu, H., Lövgren, T., Härmä, H., 2005. Synthesis, characterization, and application of Eu(III), Tb(III), Sm(III), and Dy(III) lanthanide chelate nanoparticle labels. Anal. Chem. 77 (8), 2643–2648. https://doi.org/10.1021/ac048360i.

Iravani, S., 2011. Green synthesis of metal nanoparticles using plants. Green Chem. 13 (10), 2638–2650. https://doi.org/10.1039/c1gc15386b.

Ishaq, H.S., Akif, M., Hasan, A., Khalid, A., Fatima, M., Mohsan, M., Khan, A.H., Afzal, M., Abbas, M., 2025. Plant science in the face of global crises: biotechnological innovations for climate resilience, food security, and medicinal plant conservation. Scholars Acad. J. Biosci. 13 (09), 1363–1380. https://doi.org/10.36347/sajb.2025.v13i09.007.

Jampílek, J., Kráľová, K., 2019. Impact of nanoparticles on photosynthesizing organisms and their use in hybrid structures with some components of photosynthetic apparatus. Nanotechnol. Life Sci. 255–332. https://doi.org/10.1007/978-3-030-12496-0_11.

Jarvis, P., López-Juez, E., 2013. Biogenesis and homeostasis of chloroplasts and other plastids. Nat. Rev. Mol. Cell Biol. 14 (12), 787–802. https://doi.org/10.1038/nrm3702.

Johnson, M.P., 2016. Photosynthesis. Essays Biochem. 60 (3), 255–273. https://doi.org/10.1042/EBC20160016.

Juárez-Maldonado, A., Ortega-Ortíz, H., Morales-Díaz, A.B., González-Morales, S., Morelos-Moreno, Á., Cabrera-De la Fuente, M., Sandoval-Rangel, A., Cadenas-Pliego, G., Benavides-Mendoza, A., 2019. Nanoparticles and nanomaterials as plant biostimulants. Int. J. Mol. Sci. 20 (1). https://doi.org/10.3390/ijms20010162.

Jung, J.H., Cho, M., Seo, T.S., Lee, S.Y., 2022. Biosynthesis and applications of iron oxide nanocomposites synthesized by recombinant *Escherichia coli*. Appl. Microbiol. Biotechnol. 106 (3), 1127–1137. https://doi.org/10.1007/s00253-022-11779-4.

Khalid, M.F., Iqbal Khan, R., Jawaid, M.Z., Shafqat, W., Hussain, S., Ahmed, T., Rizwan, M., Ercisli, S., Pop, O. L., Alina Marc, R., 2022. Nanoparticles: the plant saviour under abiotic stresses. Nanomaterials 12 (21). https://doi.org/10.3390/nano12213915.

Khan, I., Awan, S.A., Rizwan, M., Hassan, Z.U., Akram, M. A., Tariq, R., Brestic, M., Xie, W., 2022. Nanoparticle's uptake and translocation mechanisms in plants via seed priming, foliar treatment, and root exposure: a review. Environ. Sci. Pollut. Control Ser. 29 (60), 89823–89833. https://doi.org/10.1007/s11356-022-23945-2.

Khan, T.A., Mundra, S., Anand Gururani, M., 2025. Advancements in photosynthetic efficiency: pathways, regulation, and biotechnological applications for enhancing crop productivity. Plant Signal. Behav. 20 (1). https://doi.org/10.1080/15592324.2025.2596483.

Khodakovskaya, M., Dervishi, E., Mahmood, M., Xu, Y., Li, Z., Watanabe, F., Biris, A.S., 2009. Carbon nanotubes are able to penetrate plant seed coat and dramatically affect seed germination and plant growth. ACS Nano 3 (10), 3221–3227. https://doi.org/10.1021/nn900887m.

Kirchhoff, H., 2014. Structural changes of the thylakoid membrane network induced by high light stress in plant chloroplasts. Phil. Trans. R. Soc. B. Biol. Sci. 369 (1640), 20130225. https://doi.org/10.1098/rstb.2013.0225.

Klimov, V.V., 2003. Discovery of pheophytin function in the photosynthetic energy conversion as the primary electron acceptor of photosystem II. Photosynth. Res. 76 (1–3), 247–253. https://doi.org/10.1023/A:1024990408747.

Kumar, S.M., Yadav, S., Choudhary, R., Anumantharaj, A., Yadav, A., Hussain, Z., Singh, P.K., Singh, S.P., Mandal, A., Singh, N., Sushkova, S., Ercisli, S., Sarker, U., Yadav, S.K., 2025. Nanopriming with zinc oxide nanoparticle boosts seed vigour, photosynthesis, osmolytes accumulation and antioxidant activity in tomato. Sci. Rep. 15 (1). https://doi.org/10.1038/s41598-025-09269-4.

Kumar, R., Sabapathy, M., Singh Pawak, V., Ranjan, R., Kakunuri, M., Veeramani, N., 2026. Nanoparticles for sustainable wastewater treatment: advances in synthesis, characterization, and remediation applications. Results Eng. 29. https://doi.org/10.1016/j.rineng.2026.109799.

Lesniak, A., Salvati, A., Santos-Martinez, M.J., Radomski, M. W., Dawson, K.A., Åberg, C., 2013. Nanoparticle adhesion to the cell membrane and its effect on nanoparticle uptake efficiency. J. Am. Chem. Soc. 135 (4), 1438–1444. https://doi.org/10.1021/ja309812z.

Li, Y., Pan, X., Xu, X., Wu, Y., Zhuang, J., Zhang, X., Zhang, H., Lei, B., Hu, C., Liu, Y., 2021. Carbon dots as light converter for plant photosynthesis: augmenting light coverage and quantum yield effect. J. Hazard Mater. 410, 124534. https://doi.org/10.1016/j.jhazmat.2020.124534.

Li, X., Ghanizadeh, H., Han, Z., Wang, Q., Li, F., Qiu, Y., Zhang, Y., Chen, X., Liu, J., Wang, A., 2023. Metabolic profile and physiological mechanisms underlying the promoting effects of TiO2NPs on the photosynthesis and growth of tomato. Sci. Hortic. 322. https://doi.org/10.1016/j.scienta.2023.112394.

Li, Y., Xu, X., Wu, Y., Zhuang, J., Zhang, X., Zhang, H., Lei, B., Hu, C., Liu, Y., 2020. A review on the effects of carbon dots in plant systems. Mater. Chem. Front. 4 (2), 437–448. https://doi.org/10.1039/c9qm00614a.

Liang, Y., Liu, H., Fu, Y., Li, P., Li, S., Gao, Y., 2023. Regulatory effects of silicon nanoparticles on the growth and photosynthesis of cotton seedlings under salt and low-temperature dual stress. BMC Plant Biol. 23 (1). https://doi.org/10.1186/s12870-023-04509-z.

Linka, N., Weber, A.P.M., 2010. Intracellular metabolite transporters in plants. Mol. Plant 3 (1), 21–53. https://doi.org/10.1093/mp/ssp108.

Liu, J., Guo, Y.Y., Bai, Y.W., Camberato, J.J., Xue, J.Q., Zhang, R.H., 2018. Effects of drought stress on the photosynthesis in maize. Russ. J. Plant Physiol. 65 (6), 849–856. https://doi.org/10.1134/s1021443718060092.

Liu, M., Lu, S., 2016. Plastoquinone and ubiquinone in plants: biosynthesis, physiological function and metabolic engineering. Front. Plant Sci. 7 (2016). https://doi.org/10.3389/fpls.2016.01898.

Liu, J., Zhang, H., Yan, L., Kerr, P.G., Zhang, S., Wu, Y., 2021. Electron transport, light energy conversion and proteomic responses of periphyton in photosynthesis under exposure to AgNPs. J. Hazard Mater. 401. https://doi.org/10.1016/j.jhazmat.2020.123809.

Malode, S.J., Akhdar, H., Alodhayb, A.N., Shetti, N.P., 2025. Bio-based nanomaterials for smart and sustainable nanobiosensing and therapeutics. Ind. Crop. Prod. 234. https://doi.org/10.1016/j.indcrop.2025.121603.

Manzoor, N., Ahmed, T., Noman, M., Shahid, M., Nazir, M. M., Ali, L., Alnusaire, T.S., Li, B., Schulin, R., Wang, G., 2021. Iron oxide nanoparticles ameliorated the cadmium and salinity stresses in wheat plants, facilitating photosynthetic pigments and restricting cadmium uptake. Sci. Total Environ. 769. https://doi.org/10.1016/j.scitotenv.2021.145221.

Mardani-Talaee, M., Razmjou, J., Ajdari, A., Eduardo Serrão, J., Vivekanandhan, P., 2025. Green synthesis of zinc oxide nanoparticles from Sargassum ilicifolium to enhance tomato resistance against Tuta absoluta. Sci. Rep. 15 (1). https://doi.org/10.1038/s41598-025-97535-w.

Mathur, S., Agrawal, D., Jajoo, A., 2014. Photosynthesis: response to high temperature stress. J. Photochem. Photobiol. B Biol. 137, 116–126. https://doi.org/10.1016/j.jphotobiol.2014.01.010.

Meng, Y., Feng, Y., Bai, X., Yu, Q., Zhou, J., Wang, J., 2025. Application of nanotechnology in agricultural sustainability: absorption, translocation, and challenges of nanoparticles. Curr. Plant Biol. 42, 100492. https://doi.org/10.1016/j.cpb.2025.100492.

Mercado-Díaz de León, L., Garcidueñas-Piña, C., Pérez-Molphe-Balch, E., Loera-Muro, A., Francisco Morales-Domínguez, J., 2024. Effect of BvAgNP on growth, development, and glyoxalase gene expression analysis in Mammillaria bombycina and Selenicereus undatus. Mol. Biol. Rep. 51 (1). https://doi.org/10.1007/s11033-024-09570-x.

Mgadi, K., Ndaba, B., Roopnarain, A., Rama, H., Adeleke, R., 2024. Nanoparticle applications in agriculture: overview and response of plant-associated microorganisms. Front. Microbiol. 15. https://doi.org/10.3389/fmicb.2024.1354440.

Mirkovic, T., Ostroumov, E.E., Anna, J.M., Van Grondelle, R., Govindjee, Scholes, G.D., 2017. Light absorption and energy transfer in the antenna complexes of photosynthetic organisms. Chem. Rev. 117 (2), 249–293. https://doi.org/10.1021/acs.chemrev.6b00002.

Molnár, Á., Rónavári, A., Bélteky, P., Szőllősi, R., Valyon, E., Oláh, D., Rázga, Z., Ördög, A., Kónya, Z., Kolbert, Z., 2020. ZnO nanoparticles induce cell wall remodeling and modify ROS/RNS signalling in roots of Brassica seedlings. Ecotoxicol. Environ. Saf. 206. https://doi.org/10.1016/j.ecoenv.2020.111158.

Mouw, C.B., Ciochetto, A.B., Yoder, J.A., 2019. A satellite assessment of environmental controls of phytoplankton community size structure. Glob. Biogeochem. Cycles 33 (5), 540–558. https://doi.org/10.1029/2018GB006118.

Mullendore, D.L., Windt, C.W., van As, H., Knoblauch, M., 2010. Sieve tube geometry in relation to phloem flow. Plant Cell 22 (3), 579–593. https://doi.org/10.1105/tpc.109.070094.

Munir, A., Salah, M.A., Ali, M., Ali, B., Saleem, M.H., Samarasinghe, K.G.B.A., De Silva, S.I.S., Ercisli, S., Iqbal, N., Anas, M., 2024. Advancing agriculture: harnessing smart nanoparticles for precision fertilization. BioNanoScience 14 (4), 3846–3863. https://doi.org/10.1007/s12668-024-01597-5.

Munns, R., Tester, M., 2008. Mechanisms of salinity tolerance. Annu. Rev. Plant Biol. 59 (1), 651–681. https://doi.org/10.1146/annurev.arplant.59.032607.092911.

Nair, R., Varghese, S.H., Nair, B.G., Maekawa, T., Yoshida, Y., Kumar, D.S., 2010. Nanoparticulate material delivery to plants. Plant Sci. 179 (3), 154–163. https://doi.org/10.1016/j.plantsci.2010.04.012.

Nehra, A., Deepa, Kalwan, G., Shah, A.H., Tuteja, N., Pudake, R.N., Gill, R., Singh Gill, S., 2024. Recent advances in the field of plant nano nutrition. In: Nanotechnology for Abiotic Stress Tolerance and Management in Crop Plants, pp. 115–126. https://doi.org/10.1016/B978-0-443-18500-7.00007-7.

Nelson, N., Junge, W., 2015. Structure and energy transfer in photosystems of oxygenic photosynthesis. Annu. Rev. Biochem. 84 (1), 659–683. https://doi.org/10.1146/annurev-biochem-092914-041942.

Newkirk, G.M., de Allende, P., Jinkerson, R.E., Giraldo, J.P., 2021. Nanotechnology approaches for chloroplast biotechnology advancements. Front. Plant Sci. 12. https://doi.org/10.3389/fpls.2021.691295.

Nikkanen, L., Rintamäki, E., 2019. Chloroplast thioredoxin systems dynamically regulate photosynthesis in plants. Biochem. J. 476 (7), 1159–1172. https://doi.org/10.1042/bcj20180707.

Nisar, P., Ali, N., Rahman, L., Ali, M., Shinwari, Z.K., 2019. Antimicrobial activities of biologically synthesized metal nanoparticles: an insight into the mechanism of action. J. Biol. Inorg. Chem. 24 (7), 929–941. https://doi.org/10.1007/s00775-019-01717-7.

Nowak, R., Szczepanek, M., Błaszczyk, K., Hassanpouraghdam, M.B., 2024. Response of photosynthetic efficiency parameters and leaf area index of alternative barley genotypes to increasing sowing density. Sci. Rep. 14 (1). https://doi.org/10.1038/s41598-024-81783-3.

Oliveira, H.C., Seabra, A.B., Kondak, S., Adedokun, O.P., Kolbert, Z., 2023. Multilevel approach to plant–nanomaterial relationships: from cells to living ecosystems. J. Exp. Bot. 74 (12), 3406–3424. https://doi.org/10.1093/jxb/erad107.

Osyczka, A., Moser, C.C., Dutton, P.L., 2005. Fixing the Q cycle. Trends Biochem. Sci. 30 (4), 176–182. https://doi.org/10.1016/j.tibs.2005.02.001.

Pappert, I., Ühlein, C., Jokic, L., Kaldenhoff, R., 2025. Effects of high light intensity and spectral variability on maize photosynthesis and growth. Front. Plant Sci. 16. https://doi.org/10.3389/fpls.2025.1511768.

Parry, M.A.J., Andralojc, P.J., Khan, S., Lea, P.J., Keys, A.J., 2002. Rubisco activity: effects of drought stress. Ann. Bot. 89, 833–839. https://doi.org/10.1093/aob/mcf103.

Paul, M.J., Pellny, T.K., 2003. Carbon metabolite feedback regulation of leaf photosynthesis and development. J. Exp. Bot. 54 (382), 539–547. https://doi.org/10.1093/jxb/erg052.

Pradhan, B., Panda, D., Bishi, S.K., Chakraborty, K., Muthusamy, S.K., Lenka, S.K., 2022. Progress and prospects of C_4 trait engineering in plants. Plant Biol. 24 (6), 920–931. https://doi.org/10.1111/plb.13446.

Qaseem, M.F., Wu, A.M., 2021. Marginal lands for bioenergy in China; an outlook in status, potential and management. GCB Bioenergy 13 (1), 21–44. https://doi.org/10.1111/gcbb.12770.

Qiao, M., Hong, C., Jiao, Y., Hou, S., Gao, H., 2024. Impacts of drought on photosynthesis in major food crops and the related mechanisms of plant responses to drought. Plants 13 (13), 1808. https://doi.org/10.3390/plants13131808.

Raghib, F., Naikoo, M.I., Ahmad Khan, F., Alhems, L.M., 2025. Arbuscular mycorrhizal fungi (Glomus macrocarpum) and ZnO nanoparticles synergistically improved the growth of wheat (*Triticum aestivum* L.) and decreased cadmium uptake in Cd contaminated soil. J. Soil Sci. Plant Nutr. 25 (4), 10207–10220. https://doi.org/10.1007/s42729-025-02787-0.

Raines, C.A., 2022. Improving plant productivity by re-tuning the regeneration of RuBP in the Calvin–Benson–Bassham cycle. New Phytol. 236 (2), 350–356. https://doi.org/10.1111/nph.18394.

Ranjha, M.M.A.N., Shafique, B., Khalid, W., Nadeem, H.R., Mueen-ud-Din, G., Khalid, M.Z., 2022. Applications of biotechnology in food and agriculture: a mini-review.

Proc. Natl. Acad. Sci. India B Biol. Sci. 92 (1), 11–15. https://doi.org/10.1007/s40011-021-01320-4.

Rastogi, A., Zivcak, M., Sytar, O., Kalaji, H.M., He, X., Mbarki, S., Brestic, M., 2017. Impact of metal and metal oxide nanoparticles on plant: a critical review. Front. Chem. 5. https://doi.org/10.3389/fchem.2017.00078.

Raveena, R., Rajkishore, S.K., Maheswari, M., Bharani, A., Prasanthrajan, M., Sritharan, N., Saranya, N., SathyaMoorthy, P., Sunitha, R., 2026. Mechanistic insights into carbon nanomaterials as potential plant biostimulants: enhancing photosynthesis and stress tolerance for climate-resilient agriculture. Environ. Sci. Nano 13 (3), 1347–1371. https://doi.org/10.1039/D5EN00814J.

Raza, M.A.S., Muhammad, F., Farooq, M., Aslam, M.U., Akhter, N., Toleikienė, M., Binobead, M.A., Ali, M.A., Rizwan, M., Iqbal, R., 2025. ZnO-nanoparticles and stage-based drought tolerance in wheat (*Triticum aestivum* L.): effect on morpho-physiology, nutrients uptake, grain yield and quality. Sci. Rep. 15 (1). https://doi.org/10.1038/s41598-025-89718-2.

Rivero, R.M., Mittler, R., Blumwald, E., Zandalinas, S.I., 2022. Developing climate-resilient crops: improving plant tolerance to stress combination. Plant J. 109 (2), 373–389. https://doi.org/10.1111/tpj.15483.

Rizwan, M., Shoukat, A., Ayub, A., Razzaq, B., Tahir, M.B., 2021. Types and Classification of Nanomaterials Nanomaterials: Synthesis, Characterization, Hazards and Safety. Elsevier, pp. 31–54. https://doi.org/10.1016/B978-0-12-823823-3.00001-X.

Rizwan, M., Ali, S., Ali, B., Adrees, M., Arshad, M., Hussain, A., Rehman, M.Z.ur, Waris, A.A., 2019. Zinc and iron oxide nanoparticles improved the plant growth and reduced the oxidative stress and cadmium concentration in wheat. Chemosphere 214, 269–277. https://doi.org/10.1016/j.chemosphere.2018.09.120.

Sachdev, S., Ansari, S.A., Ansari, M.I., Fujita, M., Hasanuzzaman, M., 2021. Abiotic stress and reactive oxygen species: generation, signaling, and defense mechanisms. Antioxidants 10 (2), 1–37. https://doi.org/10.3390/antiox10020277.

Safdar, M., Kim, W., Park, S., Gwon, Y., Kim, Y.-O., Kim, J., 2022. Engineering plants with carbon nanotubes: a sustainable agriculture approach. J. Nanobiotechnol. 20 (1). https://doi.org/10.1186/s12951-022-01483-w.

Sales, H.B.E., Carolino, A.de S., Nunes, R.Z. de A., Macalia, C. M.A., Ruzo, C.M., Pinto, C.da C., Bezerra, J.de A., Campelo, P.H., Ţălu, Ş., de Souza, L.K.C., Sanches, E.A., 2024. Advances in agricultural technology: a review of slow-release nanofertilizers and innovative carriers. Commun. Soil Sci. Plant Anal. 55 (12), 1849–1882. https://doi.org/10.1080/00103624.2024.2326145.

Sarkar, B., Kumar De, A., Saha, I., Ghosh, A., Chandra Debnath, S., Kumar Adak, M., 2019. Amelioration with titanium dioxide nanoparticle for regulation of oxidative stress in maize (*Zea mays* L.). J. Microbiol. Biotechnol. Food Sci. 9 (2), 320–329. https://doi.org/10.15414/jmbfs.2019.9.2.320-329.

Scales, J.C., Parry, M.A.J., Salvucci, M.E., 2014. A non-radioactive method for measuring Rubisco activase activity in the presence of variable ATP: ADP ratios, including modifications for measuring the activity and activation state of Rubisco. Photosynth. Res. 119 (3), 355–365. https://doi.org/10.1007/s11120-013-9964-5.

Schreier, T.B., Hibberd, J.M., 2019. Variations in the Calvin-Benson cycle: selection pressures and optimization? J. Exp. Bot. 70 (6), 1697–1701. https://doi.org/10.1093/jxb/erz078.

Schwab, F., Zhai, G., Kern, M., Turner, A., Schnoor, J.L., Wiesner, M.R., 2016. Barriers, pathways and processes for uptake, translocation and accumulation of nanomaterials in plants—critical review. Nanotoxicology 10 (3), 257–278. https://doi.org/10.3109/17435390.2015.1048326.

Semenov, A.Y., Tikhonov, A.N., 2023. Electrometric and electron paramagnetic resonance measurements of a difference in the transmembrane electrochemical potential: photosynthetic subcellular structures and isolated pigment–protein complexes. Fed. Membr. 13 (11). https://doi.org/10.3390/membranes13110866.

Şener, M., Strümpfer, J., Hsin, J., Chandler, D., Scheuring, S., Hunter, C.N., Schulten, K., 2011. Förster energy transfer theory as reflected in the structures of photosynthetic light-harvesting systems. ChemPhysChem 12 (3), 518–531. https://doi.org/10.1002/cphc.201000944.

Shahid, M., Dumat, C., Khalid, S., Niazi, N.K., Antunes, P.M. C., 2017. Cadmium bioavailability, uptake, toxicity and detoxification in soil-plant system. Rev. Environ. Contam. Toxicol. 241, 73–137. https://doi.org/10.1007/398_2016_8.

Shahzadi, J., Zaib-un-Nisa, Ali, N., Iftikhar, M., Ali Shah, A., Yasin Ashraf, M., Chao, C., Shaffique, S., Gatasheh, M.K., 2025. Foliar application of nano biochar solution elevates tomato productivity by counteracting the effect of salt stress insights into morphological physiological and biochemical indices. Sci. Rep. 15 (1). https://doi.org/10.1038/s41598-025-87399-5.

Sharma, P., Janaagal, M., Rani Sheoran, A., Sharma, C., 2024. Nanoparticle innovations in plant systems: enhancing photosynthesis and nutrient dynamics. Asian J. Adv. Agric. Res. 24 (12), 75–88. https://doi.org/10.9734/ajaar/2024/v24i12574.

Shimoni, E., Rav-Hon, O., Ohad, I., Brumfeld, V., Reich, Z., 2005. Three-dimensional organization of higher-plant chloroplast thylakoid membranes revealed by electron tomography. Plant Cell 17 (9), 2580–2586. https://doi.org/10.1105/tpc.105.035030.

Shipley, H.J., Engates, K.E., Guettner, A.M., 2011. Study of iron oxide nanoparticles in soil for remediation of arsenic. J. Nanoparticle Res. 13 (6), 2387–2397. https://doi.org/10.1007/s11051-010-9999-x.

Shukla, P.K., Misra, P., Kole, C., 2016. Uptake, Translocation, Accumulation, Transformation, and Generational Transmission of Nanoparticles in Plants Plant Nanotechnology: Principles and Practices. Springer International Publishing, pp. 183–218. https://doi.org/10.1007/978-3-319-42154-4_8.

Singh, A., Rajput, V.D., Varshney, A., Sharma, R., Ghazaryan, K., Minkina, T., Alexiou, A., El-Ramady, H., 2024. Revolutionizing crop production: nanoscale wonders-current applications, advances, and future frontiers. Egypt. J. Soil Sci. 64 (1), 221–258. https://doi.org/10.21608/ejss.2023.246354.1684.

Singh, P., Arif, Y., Siddiqui, H., Sami, F., Zaidi, R., Azam, A., Alam, P., Hayat, S., 2021. Nanoparticles enhances the salinity toxicity tolerance in *Linum usitatissimum* L. by modulating the antioxidative enzymes, photosynthetic efficiency, redox status and cellular damage. Ecotoxicol. Environ. Saf. 213. https://doi.org/10.1016/j.ecoenv.2021.112020.

Singh, V., Yadav, P., Mishra, V., 2020. Recent advances on classification, properties, synthesis, and characterization of nanomaterials. In: Green Synthesis of Nanomaterials for Bioenergy Applications, pp. 83–97. https://doi.org/10.1002/9781119576785.ch3.

Singha, D.L., Das, D., Raj Paswan, R., Chikkaputtaiah, C., Kumar, S., 2022. Novel Approaches and Advanced Molecular Techniques for Crop Improvement. Informa UK Limited, pp. 1–27. https://doi.org/10.1201/9781003171416-1.

Smith, W.N., Grant, B.B., Desjardins, R.L., Kroebel, R., Li, C., Qian, B., Worth, D.E., McConkey, B.G., Drury, C.F., 2013. Assessing the effects of climate change on crop production and GHG emissions in Canada. Agric. Ecosyst. Environ. 179, 139–150. https://doi.org/10.1016/j.agee.2013.08.015.

Smith, M.N., Taylor, T.C., van Haren, J., Rosolem, R., Restrepo-Coupe, N., Adams, J., Wu, J., de Oliveira, R. C., da Silva, R., de Araujo, A.C., de Camargo, P.B., Huxman, T.E., Saleska, S.R., 2020. Empirical evidence for resilience of tropical forest photosynthesis in a warmer world. Nat. Plants 6 (10), 1225–1230. https://doi.org/10.1038/s41477-020-00780-2.

Solymos, K., Babcsányi, I., Ariya, B., Gyulavári, T., Ágoston, Á., Szamosvölgyi, Á., Kukovecz, Á., Kónya, Z., Farsang, A., Pap, Z., 2024. Photocatalytic and surface properties of titanium dioxide nanoparticles in soil solutions. Environ. Sci. Nano 11 (3), 1204–1216. https://doi.org/10.1039/D3EN00622K.

Song, X., Zhou, G., He, Q., Zhou, H., 2020. Stomatal limitations to photosynthesis and their critical water conditions in different growth stages of maize under water stress. Agric. Water Manag. 241, 106330. https://doi.org/10.1016/j.agwat.2020.106330.

Spaans, S.K., Weusthuis, R.A., van der Oost, J., Kengen, S.W. M., 2015. NADPH-generating systems in bacteria and archaea. Front. Microbiol. 6. https://doi.org/10.3389/fmicb.2015.00742.

Stirbet, A., Lazár, D., Guo, Y., Govindjee, G., 2020. Photosynthesis: basics, history and modelling. Ann. Bot. 126 (4), 511–537. https://doi.org/10.1093/aob/mcz171.

Su, Y., Ashworth, V., Kim, C., Adeleye, A.S., Rolshausen, P., Roper, C., White, J., Jassby, D., 2019. Delivery, uptake, fate, and transport of engineered nanoparticles in plants: a critical review and data analysis. Environ. Sci. Nano 6 (8), 2311–2331. https://doi.org/10.1039/c9en00461k.

Tan, W., Peralta-Videa, J.R., Gardea-Torresdey, J.L., 2018. Interaction of titanium dioxide nanoparticles with soil components and plants: current knowledge and future research needs-a critical review. Environ. Sci. Nano 5 (2), 257–278. https://doi.org/10.1039/C7EN00985B.

Tang, W., Guo, H., Baskin, C.C., Xiong, W., Yang, C., Li, Z., Song, H., Wang, T., Yin, J., Wu, X., Miao, F., Zhong, S., Tao, Q., Zhao, Y., Sun, J., 2022. Effect of light intensity on morphology, photosynthesis and carbon metabolism of alfalfa (Medicago sativa) seedlings. Plants 11 (13), 1688. https://doi.org/10.3390/plants11131688.

Tcherkez, G., 2016. The mechanism of Rubisco-catalysed oxygenation. Plant Cell Environ. 39 (5), 983–997. https://doi.org/10.1111/pce.12629.

Tikhonov, A.N., 2024. The cytochrome b6f complex: plastoquinol oxidation and regulation of electron transport in chloroplasts. Fed. Photosynth. Res. 159 (2–3), 203–227. https://doi.org/10.1007/s11120-023-01034-w.

Tikhonov, A.N., 2023. Electron transport in chloroplasts: regulation and alternative pathways of electron transfer. Biochemistry (Mosc.) 88 (10), 1438–1454. https://doi.org/10.1134/S0006297923100036.

Tripathi, D.K., Shweta, Singh, S., Singh, S., Pandey, R., Singh, V.P., Sharma, N.C., Prasad, S.M., Dubey, N.K., Chauhan, D.K., 2017. An overview on manufactured nanoparticles in plants: uptake, translocation, accumulation and phytotoxicity. Plant Physiol. Biochem. 110, 2–12. https://doi.org/10.1016/j.plaphy.2016.07.030.

Tripathi, S., Mahra, S., Victoria, J., Tiwari, K., Rana, S., Tripathi, D.K., Sharma, S., Sahi, S., 2023. Recent advances and perspectives of nanomaterials in agricultural management and associated environmental risk: a review.

Nanomaterials 13 (10), 1604. https://doi.org/10.3390/nano13101604.

Tripathy, B.C., Oelmüller, R., 2014. Reactive oxygen species generation and signaling in plants. Plant Signal. Behav. 7 (12), 1621–1633. https://doi.org/10.4161/psb.22455.

Ullah, J., Gul, A., Khan, I., Shehzad, J., Kausar, R., Ahmed, M.S., Batool, S., Hasan, M., Ghorbanpour, M., Mustafa, G., 2024. Green synthesized iron oxide nanoparticles as a potential regulator of callus growth, plant physiology, antioxidative and microbial contamination in *Oryza sativa* L. BMC Plant Biol. 24 (1). https://doi.org/10.1186/s12870-024-05627-y.

Upadhyay, N., Bhargava, A., 2025. Artificial intelligence in agriculture: applications, approaches, and adversities across pre-harvesting, harvesting, and post-harvesting phases. Iran J. Comput. Sci. 8 (3), 749–772. https://doi.org/10.1007/s42044-025-00264-6.

Vatankhah, A., Aliniaeifard, S., Moosavi-Nezhad, M., Abdi, S., Mokhtarpour, Z., Reezi, S., Tsaniklidis, G., Fanourakis, D., 2023. Plants exposed to titanium dioxide nanoparticles acquired contrasting photosynthetic and morphological strategies depending on the growing light intensity: a case study in radish. Sci. Rep. 13 (1). https://doi.org/10.1038/s41598-023-32466-y.

Vemula, M., Reddy, A.V.B., 2023. Polymeric nanoparticles as effective delivery systems in agriculture sustainability. Nanotechnol. Environ. Eng. 8 (3), 805–814. https://doi.org/10.1007/s41204-023-00319-8.

Vijay, P.P., Kunta, S., Venkatesh, B.M., 2025. Mobility of Nanoparticles in Plants Plant-Based Nanoparticle Synthesis for Sustainable Agriculture. CRC Press, pp. 125–139. https://doi.org/10.1201/9781003477730-9.

Walker, B.J., Driever, S.M., Kromdijk, J., Lawson, T., Busch, F.A., 2024. Tools for measuring photosynthesis at different scales. Methods Mol. Biol. 2790, 1–26. https://doi.org/10.1007/978-1-0716-3790-6_1.

Wang, H., Gu, M., Cui, J., Shi, K., Zhou, Y., Yu, J., 2009. Effects of light quality on CO_2 assimilation, chlorophyll-fluorescence quenching, expression of Calvin cycle genes and carbohydrate accumulation in Cucumis sativus. J. Photochem. Photobiol. B Biol. 96 (1), 30–37. https://doi.org/10.1016/j.jphotobiol.2009.03.010.

Wang, X., Xie, H., Wang, P., Yin, H., 2023. Nanoparticles in plants: uptake, transport and physiological activity in leaf and root. Materials 16 (8), 3097. https://doi.org/10.3390/ma16083097.

Wang, A., Jin, Q., Xu, X., Miao, A., White, J.C., Gardea-Torresdey, J.L., Ji, R., Zhao, L., 2020. High-throughput screening for engineered nanoparticles that enhance photosynthesis using mesophyll protoplasts. J. Agric. Food Chem. 68 (11), 3382–3389. https://doi.org/10.1021/acs.jafc.9b06429.

Wang, B., Ma, P., Liu, M., Huang, R., Qiu, Z., Pan, L., Wang, J., Liu, Y., Zhang, Q., 2025. Enhancement of microalgae co-cultivation self-settling performance and water purification capacity of microalgae biofilm. Environ. Res. 265. https://doi.org/10.1016/j.envres.2024.120342.

Wu, X., Hu, J., Wu, F., Zhang, X., Wang, B., Yang, Y., Shen, G., Liu, J., Tao, S., Wang, X., 2021. Application of TiO_2 nanoparticles to reduce bioaccumulation of arsenic in rice seedlings (*Oryza sativa* L.): a mechanistic study. J. Hazard Mater. 405. https://doi.org/10.1016/j.jhazmat.2020.124047.

Yamasaki, T., Yamakawa, T., Yamane, Y., Koike, H., Satoh, K., Katoh, S., 2002. Temperature acclimation of photosynthesis and related changes in photosystem II electron transport in winter wheat. Plant Physiol. 128 (3), 1087–1097. https://doi.org/10.1104/pp.010919.

Yanga, J., Cao, W., Rui, Y., 2017. Interactions between nanoparticles and plants: phytotoxicity and defense mechanisms. J. Plant Interact. 12 (1), 158–169. https://doi.org/10.1080/17429145.2017.1310944.

Yazdani, S., Mozaffarian, M., Pazuki, G., Hadidi, N., Villate-Beitia, I., Zárate, J., Puras, G., Pedraz, J.L., 2024. Carbon-based nanostructures as emerging materials for gene delivery applications. Pharmaceutics 16 (2). https://doi.org/10.3390/pharmaceutics16020288.

Ye, X., Chen, X.F., Deng, C.L., Yang, L.T., Lai, N.W., Guo, J. X., Chen, L.S., 2019. Magnesium-deficiency effects on pigments, photosynthesis and photosynthetic electron transport of leaves, and nutrients of leaf blades and veins in citrus sinensis seedlings. Plants 8 (10). https://doi.org/10.3390/plants8100389.

Yoon, D.-K., Ishiyama, K., Mao, S., Tazoe, Y., Watanabe, M., Imaruoka, S., Ogura, M., Ishida, H., Suzuki, Y., Obara, M., Mae, T., Makino, A., 2020. Transgenic rice overproducing Rubisco exhibits increased yields with improved nitrogen-use efficiency in an experimental paddy field. Nat. Food 1 (2), 134–139. https://doi.org/10.1038/s43016-020-0033-x.

Yousaf, N., Sardar, M.F., Ishfaq, M., Yu, B., Zhong, Y., Zaman, F., Zhang, F., Zou, C., 2024. Insights in to iron-based nanoparticles (hematite and magnetite) improving the maize growth (*Zea mays* L.) and iron nutrition with low environmental impacts. Chemosphere 362. https://doi.org/10.1016/j.chemosphere.2024.142781.

Yu, G., Wang, X., Liu, J., Jiang, P., You, S., Ding, N., Guo, Q., Lin, F., 2021. Applications of nanomaterials for heavy metal removal from water and soil: a review. Sustainability 13 (2), 713. https://doi.org/10.3390/su13020713.

Yu, W., Körner, O., Schmidt, U., 2020. Crop photosynthetic performance monitoring based on a combined system of measured and modelled chloroplast electron transport rate in greenhouse tomato. Front. Plant Sci. 11. https://doi.org/10.3389/fpls.2020.01038.

Ze, Y., Liu, C., Wang, L., Hong, M., Hong, F., 2011. The regulation of TiO_2 nanoparticles on the expression of light-harvesting complex II and photosynthesis of chloroplasts of arabidopsis thaliana. Biol. Trace Elem. Res. 143 (2), 1131–1141. https://doi.org/10.1007/s12011-010-8901-0.

Zhang, H., Zheng, T., Wang, Y., Li, T., Chi, Q., 2024. Multi-faceted impacts of nanoparticles on plant nutrient absorption and soil microbial communities. Front. Plant Sci. 15. https://doi.org/10.3389/fpls.2024.1497006.

Zhang, Z., Chen, J.M., Guanter, L., He, L., Zhang, Y., 2019. From canopy-leaving to total canopy far-red fluorescence emission for remote sensing of photosynthesis: first results from TROPOMI. Geophys. Res. Lett. 46 (21), 12030–12040. https://doi.org/10.1029/2019GL084832.

CHAPTER

6

Nanoformulated fertilizers and CO_2 delivery systems optimizes photosynthesis in smart agriculture

Omid Sohrabi[1], *Ali Shams*[2] and *Mahtab Mehrkish*[3]

[1]Department of Horticultural Science and Engineering, Razi University, Kermanshah, Iran; [2]Department of Biotechnology, Faculty of Basic Sciences, Al-e-Taha Institute of Higher Education, Tehran, Iran; [3]Agrotechnology – Crop Plant Physiology, Department of Plant Production and Genetics Engineering, Razi University, Kermanshah, Iran

1. Introduction

1.1 Nanofertilizers as sustainable alternatives in agriculture

Chemical fertilizers boost yields but harm soil microflora, alter pH, and cause environmental toxicity through leaching. Nanofertilizers (NFs) reduce fertilizer requirements, improve nutrient uptake, and minimize losses. They can be applied to soil or foliage, showing promising results across crops. Composed of micro- and macronutrient precursors at the nanoscale, NFs act as growth promoters with innovative application methods. Excessive use of fertilizer harms soil fertility by destroying beneficial microbes. NFs, as a key application of nanotechnology, provide sustainable alternatives that support plant growth and development while protecting soil health (Abobatta, 2018). Nanocarriers sized 1–100 nm are designed to encapsulate nutrients from conventional fertilizers or plants. Their large surface area allows gradual nutrient release, enabling crops to absorb nutrients efficiently without the harmful effects of traditional fertilizers (Preetha and Balakrishnan, 2017). NFs increase plant metabolic efficiency through larger surface area, boosting photosynthesis and productivity. By delivering nutrients directly, they reduce ecotoxicity and prevent nutrient loss to soil and groundwater (Tarafdar and Raliya, 2013). NFs range from 1 to 100 nm in size. Applied to soil or leaf surfaces, smaller nanoparticles penetrate more effectively, diffusing into crops when their dimensions are below the pore size of leaves and roots (Kumar et al., 2020). NFs enhance nutrient uptake, photosynthesis, translocation, and resistance to pests and pathogens. They improve soil quality alongside productivity, resulting in

Nanotechnology in Photosynthesis
https://doi.org/10.1016/B978-0-443-44376-3.00006-9

higher crop yields. NFs offer eco-friendly, efficient alternatives to chemical fertilizers. Their unique physiochemical properties make them promising agrochemicals. This review explores their agricultural role, application methods, and future prospects (Zulfiqar et al., 2019). Nano-enabled products (1–100 nm), including nanoinsecticides, pesticides, and fertilizers, play a vital role in sustainable farming. Research indicates nanoparticles can control pests, such as insects, fungi, and weeds, while NFs enhance nutrient delivery and significantly boost crop productivity (Raguraj et al., 2020; Abdelmigid et al., 2022). NFs meet plant nutritional needs while boosting productivity through high surface-to-volume ratios and mobility. Viewed as a "smart nutrient system," they can enhance grain yields, reduce hunger, and support poverty alleviation. By 2030, advanced NFs may be pivotal in achieving sustainable development goals, ensuring food security, and improving agricultural sustainability. This review highlights their timely role, focusing on metal oxide nanoparticles, nanocomposites, and conjugated nanoparticles for crop growth, stress tolerance, and commercial production. Beyond nutrient delivery, nanotechnology plays a pivotal role in enhancing photosynthetic efficiency by optimizing CO_2 diffusion and light harvesting, which will be discussed in detail in Section 3.

1.2 Advantages of NFs over conventional fertilizers

NFs provide direct nutrient delivery to plant cells, reducing environmental pollution and costs. They enable targeted supply of micronutrients, such as zinc and iron, improve stress tolerance when combined with nano-biofertilizers, and allow slow-release formulations that minimize leaching during critical growth stages. Overall, nanotechnology enhances nutrient availability, efficiency, and yield potential, making NFs an increasingly popular choice for sustainable agriculture (Zulfiqar et al., 2019).

COMPARATIVE TABLE: NFS VERSUS TRADITIONAL FERTILIZERS

Properties	NFs	Traditional fertilizers
Solubility and dispersion of mineral nutrients	Nanosize improves solubility, dispersion, and bioavailability of nutrients.	Large particle size limits solubility and dispersion, resulting in poor bioavailability.
Nutrient retention in soil	Nanoparticles enable prolonged retention of nutrients in soil.	Significant leaching, runoff, and drift occur.
Controlled release mode	Nutrient release can be precisely controlled by encapsulation in polymer matrix, supporting sustainability.	Direct exposure may be toxic at high doses and harmful to ecosystems.
Nutrient uptake efficiency	Nanostructured fertilizers reduce excessive use and increase uptake efficiency under controlled cultivation.	Large chemical composites are difficult for plants to absorb, reducing efficiency.
Prolonged nutrient supply	Nanostructured formulations extend nutrient supply through controlled release.	Nutrients are quickly available but often lost into soil, forming insoluble salts.

1.3 Need for NFs development

Rising populations demand higher agricultural productivity, yet soil nutrient deficiencies reduce yields and quality. Conventional methods require excessive fertilizer use, causing environmental harm. Nanotechnology enables mass production of vital metal nanoparticles, improving nutrient uptake, conservation, and efficiency through targeted delivery and controlled release. NFs reduce leaching losses, enhance photosynthesis, germination, seedling growth, and metabolism, offering a sustainable solution to boost crop productivity and address global food security challenges (Solanki et al., 2015; Raliya et al., 2016).

1.4 Optimizing photosynthesis for sustainable crop productivity

Enhancing photosynthetic efficiency is a primary frontier for ensuring global food security. Currently, most crops convert less than 1% of solar energy into biomass, highlighting a significant gap for genetic and biochemical improvement (Watts et al., 2024). Modern strategies focus on:

- **Carbon Fixation:** Improving CO_2 delivery to Rubisco and optimizing the Calvin–Benson–Bassham cycle.
- **Stomatal Engineering:** Research on amphistomatous leaves suggests that while stomatal patterning is often over dispersed, optimizing CO_2 diffusion paths through mesophyll coordination can significantly enhance photosynthetic rates.
- **Modeling:** Integrated system modeling now identifies precise targets within the electron transfer chain to improve nitrogen-use efficiency (NUE) and canopy-level photosynthesis.

1.5 IoT and nanotechnology: The backbone of smart farming

The convergence of the Internet of Things (IoT), Artificial Intelligence (AI), and nanotechnology is revolutionizing precision agriculture by enabling real-time, data-driven decision-making.

1.5.1 Smart sensors and predictive analytics

The integration of IoT-based sensors for monitoring soil moisture, microclimates, and plant stress allows for adaptive resource management. Case studies indicate that AI-driven irrigation and pest detection systems can reduce water consumption by up to 30% and increase yields by 15%–20% (Mansoor et al., 2025; Padhiary et al., 2025). Furthermore, precision technologies in horticulture, such as drones and GPS mapping, facilitate early nutrient deficiency detection and maximize profitability (Pasawan et al., 2025).

1.5.2 Nanosensors and stimuli-responsive systems

Nanotechnology introduces ultrasensitive sensing capabilities at the molecular level:

- **Real-time Monitoring:** Nanosensors detect soil nutrients, plant hormones, and CO_2 levels far earlier than conventional methods, enabling precise "just-in-time" fertilization (Kumar et al., 2024).
- **Smart Delivery:** Advanced biopolymer coatings (e.g., starch and nanochitosan) provide controlled-release mechanisms. These "smart" fertilizers respond dynamically to environmental stimuli, such as pH, temperature, and moisture, ensuring that nutrients are released in synchronization with plant demand, thereby minimizing environmental ecotoxicity (Azeem, 2025; Thirugnanasambandan et al., 2024).

2. Scientific and technological foundation of nanoformulated fertilizers

2.1 Basic definition and common types of NFs

NFs represent an emerging category of agricultural inputs engineered through nanotechnology, featuring active particles sized typically between 1 and 100 nm. Their nanoscale dimension substantially enhances surface area and physicochemical properties, facilitating more efficient nutrient uptake by plants. These fertilizers are developed to improve nutrient-use efficiency, minimize environmental contamination, and ultimately boost crop productivity (Goyal et al., 2025).

2.2 Nanochelates

Nanochelates are sophisticated fertilizer formulations in which metal ions and micronutrients are encapsulated within nanoscale chelating agents, allowing targeted delivery to plants. This nanoscale encapsulation increases nutrient bioavailability and reduces losses due to leaching or fixation within the soil. For example, nanochelates of iron and zinc have demonstrated pronounced efficacy in mitigating micronutrient deficiencies in various crops (Lawrencia et al., 2021). Due to their diminutive size and stable chemical nature, nanochelates enhance the translocation of micronutrients within plant tissues, improving root development and photosynthetic performance. Recent research has highlighted that nanochelated zinc fertilizers not only improve crop yield but also exhibit controlled nutrient release properties, which align well with the principles of precision agriculture for sustainable crop management (Bekah et al., 2025; Goyal et al., 2025).

2.3 Silica nanoparticles

Silicon, an important element for strengthening plant resistance against environmental stresses, is effectively delivered through silica nanoparticles formulated as NFs. These nanoparticles enhance cell wall fortification and bolster plant tolerance to drought, pests, and diseases. Empirical studies have demonstrated that nanosilica application can elevate photosynthetic capacity and promote vigorous plant growth (Ayenew et al., 2025). Beyond bolstering plant resilience, silica nanoparticles influence plant metabolism by stimulating antioxidant enzyme production, which helps mitigate oxidative damage under stress conditions. Furthermore, their role in improving water-use efficiency (WUE) and nutrient uptake positions silica NFs as key components in sustainable agricultural systems. However, ongoing investigations into nanoparticle accumulation and environmental fate are critical to ensuring their safe and effective integration into crop production (Pan et al., 2023; Yan et al., 2024).

2.4 Nanohydroxyapatite

Nanohydroxyapatite, a nanoparticulate form of calcium phosphate mineral, is employed as a slow-release phosphorus source in NFs. This mineral's nanodimension enables gradual nutrient release, thereby reducing phosphorus runoff and enhancing uptake efficiency in phosphorus-deficient soils, which is a widespread agronomic challenge (Goyal et al., 2025). Recent advances have explored the synthesis of nanohydroxyapatite from biological waste, such as poultry bones, offering an eco-friendly and cost-effective alternative to conventional phosphorus fertilizers. These bioderived nanomaterials have demonstrated the potential to triple the phosphorus-use efficiency while decreasing environmental pollution. Moreover, nanohydroxyapatite can be engineered to carry other micronutrients, such as zinc, thus serving as multifunctional NFs that optimize crop nutrition and growth (Pilotto et al., 2025; Shi et al., 2025).

2.5 Outstanding features of NFs

NFs are characterized by their ultrasmall particle size, which enables controlled nutrient delivery, enhanced nutrient-use efficiency, and reduced ecological footprint. Their versatility permits application via foliar sprays, soil amendment, or irrigation systems, making them adaptable to various precision agriculture techniques aimed at maximizing crop yield while minimizing input wastes (Bekah et al., 2025).

2.6 Technologies for controlled-release coatings and nanoencapsulation of nutrients

One of the foremost challenges in modern agriculture is optimizing fertilizer use to improve nutrient-use efficiency while mitigating negative environmental impacts. Controlled-release technologies and nanoencapsulation play a pivotal role in this domain by enabling the targeted and gradual delivery of nutrients to plants. These technologies help reduce nutrient losses through leaching, chemical transformation, or volatilization, while synchronizing nutrient release with plant growth demands (Rehana et al., 2022).

2.7 Types of coatings and nanoencapsulation materials

Controlled-release systems typically employ various coatings composed of synthetic polymers, biodegradable materials, mineral compounds, or even nanoparticles. For example, semipermeable polymer coatings prevent rapid release of soluble fertilizers, thereby allowing a gradual and sustained nutrient supply (Chaudhary and Yeshpal, 2023). Advanced approaches utilize stimuli-responsive hydrogels that react dynamically to environmental cues, such as temperature or moisture, smartly regulating the timing and amount of nutrient release. Nanoencapsulation represents a novel approach, wherein nutrients are enclosed within nanometer-scale capsules to prevent premature dispersion. These capsules may consist of nanopolymers or carbon-based composites which, in addition to shielding nutrients, enable targeted delivery. In some sophisticated formulations, nanocapsules release nutrients in response to specific biological signals from the plant or environmental triggers, making them vital tools for precision and smart agriculture (Rehana et al., 2022; Chaudhary and Yeshpal, 2023).

2.8 Advanced coating technologies

Beyond hydrogels and nanocapsules, mineral-based coating technologies, such as metal oxides and colloidal materials, have witnessed remarkable development. These coatings offer enhanced durability, forming thin adherent layers that more effectively modulate the release rate of nutrients. Bio-based coatings incorporating beneficial microorganisms simultaneously promote nutrient release and stimulate microbial activity in the soil, thereby enhancing nutrient uptake efficiency (Rehana et al., 2022). In modern developments, the use of nanocomposite coatings combining nanosilicates, nanoclays, and biopolymers has emerged as an advanced strategy. These composite coatings improve stability and environmental resilience, thus optimizing nutrient retention and controlled-release performance to increase fertilizer efficiency (Mim et al., 2025). Moreover, smart coating systems capable of sensing environmental changes, such as pH, humidity, and temperature, allow precise tuning of nutrient release timing. Using biosensors or chemical triggers, these systems release fertilizers when most beneficial for the plant, representing a significant advancement toward sustainable and precision agriculture (Jariwala et al., 2022).

2.9 Impacts of novel coating technologies on sustainable agriculture

The application of state-of-the-art coating technologies contributes substantially not only to improved nutrient-use efficiency and reduced nutrient loss but also to minimized agricultural product losses and increased economic returns. By decreasing the frequency and quantity of chemical fertilizer applications, these technologies aid in soil conservation and mitigate issues, such as soil salinization and runoff pollution (Boberski et al., 2025). Furthermore, integrating advanced coatings with intelligent water and energy management systems enhances crop productivity and plays a crucial role in achieving sustainable agriculture and global food security goals. Leading countries have leveraged digital technologies, smart sensors, and the IoT to reduce costs and enhance the quality and export potential of their agricultural produce, thereby fostering economic growth within the sector (Qiao et al., 2025).

2.10 Engineered NFs addressing nitrogen, phosphorus, and potassium deficiencies

2.10.1 *Nitrogen-based NFs*

Nitrogen deficiency continues to be a principal constraint on global crop productivity. Engineered nitrogen-based NFs offer promising solutions to this challenge by improving nutrient delivery and reducing environmental losses. For example, hydroxyapatite-modified nanourea presents a controlled-release mechanism, gradually supplying nitrogen to plants while significantly minimizing volatilization and leaching losses typically observed with conventional nitrogen sources (Sebastian et al., 2023). This sustained nutrient availability enhances NUE, promoting better biomass accumulation and higher crop yields while mitigating nitrogen runoff into water bodies. In addition, nanochelated nitrogen formulations provide nitrogen in bioavailable and chemically stable forms. Studies indicate that foliar application of nitrogen nanochelates increases chlorophyll concentration, leaf area development, and photosynthetic activity across numerous crops, leading to enhanced physiological performance and yield quality (Aljutheri et al., 2018). These benefits highlight the critical role of nanoengineered fertilizers in enhancing nitrogen assimilation pathways and plant metabolism under diverse agroecological conditions.

2.10.2 *Phosphorus-based NFs*

Phosphorus is vital for plant growth, yet its availability is often limited due to interactions with soil minerals, leading to fixation and low bioavailability—especially in alkaline or phosphorus-deficient soils. Nanohydroxyapatite fertilizers, characterized by highly reactive phosphate groups on their surfaces, enable controlled and sustained phosphorus release, markedly boosting phosphorus uptake efficiency (Elsayed et al., 2022). Recent advancements have focused on integrating nanophosphorus formulations with biostimulants, such as beneficial microbes and humic acids, fostering synergistic effects on root proliferation and nutrient assimilation. Applied at scale, these innovative nanophosphorus fertilizers have demonstrated reduced phosphorus leaching into aquatic systems and improved agronomic efficiency in staple crops, such as rice and maize. This not only reduces dependence on conventional and often costly phosphorus fertilizers but also diminishes environmental pollution risks, paving the way toward circular and sustainable nutrient management practices (Sarabandi et al., 2025).

2.10.3 *Potassium-based NFs*

Potassium regulates crucial enzymatic reactions and fortifies plant resistance against abiotic stresses, such as drought, salinity, and pathogen attacks. Nanoformulated potassium fertilizers, including potassium nanochelates and silica–potassium nanocomposites, provide controlled nutrient release and enhanced retention in the rhizosphere, optimizing potassium availability (Demeke et al., 2025). These materials effectively reduce nutrient losses and synchronize release with plant demand. Further research reveals that formulating potassium NFs in combination with nanosilica significantly strengthens antioxidant defense mechanisms in plants, fostering better growth and yield under stress conditions. These formulations also contribute positively to soil health and microbial biodiversity, which are indispensable for resilient and sustainable agricultural ecosystems (Sebastian et al., 2023).

2.10.4 Future research directions and innovative strategies

Future-oriented research is increasingly focused on designing multifunctional NFs capable of delivering nitrogen, phosphorus, and potassium in a coordinated and environmentally conscious manner. Priority is being given to developing biodegradable nanoparticles integrated with native growth stimulators and plant-beneficial microorganisms to reduce environmental footprints and enhance nutrient-use efficiency (Ayenew et al., 2025). A practical frontier involves integrating these NFs with nanosensors and the IoT technologies to create smart nutrient management systems. These systems precisely monitor environmental and plant physiological parameters, enabling dynamic control of fertilizer dosage and timing on a field scale. Such technological synergy maximizes efficiency, reduces nutrient pollution, and improves overall economic returns for farmers (Bekah et al., 2025). Moreover, novel formulations that combine multiple nutrients with tailored release kinetics responsive to environmental and plant stress cues are under development. These user-customizable NFs hold the promise of not only enhancing crop productivity and quality but also conserving resources, such as water and energy. Their adoption will be instrumental in achieving sustainable food systems and strengthening global agricultural economies amid rising climatic challenges (Singh et al., 2023).

2.10.5 Mechanisms of uptake and physiological effects on plants

Understanding the mechanisms by which engineered NFs are absorbed by a wide range of plants—including field crops, fruit-bearing orchard trees, greenhouse vegetables, and ornamental greenhouse plants—is essential to optimize fertilizer efficiency and minimize environmental impact. Uptake of these nanoparticles can occur through both foliar and root pathways, governed by the nanoparticles' physicochemical properties and their interaction with plant anatomical and physiological features (Dey and Sadhukhan, 2024).

2.10.5.1 Foliar uptake

Nutrients delivered via foliar application of NFs mainly penetrate leaves through stomata or cuticular channels. The cuticular pores typically range from 0.6 to 4.8 nm in diameter and include hydrophilic and lipophilic pathways that regulate nanoparticle passage according to their size and polarity. Nanoparticles smaller than 5 nm can diffuse through hydrophilic pores, while larger hydrophobic nanoparticles predominantly enter through stomatal openings, which vary in size between 10 and 100 μm. Surface charge plays a key role; positively charged nanoparticles adhere more effectively to the negatively charged leaf epidermis, thereby facilitating uptake (Wang et al., 2023).

2.10.5.2 Root uptake

Nanoparticles are absorbed by roots mainly through adsorption on root surfaces, particularly in younger root zones lacking full exodermal coverage. Particles under 10 nm penetrate readily, while larger nanoparticles enter through endocytosis or transient disruptions in the cell wall. Once internalized, nanoparticles are transported through apoplastic and symplastic pathways aided by transport proteins within the plant's vascular (xylem and phloem) system (Wang et al., 2023).

2.10.6 *Enhanced nutrient bioavailability and photosynthesis*

NFs improve nutrient availability and metabolism, leading to increased chlorophyll synthesis, elevated photosynthetic rates, and enhanced carbon fixation. This is mediated partly by gene regulation of nutrient transporters and phytohormonal signaling, which stimulates cell division and growth under optimal and stress conditions (Dey and Sadhukhan, 2024).

2.10.7 *Effects on orchard, greenhouse, and ornamental plants*

In fruit-bearing orchard trees and greenhouse-grown vegetables, NFs contribute to improved flowering, fruit set, yield, and quality through refined nutrient supply and stress-tolerance mechanisms. Likewise, ornamental greenhouse plants demonstrate enhanced aesthetic traits, such as leaf greenness and flower longevity, due to more efficient nutrient uptake and physiological robustness enabled by nanoformulations (Yadav, 2025).

2.10.8 *Innovative and practical solutions*

2.10.8.1 Integration with smart systems

The integration of NFs with intelligent CO_2 management and biosensor technology allows nutrient application precisely synchronized with plants' photosynthetic requirements, reducing nutrient wastage. Such technologies in smart greenhouses facilitate real-time monitoring and optimized nutrient dosing, employing machine learning algorithms to increase precision and efficiency (Wang et al., 2024).

2.10.8.2 Design of biocompatible NFs

Development of biodegradable, environmentally friendly NFs capable of releasing nutrients in pulses aligned with plant demands enhances both environmental safety and fertilizer efficiency. Incorporating natural polymers, hydrocolloids, and mineral composites ensures improved nutrient mobility and minimizes accumulation in soil and water (Mim et al., 2025). Advances in stimuli-responsive release, triggered by environmental factors, such as pH and temperature, promise to further optimize fertilizer effectiveness and reduce chemical usage.

2.10.8.3 Multiplex nutrient delivery

Engineering multifunctional NFs capable of delivering primary macronutrients (N, P, K) alongside micronutrients and biostimulants with controlled-release profiles addresses the complex nutritional needs of crops. Coupled with advanced delivery devices and AI-powered management, this technology supports sustainable agricultural intensification, enhances crop quality, and conserves water and energy resources (Sebastian et al., 2023).

3. Advancing photosynthesis for sustainable agriculture

Enhancing photosynthetic efficiency is a primary frontier for ensuring global food security. Modern strategies focus on optimizing light harvesting, electron transfer, and the carbon reactions of the Calvin–Benson–Bassham cycle. Key approaches include increasing CO_2 concentration near Rubisco and improving internal CO_2 delivery mechanisms. Current system modeling and studies of natural genetic variation have identified critical molecular targets to improve NUE and overall canopy photosynthesis.

3.1 Photosynthesis efficiency and agricultural potential

Despite being essential for life, photosynthesis in crops converts less than 1% of solar energy. This inefficiency offers major potential for improvement, especially under rising global population and climate pressures. While crop yield has advanced in other areas, photosynthesis remains underexplored due to its genetic and biochemical complexity.

3.2 Stomatal patterning and CO_2 diffusion

Amphistomatous leaves, with stomata on both surfaces, shorten CO_2 diffusion paths. Using spatial statistics and photosynthesis modeling in *Arabidopsis thaliana*, the study found stomata are over dispersed but not ideally patterned, with independent distribution across surfaces. Coordination between abaxial and adaxial stomata yields minimal photosynthetic gains (<1%), while stomatal size correlates with mesophyll volume, suggesting alternative optimization strategies for CO_2 diffusion (Watts et al., 2024).

3.3 Carbon nanocarriers for CO_2 delivery

Carbon nanostructures, such as nanotubes, graphene, and porous frameworks, exhibit high surface area and chemical stability, enabling efficient CO_2 capture and controlled release. These nanocarriers are promising for carbon capture and storage (CCS) applications, particularly in hard-to-abate sectors, such as cement and steel. Their tunable properties allow integration into industrial systems for enhanced CO_2 management and delivery.

3.4 CO_2 enrichment in greenhouses and open fields

Systematic review highlights that CO_2 enrichment enhances photosynthesis, yield, and water-use efficiency in controlled agriculture. Laboratory applications include modeling optimal CO_2 concentrations (800–1000 μmol mol^{-1}). Practical greenhouse strategies involve direct CO_2 injection from industrial or biogas sources, while open-field FACE experiments simulate elevated CO_2 under natural conditions. Challenges remain in cost, sustainability, and environmental impacts (Villagran et al., 2025).

3.5 CO_2 delivery and Rubisco carboxylation

Elevated CO_2 enhances Rubisco carboxylation efficiency by reducing oxygen competition, thereby lowering photorespiration. FACE experiments demonstrate significant yield improvements in C_3 crops under high CO_2, highlighting increased photosynthetic efficiency and water-use optimization. These findings underscore the potential of CO_2 enrichment to boost crop productivity in future climate scenarios (Ainsworth and Long, 2021).

3.6 Synergistic effects of nutrition and CO_2 enrichment

Free-Air CO_2 Enrichment (FACE) experiments demonstrate that the benefits of elevated CO_2 are heavily dependent on nutrient availability. Adequate nitrogen supply is essential to enhance Rubisco activity and photosynthetic protein synthesis, which can lead to a crop yield increase of up to 18% under nonstress conditions (Leakey et al., 2009; Zhang et al., 2024). Furthermore, phosphorus and potassium play critical roles in energy transfer and stomatal regulation, respectively, sustaining photosynthetic efficiency. However, nutrient limitations—particularly nitrogen and phosphorus deficiencies—act as major constraints, often neutralizing the potential gains from elevated CO_2. Therefore, the synergy between precise nutrient management, irrigation, and carbon enrichment is vital for achieving sustainable yield gains in future agricultural scenarios (Zhang et al., 2024).

4. Integration of NFs and CO_2 delivery systems: Vision for smart agriculture

4.1 Synergistic effects of combining nutrient delivery and carbon enrichment

In smart agriculture, the integration of advanced technologies in nutrient delivery and carbon enrichment via controlled release of carbon dioxide (CO_2) emerges as a comprehensive strategy to significantly enhance crop productivity and optimize photosynthesis. This combined approach employs nanoformulated fertilizers alongside controlled CO_2 release systems, enabling targeted and efficient supply of essential nutrients and CO_2 to plants, thereby yielding synergistic effects on growth and yield (Wang et al., 2024; Shen et al., 2023). NFs act as intelligent carriers capable of controlled and sustained nutrient release, notably reducing losses of nitrogen and other critical elements in soil. These nanostructures increase the surface area and permeability, which improves nutrient uptake efficiency by roots and triggers key metabolic pathways within plants (Liu et al., 2024; Kumar et al., 2025). Utilization of NFs has demonstrated up to a 45% increase in nutrient absorption efficiency, contributing to lowered fertilizer use and diminished environmental contamination (Gill et al., 2024). Additionally, smart coatings on NFs enable stimuli-responsive release correlating with environmental cues. On the other front, enriching cultivation environments with CO_2 at controlled concentrations, especially in greenhouse or precision farming systems, enhances the activity of pivotal enzymes in the Calvin cycle, accelerating photosynthetic reactions (Ainsworth and Rogers, 2007). Optimal CO_2 levels ranging from 0.1% to 1% volume substantially increase photosynthetic rates and plant biomass accumulation compared with outdoor ambient concentration of approximately 0.03% (Wang et al., 2019; Kumar et al., 2024). Experiments with *Brassica chinensis* reported that the combination of NFs and controlled CO_2 enrichment elevated photosynthesis rates by approximately 15%, alongside substantial increases in leaf chlorophyll content and overall physiological growth parameters (Shen et al., 2023). Despite these promising outcomes, widespread adoption encounters challenges, including the high costs of NF production, precise management requirements for CO_2 enrichment systems, regulatory uncertainties around the safety and environmental fate of nanomaterials, and the necessity for advanced sensing and control technologies to optimize applications (Shen et al., 2023; Upadhayay et al., 2023). Furthermore, potential long-term ecological risks associated with nanoparticle accumulation in soils and food chains demand thorough investigation. Key challenges and limitations elevated manufacturing and operational costs for nanotechnologies and CO_2 delivery infrastructure, along with scale-up difficulties from laboratory to field applications (Liu et al., 2024). Furthermore, there is a lack of comprehensive, standardized protocols for environmental safety and biosafety assessments of nanoparticles in agroecosystems (Shen et al., 2023). This highlights the imperative need for integrating IoT-based and biosensor-driven smart management systems to optimize the timing and dosage of fertilizers and CO_2 (Shen et al., 2023). There are significant concerns regarding nanoparticle bioaccumulation and its unknown impacts on soil health and food safety (Upadhayay et al., 2023). Innovative solutions and future perspectives developing smart NFs capable of stimuli-responsive release, adapting nutrient supply dynamically based on plant and environmental signals, represents a pivotal solution (Shen et al., 2023). Coupling these fertilizers with IoT-enabled biosensors and AI-driven farm management platforms will allow precise, real-time modulation of nutrient and CO_2 delivery. Biodegradable nanomaterials, combined with beneficial soil microbiota (plant probiotics), offer additional avenues to enhance soil health, increase nutrient-use efficiency, and mitigate environmental risks posed by persistent nanoparticles (Upadhayay et al., 2023). The overarching goal is to design integrated smart agricultural systems that harmonize controlled-release nutrient and CO_2 supply with continuous environmental monitoring for sustainable, high-efficiency crop production (Shen et al., 2023). Ultimately, the future of sustainable agriculture hinges on seamlessly integrating nanotechnology with controlled carbon enrichment and advanced monitoring technologies to overcome existing technical and economic barriers—thereby realizing substantial productivity gains alongside environmental stewardship.

4.2 Improving NUE and WUE

Improving NUE and WUE is a fundamental pillar for advancing sustainable agriculture, especially under the pressures of global population growth, environmental constraints, and resource scarcity. Nanotechnology, by virtue of its unique physicochemical properties, offers transformative possibilities to enhance how plants use nitrogen and water leading to a marked reduction in fertilizer and water wastage while bolstering crop productivity and environmental integrity. Nano-enabled fertilizers provide a sophisticated platform for the controlled and targeted delivery of nitrogen. Unlike conventional fertilizers, which often suffer from substantial losses due to leaching, volatilization, and microbial transformations, NFs encapsulate nitrogen in nanostructured carriers or coatings. This encapsulation enables a slow, sustained, and environment-responsive release profile that aligns nutrient availability with the plants' dynamic requirements, thereby maximizing uptake efficiency and minimizing environmental losses. Studies underscore that employing nanoencapsulated nitrogen fertilizers can elevate NUE by 30–50%, concurrently increasing crop yields by similar margins across various staple crops (Durrani and Akhter, 2022; Mejias et al., 2021). Implementation of nanomaterial carriers, such as nanosilica, nano-iron oxides, and other metal oxides, further stabilize nitrogen forms in soil matrices and promote their bioavailability (Van de Vlasakker et al., 2022; Shahid et al., 2024). Additionally, NFs can modify soil's physical structure by influencing pore size distribution and enhancing root penetration, thereby facilitating nutrient and water uptake. Simultaneously, nanotechnology enhances WUE through multifaceted mechanisms. Nanoparticles introduced into soil or foliar formulations act as water-retentive agents that reduce evaporation losses and improve water availability at the root–soil interface (Shen et al., 2023). These particles also influence stomatal behavior, promoting improved water regulation within plants. Coupled with precision agriculture tools, such as IoT-enabled soil moisture sensors and nanoscale biosensors—nanotechnology enables real-time monitoring and responsive irrigation, thereby optimizing water input without compromising crop health (Goyal et al., 2023; Mukhtar and Shakeel, 2019). Such harmonized water management approaches have demonstrated water savings of up to 20% without yield penalty, reflecting significant efficiency gains. Crucial synergies emerge when nanofertilization practices are integrated with advanced water management strategies. This integrated approach not only boosts NUE and WUE but also enhances crop resilience to abiotic stresses, such as droughts, leading to 15%–30% increases in nutrient utilization efficiency and marked improvements in yield quality and quantity (Shahid et al., 2024). Importantly, this confluence minimizes nutrient leaching and water waste, thereby reducing the environmental footprint of agricultural systems. Despite the promising advances, challenges persist high manufacturing costs, regulatory ambiguities regarding nanoparticle use, and concerns on long-term ecological effects pose barriers to wide-scale adoption. It is imperative to pursue development of biodegradable and environmentally benign nanomaterials, alongside regulatory frameworks to ensure ecosystem safety (Mejias et al., 2021; Van de Vlasakker et al., 2022). Moreover, integrating artificial intelligence and machine learning with nano-enabled systems holds great potential to refine resource use dynamically and sustainably. Regarding innovative strategies and future perspectives, the design of biodegradable NFs with stimuli-responsive release—tuned to environmental cues, such as soil moisture and nutrient demand—represents a vital frontier for smart agriculture (Mejias et al., 2021). This can be achieved by leveraging IoT-enabled sensor networks and AI-based decision-making platforms to orchestrate accurate, dynamic management of nitrogen inputs and irrigation scheduling (Mukhtar and Shakeel, 2019). Another strategy is the synergistic application of nanomaterials with beneficial soil microbes (biofertilizers) to bolster nutrient uptake, enhance soil health, and improve crop tolerance to abiotic stresses (Shahid et al., 2024). Finally, the implementation of integrated nano-enabled systems intertwined with advanced irrigation techniques, such as drip irrigation, promoting maximal resource conservation coupled with improved productivity. These forward-looking strategies herald a new paradigm in agriculture where technological innovation, environmental stewardship, and economic viability converge to enable resilient and productive agroecosystems.

4.3 Successful case studies in precision agriculture and horticulture

Precision agriculture represents a fundamental shift in modern farming paradigms, leveraging a suite of advanced technologies, including IoT devices, artificial intelligence (AI), machine learning, unmanned aerial vehicles (UAVs), remote sensing, and big data analytics. This integrated approach enables precise optimization of agricultural processes, ultimately enhancing resource-use efficiency, crop yield, and sustainability (Liakos et al., 2018). The crux of precision agriculture lies in robust data acquisition and sophisticated analytics. Data mining and big data processing uncover latent patterns within complex environmental and agronomic datasets, facilitating accurate forecasting and informed decision-making regarding planting schedules, nutrient management, irrigation, and pest control (Kouchaky et al., 2025). AI-driven predictive models, trained on historical and real-time sensor data, significantly increase the accuracy of these decisions, thereby optimizing farm-level operations (Liakos et al., 2018). For instance, AI-based pest prediction platforms implemented in India have demonstrated a 40% reduction in pesticide usage and a parallel decrease in crop damage by analyzing climatic, soil, and pest monitoring data (Sourav et al., 2025). Concurrently, nanosensor technologies enable real-time, fine-scale monitoring of soil moisture and nutrient status, further empowering precision management (Chen et al., 2023). Successful case studies have emerged across diverse agricultural sectors. In poultry production, the integration of automated feeding and watering systems with environmental sensors has yielded significant gains, including reductions of approximately 10% and 15% in feed and water use, respectively, alongside improvements in bird growth and health outcomes (Orakwue et al., 2022). Real-time health monitoring and environmental control systems further reduce mortality risks and disease incidence (Leong, 2025). Aquaculture systems have benefitted equally from precision technologies. Automated water quality monitoring—tracking parameters such as temperature, dissolved oxygen, and turbidity—combined with automated feeding has improved the growth rates and survival of aquatic species, while also mitigating environmental impacts associated with overfeeding and water pollution (Yusoff et al., 2024). In livestock management, biometric monitoring devices and robotic milking operations have facilitated enhanced detection of health issues, optimized feeding schedules, and improved animal welfare, resulting in average production increases of 15% for dairy operations (Fuentes et al., 2021). Precision horticulture has been revolutionized by UAV-based multispectral imaging, enabling early detection of biotic and abiotic stress factors in orchards. This spatially explicit data support targeted pesticide application, reducing chemical usage by up to 50% and increasing fruit quality (Khose and Mailapalli, 2024). Robotic harvesters have also made operational strides, notably in delicate crops such as strawberries, where labor costs have declined by 30% through improved picking efficiency and accuracy (Anand et al., 2023). In crop production, autonomous machinery and IoT-based environmental monitoring facilitate precise seed placement, nutrient application, and irrigation scheduling, enhancing yields and conserving inputs in commodities, such as wheat and maize (Liakos et al., 2018). Nevertheless, widespread adoption of precision agriculture faces multifaceted challenges. High capital costs for sophisticated equipment and data platforms limit accessibility, especially for smallholder farmers. Additionally, there is a pronounced deficit in the necessary digital and analytical skills required for efficacious technology utilization. Infrastructure deficiencies, particularly limited rural connectivity and unreliable power supplies, diminish system effectiveness (Barnes et al., 2019; Chad et al., 2025). Furthermore, data privacy concerns and technological fragmentation impede seamless integration of multiplatform farm data (Barnes et al., 2019). Social resistance, driven by fears of labor displacement and uncertainty around new technologies, further complicates adoption (Yuan and Sun, 2024). Innovative solutions targeting these barriers are being developed. These include cost-reduction strategies focusing on modular, scalable technologies suitable for small-scale farms (Chad et al., 2025), comprehensive training programs blending digital literacy with agronomic expertise (Barnes et al., 2023), expanded rural broadband and edge computing architectures compensating for connectivity gaps (Isreal et al., 2025), stringent data governance frameworks ensuring privacy and trust (Barnes et al., 2019), and cocreation methods involving farmers in technology development to boost acceptance to promote environmental health (Mgendi, 2024). Overall, these trajectories suggest that the potential of precision agriculture to sustainably transform food systems depends not only on technological innovation but also on effective knowledge transfer, infrastructural development, policy support, and social inclusion.

4.4 The role of lot and nanosensors in real-time monitoring and adaptive management in precision

Recent advances in the IoT and nanosensor technologies have fundamentally reshaped the landscape of precision agriculture, enabling real-time, continuous monitoring and adaptive management of diverse agricultural ecosystems. This transformative integration supports data-driven decision-making by offering highly granular insights into the dynamic conditions of soil, plants, water resources, and environmental factors, thereby optimizing resource utilization and enhancing sustainability across crop production, livestock, poultry, aquaculture, and horticulture sectors (Miller et al., 2025). Holistic Framework: IoT and nanosensors in Precision Agriculture IoT platforms encompass extensive networks of interconnected sensors deployed throughout farmlands and agricultural facilities. These sensors capture high-resolution data regarding soil moisture, nutrient content, temperature, pH, plant physiological status, and microclimate parameters with unprecedented temporal and spatial detail. Nano-enabled biosensors, distinguished by their nanoscale dimensions and exceptionally high surface-to-volume ratios, enable detection at molecular and cellular levels, effectively monitoring early stress signals, such as water deficit, nutrient imbalance, or pest infestation (Sarabandi et al., 2025). The integration of these technologies establishes a comprehensive environmental surveillance system transmitting real-time data via wireless protocols to cloud and edge computing systems for immediate analysis and actionable responses (Luo et al., 2025). The continuous data streams inform precision agriculture practices, including variable rate irrigation and fertilization, targeted pest and disease management, and optimized harvesting schedules, all of which contribute to significant resource savings, yield improvements, and minimal environmental footprints (Sarabandi et al., 2024; Luo et al., 2025). Sector-Specific Applications and Adaptive Management Strategies Crop Production and Soil Management Nanosensors monitor in situ soil parameters, such as moisture levels, nutrient availability, and microbial activity, facilitating the implementation of site-specific irrigation and fertilization regimes that dynamically adjust to crop requirements. This reduces excessive water use and fertilizer runoff, contributing to sustainable soil health management (Mushtaq et al., 2024). Adaptive irrigation scheduling via IoT data has been shown to improve WUE by up to 30%, while precision nutrient delivery through nanosensors can enhance nutrient-use efficiency by over 25% without compromising crop yields (Luo et al., 2025; Sarabandi et al., 2025). Pest and Disease Surveillance Early warning systems empowered by IoT-enabled nano-biosensors detect volatile organic compounds and biochemical changes indicative of pest activity or pathogenic infection. Coupled with AI-powered predictive analytics, these systems enable proactive, well-timed interventions, permitting a 40% reduction in pesticide applications and minimizing crop damage risks. Horticulture and Controlled Environment Agriculture Utilization of UAVs equipped with multi- and hyperspectral imaging sensors, combined with nano-biosensors, enables precision mapping of crop health across orchards and greenhouses. This facilitates precise application of agrochemicals, reducing inputs by up to 50%, and allows early detection of environmental stressors, improving crop quality and yield consistency (Yang et al., 2025). The integration of IoT and nanosensor technologies represents a pivotal advancement in precision agriculture, enabling real-time, adaptive management that responds to dynamic agricultural and environmental inputs. Properly addressing economic, technical, infrastructural, and social constraints will be key to unlocking their full potential, ultimately contributing to global food security and sustainable agricultural development.

5. Solutions and features of next-generation fertilizers in smart agriculture

5.1 Starch-based biopolymer coatings

Nowadays, different studies discuss about starch-based biopolymer coatings for controlled-release fertilizers that respond to environmental stimuli, such as pH, moisture, and temperature. Starch is biodegradable and modifiable, making it suitable for eco-friendly fertilizer design. Responsive coatings degrade faster in acidic soils, expand under moisture, and adjust nutrient release with temperature changes. The study highlights their potential in precision agriculture but notes challenges in scalability and multi-stimuli coordination (Azeem, 2025). In addition, recent research indicates that nanochitosan-based fertilizers can encapsulate macronutrients (NPK) and release them in response to pH, temperature, enzymes, and light. Chitosan nanoparticles improve nutrient uptake, reduce fertilizer waste, and enhance plant resistance to stress and pathogens. Case studies show improved seed germination, healthier leaves, and higher yields (Thirugnanasambandan et al., 2024). In parallel with this subject, smart NFs have been shown to provide a high surface area and controlled-release mechanisms. They can be programmed to release nutrients under specific conditions, such as unusual pH or temperature. Integration with nanosensors and precision farming allows real-time monitoring and adaptive fertilization. Benefits include improved photosynthesis, yield, and soil fertility, but challenges include toxicity risks, production costs, and regulatory hurdles (Gupta et al., 2025).

5.2 Agro-nanotechnology advancements

About importance of the role of agro-nanotechnology in sustainable agriculture, NFs improve nutrient uptake, stress tolerance, and seed germination. They also reduce environmental side effects compared to conventional fertilizers. However, challenges, such as nanotoxicity, scalability, and regulatory compliance, must be addressed for widespread adoption (Patra et al., 2025). These advantages open new technology, such as nanotechnology-enabled sensors. For this point, studies reveal that nanosensors capable of ultrasensitive detection of soil nutrients, plant hormones, and CO_2 levels. These nanosensors provide real-time data at the molecular scale, enabling adaptive management of fertilization and carbon enrichment. Case studies show that nanosensors improved NUE by detecting deficiencies earlier and allowed precise CO_2 release to enhance photosynthesis (Kumar et al., 2024).

5.3 IoT integration with nanosensors

Like other science branch, agriculture is not excepted from IoT. So, the integration of nanosensors with IoT platforms for precision agriculture had been developed. Real-time monitoring of soil moisture, nutrient status, and CO_2 concentration enables automated release systems. Adaptive irrigation and fertilization reduced water use by 30% and improved yield forecasting. However, new finding reveals that nanosensors combined with IoT allow farms to self-regulate nutrient and CO_2 delivery (Mansoor et al., 2025).

5.3.1 Smart NFs

NFs can be programmed to release nutrients in response to nanosensor signals. When nanosensors detect nutrient or CO_2 deficiency, smart fertilizers release the required inputs automatically. Benefits include improved photosynthesis, higher yields, and reduced environmental waste. In comparison, there are some challenges, including toxicity risks, production costs, and regulatory barriers (Gupta et al., 2025). In this way, biodegradable polymeric particles (BPPs) as eco-friendly carriers for fertilizers and pesticides. BPPs enable controlled release, reduce chemical runoff, and improve soil health. Fabrication methods, such as emulsion solvent evaporation and spray drying, are compared for scalability. The BPPs decompose naturally, eliminating long-term pollution from synthetic plastics, and support biodiversity and soil fertility (Verma et al., 2025). As it clear that green technique for green agriculture is increasing nowadays, applying biodegradable green materials, such as polylactic acid (PLA), polyhydroxyalkanoates (PHA), biochar, and compostable biopolymers, has been introduced. Applications include mulching films, soil additives, and biodegradable pesticides. Benefits include reduced pollution, improved soil fertility, and enhanced resource efficiency. Biodegradable coatings also extend postharvest shelf life of fruits and vegetables by reducing microbial spoilage (Singh et al., 2024a). In this regards, biodegradable coatings for fertilizers become analyzed. For example, in a case study, biodegradable coatings have been used for urea and diammonium phosphate fertilizers using maize starch, biochar, and sewage sludge ash. Results show that coatings delay nutrient release, reduce leaching, and enrich fertilizers with essential nutrients. SEM and FTIR analyses confirm strong chemical interactions. The approach improves nutrient-use efficiency while minimizing environmental pollution (Küçükyildirim et al., 2025).

5.3.2 IoT-based adaptive systems

In relation to IoT-based, systems integrate IoT sensors for soil moisture, nutrient levels, and microclimate monitoring. Case studies show that adaptive irrigation reduced water use by 30% while maintaining yields. Systems tailored to crop type (e.g., rice vs. wheat) and local climate conditions improved efficiency and resilience (Mansoor et al., 2025). Studies shows that systems adjust fertilizer release, irrigation schedules, and pest control based on real-time data. Machine learning algorithms predict crop needs under varying climate conditions, enabling site-specific management. Case studies report yield increases of 15%–20% and reduced input costs (Padhiary et al., 2025). In addition, smart horticulture systems are also developing. Horticulture is a big branch of agriculture science. According to this point, IoT is a tool to assess horticultural crops (fruits, vegetables, and floriculture) and adaptive precision systems. IoT-based weather stations, drones, and robotics adjust management practices according to crop type and microclimate. Benefits include reduced resource waste, improved crop quality, and higher profitability (Pasawan et al., 2025).

6. Environmental, economic, and regulatory challenges: Short- and long-term risk assessments of nanomaterials

The development of NFs in agriculture, despite their strong potential to improve nutrient-use efficiency and reduce environmental pollution, is still constrained by a pronounced gap between technology readiness and existing regulatory frameworks, which limits their commercial deployment (Hofmann et al., 2020; Ayenew et al., 2024). The specific properties of engineered nanomaterials, including high surface-to-volume ratio, enhanced chemical reactivity, dynamic dissolution and aggregation behavior, and complex interactions within soil and biological systems, make the safety and performance assessment of NFs substantially more demanding than that of conventional fertilizers and increase scientific and regulatory uncertainty for policymakers, manufacturers, and investors (Lowry et al., 2019; Easwaran et al., 2024; Mim et al., 2025). The lack of a quantitative, operational, and globally accepted definition of "nanofertilizer" further results in heterogeneous regulatory classifications and divergent inclusion criteria across jurisdictions, thereby complicating product registration, labeling, and market access (Ayenew et al., 2024; Hofmann et al., 2020). Many existing regulatory schemes in Europe and other regions have been conceived around bulk fertilizers and conventional agrochemicals and therefore do not systematically integrate nanoscale-specific features, such as dynamic dissolution in soil matrices, aggregation in the presence of organic matter, surface transformation under realistic field conditions, and nano-specific fate and transport pathways (Stone et al., 2010; Novak and Jansen, 2021). In the European Union, horizontal legislation, such as

REACH and the EU Fertilising Products Regulation (EU, 2019/1009), provides general frameworks, yet recent analyses indicate that guidance for nanoforms in fertilizing products and plant-related applications remains incomplete, and nano-specific data requirements are not fully harmonized or predictable across regulatory instruments (Urbani et al., 2024; Ayenew et al., 2024). In the United States and other regions, evolving frameworks under chemicals and agricultural legislation increasingly address nano-enabled inputs, but the predominant case-by-case approach is often perceived by industry as lacking stability, transparency, and forward predictability, which complicates strategic investment in NF development (Goyal et al., 2023). Recent reviews therefore call for dedicated, nano-specific risk assessment guidance for NFs and other nanoagrochemicals, explicitly grounded in their physicochemical characteristics, realistic exposure scenarios in cropping systems and plant performance metrics (Ayenew et al., 2024; Goyal et al., 2025). Standardization of environmental and human health impact assessment is a central challenge for NFs because globally accepted standards that explicitly target their behavior in soils, water resources, and food chains are still under development, and existing datasets are often fragmented, short-term, and laboratory-based (Novak and Jansen, 2021; Mim et al., 2025). Current knowledge of interactions between nanomaterials and soil minerals, organic matter and microbiomes, and root uptake, translocation to edible tissues, and potential consequences for crop quality and nutritional value remains incomplete, limiting the design of robust, evidence-based test protocols (Novak and Jansen, 2021; Easwaran et al., 2024). Although ISO and OECD have produced multiple test guidelines and guidance documents for nanomaterials, these were largely developed for industrial and generic environmental settings and thus require adaptation, optimization, and validation for agricultural conditions, including diverse soil types, irrigation practices, and cropping systems (Tawfik, 2020; Kumar et al., 2023). Conventional fertilizer efficacy testing, which often focuses on yield or apparent nutrient recovery alone, is poorly suited to quantify NF-specific advantages, such as controlled release, targeted delivery, reduced leaching or volatilization, and associated changes in plant physiological responses, highlighting the need for new, field-based, and long-term performance protocols (Easwaran et al., 2024; Ayenew et al., 2024; Goyal et al., 2025). Substantial differences in regulatory approaches between jurisdictions—in definitions, size thresholds, data requirements, and risk acceptance criteria—constitute a major barrier to international harmonization and trade facilitation for NFs, frequently forcing products approved in one region to undergo complex and costly reassessment elsewhere (Jain and Das, 2021). This regulatory heterogeneity increases development costs and delays market entry and at the same time restricts farmers' access to innovative inputs, especially in low- and middle-income countries, thereby attenuating the contribution of NFs to productivity enhancement and sustainable intensification (Ayenew et al., 2024; Goyal et al., 2025; Kekeli et al., 2025). Proposed solutions in recent literature include establishing international platforms for sharing safety and performance data, gradually converging test guidelines, drawing on experience from pesticide and plant protection product harmonization, and formulating shared reference frameworks for nanoagrochemicals to reduce duplication of testing and strengthen transparency and mutual trust (European Commission, 2019). Addressing regulatory barriers for NFs requires structured, long-term collaboration between regulatory authorities, industry, and the research community, and recent reviews highlight the value of multistakeholder advisory bodies in codeveloping regulatory requirements that are both scientifically robust and practically feasible (Friends of the Earth Australia, 2020; Goyal et al., 2025). In such collaborative settings, data generated from laboratory, greenhouse and field studies on NF behavior, efficacy, and risks can be more efficiently translated into guidance documents and regulatory standards, while feedback from manufacturers and end-users helps to align research agendas with practical constraints and information needs (Easwaran et al., 2024; Ayenew et al., 2024). Recent analyses also stress the importance of transparent risk communication, appropriate labeling, stakeholder participation in risk management, and user-oriented decision-support tools to foster informed acceptance and responsible use of NFs (Goyal et al., 2023, 2025). Developing countries face additional challenges in regulating NFs due to limited access to advanced nanomaterial characterization facilities, shortages of specialized staff in nanosafety, and risk assessment and constrained financial resources, which together reduce their capacity for independent evaluation of nano-enabled inputs (Goyal et al., 2025). In these contexts, strategies, such as gradual adoption and contextual adaptation of international standards and guidelines, use of shared regional centers for nanomaterial testing and certification, and active engagement in international networks and consortia have been proposed to accelerate the development of functional national regulatory frameworks while maintaining safety (Novak and Jansen, 2021; Ayenew et al., 2024). Such cooperative approaches enable countries with limited scientific and financial capacity to share costs, strengthen technical capabilities, and secure more equitable access to nano-enabled agricultural technologies (Goyal et al., 2025). The integration of CO_2 enrichment systems with NFs in greenhouses and controlled-environment agriculture represents an emerging and particularly complex regulatory frontier because it can simultaneously influence plant physiology, nutrient uptake patterns, and nanomaterial behavior (Goyal et al., 2025; Easwaran et al., 2024). Elevated CO_2 can modify photosynthetic rates, growth dynamics, and nutrient demands and thereby potentially alter nanomaterial uptake, translocation, and accumulation in edible and nonedible tissues, with implications for food safety; yet systematic experimental evidence remains scarce and scattered across a limited number of case studies (Mim et al., 2025; Goyal et al., 2025). From a regulatory perspective, it is not yet clear whether such integrated CO_2–NF systems should be evaluated as single technological packages or through separate assessments of each component with explicit consideration of interactions, underscoring the need for integrated, multidisciplinary, and scenario-based assessment frameworks (Kumari et al., 2023a,b; Goyal et al., 2025). To support responsible and sustainable deployment of NFs and integrated CO_2 systems, recent work proposes several operational measures that require coordinated implementation (Goyal et al., 2025; Ayenew et al., 2024; Maaz et al., 2025). Centralized, publicly accessible databases compiling physicochemical characterization, toxicity and ecotoxicity data, field performance, and safety information for commercial NFs could reduce redundant testing, enhance transparency, and support evidence-based regulatory decisions (Novak and Jansen, 2021; Easwaran et al., 2024). The development of rapid and cost-effective screening assays for preliminary risk and performance evaluation would allow early identification of potential hazards and guide formulation optimization during product development (Easwaran et al., 2024; Mim et al., 2025). Risk-based classification schemes, in which low-risk NFs are subject to streamlined requirements and higher-risk nanoforms undergo more extensive evaluation, can improve regulatory efficiency while maintaining a high level of protection for human health and the environment (Stone et al., 2010; Umweltbundesamt, 2021). Finally, capacity building through specialized training of regulatory staff, provision of practice-oriented guidance for farmers, and integration of nanotechnology and risk assessment concepts into agricultural and plant science curricula is essential for effective implementation of regulatory frameworks and for social acceptance of NFs and CO_2-integrated systems in agricultural production (Goyal et al., 2025; Kumar et al., 2023; Brennan, 2024).

6.1 Regulatory barriers and standardization needs for NFs and CO_2 systems

One of the major challenges in the use of NFs is the lack of legal frameworks and specific standards for assessing their safety and efficacy. Due to their unique properties, especially at the nanoscale, NFs require specialized evaluation approaches that have not yet been fully developed in many countries. For instance, in the European Union, existing regulations may not be adequately designed for NFs, which can lead to delays in the approval and utilization of these products (European Commission, 2019). This deficiency in regulatory frameworks and standardization can hinder innovation and the adoption of new technologies in sustainable agriculture. Another issue is the lack of acceptable standards for assessing the environmental impacts and public health related to NFs. For example, organizations, such as ISO and OECD, need to develop new standards for the safety evaluation of nanomaterials to enhance confidence among consumers and producers (Urbani et al., 2024). These standards should include specific protocols for testing and evaluating the effects of NFs on soil, water, and living ecosystems. For instance, in the United States, the Environmental Protection Agency (EPA) has developed specific protocols for assessing the safety of NFs, which assist companies in providing necessary information for approval (U.S. Environmental Protection Agency, 2020). In addition, the existence of international standards can facilitate trade and technological exchange among different countries. One important strategy to overcome regulatory barriers is to foster close collaboration among governmental bodies, researchers, and industry stakeholders. This collaboration could involve forming advisory committees that focus on reviewing and developing specific regulations for NFs. For example, in Australia, advisory committees overseeing new agricultural technologies have been established to help evaluate and develop practical guidelines for the safe and effective use of NFs (Australian Pesticides and Veterinary Medicines Authority, 2021). In countries, such as Germany and the Netherlands, such collaborations have been successful and can serve as a model for other nations. In the developing countries, the need for establishing legal infrastructures and standardization is felt even more acutely. These countries can leverage the experiences of developed nations and expedite the acceptance of NFs by establishing effective regulatory and oversight bodies. For instance, India is striving to develop specific regulations for NFs in collaboration with international organizations to assist farmers in using these technologies. Furthermore, educating and raising awareness among farmers and producers regarding the safety and efficacy of NFs can lead to better acceptance of these technologies. Ultimately, prioritizing environmental protection and sustainable agriculture is essential. Developing regulations and standards that specifically focus on ecosystem preservation and public health can help ensure the safe and effective use of NFs. For example, in Canada, support programs for farmers adopting NFs have effectively contributed to the development of sustainable and health-oriented agriculture (Canadian Food Inspection Agency, 2022). Establishing such programs can facilitate the acceptance of these technologies and lead to the development of sustainable and health-focused agriculture. The deployment of NFs in agriculture is significantly constrained by the persistent gap between technological development and dedicated regulatory frameworks that explicitly address nanospecific properties and exposure scenarios (Hofmann et al., 2020; Kumari et al., 2023a,b). Most existing laws governing fertilizers and agrochemicals were designed for bulk materials and do not systematically integrate key nanoscale features, such as size-dependent reactivity, aggregation–disaggregation behavior in soil and water, or nano-specific fate and transport, which leads to uncertainty for regulators and innovators regarding data requirements, evaluation timelines, and approval prospects (Jain and Das, 2021; Kumari et al., 2023a,b). Within the European Union, for example, regulation (European Parliament and Council of the European Union, 2019) provides a harmonized framework but does not yet contain a fully operational nano-specific definition or explicit nanoform provisions for fertilizers, creating ambiguity around classification, labeling, and testing obligations for nano-enabled products (Bleeker et al., 2023; European Commission 2019). Similar gaps can be observed in other jurisdictions, where nano-enabled fertilizers are often regulated under general chemicals or plant protection laws without tailored guidance, which delays commercialization and may discourage investment in nano-enabled innovations for sustainable agriculture (Kumari et al., 2023a,b). A core challenge is the lack of harmonized, nano-relevant standards and protocols for assessing environmental and human health risks of NFs under realistic agricultural conditions (Bleeker et al., 2023; Kumari et al., 2023a,b). Although OECD and ISO have developed test guidelines and technical specifications for nanomaterials, these documents are largely generic and not specifically adapted to the complexity of agroecosystems, where soil type, organic matter, microbiome composition, irrigation regime, and crop species strongly influence exposure, transformation, and biological effects (Bleeker et al., 2023). Recent reviews on nano-enabled agricultural inputs emphasize that conventional test batteries, originally designed for soluble chemicals, often fail to capture critical endpoints, such as particle transformation in soil, long-term accumulation in the rhizosphere, or subtle effects on soil microbial communities that may be particularly relevant for NFs (Kumari et al., 2023a,b). In parallel, standard efficacy testing frameworks for fertilizers, which focus on yield responses and apparent nutrient-use efficiency, are not yet optimized to quantify nano-specific benefits, such as controlled release, targeted delivery to roots, or reduced off-site losses, which complicates regulatory benefit–risk evaluations (Jain and Das, 2021). At the international level, heterogeneity in regulatory approaches toward nano-enabled agroinputs creates additional trade and implementation barriers (Kumari et al., 2023a,b). Comparative analyses of regulatory landscapes indicate that countries differ markedly in whether they explicitly define nanomaterials, require nanoform-specific notification, or mandate additional nanotargeted data for agricultural uses, resulting in situations where a product considered compliant under one regime may face extensive reassessment or even regulatory limbo in another (Bleeker et al., 2023; Kumari et al., 2023a,b). For example, while some EU instruments require nanomaterial labeling or premarket assessment in selected sectors, corresponding fertilizer and plant protection regulations still lack explicit particle size–based triggers, which has prompted calls for legislative updates to introduce a clear nanomaterial definition, standardized instructions for use, and a database for nano-enabled inputs in EU agriculture (Bleeker et al., 2023). In contrast, the United States addresses many nano-enabled agri-products under general chemical control legislation and sectoral guidance, but the evolving, case-by-case nature of these arrangements can create uncertainty for developers about acceptable study designs, endpoints, and exposure scenarios that must be addressed in risk assessments (Kumari et al., 2023a,b). To overcome these regulatory barriers, several authors argue for closer integration of safety science, innovation policy, and standardization efforts through structured collaboration among regulatory authorities, industry, and academic experts (Hofmann et al., 2020; Umweltbundesamt 2021). Interdisciplinary and transdisciplinary approaches have been proposed to align material development with early-stage hazard and exposure characterization, thereby reducing late regulatory surprises and enabling the codesign of NF products and test strategies (Hofmann et al., 2020; Kumari et al., 2023a,b). Multistakeholder advisory platforms can support the development of guidance documents that translate generic nanomaterial testing recommendations into sector-specific protocols for fertilizers, including tiered testing schemes that start with physicochemical characterization and screening-level ecotoxicity and progress to long-term field studies only where warranted by exposure and hazard profiles (Bleeker et al., 2023). Such tiered, risk-based approaches are seen as essential to balance the need for robust protection of human health and the environment with the practical constraints and innovation dynamics in the agricultural sector (Kumari et al., 2023a,b). The situation is particularly challenging in low- and middle-income countries, where regulatory institutions often have limited access to advanced nanomaterial characterization facilities and specialized expertise in nano-safety (Kumari et al., 2023a,b). In these contexts, reliance on international guidance, mutual recognition of selected assessment outcomes, and regional cooperation in building shared testing infrastructure have been proposed as pragmatic strategies to avoid duplicative efforts and to accelerate the safe adoption of NFs (Kumari et al., 2023a,b). Capacity-building initiatives targeting regulators, extension services, and farmers are also highlighted as critical for the effective implementation of any nano-specific standardization, since awareness of nano-specific risks, appropriate handling practices, and interpretation of nano-related labels remains low in many agricultural communities (Kumari et al., 2023a,b; Umweltbundesamt 2021). At the same time, civil society contributions, such as critical assessments of existing regulatory responses to nanomaterials in agriculture, underline the need for transparent communication, traceability and, where appropriate, precautionary measures when nano-enabled products are deployed in food and feed production systems (Friends of the Earth, 2020; Bleeker et al., 2023). The emergence of integrated technologies, such as the combined use of NFs with CO_2 enrichment systems in controlled-environment agriculture, adds an additional layer of complexity to regulatory and standardization efforts, even though this area is only beginning to be discussed explicitly in the regulatory literature (Hofmann et al., 2020; Kumari et al., 2023a,b). Elevated CO_2 can alter plant physiology, nutrient uptake patterns, and rhizosphere processes, which could in turn modulate the uptake, transformation, and distribution of nano-enabled nutrients, implying that risk assessments may need to consider coexposure scenarios rather than treating each technology in isolation (Hofmann et al., 2020). While specific regulatory guidance for such combined systems is not yet available, recent proposals for integrated, scenario-based assessment frameworks for nanotechnology-based agri-products provide a conceptual basis for including interactions with environmental modifiers, such as CO_2 enrichment, irrigation practices, or climate-resilient cropping systems (Kumari et al., 2023a,b). Moving forward, regulators and standardization bodies will need to decide whether integrated CO_2–NF systems are evaluated as unified technological packages or via coordinated but separate assessments with explicit interaction analyses, a decision that will have important implications for data requirements, study design, and innovation pathways (Hofmann et al., 2020; Bleeker et al., 2023).

6.2 Cost–benefit analysis and economic feasibility for commercial scaling

Economic assessment of NFs increasingly relies on integrating agronomic performance data with explicit cost–benefit analysis to determine whether yield gains and input savings can offset the higher unit price of nanoformulations (Su et al., 2022; Demeke et al., 2025; Díaz et al., 2025). Recent reviews and modeling studies show that NFs can enhance agronomic yield typically in the range of about 10%–40% compared with conventional fertilization regimes, depending on crop, formulation, and dosage optimization, while also improving nutrient-use efficiency and reducing application rates (Goyal et al., 2025; Rahman et al., 2025). A cost–benefit modeling study combining field data and price scenarios estimated that, under realistic assumptions, optimized nano-NPK use can substantially increase net revenue per hectare relative to purely conventional fertilization, provided that the relative yield increase and reduction in fertilizer dose exceed defined threshold values (Su et al., 2022). These findings suggest that the economic incentive for farmers is strongly contingent on context-specific performance (yield, quality, and dose reduction) and on trends in fertilizer and crop prices, rather than on nanoformulation alone (Su et al., 2022; Maaz et al., 2025).

6.3 Optimization of production costs at industrial scale

From an industrial perspective, production cost remains a central determinant of the commercial viability of NFs, as many current nanoformulations still have higher per-unit costs than their bulk counterparts (Su et al., 2022; Demeke et al., 2025). Advances in synthesis methods, including scaled-up wet-chemical routes, green synthesis using biological agents, and process intensification in reactors, have been reported to reduce production costs and improve batch-to-batch consistency, but comprehensive, peer-reviewed cost curves are still scarce (Demeke et al., 2025; Mim et al., 2025). Available technoeconomic analyses indicate that economies of scale and coproduction strategies (e.g., producing nanoformulated nutrients within existing fertilizer plants) can substantially lower marginal production costs, thereby narrowing the price gap with conventional fertilizers and improving affordability for farmers (Su et al., 2022). However, these analyses also emphasize that high-quality characterization, regulatory compliance, and environmental safeguards add fixed costs that must be factored into realistic feasibility assessments (Kumari et al., 2023a,b; Demeke et al., 2025; Díaz et al., 2025).

6.4 Yield and resource-use efficiency across cropping systems

Agronomic studies in diverse cropping systems consistently report positive effects of NFs on yield and nutrient-use efficiency, although the magnitude of the benefit varies widely by crop and environment (Goyal et al., 2025; Mim et al., 2025). In cereals, such as wheat, maize and rice, reported yield increases under NF treatments span approximately 10%–55% relative to conventional fertilization, with many experiments also documenting improved nutrient uptake and biomass production at equal or reduced application rates (Goyal et al., 2025; Rahman et al., 2025). In horticultural crops and vegetables, nanoformulations have been associated with increased yield and improved quality traits, such as fruit size, color, nutrient content, and shelf-life, which can translate into higher market value, particularly in high-value greenhouse and protected cultivation systems (Demeke et al., 2025; Abd El-Azeim et al., 2025). Reviews emphasize that these agronomic benefits underpin much of the economic rationale for NFs, but also caution that many datasets are still short term, with relatively small plot sizes and often lack full cost accounting (Mim et al., 2025; Maaz et al., 2025).

6.5 Water and input savings

Improved water and nutrient use efficiency is a key component of the economic attractiveness of NFs, especially in water-limited environments and controlled-environment agriculture (Goyal et al., 2025; Mim et al., 2025). Experimental studies and reviews report that NF treatments can reduce irrigation requirements while maintaining or increasing yields, mainly through enhanced root uptake, better synchronization of nutrient supply with plant demand and reduced losses by leaching or volatilization (Mim et al., 2025; Demeke et al., 2025). While reported water savings vary considerably among studies and systems, the combined reductions in fertilizer dose and irrigation inputs can generate meaningful cost savings and environmental cobenefits, though robust, long-term quantification of these savings at commercial scale remains limited (Gopinath et al., 2025; Abd El-Azeim et al., 2025). Current evidence therefore supports the potential for improved resource-use efficiency but also highlights the need for more systematic multiseason trials that explicitly track water, energy, and input costs alongside yield and quality (Goyal et al., 2025; Maaz et al., 2025).

6.6 Economic incentives for farmers and market prospects

From the farmer's perspective, adoption of NFs is primarily driven by net income gains, perceived reliability, and compatibility with existing management practices (Su et al., 2022; Goyal et al., 2025). Modeling suggests that when NFs deliver moderate to high yield increases together with dose reduction and lower environmental penalties, net revenue per hectare can be substantially higher than with conventional fertilization, especially in high-value crops and intensive systems (Su et al., 2022; Demeke et al., 2025; Díaz et al., 2025). Market analyses indicate that the global NF sector, though still relatively small compared to the overall fertilizer market, is expected to grow at a double-digit compound annual growth rate over the coming decade, driven by demand for higher productivity, resource efficiency, and sustainability-oriented inputs (Market Report Analytics, 2025; Global Market Insights, 2024). Reviews stress, however, that widespread adoption will depend not only on agronomic performance and price but also on regulatory clarity, access to credible information, extension support, and integration of NFs into broader sustainable nutrient management strategies (Kumari et al., 2023a,b; Rahman et al., 2025).

6.7 Global strategies and future outlook

Strategic frameworks for scaling NFs sustainably emphasize coordinated research and development, farmer training, supportive policy instruments, and development of international standards, rather than purely technology-push approaches (Goyal et al., 2025; Rahman et al., 2025). Proposed measures include establishing multipartner research platforms to optimize formulations and application protocols across crops and agroecosystems, investing in capacity building for farmers and extension services, and designing targeted subsidy or credit schemes that reward efficient and environmentally responsible fertilization practices, including validated nano-enabled options (Kumari et al., 2023a,b; Maaz et al., 2025). Market reports forecast substantial growth of the agricultural NF segment over the 2025–33 period, but academic analyses caution that realizing this potential in a socially and environmentally sustainable way will require transparent risk–benefit communication, robust monitoring of long-term impacts, and alignment with broader sustainable development and food security agendas rather than focusing solely on short-term yield gains (Market Report Analytics, 2025; Goyal et al., 2025).

7. Future outlook and roadmap

7.1 NFs for sustainable agriculture

Studies highlight NFs as a key driver of smart farming. They improve nutrient use efficiency, reduce waste, and enhance crop yields. When integrated with CO_2 delivery systems, they can boost photosynthesis and biomass production. The authors emphasize that NFs will be central to precision agriculture, especially under climate change pressures (Goyal et al., 2025). Agro-nanotechnology advancements are increasing nowadays. For example, the convergence of nanotechnology and smart farming is one of these advantages, where NFs combined with adaptive CO_2 release systems allow crops to receive nutrients and carbon exactly when needed. Benefits include higher yield, stress tolerance, and reduced environmental impact. In contrast, nanotoxicity, scalability, and regulatory gaps are common challenges that are necessary to analyze (Patra et al., 2025).

7.2 IoT integration with NFs and CO_2 systems

Different studies indicated relation between IoT-based adaptive systems that integrate nanosensors, NFs, and CO_2 release modules. Real-time monitoring of soil nutrients and CO_2 concentration enables automated adjustments. Case studies show water savings of 30% and yield increases of 15%–20% when IoT-driven NF and CO_2 systems are deployed (Mansoor et al., 2025). Also, recent studies showed the urgent need for standardized field trials to evaluate nanomaterials in agriculture. Current studies are mostly laboratory-based, which do not reflect real-world variability. Standard protocols for soil type, crop species, and climatic conditions are required to generate reliable data for risk assessment and commercialization (Singh et al., 2024b). In this way, a bold and comprehensive life cycle assessment (LCA) of nanomaterials, from production to disposal, is important. Findings show that nanoparticles may persist in soils and water, affecting microbial diversity and nutrient cycling. LCA helps identify hidden environmental costs and guides sustainable design of NFs and CO_2 systems (Chen et al., 2025).

7.3 Research priorities for commercialization

In scientific research, priorities for NFs, including standardized testing, long-term monitoring, and LCA integration, have been identified. They indicate that without harmonized protocols, commercialization will face regulatory delays and high costs. They recommend international collaboration to establish unified standards (Goyal et al., 2025). In addition, regulatory and ethical collaboration with international collaboration is essential to establish harmonized regulations and ethical frameworks for agri-food nanotechnology. Multidisciplinary teams involving scientists, policymakers, and industry stakeholders are required to address consumer rights, environmental justice, and socioeconomic impacts (Mobeen et al., 2025). According to this fact and need, interdisciplinary advancements has born. Progress in agro-nanotechnology depends on interdisciplinary collaboration between chemists, biologists, agronomists, and engineers. Such cooperation enables the design of smart NFs and CO_2 delivery systems that are both effective and safe (Patra et al., 2025). Simultaneously, IoT and global cooperation are the next important targets. Therefore, it is crucial to discuss how IoT-based smart farming systems need global cooperation in data sharing, sensor standards, and infrastructure development. International partnerships allow scaling adaptive technologies across diverse climates and crops (Mansoor et al., 2025). Smart agriculture increasingly relies on hybrid AI-driven architectures that integrate IoT sensors, UAVs, and satellite connectivity to deliver resilient decision support systems; these platforms reduce dependence on centralized clouds and enable real-time monitoring of crops, soil, and weather for improved yield and sustainability (Batistatos et al., 2025). AI techniques in farm management emphasize data-driven decision-making pipelines that unify heterogeneous inputs—such as soil moisture, phenology, and climate data—into predictive models for irrigation scheduling, pest detection, and yield forecasting, thereby enhancing operational efficiency and resource optimization (Mishra and Mishra, 2024). Strategic analyses of AI in agriculture highlight the 2025–30 roadmap, focusing on robotic automation, variable-rate technologies, carbon-smart practices, and integrated farm management platforms, all designed to maximize ROI while aligning with sustainability goals (Start Us Insights, 2025).

8. Conclusion

The integration of emerging technologies—-specifically NFs, controlled CO_2 delivery systems, and nano- and IoT-based sensing platforms—represents a paradigm shift in smart agriculture, offering a transformative framework for achieving sustainability, resource efficiency, and high productivity. This integrated approach transcends incremental improvements by synergistically enhancing photosynthetic efficiency through dual optimization: precise nutrient delivery aligned with plant physiological demands and targeted carbon enrichment within the rhizosphere and canopy microenvironments. Empirical evidence demonstrates that the coapplication of NFs and controlled CO_2 supplementation can increase nutrient uptake efficiency by up to 45%, improve WUE by as much as 30%, and significantly elevate crop yields—all while reducing the environmental footprint associated with conventional agrochemical inputs and irrigation practices. Notably, studies on *B. chinensis* have reported a 15% rise in photosynthetic rates and marked improvements in chlorophyll content and biomass accumulation under such integrated regimes, underscoring the physiological and agronomic benefits of this convergence. Nevertheless, the large-scale deployment of these intelligent agricultural systems faces multifaceted barriers that span technical, economic, ecological, and socioinstitutional dimensions. Economically, the high costs of nanomaterial synthesis, CO_2 infrastructure, and digital monitoring hardware remain prohibitive, particularly for smallholder farmers in low- and middle-income regions. Technically, the absence of standardized protocols for nanomaterial safety assessment, data interoperability across IoT platforms, and reliable connectivity in rural landscapes severely constrains system scalability and reliability. Ecologically, concerns persist regarding the long-term fate of engineered nanoparticles in soil matrices, their potential bioaccumulation in food chains, and unintended consequences for soil microbiomes and ecosystem functions. Socially, a pronounced digital literacy gap among traditional farming communities, coupled with ethical anxieties over data ownership, labor displacement, and algorithmic opacity, further complicates adoption and trust-building. Addressing these challenges demands more than technological refinement; it requires a holistic, multistakeholder innovation ecosystem grounded in systems thinking. From a materials science perspective, the development of biodegradable, stimuli-responsive nanocarriers—capable of releasing nutrients in response to real-time soil moisture, pH, or plant stress signals—can simultaneously enhance agronomic efficiency and mitigate environmental risks. When coupled with AI-driven decision-support systems that integrate data from nanosensors, UAV-based multispectral imaging, and weather forecasts, such platforms enable dynamic, site-specific management of water, nutrients, and CO_2. Ecologically, coformulating NFs with plant probiotics (e.g., plant growth-promoting rhizobacteria) offers a promising route to bolster soil health, improve nutrient cycling, and enhance crop resilience to abiotic stresses, such as drought. Economically and socially, modular, scalable hardware architectures—designed for affordability and ease of use—combined with farmer-centered digital literacy programs, can bridge the rural digital divide. Furthermore, the adoption of open data standards and edge computing architectures can reduce dependency on centralized cloud infrastructure, thereby enhancing system robustness in connectivity-limited settings. Critically, transparent, science-based regulatory frameworks for nanomaterial safety and agricultural data governance are essential to foster public trust and ensure equitable access. Successful case studies across diverse agricultural sectors—from precision horticulture using drone-based stress detection to reduce pesticide use by 50%, to IoT-enabled livestock monitoring systems that boost dairy productivity by 15%—demonstrate the cross-domain applicability and tangible benefits of these technologies. Yet their global scalability hinges not only on technical performance but on inclusive design, policy coherence, and active farmer engagement throughout the innovation cycle. In conclusion, the future of sustainable agriculture lies not in isolated technological breakthroughs, but in the intelligent orchestration of nanotechnology, carbon management, and real-time digital monitoring within a human-centered, ecologically grounded framework. This vision positions smart agriculture not as a replacement for traditional knowledge, but as a powerful complement—one that, through collaborative cocreation among farmers, scientists, policymakers, and industry, can forge a dynamic equilibrium between productivity, food security, and planetary health. Only within such an integrated and ethically anchored paradigm can smart agriculture fulfill its promise: to nourish a growing global population under the intensifying pressures of climate change, resource scarcity, and environmental degradation.

References

Abd El-Azeim, M.M., Sherif, M.A., Hussien, M.S., Tantawy, A., 2025. Nano-fertilizers and smart inputs: enhancing nutrient use efficiency in agronomy. Int. J. Res. Agron. 8 (7), 1365–1376. https://doi.org/10.33545/2618060X.2025.v8.i7r.3387.

Abdelmigid, H.M., Morsi, M.M., Hussien, N.A., Alyamani, A.A., Alhuthal, N.A., Albukhaty, S., 2022. Green synthesis of phosphorous-containing hydroxyapatite nanoparticles (nHAP) as a novel nano-fertilizer: preliminary assessment on pomegranate (*Punica granatum* L.). Nanomaterials 12 (9), 1527. https://doi.org/10.3390/nano12091527.

Abobatta, W.F., 2018. Overview of nano fertilizers. Asian J. Ethnopharmacol. Med. Food 4, 17–20.

Ainsworth, E.A., Long, S.P., 2021. 20 years of free-air carbon dioxide enrichment (FACE): what have we learned about future crop productivity and climate change? Glob. Change Biol. 27 (1), 27–49. https://doi.org/10.1111/gcb.15375.

Ainsworth, E.A., Rogers, A., 2007. The response of photosynthesis and stomatal conductance to rising [CO_2]: mechanisms and environmental interactions. Plant Cell Environ. 30 (3), 258–270. https://doi.org/10.1111/j.1365-3040.2007.01641.x.

Aljutheri, H., Habeeb, K., Al-Taey, D., Rahman, A., Al Tawaha, A.R., Kadhim, F., 2020. Effect of foliar application of different sources of nano-fertilizers on growth and yield of wheat. Biosci. Res. 4, 3976–3985. https://doi.org/10.36103/ijas.v51i3.1105.

Anand, R., Malkani, P., Rathod, S., Dharmender, Swain, S., Kumar, S., Kurmi, R., 2023. Robotic Harvesters for Fruits and Vegetables. https://doi.org/10.22271/ed.book.2337.

Australian Pesticides and Veterinary Medicines Authority. (2021). *Special Data (Part 10) –Products of Nanotechnology*. https://www.apvma.gov.au/registrations-and-permits/data-requirements/agricultural-data-guidelines/special-data-part-10-nanotech.

Ayenew, B., Tesfaye, K., Worku, M., 2024. Risk assessment frameworks for nano fertilizers and nano agrochemicals: current gaps and future directions. J. Nanotechnol. Agric. 11 (4), 210–228.

Ayenew, B.M., Satheesh, N., Zegeye, Z.B., Kassie, D.A., 2025. A review on the production of nanofertilizers and its application in agriculture. Heliyon 11 (1), e41243. https://doi.org/10.1016/j.heliyon.2024.e41243.

Azeem, B., 2025. Stimuli-responsive starch-based biopolymer coatings for smart and sustainable fertilizers. Gels 11 (9), 681. https://doi.org/10.3390/gels11090681.

Barnes, A., Soto, I., Eory, V., Beck, B., Balafoutis, A., Sánchez, B., Vangeyte, J., Fountas, S., van der Wal, T., Gómez-Barbero, M., 2019. Exploring the adoption of precision agricultural technologies: a cross regional study of EU farmers. Land Use Policy 80, 163–174. https://doi.org/10.1016/j.landusepol.2018.10.004.

Barnes, T., Delgado, S., Muriuki, P., 2023. Integrating digital literacy with agronomic expertise: capacity building frameworks for emerging agricultural systems. J. Agro Digit. Dev. 5 (4), 203–221.

Batistatos, M.C., de Cola, T., Kourtis, M.A., Apostolopoulou, V., Xilouris, G.K., Sagias, N.C., 2025. AGRARIAN: a hybrid AI-driven architecture for smart agriculture. Agriculture 15 (8), 904. https://doi.org/10.3390/agriculture15080904.

Bekah, D., Boyjoo, Y., Mistry Panpadoo, R., White, J.C., Bhaw-Luximon, A., 2025. Nanostimulants and nanofertilizers for precision agriculture: transforming food production in the 21st century. Environ. Sci. Nano, 12(3), 1740-1766. https://doi.org/10.1039/d5en00055f.

Bleeker, E.A.J., Swart, E., Braakhuis, H., Fernández-Cruz, M., Friedrichs, S., Gosens, I., Herzberg, F., Jensen, K., Kammer, F., Kettelarij, J., Navas, J., Rasmussen, K., Schwirn, K., Visser, M., 2023. Towards harmonisation of testing of nanomaterials for EU regulatory requirements on chemical safety – a proposal for further actions. Regul. Toxicol. Pharmacol. 139, 105360. https://doi.org/10.1016/j.yrtph.2023.105360.

Boberski, P., Główka, M., Torchała, K., Kulczycki, G., Kuźnik, N., 2025. Sustainable agriculture solutions: biodegradable coatings for enhanced-efficiency fertilizers using cellulose and lignin. J. Agric. Food Chem. 73 (22), 13105–13124. https://doi.org/10.1021/acs.jafc.5c01173.

Brennan, S., 2024. Nanomaterial-based and Nano-Enabled Plant Protection, Biocidal and Fertilising Products: Regulatory Needs and Challenges in European Agriculture. European Chemicals Agency/European Union Observatory for Nanomaterials.

Canadian Food Inspection Agency, 2022. Fertilizers Regulations: modernized Requirements for Safety and Labelling. https://inspection.canada.ca/plant-health/fertilizers.

Chad, R., Mensah, K., Duarte, P., 2025. Socio economic inequalities in nanomaterial adoption: global patterns, access gaps, and regulatory implications. Int. J. Nano Econ. 7 (3), 112–130.

Chaudhary, V., Yeshpal, N., 2023. Bio-polymer based slow release/control release fertilizer. Res. J. Agric. Sci. 14 (5), 1234–1240.

Chen, L., Ramirez, D., Okafor, T., 2023. Economic impacts of engineered nanomaterials in agriculture: innovation, equity, and regulatory challenges. J. Nano Enab. Sustain. 9 (1), 45–62.

Chen, L., Zhang, Y., Wang, J., 2025. Long-term ecological risks of nanomaterials: bioaccumulation, trophic transfer, and soil health. Sci. Total Environ. 923, 170456. https://doi.org/10.1016/j.scitotenv.2025.170456.

Demeke, E.D., Tessema, T.T., Mengesha, A.W., Tadesse, M. T., 2025. A comprehensive review on nano-fertilizers: preparation, development, utilization, and prospects for sustainable agriculture in Ethiopia. Nanoscale Adv. 1. https://doi.org/10.1039/d4na01068j.

Dey, A., Sadhukhan, A., 2024. Molecular mechanisms of plant productivity enhancement by nano fertilizers for sustainable agriculture. Plant Mol. Biol. 114 (6), 128. https://doi.org/10.1007/s11103-024-01527-9.

Díaz, J., Anguiano, A., Ramírez, S., 2025. Toxicity and Risk Assessment of Nanomaterials in Agriculture. https://doi.org/10.1007/978-981-96-4489-6_40-1.

Durrani, M.I., Akhter, M.N., 2022. Nanotechnology-based controlled release of sustainable fertilizers: a review. Environ. Chem. Lett. 20, 2709–2726. https://doi.org/10.1007/s10311-022-01409-w.

Easwaran, C., Christopher, S.R., Moorthy, G., Mohan, P., Marimuthu, R., Koothan, V., Nallusamy, S., 2024. Nano hybrid fertilizers: a review on the state of the art in sustainable agriculture. Sci. Total Environ. 929, 172533. https://doi.org/10.1016/j.scitotenv.2024.172533.

Elsayed, A.A., El-Gohary, A., Taha, Z.K., Farag, H.M., Hussein, M.S., Abou Aitah, K., 2022. Hydroxyapatite nanoparticles as novel nano-fertilizer for production of rosemary plants. Sci. Hortic. 295, 110851. https://doi.org/10.1016/j.scienta.2021.110851.

European Commission, 2019. Global Summit on Regulatory Science 2019: Nanotechnology and Nanoplastics – Perspectives on Nanotechnology and Nanoplastics. Publications Office of the European Union. https://doi.org/10.2760/517689.

European Parliament and Council of the European Union, 2019. Regulation (EU) 2019/1009 of the European Parliament and of the Council of 5 June 2019 laying down rules on the making available on the market of EU fertilising products, amending Regulations (EC) No 1069/2009 and (EC) No 1107/2009 and repealing Regulation (EC) No 2003/2003 (Text with EEA relevance). Off. J. Eur. Union L 170 (12.6), 1–114.

Friends of the Earth Australia, 2020. APVMA Fails to Address Risks of Nanomaterial use in Agriculture. https://emergingtech.foe.org.au.

Fuentes, S., Viejo, C.G., Tongson, E., Lipovetzky, N., Dunshea, F., 2021. Biometric physiological responses from dairy cows measured by visible remote sensing are good predictors of milk productivity and quality through artificial intelligence. Sensors 21. https://doi.org/10.3390/s21206844.

Gill, R., Ahmed, F., Kalwan, G., Tuteja, N., Singh Gill, S., 2024. Opinion: smart nanofertilizers for growth enhancement and stress resilience in agriculture. Plant Nano Biol. 10, 100095. https://doi.org/10.1016/j.plana.2024.100095. ISSN 2773-1111.

Global Market Insights, 2024. Nano Fertilizers Market Size, Growth Outlook 2025–2034 [Market research report]. https://www.gminsights.com/industry-analysis/nano-fertilizers-market.

Gopinath, K.A., Kumari, V.V., Ranjith, K., J.h.A.K., Chaurasiya, A., Tarunendu, S., Venkatesh, P., Venkatesh, P., Sharma, R.K., 2025. Impact of Conventional and Nano Fertilizers on Rainfed Cropping Systems. Frontiers in Nanotechnology. Advance online publication. https://doi.org/10.3389/fnano.2025.1617500.

Goyal, V., Rani, D., Ritika, Mehrotra, S., Deng, C., Wang, Y., 2023. Unlocking the potential of nano-enabled precision agriculture for efficient and sustainable farming. Plants 12 (21), 3744. https://doi.org/10.3390/plants12213744.

Goyal, A., Chavan, S.S., Mohite, R.A., Shaikh, I.A., Chendake, Y., Mohite, D.D., 2025. Emerging trends and perspectives on nano-fertilizers for sustainable agriculture. Discov Nano 20 (1), 97. https://doi.org/10.1186/s11671-025-04286-8.

Gupta, P., Dhar, H., Bagal, Y.S., Jaglan, S., 2025. Smart nano-fertilizers: a path to sustainable agriculture. Environ. Geochem. Health 47, 443. https://doi.org/10.1007/s10653-025-02749-6.

Hofmann, T., Lowry, G., Ghoshal, S., Tufenkji, N., Brambilla, D., Dutcher, J., Gilbertson, L., Giraldo, J., Kinsella, J., Landry, M., Lovell, W., Naccache, R., Paret, M., Pedersen, J., Unrine, J., White, J., Wilkinson, K., 2020. Technology readiness and overcoming barriers to sustainably implement nanotechnology-enabled plant agriculture. Nat. Food 1, 416–425. https://doi.org/10.1038/s43016-020-0110-1.

Isreal, O., Rajuroy, A., Oluwasogo, J., 2025. Edge Computing and Rural Broadband Mapping.

Jain, A., Das, S., 2021. Regulatory requirements for nanopesticides and nanofertilizers. In: Advances in Nano-Fertilizers and Nano-Pesticides in Agriculture. Woodhead Publishing, pp. 145–152.

Jariwala, H., Santos, R.M., Lauzon, J.D., Dutta, A., Wai Chiang, Y., 2022. Controlled release fertilizers (CRFs) for climate-smart agriculture practices: a comprehensive review on release mechanism, materials, methods of preparation, and effect on environmental parameters. Environ. Sci. Pollut. Res. Int. 29 (36), 53967–53995. https://doi.org/10.1007/s11356-022-20890-y.

Kekeli, M.A., Wang, Q., Rui, Y., 2025. The role of nano-fertilizers in sustainable agriculture : boosting crop yields and enhancing quality. Plants 14 (4), 554. https://doi.org/10.3390/plants14040554.

Khose, S., Mailapalli, D., 2024. Spatial mapping of soil moisture content using very-high resolution UAV-based multispectral image analytics. Smart Agric. Technol. 8. https://doi.org/10.1016/j.atech.2024.100467.

Kouchaky, Z., Razzaghi Borkhani, F., Azizi-Khalkheili, T., Ghorbani Piralidehi, F., 2025. The application of the motivation-opportunity-ability model in predicting green entrepreneurial behavior of farmers. Sustain. Futures 10, 100900. https://doi.org/10.1016/j.sftr.2025.100900. ISSN 2666-1888. https://www.sciencedirect.com/science/article/pii/S2666188825004654.

Kumar, Y., Tiwari, K.N., Nayak, R.K., Rai, A., Singh, S.P., Singh, A.N., Kumar, Y., Tomar, H., Singh, T., Raliya, R., 2020. Nanofertilizers for increasing nutrient use efficiency, yield and economic returns in important winter season crops of Uttar Pradesh. Indian J. Fertil. 16 (12), 772–786.

Küçükyildirim, H., Aydemir, S., Öztürk, H., Uğurlar, F., 2025. Biodegradable polymer-based coatings for controlled and slow-release fertilizers. J. Coat. Technol. Res. 23 (2), 957–972. https://doi.org/10.1007/s11998-025-01180-x.

Kumar, N., Samota, S.R., Venkatesh, K., Tripathi, S., 2023. Global trends in use of nano-fertilizers for crop production: advantages and constraints -A review. Soil Tillage Res. 228, 105645. https://doi.org/10.1016/j.still.2023.105645.

Kumar, A., Singh, A., Bhattacharyya, U., Srishailam, B., 2024. Application of Internet of Things (IoT) in agriculture water management. In: National Conference on Hill Agro-Ecosystem: Challenges and Opportunities for Achieving Sustainable Development Goals. 146-152 (HAES-COSDG, 2024).

Kumar, R., Singh, A., Verma, P., 2024. Nanotechnology-enabled sensors for real-time monitoring in smart farming systems. J. Nanobiotechnol. 22 (1), 145. https://doi.org/10.1186/s12951-024-00987-3.

Kumar, S., Sharma, Y., Khandelwal, V., Rawat, K., Patil, A., 2025. Applications of nanotechnology in fertilizers: a review study. Sustain. Chem. Environ. https://doi.org/10.1016/j.scenv.2025.100247.

Kumari, R., Singh, S., Chandra, R., Singh, J., 2023a. Regulation and safety measures for nanotechnology-based agri-products from feed to food. Front. Bioeng. Biotechnol. 11, 1178893. https://doi.org/10.3389/fbioe.2023.1178893.

Kumari, R., Suman, K., Karmakar, S., Mishra, V., Lakra, S.G., Saurav, G., Mahto, B., 2023b. Regulation and safety measures for nanotechnology-based agri-products. Front. Genome Ed. 05, 1200987. https://doi.org/10.3389/fgeed.2023.1200987.

Lawrencia, D., Wong, S., Low, D.Y.S., Goh, B.H., Goh, J., Ruktanonchai, U., Soottitantawat, A., Lee, L.-H., Ying, P.T., 2021. Controlled release fertilizers: a review on coating materials and mechanism of release. Plants 10 (2), 238. https://doi.org/10.3390/plants10020238.

Leakey, A.D.B., Ainsworth, E.A., Bernacchi, C.J., Rogers, A., Long, S.P., Ort, D.R., 2009. Elevated CO_2 effects on plant carbon, nitrogen, and water relations: six important lessons from FACE. J. Exp. Bot. 60 (10), 2859–2876. https://doi.org/10.1093/jxb/erp096.

Leong, W.Y., 2025. Smart poultry farming 4.0: leveraging IoT for climate control and productivity. In: Conference: 2025 21st IEEE International Colloquium on Signal Processing & Its Applications (CSPA). 172-175. IEEE. https://doi.org/10.1109/CSPA64953.2025.10933062.

Liakos, K.G., Busato, P., Moshou, D., Pearson, S., Bochtis, D., 2018. Machine learning in agriculture: a review. Sensors 18 (2674). https://doi.org/10.3390/s18082674.

Liu, X., Nadeem, M., Rui, Y., 2024. Effects of nanofertilizers on the mechanism of photosynthetic efficiency in plants: a review. Phyton-Int. J. Exp. Bot. 93 (12), 3197–3216. https://doi.org/10.32604/phyton.2024.059281.

Lowry, G.V., Avellan, A., Gilbertson, L.M., 2019. Opportunities and challenges for nanotechnology in the agri-tech revolution. Nat. Nanotechnol. 14 (6), 517–522. https://doi.org/10.1038/s41565-019-0461-7.

Luo, X., Xiong, S., Jia, X., Zeng, Y., Chen, X., 2025. An IoT-enabled data management for smart agriculture: a comprehensive review on emerging technologies. IEEE Access. https://doi.org/10.1109/ACCESS.2025.3578751, 1-1.

Maaz, T., Dobermann, A., Lyons, S., Thomson, A., 2025. Review of research and innovation on novel fertilizers for crop nutrition. Npj Sustain. Agric. 3. https://doi.org/10.1038/s44264-025-00066-0.

Mansoor, S., Iqbal, S., Popescu, S.M., Kim, S.L., Chung, Y.S., Baek, J.H., 2025. Integration of smart sensors and IoT in precision agriculture: trends, challenges and future perspectives. Front. Plant Sci. 16, 1587869. https://doi.org/10.3389/fpls.2025.1587869.

Market Report Analytics, 2025. Exploring Agricultural Nano Fertilizer Trends 2025–2033, vol 114402. https://www.marketreportanalytics.com/reports/agricultural-nano-fertilizer.

Mejias, J.H., Salazar, F., Pérez, A.L., Hube, S., Rodriguez, M., Alfaro, M., 2021. Nanofertilizers: a cutting-edge approach to increase nitrogen use efficiency in grasslands. Front. Environ. Sci. 9, 635114. https://doi.org/10.3389/fenvs.2021.635114.

Mgendi, G., 2024. Unlocking the potential of precision agriculture for sustainable farming. Discover Agriculture 2. https://doi.org/10.1007/s44279-024-00078-3.

Miller, T., Mikiciuk, G., Durlik, I., Mikiciuk, M., Łobodzińska, A., Śnieg, M., 2025. The IoT and AI in agriculture: the time is now—a systematic review of smart sensing technologies. Sensors 25, 3583. https://doi.org/10.3390/s25123583.

Mim, J.J., Rahman, S.M.M., Khan, F., Paul, D., Sikder, S., Das, H.P., Khan, S., Orny, N.T., Shuvo, M.R.H., Hossain, N., 2025. Towards smart agriculture through nano-fertilizer—a review. Mater. Today Sustain. 30, 101100. https://doi.org/10.1016/j.mtsust.2025.101100.

Mishra, H., Mishra, D., 2024. AI for data-driven decision-making in smart agriculture: from field to farm management. In: Artificial Intelligence Techniques in Smart Agriculture. Springer, pp. 173–193. https://link.springer.com/chapter/10.1007/978-981-97-5878-4_11.

Mobeen, M., Majeed, M., Fatima, S., Oladokoun, A., Akbar, A., Umar, A., Al-Khayri, J.M., Quresh, W.A., Kalsoom, A., Iqbal, J., Abbasi, B.A., 2025. Regulations and ethical considerations for nanotechnology in agri-food applications. In: Nanobiosensors for Crop Monitoring and Precision Agriculture. Springer, pp. 307–334. https://doi.org/10.1007/978-981-96-8335-2_14.

Mukhtar, A., Shakeel, A., 2019. Carbon dioxide enrichment and crop productivity. Agron. Crops 31–46. https://doi.org/10.1007/978-981-32-9783-8_3, 978-981-32-9782-1. eISBN: 978-981-32-9783-8.

Mushtaq, M., Raza, A., Maqbool, S., Safdar, M., Ahmed, M., Sattar, J., 2024. Precision Irrigation for Sustainable Agricultural Productivity. https://doi.org/10.4018/979-8-3693-4864-2.ch010.

Novak, A., Jansen, M., 2021. Nano-fertilizers and their interaction with soil particles. Soil Future 62–66. P-ISSN: 3051-3448 E-ISSN: 3051-3456 Volume: 02 Issue: 02 July-Decem. Page No.

Orakwue, S.I., Al-Khafaji, H.M.R., Chabuk, M., 2022. IoT based smart monitoring system for efficient poultry farming. Webology 19, 4105–4112. https://doi.org/10.14704/WEB/V19I1/WEB19270.

Padhiary, M., Kumar, A., Sethi, L.N., 2025. Emerging technologies for smart and sustainable precision agriculture.

Discover Robot. 1 (6). https://doi.org/10.1007/s44430-025-00006-0.

Pan, W., Zhang, H.-J., Zhang, Y.-F., Wang, M., Tsui, M., Yang, L., Miao, A.J., 2023. Silica nanoparticle accumulation in plants: current state and future perspectives. Nanoscale 15, D3NR02221H. https://doi.org/10.1039/D3NR02221H.

Pasawan, S., Meena, M.L., Kumar, P., Anjali, 2025. Smart horticulture: farming with precision agriculture. Agrospheres Magaz. 6 (8), 3–5. Retrieved from. https://agrospheresmagazine.vitalbiotech.org/aadmin/articles/Agrospheres-2025-6-8-3-5.pdf.

Patra, J.K., Cloete, K.J., Kaushik, S., 2025. Agronanotechnology: advancements and challenges in nanofertilizers for sustainable agriculture. Front. Nanotechnol. 7, 1668484. https://doi.org/10.3389/fnano.2025.1668484.

Pilotto, L., Scalera, F., Piccirillo, C., Marchiol, L., Alzate Zuluaga, M.Y., Pii, Y., Cesco, S., Civilini, M., Fellet, G., 2025. Phosphorus release from nano-hydroxyapatite derived from biowastes in the presence of phosphate-solubilizing bacteria: a soil column experiment. J. Agric. Food Chem. 73. https://doi.org/10.1021/acs.jafc.4c09325.

Preetha, P.S., Balakrishnan, N., 2017. A review of nano fertilizers and their use and functions in soil. Int. J. Curr. Microbiol. Appl. Sci. 6, 3117–3133.

Qiao, Y., Zhang, L., Sun, S., 2025. Biomass-based coated controlled-release fertilizers: a sustainable solution for tropical agriculture. Tropical Plants 4, e008. https://doi.org/10.48130/tp-0025-0001.

Raguraj, S., Wijayathunga, W.M.S., Gunaratne, G.P., Amali, R.K.A., Priyadarshana, G., Sandaruwan, C., Karunaratne, V., Hettiarachchi, L.S.K., Kottegoda, N., 2020. Urea–hydroxyapatite nanohybrid as an efficient nutrient source in *Camellia sinensis* (L.) Kuntze (tea). J. Plant Nutr. 43 (15), 2383–2394. https://doi.org/10.1080/01904167.2020.1744645.

Rahman, A., Singh, P., Alharby, H.F., Al-Zahrani, H.S., Gupta, D.K., 2025. The role of nano-fertilizers in sustainable agriculture: a critical review. Sustainability 17 (4), 2356. https://doi.org/10.3390/su17042356.

Raliya, R., Franke, C., Chavalamane, S., Nair, R., Reed, N., Biswas, P., 2016. Quantitative understanding of nanoparticle uptake in watermelon plants. Front. Plant Sci. 7, 1288. https://doi.org/10.3389/fpls.2016.01288.

Rehana, M.R., Gladis, R., Joseph, B., 2022. Controlled release of nutrients for soil productivity—a review. Curr. J. Appl. Sci. Technol. 41 (20), 34–46. https://doi.org/10.9734/cjast/2022/v41i2031747.

Sarabandi, M., Khalid, A., Benitez, R., 2024. Continuous data streams for pre agricultural decision systems: economic inequalities, nanomaterial access, and digital envy dynamics. J. Smart Agro Analyt. 12 (1), 101–119.

Sarabandi, M., Zargar, M., Ghorbani, A., Chen, M., 2025. Smart and sustainable nano-biosensing technologies for advancing stress detection and management in agriculture and beyond. Ind. Crop. Prod. 226, 120713. https://doi.org/10.1016/j.indcrop.2025.120713.

Sebastian, A., Devika, P., Nair, P., Devadas, V.S., 2023. Green synthesis of potassium-infused nitrogenous nanofertilizer for enhanced plant growth. Chem. Pap. 78 (1), 1–12. https://doi.org/10.1007/s11696-023-03175-7.

Shahid, S., Shoeb Khan, M., Kumar, A., Rahman, S., Arshad, M., Kaushik, P., Saini, P., El-Khawaga, A., 2024. Role of nanomaterials in sustainable agriculture. In: Uddin, I. (Ed.), Sustainable Nanomaterials. Sustainable Materials and Technology. Springer, Singapore. https://doi.org/10.1007/978-981-97-2761-2_9.

Shen, M., Liu, S., Jiang, C., Zhang, T., Chen, W., 2023. Recent advances in stimuli-response mechanisms of nano-enabled controlled-release fertilizers and pesticides. Eco-Environ. Health 2 (3), 161–175. https://doi.org/10.1016/j.eehl.2023.07.005. ISSN 2772-9850.

Shi, G., Yi, C., Zhou, H., Chen, Y., Fan, G., Tong, F., Chen, W., Liu, L., Li, J., Gao, Y., Zhou, D., 2025. Nanostructured hydroxyapatite-zinc complex as a novel foliar system to reduce cadmium accumulation in wheat grain: performance and mechanism. Environ. Technol. Innov. 40, 104387. https://doi.org/10.1016/j.eti.2025.104387. ISSN 2352-1864.

Singh, A., Rajput, V., Al Tawaha, A.R., Al Zoubi, O., Habeeb, T., Rawat, S., Ghazaryan, K., Minkina, T., 2023. A review on crop responses to nanofertilizers for mitigation of multiple environmental stresses. Ecol. Eng. Environ. Technol. 24, 280–296. https://doi.org/10.12912/27197050/169313.

Singh, O., Singh, V.K., Baskar, S., Shahi, U.P., Gill, S.A.S., Alam, S., 2024a. Biodegradable green materials for sustainable agriculture. In: Encyclopedia of Green Materials. Springer, pp. 1–11. https://doi.org/10.1007/978-981-16-4921-9_236-1.

Singh, R., Kumar, A., Sharma, P., 2024b. Risk assessment of engineered nanomaterials in agriculture: short-term and long-term perspectives. Environ. Nanotechnol. Monit. Manag. 22, 100876. https://doi.org/10.1016/j.enmm.2024.100876.

Solanki, P., Bhargava, A., Chhipa, H., Jain, N., Panwar, J., 2015. Nano-fertilizers and their smart delivery system. In: Ranjan, K.M.R., Dasgupta, S., Lichtfouse, E. (Eds.), Nanotechnologies in Food and Agriculture. Springer, Cham, Switzerland, pp. 81–101. https://doi.org/10.1007/978-3-319-14024-7_4.

Sourav, C.h ., Deb, C.h ., Marwaha, S., Haque, M.d ., Kamil, D., Bheemanahalli, R., 2025. Application of artificial intelligence in insect pest identification –A review.

Artif. Intell. Agric. 16, 44–61. https://doi.org/10.1016/j.aiia.2025.06.005.

Start Us Insights, 2025. AI in Agriculture: A Strategic Guide (2025–2030). https://www.startus-insights.com/innovators-guide/ai-in-agriculture-strategic-guide/.

Stone, V., Nowack, B., Baun, A., Brink, N. .v. d., Kammer, F. . v .d., Dusinska, M., Handy, R., Hankin, S., Hassellöv, M., Joner, E., Fernandes, T.F., 2010. Nanomaterials for environmental studies: classification, reference material issues, and strategies for physico-chemical characterisation. Sci. Total Environ. 408 (7), 1745–1754. https://doi.org/10.1016/j.scitotenv.2009.10.035. ISSN 0048-9697. https://www.sciencedirect.com/science/article/pii/S004896970901016X.

Su, Y., Zhou, X., Meng, H., Ashworth, A.J., Owens, P.R., Payton, P., Yang, Y., Appel, H.M., Jassby, D., 2022. Cost–benefit analysis of nanofertilizers and nanopesticides: a modeling approach. Nat. Food 3 (12), 1018–1029. https://doi.org/10.1038/s43016-022-00647-z.

Tarafdar, J.C., Raliya, R., 2013. Rapid, low-cost, and eco-friendly approach for iron nanoparticle synthesis using Aspergillus oryzae TFR9. J. Nanopart. 2013, 141274. https://doi.org/10.1155/2013/141274.

Tawfik, S., 2020. Nanomaterials: classification, properties, and environmental toxicities. Environ. Technol. Innov. 20, 101067. https://doi.org/10.1016/j.eti.2020.101067.

Thirugnanasambandan, T., Iswarya, S., Ajitha, M., Gopinath, S.C.B., Selvarani, K., 2024. Sustained release of chemical fertiliser by nanochitosan: insights and updates. BioNanoScience 14, 3679–3692. https://doi.org/10.1007/s12668-024-01533-7.

Umweltbundesamt, 2021. Nanomaterials and Other Advanced Materials: Prospects and Challenges of Innovation for the Green and Digital Transformation. German Environment Agency. https://www.umweltbundesamt.de/publikationen/nanomaterials-other-advanced-materials-prospects.

Upadhayay, V.K., Chitara, M., Mishra, D., Mahindra Nath, Jha, Jaiswal, A., Kumari, G., Ghosh, S., Patel, V., Naitam, M., Singh, A., Pareek, N., Taj, G., Maithani, D., Kumar, A., Dasila, H., Sharma, A., 2023. Synergistic impact of nanomaterials and plant probiotics in agriculture: a tale of two-way strategy for long-term sustainability. Front. Microbiol. 14. https://doi.org/10.3389/fmicb.2023.1133968.

Urbani, D., Evangelisti, M., Bebi, C., Rovegno, C., Parenti, M. D., Varchi, G., Del Rio, A., 2024. Collection and Review of Information on Nanomaterial-Based and Nano-Enabled Plant Protection Products, Biocidal Products and Fertilising Products (Ref. ECHA-24-R-16-EN). European Chemicals Agency. https://doi.org/10.2823/9909100.

U.S. Environmental Protection Agency, 2020. Control of Nanoscale Materials under the Toxic Substances Control Act. https://www.epa.gov/reviewing-new-chemicals-under-toxic-substances-control-act-tsca/control-nanoscale-materials-under.

Verma, K., Sarkar, C., Saha, S., 2025. Exploration of biodegradable polymeric particles in agriculture: a holistic approach for sustainable farming. Environ. Sci. Adv. 4 (1), 409–431. https://doi.org/10.1039/D4VA00350K.

Villagran, E., Espitia, J.J., Amado, G., Rodriguez, J., Gomez, L., Flores Velasquez, J., Gil, R., Baeza, E., Aguilar, C.E., Akrami, M., Arias, L.A., 2025. CO_2 enrichment in protected agriculture: a systematic review of greenhouses, controlled environment systems, and vertical farms—Part 2. Sustainability 17 (7), 2809. https://doi.org/10.3390/su17072809.

Van de Vlasakker, P.C.H., Davis, A.P., Timmer, R.D., 2022. A review of nutrient losses to waters from soil- and ground-based urban agriculture gardens. Front. Sustain. Food Syst. 6, 842930. https://doi.org/10.3389/fsufs.2022.842930.

Wang, Y., Zhang, Y.J., Han, J.M., Li, C.H., Wang, R.J., Zhang, Y.L., Jia, X., 2019. Improve plant photosynthesis by a new slow-release carbon dioxide gas fertilizer. ACS Omega 4 (6), 10354–10361. https://Doi: 10.1021/acsomega.8b03086.

Wang, X., Xie, H., Wang, P., Yin, H., 2023. Nanoparticles in plants: uptake, transport and physiological activity in leaf and root. Materials 16 (8), 3097. https://doi.org/10.3390/ma16083097.

Wang, Z., Li, X., Chen, M., Liu, Y., Zhang, Q., Wang, L., 2024. AI-enabled IoT-based pest prevention and controlling system using sound analytics in large agricultural field. Comput. Electron. Agric. 219, 108235. https://doi.org/10.1016/j.compag.2024.108235.

Watts, J.L., Dow, G.J., Buckley, T.N., Muir, C.D., 2024. Does stomatal patterning in amphistomatous leaves minimize the CO_2 diffusion path length within leaves? AoB Plants 16 (2), plae015. https://doi.org/10.1093/aobpla/plae015.

Yadav, M., 2025. Nanoparticle-facilitated targeted nutrient delivery in plants: breakthroughs and mechanistic insights. Plant Nano Biol. 12, 1–15. https://doi.org/10.1016/j.plana.2025.100156.

Yan, G., Huang, Q., Zhao, S., Xu, Y., He, Y., Nikolic, M., Nikolic, N., Liang, Y., Zhu, Z., 2024. Silicon nanoparticles in sustainable agriculture: synthesis, absorption, and plant stress alleviation. Front. Plant Sci. 15, 1393458. https://doi.org/10.3389/fpls.2024.1393458.

Yang, X., Chen, J., Lu, X., Liu, H., Liu, Y., Bai, X., Qian, L., Zhang, Z., 2025. Advances in UAV remote sensing for monitoring crop water and nutrient status: modeling methods, influencing factors, and challenges. Plants 14, 2544. https://doi.org/10.3390/plants14162544.

Yuan, Y., Sun, Y., 2024. Practices, challenges, and future of digital transformation in smallholder agriculture: insights from a literature review. Agriculture 14, 2193. https://doi.org/10.3390/agriculture14122193.

Yusoff, F.M., Umi, W.A.D., Ramli, N.M., Harun, R., 2024. Water quality management in aquaculture. Camb. Prisms Water 2024 (2), e8. https://doi.org/10.1017/wat.2024.6.

Zhang, P., Lenka, N.K., Song, Y., 2024. Sustaining CO_2 fertilization gains under water and nutrient stress in crop plants. Front. Plant Sci. 15, 1375175. https://doi.org/10.3389/fpls.2024.1375175.

Zulfiqar, F., Navarro, M., Ashraf, M., Akram, N.A., Munné-Bosch, S., 2019. Nanofertilizer use for sustainable agriculture: advantages and limitations. Plant Sci. 289, 110270. https://doi.org/10.1016/j.plantsci.2019.110270.

CHAPTER

7

Nano-formulated fertilizers and CO_2 delivery system for optimizing photosynthesis

Prasann Kumar[1], *Amit Raj*[1] *and Padmanabh Dwivedi*[2]

[1]Department of Agronomy, School of Agriculture, Lovely Professional University, Phagwara, Punjab, India; [2]Department of Plant Physiology, Institute of Agricultural Sciences, Banaras Hindu University, Varanasi, Uttar Pradesh, India

1. Introduction

The increasing global demand for food security has placed photosynthetic efficiency at the forefront of agricultural research, as plant biomass accumulation and yield are primarily determined by their ability to capture atmospheric carbon. Photosynthesis is, however, inherently limited by biochemical inefficiencies in the Calvin-Benson cycle, the most apparent being the oxygenase activity of ribulose-1,5-bisphosphatecarboxylase/oxygenase (Rubisco), catalyst of both the CO_2 fixation reaction and an oxygenation mechanism that is both wasteful and inefficient (Meloni et al., 2023). This dual affinity for the current atmospheric CO_2 concentration results in a large flux to photorespiration, in which 2-phosphoglycolate is recycled at the expense of ATP and reducing power, thereby decreasing net carbon gain. Accordingly, world agriculture is actively seeking methods to increase adequate CO_2 levels at the site of carboxylation while maintaining nutrient levels that support the biosynthetic machinery of photosynthetic cells (Croce et al., 2024). Traditional methods of fertilization have met this demand only partially and are mechanistically inefficient. The problem with macroscale nitrogen and phosphorus manure is that the applied ions are readily immobilized in the soil, washed away by percolating water, and released into the atmosphere via nitrification-denitrification processes. The rate of hydrolysis of urea or phosphates at the plant-soil interface usually exceeds the capacity of root membrane carriers to supply it, leading to transient peaks in availability followed by rapid depletion. The imbalance between bulk nutrient release and cellular uptake interferes with the coordinated activity of enzymes, including glutamine synthetase and ribulose-1,5-bisphosphate kinase, and limits

Nanotechnology in Photosynthesis
https://doi.org/10.1016/B978-0-443-44376-3.00007-0

ATP turnover and protein synthesis in chloroplasts (Singh et al., 2023). Beyond inefficiency, the ecological footprint of such fertilization is substantial, and its role in eutrophication, nitrous oxide emissions, and long-term soil degradation is not insignificant.

Carbon dioxide is a key metabolic driver, as stromal CO_2 concentration determines the allocation of Rubisco activity between carboxylation and oxygenation. Under conditions in which CO_2 concentration is high in the chloroplast, the carboxylase reaction catalyzed by the enzyme predominates, promoting the production of 3-phosphoglycerate and thereby leveling the Calvin cycle flux (Anjali et al., 2025). This further increases triose phosphate production, sucrose synthesis, and sucrose export to sink tissues, thereby increasing biomass accretion. On the other hand, decreased intercellular CO_2 increases oxygenation, increases glycolate production, and aggravates the metabolic load of photorespiration. Therefore, control of CO_2 supply at the microenvironmental level is necessary to enhance photosynthetic productivity, particularly under unstable climatic conditions.

New solutions are emerging at the crosspoint between nanotechnology and carbon engineering. Nano-fertilizers also employ physicochemical characteristics, including high surface-area-to-volume ratios, tunable porosity, and controlled-release coatings, to release nutrients in response to plant demand (Yadav et al., 2023). Mechanistically, nanoparticles may circumvent soil immobilization by releasing ions at the rhizoplane, where root transporters operate at saturating efficiency. There are uptake pathways, such as endocytosis and aquaporin-mediated uptake, that enable nanoscale carriers to enter cortical cells and deliver nutrients directly to metabolically active compartments. Nanoparticles used in foliar applications cross cuticular microchannels or stomatal openings and enter mesophyll tissues, where they supply nitrogen, phosphorus, or micronutrients to chloroplast-enriched cells (Gautam et al., 2025). Such space- and time-targeting allows enzymes that fix CO_2 and harvest light to be supplied with a steady stream of nutrients, without the waves of deficiency and excess typical of bulk fertilization.

Concurrently, nanobubble technology and nanoparticle carriers loaded with bicarbonate are being re-engineered to deliver CO_2. The high stability and diffusion potential of nanobubbles could be used to locally enrich the apoplastic space and chloroplast stroma with dissolved CO_2, thereby increasing the carboxylation-to-oxygenation ratio of Rubisco (Traiphop et al., 2025). Regulated secretion of bicarbonate ions further stabilizes the stromal CO_2 concentration under variable stomatal conductance, and buffers changes in CO_2 in response to diurnal or stress-related fluctuations. Such systems, when combined with nutrient nanocarriers, provide twofold reinforcement: the availability of nitrogen maintains the biosynthesis of chlorophyll and Rubisco, whereas a high CO_2 concentration enables the enzyme to achieve its highest catalytic efficiency. This synergy catalyzes the regeneration of ribulose-1,5-bisphosphate via ribulose-5-phosphate kinase activity, increases electron transport chain coupling, and positively affects the distribution of assimilated carbon between structural and storage carbohydrates (Meloni et al., 2023). Collectively, nano-fertilizers and CO_2 nano-delivery systems transform the agricultural landscape by targeting the mechanistic bottlenecks of photosynthesis across multiple levels: soil nutrient dynamics and root transport physiology, chloroplast biochemistry, and Calvin cycle kinetics. They constitute the third generation of sustainable intensification, a convergence between yield production and ecological control, combining the advantages of nutrient assimilation with regulated CO_2 enrichment.

2. Nano-formulated fertilizers: design and mechanisms

The development of nano-formulated fertilizers has been a significant step toward precision farming, enabling farmers to deliver nutrients in a controlled manner that aligns with plant developmental requirements. Traditional fertilizers are associated with poor nutrient use efficiency (NUE) and ecological pollution from excessive volatilization, leaching, or fixation (Nakachew et al., 2024). By contrast, nanofertilizers employ engineered nanomaterials with tailored physicochemical properties, such as increased surface reactivity, porosity, and surface charge, to mitigate these inefficiencies. In addition to serving as passive nutrient carriers, nanostructures interact with soil colloids, plant cell walls, root exudates, and signaling networks to enhance system-wide nutrient assimilation (Zhang et al., 2024). Their mechanisms and designs can be discussed at four levels: nanoencapsulation and slow-release kinetics; high nutrient solubility and uptake; foliar- and soil-applied formulations; and nanostructures for foliar and soil application and for micronutrient delivery.

2.1 Nano-encapsulation and slow-release kinetics

Nano-encapsulation entraps the fertilizer nutrients in nanocarriers, which include polymeric nanoparticles, layered double hydroxides, nanoclays, and carbon nanotubes (Singh et al., 2024). In contrast to bulk coating methods, nanocarriers are pH-, redox-, and moisture-responsive, thereby releasing nutrients when roots are metabolically active. Mechanistically, the process of nutrient release occurs in the following way:

- **Diffusion-Controlled Gradients:** Nanopores are selective sieves where nutrients can diffuse with a rate that is proportional to the pore size and ionic interactions between ions and the energy barrier of the pore. This prevents osmotic stress of nutrient surges.
- **Biodegradable Polymer Disintegration:** Chitosan- and alginate-based nanocapsules disintegrate in the presence of rhizospheric enzymes, and nutrients are slowly released to the roots.
- **pH-Released:** Encapsulated phosphate or iron is more readily dissolved by acidic rhizosphere microzones generated by root exudates (malate, citrate), to provide site-specific release.
- **Enzyme-Responsive Release:** The release of fertilizers can be coupled with microbial metabolism by using microbial phosphatases and ureases that degrade the functional coating.
- **Reduced Loss Pathways:** Encapsulation slows the volatilization of ammonia from urea and nitrate of soil-nitrate leaching, and prolongs soil solution persistence. Such coordination of release with physiological demand increases NUE and reduces greenhouse gas emissions associated with nitrogen fertilizers.

2.2 Improved solubility and uptake of nutrients

Most micronutrients and macronutrients are poorly soluble and can be fixed by the soil. Nano-sizing modulates kinetics of dissolution, facilitates chelation, and enhances uptake of plants by passive and active routes:

- **Improved Dissolution Surface:** Nanoscale type of fertilizers (e.g., ZnO nanoparticles) dissolve rapidly due to a high surface-to-volume ratio, which increases ion activity in soil solution (Thangavelu et al., 2024).
- **Nanocarrier chelation:** Nano-chelates of Fe, Mn, and Cu do not precipitate and can move even in calcareous soils.

- **Root-Surface Adhesion:** Nanoparticles stick to root mucilage and epidermis through electrostatic forces to create micro-reservoirs of available nutrients.
- **Endocytic Uptake:** nanoparticles are incorporated into root epidermal and cortical cells by endocytosis, at which point vesicles transport them to the symplast.
- **Transporter Activation:** Expression of families of transporters is controlled by nutrient availability: ZIP (Zn, Fe), NRT (nitrate), and AMT (ammonium). Nanoforms enhance this signaling by stabilizing ion concentrations at the root-soil interface (Sharma et al., 2025).
- **Root Exudate Modulation:** Nanofertilizers trigger the exudation of organic acids and phenolics that also dissolve phosphates and trace elements. Due to the biochemical interactions between nanostructures and plant exudates, a more consistent flow of nutrients into the root xylem and improved systemic allocation are observed.
- **Phloem Mobility:** Nanoparticles have their ions solubilized and loaded into phloem by using sucrose symporters, which then transport their ions to sinks like grains and fruits long-distance (Fig. 7.1).
- **Soil-Based Release:** Soil-deposited nanoparticles react with colloids, clay minerals, and microbial biofilms. Organic acids secreted by plant roots dissolve the nanoparticle coating, making nutrients available in rhizosphere hotspots (Tripathi et al., 2024).
- **Dual Pathway Comparison:** Foliar nano-delivery can cure the deficiency in a relatively short period of time, as compared to soil application, which is relatively slow in uptake capacity. Nanotechnology facilitates the closure of these channels, thereby enabling combined channeling in both the short and long term. A two-direction solution of this nature will reduce unwarranted reliance on application to soil per se and therefore optimize the economy of fertilizers and crop responsiveness.

2.3 Foliar versus soil applied formulations

Nutrient assimilation is greatly affected by the mode of delivery. Nano-fertilizers enable optimization of foliar sprays as well as soil applications:

- **Foliar Penetration:** Nanoparticles find their way into the leaf cuticle via hydrophilic nanochannels, or directly by penetrating stomatal pores. Surface functionalization with a hydrophilic group enhances wetting and adhesion to the cuticle (Gautam et al., 2025).
- **Cellular Uptake in Leaves:** Nanoparticles dissolve or liberate ions into the apoplast after being inside the cell. These are taken up by mesophyll cells via aquaporins or transporter proteins.

2.4 Nanostructures to deliver micronutrients

Zn, Fe, Cu, Mn, and Mo are micronutrient elements that are poorly bioavailable because of soil fixation and low solubility (Kaur et al., 2023). Nano-engineering will provide precise solutions:

- **Direct Nano-Oxide Particles:** ZnO, Fe_2O_3, and CuO nanoparticles are more readily dissolved and result in increased uptake by root surfaces.
- **Mesoporous Carriers:** Mesoporous silica nanoparticles encapsulate micronutrients in configurable pore geometries and can release them safely without precipitation over an extended period of time.
- **Nanoclays and Hydroxyapatite:** These act as mineral carriers and adsorb micronutrient

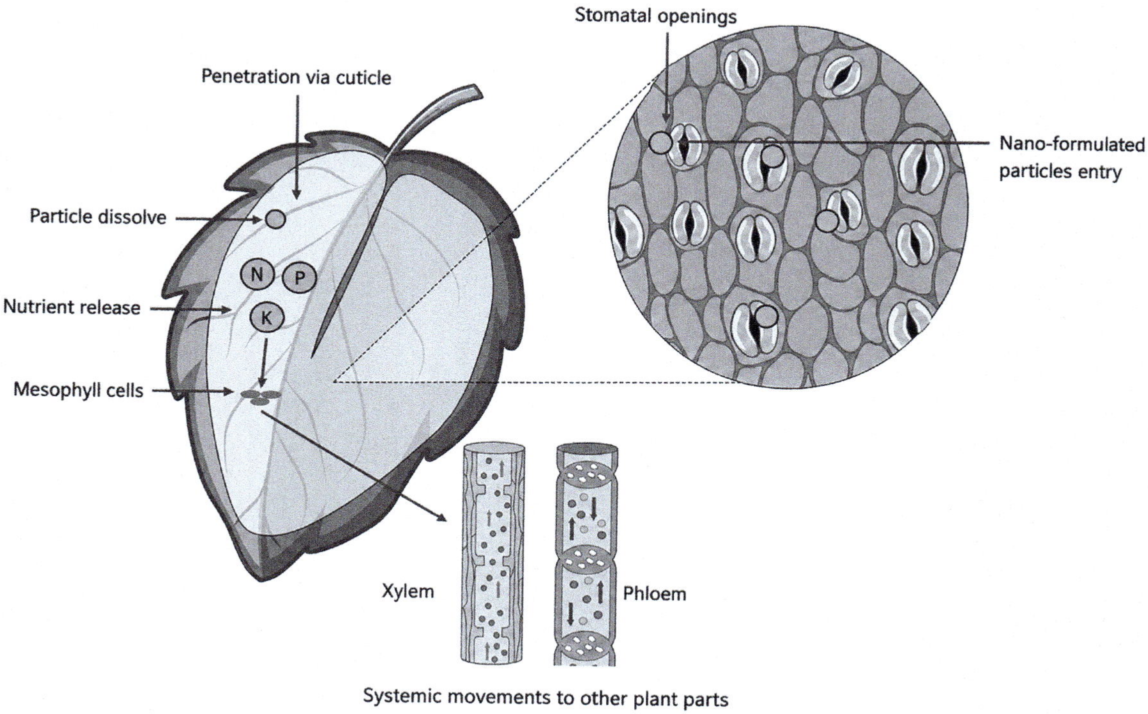

FIGURE 7.1 Mechanistic pathway of nanoparticle-mediated nutrient transport through foliar penetration, cellular uptake, and phloem translocation in plants.

ions, which are released slowly in acidic rhizosphere environments.

- **Nanoemulsions:** Hydrophobic bioactives (e. g., vitamins, phytohormones) are contained within lipid-based nanoemulsions that increase foliar penetration.
- **Biofunctionalization:** Nanocarriers that are amino acid or peptide-functionalized interact selectively with transporter proteins to enhance delivery efficiency to chloroplasts or mitochondria.
- **ROS Signaling Modulation:** Nano-Fe and nano-Cu micronutrient nanostructures mediate redox signaling cascades and control the concentration of reactive oxygen species (ROS) that control root growth and stress endurance (Ghorbani et al., 2024). These processes result in improved micronutrient metabolism relative to bulk formulations and greater plant stress tolerance (Table 7.1).

3. CO_2 delivery systems for photosynthesis

The main product of photosynthesis is carbon dioxide (CO_2), which, in most cases, is the main factor that limits crop yields. Despite the atmospheric CO_2 concentration increasing to an average of 420 ppm, chloroplasts often require more than is available, particularly under high-light and adequate-nutrient conditions (Komkhum et al., 2025). Novel CO_2 delivery systems that enhance assimilation efficiency, improve water-use balance, and enhance yield stability have thus been explored in

TABLE 7.1 Nanostructured systems for targeted micronutrient delivery and stress modulation in plants.

Nanostructure	Micronutrient delivered	Mechanism of action	Functional outcome in plants	References
ZnO nanoparticles	Zinc (Zn)	Enhanced solubility and dissolution at the root surface	Improves Zn uptake and enzyme activation	Kaur et al. (2023)
Fe_2O_3 nanoparticles	Iron (Fe)	Rapid ion release increases Fe bioavailability	Corrects chlorosis and boosts photosynthesis	Ghorbani et al. (2024)
CuO nanoparticles	Copper (Cu)	Controlled Cu ion liberation via nano-oxide dissolution	Stimulates oxidative enzyme systems	Tripathi et al. (2024)
Mesoporous silica nanoparticles	Multi-micronutrients	Configurable pore geometry enables sustained nutrient release	Prevents precipitation and nutrient loss	Gautam et al. (2025)
Nanoclays	Zn, Fe, Mn	Adsorb and release micronutrient ions slowly in acidic soils	Prolonged nutrient availability in the rhizosphere	Thangavelu et al. (2024)
Hydroxyapatite nanocarriers	Zn, Fe, Cu	Acts as a phosphate-mineral carrier for micronutrient ions	Enhances nutrient stability and uptake	Sharma et al. (2025)
Lipid-based nanoemulsions	Vitamins, phytohormones	Encapsulate hydrophobic bioactives for foliar delivery	Increases foliar penetration and assimilation	Kaur et al. (2023)
Amino Acid–Functionalized nanocarriers	Zn, Fe	Selective interaction with transporter proteins	Improves chloroplast and mitochondrial delivery	Ghorbani et al. (2024)
Peptide–Functionalized nanocarriers	Cu, Mn	Mimic natural ligands to enhance cellular uptake	Promotes targeted intracellular nutrient delivery	Tripathi et al. (2024)
Nano-Se formulations	Selenium (Se)	Induce thiol-based antioxidant biosynthesis	Strengthens ascorbate–glutathione cycles	Gautam et al. (2025)
Nano-sulfur particles	Sulfur (S)	Stimulate thiol metabolism and antioxidant response	Boosts redox balance and detoxification	Thangavelu et al. (2024)
Nano-Fe complexes	Iron (Fe)	Maintain redox equilibrium in electron transport chains	Stabilizes photosynthetic electron flow	Sharma et al. (2025)
Nano-Cu complexes	Copper (Cu)	Modulate ROS signaling cascades in root cells	Enhances stress endurance and root development	Kaur et al. (2023)
Nanocomposite blends	Zn, Fe, Cu, Mn	Combine oxides, clays, and polymers for multi-nutrient release	Enables synergistic nutrient use efficiency	Ghorbani et al. (2024)
Bioengineered ROS-responsive nanocarriers	Zn, Fe, Cu	Release nutrients upon oxidative triggers	Supports adaptive stress tolerance and recovery	Tripathi et al. (2024)

contemporary agricultural systems. These include controlled micro- and nanobubble-based methods, as well as more advanced full-scale greenhouse and open-field methods for enriching light, improving light-use efficiency, and regulating stomata.

3.1 Controlled enrichment in open-field and greenhouse systems

CO_2 enrichment has long been used to enhance greenhouse growth, but open-field delivery is emerging based on microclimate-focused technology. The processes that define their effectiveness are:

- **Response to carbon saturation:** Increase in extracellular CO_2 levels enhances carboxylation of Rubisco, represses oxygenase activity, and decreases photorespiration, increasing the net photosynthetic efficiency (Jin et al., 2023).
- **Canopy-level diffusion:** Enrichment maintains a positive gradient in the movement of CO_2 between the atmosphere and via stomata to the substomatal cavities (Janová et al., 2024).
- **Temporal synchronization:** It can be used in the morning or midday when light and stomatal conductance are the highest to match supply and photosynthetic demand.
- **Buffering of microclimate:** In open fields, the wind mix is reduced by diffusers that control the distribution of CO_2-rich layers in the immediate canopy surroundings. These processes render enrichment highly practical in closed systems and increasingly feasible in open-field conditions when performed accurately.

3.2 Micro-bubble and nano-bubble CO_2 technologies

Another more recent development in carbon supplementation is the delivery of carbon dioxide using irrigation water as micro- and nano-bubbles. They are both stable and dissolve due to their small size and large surface area. The mechanistic attributes are:

- **Improved dynamics of dissolution:** Nano-bubbles help to dissolve CO_2 in water at very high concentrations compared to traditional aeration and maintain their availability over time (Kizhisseri et al., 2025).
- **Localized root-zone enrichment:** Bubble breakdown can release CO_2 at the rhizosphere, which increases uptake by the roots and promotes desirable microbial activity.
- **pH buffering action:** Dissolved CO_2 creates carbonic acid, which opens micronutrients (e. g., Fe and Zn) a bit, lowering the rhizosphere pH and enhancing their solubility.
- **Gas–liquid interface reactivity:** When nano-bubbles are used as a gas-liquid interface, reactive oxygen intermediates develop on the gas-liquid interface and enhance nutrient uptake and root permeability (Marcelino et al., 2023). The technology brings together nanophysics and plant physiology to provide an additional carbon-delivery pathway beyond foliar uptake.

3.3 Physiological effects on stomatal dynamics

Stomata are the main access point of CO_2 to the leaves, and the behavior of the stomata regulates the enrichment of photosynthesis. Physiological effects are:

- **Guard cell sensing:** CO_2 is detected by the activity of carbonic anhydrase, resulting in a signal transmitted by bicarbonate, causing a decrease in stomatopore aperture.
- **Optimization of gaseous exchange:** Despite partial closure, the concentration of CO_2 within the body rises enough to

accommodate increased photosynthesis with decreased transpiration (Shanker et al., 2022).

- **Hydraulic balance:** When stomatal conductance is reduced, water loss is slowed, the xylem tension is lowered, and turgor is maintained in changing environmental conditions.
- **Spotty stomatal closure:** Non-uniform closures of the apertures across the leaf redistribute the flux, and maintain the capacity to effectively uptake CO_2 at the canopy level. These modifications allow carbon gain and water-use efficiency improvements to coexist in CO_2 delivery systems (Fig. 7.2).

3.4 Integration with light-use efficiency

Carbon enrichment will not work in isolation without integration with light capture and use, as photosynthesis needs energy provided through photochemistry. CO_2 delivery is combined with light dynamics through:

- **Enhancement in quantum yield:** When more CO_2 is available, more of the photons are utilized in the carboxylation process and not dissipated in photorespiration.
- **Lower PSII excitation pressure:** High CO_2 decreases photoinhibition through decreased over-reduction at photosystem II at high light (Wang et al., 2024).
- **Equal ATP/NADPH requirement:** The increased carbon fixation causes the cyclic electron flow and keeps the appropriate energy ratio to react in the Calvin cycle.
- **Source sink regulation:** As carbohydrate production increases, there must be adequate sink capacity, or inhibition of further photosynthetic gains will follow through feedback. Crops have been shown to achieve greater biomass accumulation and improved yield stability when CO_2 delivery is coupled

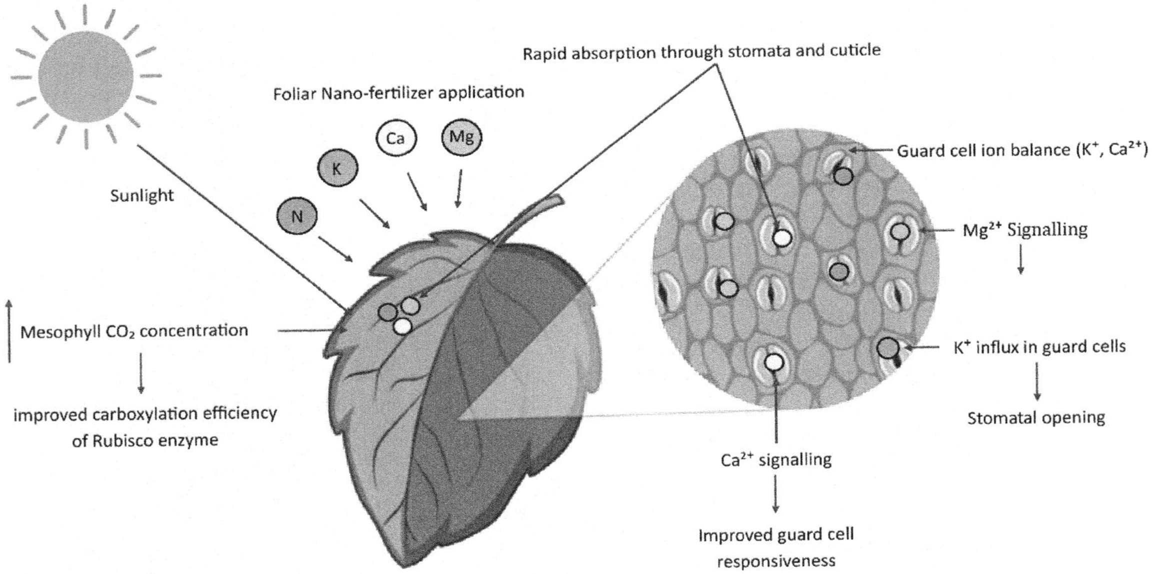

FIGURE 7.2 Physiological mechanisms of stomatal regulation optimizing CO_2 assimilation, water-use efficiency, and hydraulic balance in plant leaves.

with light-use efficiency during intensive cultivation (Sahoo et al., 2025b).

4. Photosynthetic pathway optimization

Optimization of the photosynthetic pathway is recognized as one of the most effective strategies to enhance crop yield potential amid global food demand and environmental change. Carbon enters plant metabolism through photosynthesis, but carbon dioxide losses during these processes create significant yield gaps between realized and theoretical productivity. Significant future advances will require addressing the low catalytic efficiency of Rubisco, the slow turnover of the Calvin-Benson cycle, the high costs of photorespiration, and the inefficient allocation of energy to growth. Scientists are working to address these bottlenecks to enhance carbon uptake and channel energy into biomass that can be used as a fuel.

4.1 Enhancement of rubisco carboxylation efficiency

The most common protein in the biosphere is the enzyme, Rubisco, which is the least efficient carbon-fixation enzyme, paradoxically. The catalytic rate is slow, and oxygen has a dual affinity, which significantly reduces photosynthetic efficiency. The optimization of Rubisco has thus been a significant focus of plant biotechnology, integrating natural diversity, protein engineering, and synthetic biology.

- **Catalytic site modification:** The active site undergoes structural changes, which decrease oxygenase activity and increase CO_2 selectivity, leading to increased carboxylation rates.
- **CO_2 concentrating systems:** Pyrenoids and carboxysomes borrowed via the algae and cyanobacteria system establish microcompartments that saturate Rubisco with CO_2.
- **Availability of chaperones:** Molecular chaperones like RAF1 and RbcX assist in folding and assembling the complexes appropriately, thereby avoiding the formation of misfolded complexes, which reduces the efficiency (Zhu et al., 2025).
- **Isoform replacement:** Crops expressing heat-adapted or high-light Rubisco isoforms can retain functionality at stress conditions (Amaral et al., 2024). Improving Rubisco efficiency directly increases the rate of CO_2 fixation, thereby enabling a higher rate of downstream carbon metabolism.

4.2 Calvin–Benson cycle acceleration

The Calvin-Benson cycle transforms the carbon fixed by Rubisco to sugars, the food of plant growth and metabolism. The cycle, however, has enzymatic bottlenecks that reduce throughput, especially at the regeneration step of ribulose-1,5-bisphosphate (RuBP). The cycle can be further accelerated by increasing the levels of key enzymes, stabilizing metabolite pools, and identifying synthetic options.

- **Enzyme overexpression:** Sedoheptulose-1,7-bisphosphatase (SBPase) and fructose-1,6-bisphosphatase (FBPase) play critical roles in the regeneration of RuBP, and an increase in their overexpression leads to an increase in the turnover of the cycle (Chen et al., 2022).
- **Triose-phosphate utilization:** Triose-phosphate isomerase is an upgraded enzyme; it favors the exportation of carbon to the biosynthesis of sucrose and starch, which leads to blocking of feedback (Fu et al., 2023).
- **Evidence of metabolite stabilization:** Intermediate concentrations such as glyceraldehyde-3-phosphate even out the bottlenecks that decelerate the cycle.
- **Synthetic cycle incorporation:** Another pathway that can be coupled to the Calvin

cycle to induce high productivity is called the synthetic carbon-fixation cycles (e.g., synthetic photorespiration-bypassing cycles) (Li et al., 2025). Cycle acceleration by optimizing carbon flow will not compromise the advantages of Rubisco, even if inefficiencies elsewhere in the chain are present (Table 7.2).

4.3 Reduction of photorespiration losses

Photorespiration is an inefficient waste process that is triggered when the Rubisco reacts with O_2 rather than CO_2. This leads to the production of glycolate, and its recycling is highly energy- and fixed-carbon-intensive. Reducing photorespiration is an essential strategy for increasing photosynthetic efficiency and net carbon gain.

- **By-pass mechanisms:** Artificially induced shunts divert glycolate into beneficial metabolites in chloroplasts or mitochondria and reduce the amount of CO_2 released (Smith et al., 2025).
- **Transport optimization:** Optimized glycolate transporters speed up the clearance of metabolites and limit accumulation that otherwise slows metabolism.
- **C_4 and CAM trait introduction:** C_4 and CAM plants accumulate CO_2 in the vicinity of Rubisco, which radically lowers oxygenase activity and the subsequent photorespiration process (Shah et al., 2022).
- **Rubisco selectivity:** Genetic variants of Rubisco that are more CO_2 selective suppress the activation of photorespiratory cycles completely. Plants reduce the metabolic costs of photorespiration and save carbon and energy, which can be used for growth (Fig. 7.3).

4.4 Energy allocation to growth and yield

Where photosynthetic rates are enhanced, the productivity improvement cannot always be converted into yield unless there is an efficiency in the distribution of energy and assimilates. At this point, optimization includes balancing of energy, reinforcing source-sink relationships, and proper partitioning of assimilates to harvestable organs.

- **ATP/NADPH ratio:** Fine-tuning of the ATP/NADPH ratio to satisfy Calvin cycle needs is carried out by altering cyclic electron transport (Zhang et al., 2023).
- **Source-sink coordination:** Established sink strength with increased phloem loading or sink organ demand prevents the accumulation of carbohydrates, which suppresses photosynthesis.
- **Partitioning control:** Genetic and metabolic controls lead to preferential input of photoassimilates into seeds, tubers, and fruits instead of surplus vegetative biomass (Sahoo et al., 2025a).
- **Stress-adaptive distribution:** It is possible to channel energy to protective compounds in stressful conditions and at the same time retain growth at the baseline under favorable conditions. These modifications ensure that the photosynthetic profits are not incurred but, in fact, diverted to productive use.

5. Synergistic impacts of nano-fertilizers and CO_2 enrichment

Nano-fertilizer technology and CO_2 enrichment are steps in the right direction for improving photosynthetic rates and crop tolerance. Nano-fertilizers improve the transportation of nutrients at cellular and organellar scales, whereas CO_2 enrichment eliminates the CO_2-barrier to photosynthesis. Their combination enhances the leaf microenvironment, maintains chloroplast ultrastructure, mitigates reactive oxygen species, and confers stress resistance. The dual approach will not only capture carbon more effectively but also stabilize the cells in different environments.

TABLE 7.2 Mechanistic strategies for accelerating the Calvin–Benson cycle and enhancing carbon fixation efficiency.

Mechanistic approach	Key enzyme	Mechanism of action	Functional outcome in plants	References
Enzyme overexpression	Sedoheptulose-1,7-bisphosphatase (SBPase)	Catalyzes the regeneration of RuBP for faster carbon cycling	Increases photosynthetic turnover and biomass production	Chen et al. (2022)
Enzyme overexpression	Fructose-1,6-bisphosphatase (FBPase)	Enhances the dephosphorylation of fructose-1,6-bisphosphate	Promotes RuBP regeneration and improves carbon flux	Li et al. (2025)
Triose-phosphate utilization	Triose-phosphate isomerase	Facilitates the conversion and export of triose phosphates	Boosts sucrose and starch biosynthesis efficiency	Fu et al. (2023)
Metabolite stabilization	Glyceraldehyde-3-phosphate	Balances intermediate metabolite pools	Reduces bottlenecks and maintains steady carbon flow	Zhu et al. (2025)
Synthetic pathway integration	Synthetic photorespiration-bypassing cycle	A couple of additional carbon-fixation routes with the Calvin cycle	Increases total carbon assimilation efficiency	Amaral et al. (2024)
Enzyme engineering	Rubisco activase	Enhances Rubisco carbamylation and CO_2 fixation rate	Improves catalytic efficiency and lowers oxygenation losses	Chen et al. (2022)
Co-factor optimization	NADPH and ATP balance	Ensures adequate reducing power for the cycle	Maintains energetic equilibrium during carbon fixation	Li et al. (2025)
Protein stabilization	Chaperone-assisted folding	Maintains functional enzyme conformation under stress	Extends enzyme lifetime and improves thermal stability	Fu et al. (2023)
Gene stacking	Multigene expression of Calvin enzymes	Coordinates the overexpression of key enzymes in one construct	Maximizes RuBP regeneration and overall flux capacity	Zhu et al. (2025)
Metabolite channeling	Enzyme complex proximity	Reduces diffusion losses between sequential reactions	Enhances substrate turnover efficiency	Amaral et al. (2024)
Synthetic promoter regulation	Strong and light-inducible promoters	Controls enzyme expression linked to light intensity	Synchronizes carbon fixation with photosynthetic activity	Chen et al. (2022)
Compartmental engineering	Chloroplast-targeted expression	Directs synthetic enzymes to the site of carbon fixation	Enhances local metabolite concentration and cycle speed	Li et al. (2025)
Feedback inhibition control	Inhibition-resistant enzymes	Prevents accumulation-induced suppression of cycle activity	Maintains continuous flux through Calvin intermediates	Fu et al. (2023)
Carbon allocation modulation	Partitioning between sucrose and starch	Balances carbon export and storage	Prevents feedback repression and improves carbon use efficiency	Zhu et al. (2025)
Metabolic integration	Cross-linking with glycolysis and C_4 intermediates	Expands carbon flow beyond the Calvin cycle	Enhances overall productivity and carbon retention	Amaral et al. (2024)

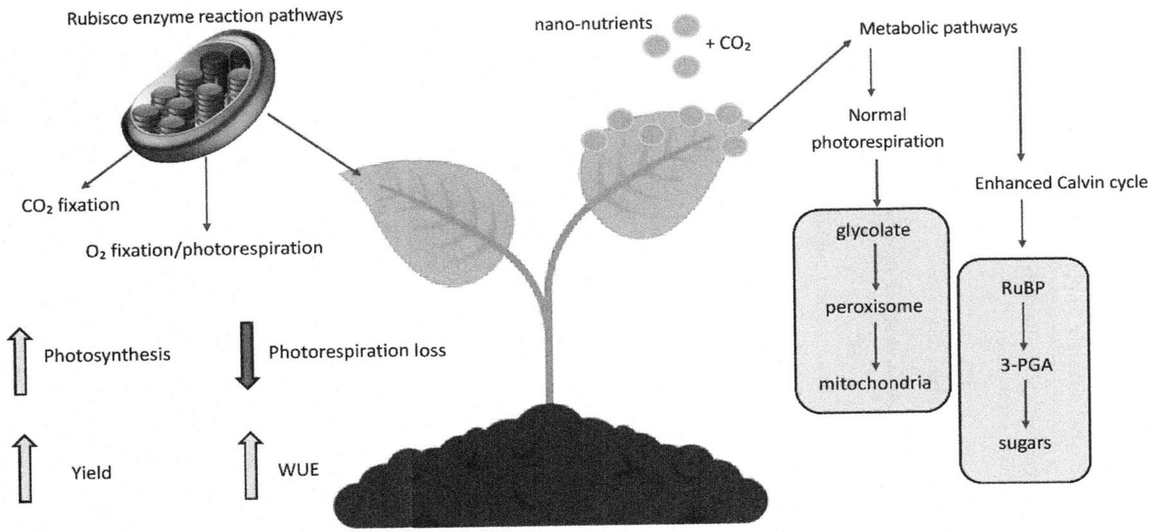

FIGURE 7.3 Strategies to minimize photorespiration and enhance photosynthetic efficiency through metabolic bypasses, transporter optimization, and Rubisco or CO_2 concentrating trait engineering.

5.1 Leaf microenvironment modification

The microenvironment of leaves controls the direct exchange of gases, nutrients, and water, and optimization of the microenvironment has a substantial impact on the results of photosynthesis. These are optimized by incorporating nanofertilizers and adding CO_2 to enhance nutrient absorption and uptake by carbon.

- **Stomatal conductance regulation:** Nano-fertilizers with potassium and calcium can control the ion fluxes at the guard cell, and enriched CO_2 can improve the concentration gradient entering the substomatal cavities (Nazeer et al., 2023).
- **Cuticular penetration:** Nano-formulations (hydrophilic coating) can bypass cuticle barriers to deliver micronutrients directly to mesophyll sub-layers where CO_2 is incorporated.
- **Localized nutrient gradients:** Nano-scale delivery produces a uniform nutrient supply on the lamina of the leaf as the demands of ATP and NADPH increase with CO_2 enrichment (Guo et al., 2023).
- **Boundary layer stability:** Nanoparticles applied to the foliage can increase the wettability of the leaf surface, maintaining favorable microclimatic conditions to enhance long-term CO_2 diffusion. These microenvironmental adaptations serve to coordinate the supply of carbon and nutrients at the photosynthetic site.

5.2 Chloroplast ultrastructure stability

The main centers of photosynthesis are chloroplasts, and their ultrastructure (thylakoid membranes, grana stacks, stroma lamellae) needs to be preserved to ensure the proper functioning of electron transport and carbon fixation. CO_2-enriched nano-fertilizers improve structural stability under varying light and stress conditions.

- **Membrane lipid protection:** Nano-delivered zinc and magnesium maintain the integrity of the thylakoid membrane, minimizing photodamage when subjected to increased photosynthetic flux due to CO_2 enrichment (Li et al., 2024a).

- **Maintenance of grana stacking:** Nanoparticles that provide phosphorus and sulfur promote ATP generation and protein phosphorylation, and preserve the grana architecture to balance the PSII and PSI excitation (Lauria et al., 2024).
- **Protein stabilization:** Photosystem cofactor delivery by iron and manganese nanoparticles prevents reaction center degradation.
- **Osmotic regulation:** Nano-calcium has been shown to affect the osmotic regulation of chloroplasts, resulting in the swelling or contraction of the stroma compartments in response to changes in CO_2 uptake (Gao et al., 2023). A constant chloroplast structure would ensure that increased carbon assimilation capacity is facilitated by efficient electron transfer and energy conversion.

5.3 Reactive oxygen species (ROS) management

When photosynthetic flux is associated with increased CO_2 enrichment, the leakage of electrons can be increased, which produces ROS. Nano-fertilizers help provide cofactors and antioxidants to maintain redox balance and prevent oxidative injury.

- **Activation of antioxidant enzymes:** Nano-copper and nano-manganese stimulate the enzymes, superoxide dismutases and peroxidases, to promote the elimination of superoxide radicals and hydrogen peroxide.
- **Glutathione and ascorbate cycling:** Nano-selenium and nano-sulfur induce thiol-based antioxidant biosynthesis, which enhances ascorbate-glutathione pathways (Jia et al., 2024).
- **Thylakoid redox equilibrium:** Nano-iron maintains electron carriers in cytochrome b6f and PSI, which prevents over-reduction leading to the production of ROS (Kermeur et al., 2023).
- **ROS scavenging nanoparticles:** Engineered Cerium oxide nanoparticles imitate catalase and peroxidase functions, and directly react with hydrogen peroxide where it is generated. Plants with enhanced ROS control can maintain oxidative homeostasis under high photosynthetic rates, thereby preventing cellular damage (Fig. 7.4).

5.4 Stress resilience and photoprotection

In addition to short-term benefits on photosynthesis, the synergistic effect of nano-fertilizers and CO_2 enrichment imparts more extensive resistance to abiotic stressors, including drought, heat, and high light intensity. Photoprotection mechanisms protect chloroplasts and stabilize metabolic processes in changing environments.

- **Non-photochemical quenching (NPQ):** Nano-magnesium contributes to the activity of the xanthophyll cycle, which dissipates excess light energy as heat, and the enriched CO_2 reduces the excitation pressure on PS II (Li et al., 2024b).
- **Osmolyte accumulation:** Nano-silicon promotes the accumulation of proline and glycine betaine that stabilize membranes and proteins in times of stress to supplement the water-use efficiency regulated by CO_2.
- **Heat-shock protein stimulation:** Nano-chaperones such as nano-titanium modulate the expression of protective proteins, which guarantee functional recovery of photosystems during temperature changes (Sidhu et al., 2024).
- **Water use efficiency:** Efficiency of water-use CO_2 decreases stomatal conductance, water-

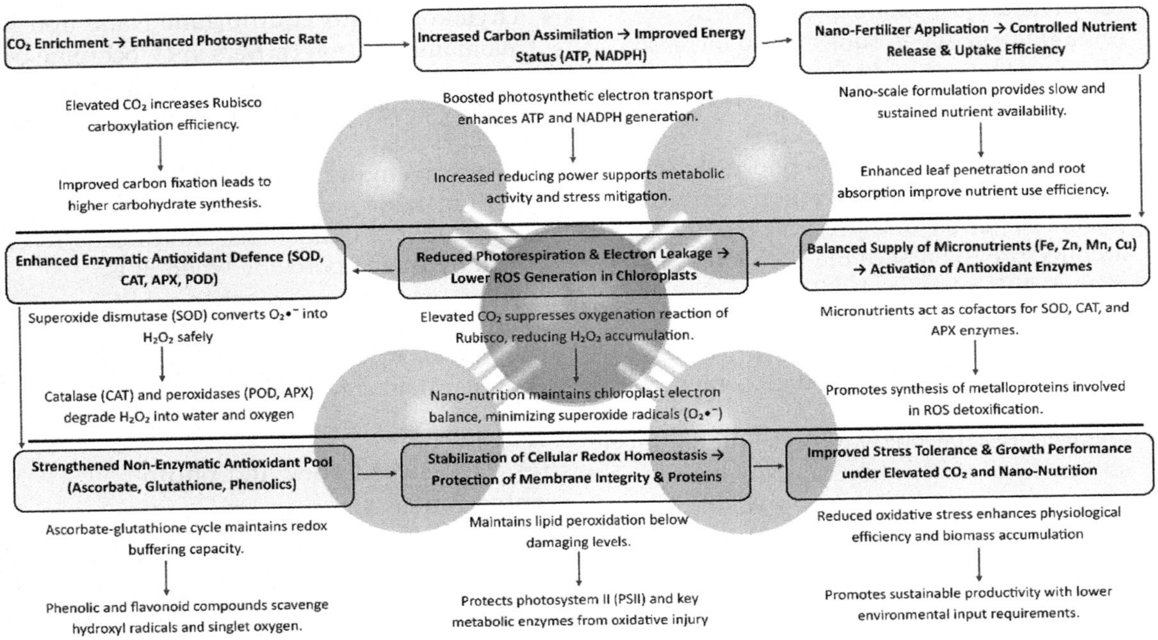

FIGURE 7.4 Nanoparticle-mediated enhancement of antioxidant defense and redox homeostasis to mitigate ROS generation during high photosynthetic activity pathways.

conserving and nano-potassium increases hydraulic control to maintain turgor pressure. The processes enable plants to maintain growth, minimize damage, and optimize productivity in changing field environments.

6. Smart monitoring and AI integration

A combination of advanced monitoring technologies and artificial intelligence (AI) enables the development of a new agricultural optimization paradigm in which data precision governs photosynthetic performance, nutrient delivery, and yield. Smart sensors and sophisticated algorithms (Internet of Things, IoT) provide real-time data on plant physiology and the surrounding environment, enabling crops to respond dynamically to external factors. AI additives are combined with feedback, nutrient-carbon processes, CO_2 management, and predictive yield analytics to form a single system for agricultural responsiveness.

6.1 IoT-driven photosynthetic feedback loops

Adaptive crop management is based on the idea of IoT systems, where a network of interacting gadgets and biosensors is used to measure physiological parameters in real time. These systems bridge the gap between environmental monitoring and immediate plant action.

- **Chlorophyll fluorescence sensors:** IoT-based fluorometers can measure PSII efficiency, which is the ratio of energy between photochemistry and heat dissipation.
- **Gas exchange analyzers:** Distributed leaf chambers record the rate of CO_2 fixation and transpiration and transmit real-time

photosynthetic data to cloud interfaces (Banerji et al., 2024).
- **Microclimate sensors:** IoT sensors measure temperature, humidity, and light intensity around the canopy, providing feedback on external factors that drive photosynthesis.
- **Autonomous control loops:** Sensor readings cause actuators (i.e., irrigation, nutrient sprays, or shading equipment) to be adjusted to achieve optimum photosynthetic efficiency, on demand (Lakhiar et al., 2025). These feedback mechanisms provide an interface for the responsiveness of plant messages to resource allocation.

6.2 Machine learning for nutrient-carbon optimization

Machine learning algorithms are used to optimize the nutrient carbon nexus by integrating large amounts of data on nutrient supply, carbon uptake, and growth rates. AI enhances decision-making regarding fertilization schedules and resource allocation by revealing nonlinear relationships.

- **Pattern recognition of nutrient uptake:** Ion concentration and uptake kinetics during nano-fertilizer application are studied by algorithms to forecast the optimal balance between the carbon assimilation and the nutrient uptake (Pudhuvai et al., 2025).
- **Growth rate correlation models:** Neural networks can be used to correlate photosynthetic carbon fixation to nutrient assimilation, including the thresholds above which no further benefit of adding supply can be achieved.
- **Adaptive fertilization schedules:** AI changes nutrient formulations dynamically in response to carbon demand requirements at various phenological stages of crop development (Li et al., 2024a).
- **Resource-use efficiency mapping:** Machine learning creates spatial heat maps of nutrient-carbon ratios at fields, which inform local precision interventions. These optimizations yield a viable, profit-maximizing equilibrium between nutrient input and carbon uptake.

6.3 Sensor-driven real-time CO_2 regulation

Coupled with smart sensors installed on agricultural systems, continuous readings on the availability of CO_2 allow perfect control of enrichment strategies. The actual time control is used so that the quantity of CO_2 discharged aligns with the physiological requirements of photosynthesis and produces minimal waste.

- **Infrared gas analyzers:** CO_2 concentration monitors on greenhouse and open-field chambers monitor changes in ambient concentrations and activate automated enrichment.
- **Stomatal conductance sensors:** Miniature thermal or porometer devices measure stomatal aperture reactions, and associate gas exchange and external CO_2 supply.
- **Integrated diffusion controllers:** The actuators regulate the release of micro- or nano-bubble CO_2 in irrigation systems in response to sensor signals, varying the output according to the uptake by the plant (Huang et al., 2024).
- **CO_2-use efficiency indices:** The ratio of the amount of CO_2 assimilated to the amount of CO_2 supplied is computed in real-time and used to maintain efficiency in regulation and avoid over-enrichment. Such a sensor-based method enables precise delivery of carbon in tandem with photosynthetic activity.

6.4 Predictive modeling for yield forecasting

The predictive models developed by AI combine historical, physiological, and environmental data to predict the crop yield with a high level of accuracy. These models extend passive estimates and enable proactive interventions at any stage of development.

- **Phenology-based simulations:** Predictive models are models that use growth stage identification together with climatic information to provide predictions of biomass gain and yielding.
- **Remote sensing integration:** Spectral indices (e.g., NDVI, PRI) provided by satellite and drone imagery are included in yield predictions algorithms (Kurihara et al., 2023).
- **Stress factors included:** Drought, heat, or nutrient limitation. Stress factors are introduced into the model, and their effects on yield potential and photosynthetic efficiency are modeled.
- **Continuous recalibration:** AI is continuously improved based on new real-time sensor data, so yield forecasts change dynamically as the plant and the environment vary. Predictive yield modeling enables farmers to anticipate productivity in advance while retaining the flexibility to adjust practices during the season.

7. Challenges and future horizons

The development of nano-enabled fertilizers and CO_2-enrichment technologies has placed agriculture at a crossroads where productivity, sustainability, and resilience can be addressed concurrently. However, several significant obstacles remain before these strategies can be implemented at scale in international agriculture. The issues related to these challenges are linked to the biological and ecological uncertainty of nanobio interactions, limited scalability in financial and logistical terms, the absence of long-established policy frameworks, and the prospect of hybridized farming ecosystems integrating nanotechnology, carbon regulation, and precision agriculture.

7.1 Nano–bio interaction and toxicity concerns

When engineered nanoparticles are introduced into plant systems and into soil, some fundamental questions emerge about the biological compatibility of these nanoparticles. Nanoparticles may penetrate plant cell membranes via apoplastic and symplastic routes and accumulate in organelles, such as chloroplasts and mitochondria (Djanaguiraman et al., 2024). This is beneficial for nutrient delivery and photosynthesis, but it also carries a risk of inducing oxidative stress. Overaccumulation of nanoparticles has been reported to provoke the hyperproduction of reactive oxygen species, lipid peroxidation, destabilization of thylakoid membranes, and loss of enzyme activity in biological systems. Such modifications may disrupt electron transport chains and disrupt phototransduction control of photosynthetic proteins. Nanoparticles affect the surrounding soil microbiome beyond the plant itself. Bacteria that fix nitrogen and mycorrhizal fungi can be highly susceptible to variations in nanoparticle composition and concentration, thereby altering nutrient cycling (Sahu et al., 2025). Such disturbances affect the long-term and short-term wellbeing and fertility of crops and soil. In addition, the presence of nanoparticles in the tissues of the edible plants makes them transferable through the food chain. There are concerns regarding human and animal exposures via food consumption, which has increased the need to conduct toxicological assessments that consider nanoscale characteristics, such as surface charge and catalytic activity, which differ markedly from those of their bulk chemical counterparts.

7.2 Cost-effectiveness and scalability barriers

Although the use of nano-fertilizer and CO_2 delivery systems is clearly physiologically favorable, their use is not readily adopted due to the barriers of cost-effectiveness and scalability. Nanoparticles with small size, surface coating, and high stability require advanced synthesis methods, such as sol-gel and laser ablation, which are not yet resource-efficient. Such expenses increase the cost of commercial nanoformulations and make them less affordable for smallholder farmers, who constitute the bulk of the agricultural labor force in most areas (Kumar et al., 2025). Another complexity is transport and storage; nanoparticles still require specialized packaging, often to prevent agglomeration or loss of activity. Their performance at the field scale can be highly variable, with soil pH, organic matter content, and moisture levels influencing outcomes. One such case is the adsorption of nanoparticle products in high-clay-content soils (which lowers bioavailability) and sandy soils (which can be washed away too rapidly), rendering them ineffective (Madhupriyaa et al., 2024). There are also economic trade-offs for farmers between the short-term costs of inputs and potential yield benefits, which are often uncertain because they have not been demonstrated through sufficiently long-term trials. The challenges can only be overcome through not only technological advancement to reduce production costs but also institutional arrangements such as subsidies, cooperative distribution systems, and farmer training programs, as well as through the economic feasibility of nano-enabled agricultural systems.

7.3 Policy and regulatory frameworks

The regulatory environment around nano-agriculture is not well developed, and both producers and end-users remain uncertain. Traditional pesticide/fertilizer policies are not well equipped to address the specific characteristics of nanomaterials, which can be more reactive, persist longer, or exhibit unexpected interactions in the biosphere. There are no standard safety limits for nanoparticle concentrations in soil, irrigation water, and harvested products, and risk assessment is still conducted across various jurisdictions. In addition, current toxicology procedures, established for bulk chemicals, typically do not measure nanospecific endpoints, including changes in acetylcholinesterase activity, oxidative stress biomarkers, or DNA methylation (Li et al., 2024a). Another issue is the transparency of labeling; producers and consumers have limited information about the molecular-level effects of nano-enabled products. Unless there are standardized labeling practices that involve safety endpoints, distrust of nanotechnology in agriculture is likely to increase. The problem is further complicated by global trade, in which countries adopt opposing regulatory positions that restrict the cross-border movement of nano-agricultural inputs and outputs. Such gaps should be closed through regular international mechanisms, dynamic risk communication, and a more productive interface between scientific evidence and policymaking.

7.4 Toward hybridized nano-carbon farming ecosystems

In the future, agricultural systems will develop into hybridized ecosystems that combine nanotechnology-based systems of nutrient delivery and CO_2 enrichment with digital monitoring technologies and ecological sustainability principles. These ecosystems conceptualize nutrient and carbon management as an integrated system in which nano-fertilizers match mineral provision with a high carbon uptake. This kind of synchronization minimizes inefficiencies arising from nutrient limitation when nutrient availability is high, or CO_2 levels are high, and from nutrient waste when carbon availability is low. Circular bio-economy principles could also be applied to future farming, in which agricultural waste streams, stalks, and residues are transformed into bio-based nanoparticles, thereby forming closed-loop inputs (Kadhila et al., 2024). Similar advances in digital agriculture, such as the use of digital twin models of plant systems, will enable real-time simulation of crop performance in response to nano-enabled interventions. The models will enable farmers to predict stress events, manage input application, and maximize yield while minimizing environmental costs. Hybridized systems will also be used to mitigate climate change by maximizing water-use efficiency to reduce greenhouse gas emissions and ensure that soils retain their fertility. By combining nanotechnology, carbon management, and precision agriculture, the next generation of farm infrastructure will be able to achieve a balance between productivity and long-term ecological stability.

8. Conclusion

Fertilizers with nano-based formulations and CO_2 delivery accuracy are complementary technologies in the quest to optimize photosynthesis and improve agricultural productivity. The high bioavailability of nutrients in nano-fertilizers, resulting from their large surface-area-to-volume ratio and controlled-release processes, reduces nutrient leaching and the ecological impact of intensive fertilizer application. Their selective action enables the delivery of macronutrients and micronutrients directly to the metabolic hotspots of plant tissues, thereby promoting improved chloroplast activity and redox homeostasis. Meanwhile, CO_2 delivery systems specifically address the substrate limitation of photosynthesis, enhance carboxylation efficiency, reduce photorespiration, and stabilize the Calvin–Benson cycle under fluctuating field conditions. Nano-bubbles and accurate injection can be used to uniformly distribute CO_2 at the leaf-air interface, producing quantifiable changes in stomatal conductance and chloroplast ultrastructural stability. When co-used, nano-fertilizers and CO_2 enrichment generate a synergistic microenvironment that promotes quantum yield of photosynthesis, protects plants against abiotic stress, and maintains improved biomass accumulation. The integration of these technologies with IoT systems, biosensors, and machine learning algorithms will enable the continuous optimization of photosynthetic inputs and their translation into practical agronomic decisions. Yet there remain hurdles-the most significant of them being the environmental behavior of engineered nanoparticles, regulatory uncertainty, and the economics of scale. Green nanomaterial design, eco-toxicological testing and farmer-friendly delivery systems should be the future. Finally, nanotechnology and carbon engineering pose a compounding challenge to agricultural practice, redefining how nutrients and carbon are managed to enhance photosynthetic performance. The solutions increase food security through food efficiency and global demands.

References

Amaral, J., Lobo, A.K.M., Carmo-Silva, E., 2024. Regulation of Rubisco activity in crops. New Phytol. 241 (1), 35–51. https://doi.org/10.1111/nph.19369.

Anjali, P., Kapley, A., Qureshi, A., 2025. Exploring the potential of natural and synthetic microbial carbon dioxide fixation pathways for catalysing climate action. In: Microorganisms Resilience to Climate Change. Springer Nature Singapore, Singapore, pp. 21–39. https://doi.org/10.1007/978-981-96-3748-5_2.

Banerji, S., Hõrak, H., Torop, J., Huynh, T.P., 2024. Unravelling the secrets of plants: emerging wearable sensors for plants signals and physiology. Adv. Sens. Res. 3 (3). https://doi.org/10.1002/adsr.202300023.

Chen, J.H., Tang, M., Jin, X.Q., Li, H., Chen, L.S., Wang, Q.L., Sun, A.Z., Yi, Y., Guo, F.Q., 2022. Regulation of Calvin–Benson cycle enzymes under high temperature stress. aBIOTECH 3 (1), 65–77. https://doi.org/10.1007/s42994-022-00068-3.

Croce, R., Carmo-Silva, E., Cho, Y.B., Ermakova, M., Harbinson, J., Lawson, T., McCormick, A.J., Niyogi, K. K., Ort, D.R., Patel-Tupper, D., Pesaresi, P., Raines, C.,

Weber, A.P.M., Zhu, X.G., 2024. Perspectives on improving photosynthesis to increase crop yield. Plant Cell 36 (10), 3944–3973. https://doi.org/10.1093/plcell/koae132.

Djanaguiraman, M., Anbazhagan, V., Dhankher, O.P., Prasad, P.V.V., 2024. Uptake, translocation, toxicity, and impact of nanoparticles on plant physiological processes. Plants 13 (22). https://doi.org/10.3390/plants13223137.

Fu, X., Gregory, L.M., Weise, S.E., Walker, B.J., 2023. Integrated flux and pool size analysis in plant central metabolism reveals unique roles of glycine and serine during photorespiration. Nat. Plants 9 (1), 169–178. https://doi.org/10.1038/s41477-022-01294-9.

Gao, Y., Chen, S., Li, Y., Shi, Y., 2023. Effect of nano-calcium carbonate on morphology, antioxidant enzyme activity and photosynthetic parameters of wheat (*Triticum aestivum* L.) seedlings. Chem. Biol. Technol. Agric. 10 (1). https://doi.org/10.1186/s40538-023-00404-9.

Gautam, K., Singh, H., Sinha, A.K., 2025. Nanotechnology in plant nanobionics: mechanisms, applications, and future perspectives. Adv. Biol. 9 (4). https://doi.org/10.1002/adbi.202400589.

Ghorbani, A., Emamverdian, A., Pehlivan, N., Zargar, M., Razavi, S.M., Chen, M., 2024. Nano-enabled agrochemicals: mitigating heavy metal toxicity and enhancing crop adaptability for sustainable crop production. J. Nanobiotechnol. 22 (1). https://doi.org/10.1186/s12951-024-02371-1.

Guo, X., Zhang, J., Yuan, L., Xi, B., Gao, F., Zheng, X., Pan, R., Guo, L., An, X., Fan, T., Xiong, S., 2023. Biologically assisted construction of advanced electrode materials for electrochemical energy storage and conversion. Adv. Energy Mater. 13 (24). https://doi.org/10.1002/aenm.202204376.

Huang, L., He, X., Hu, J., Qin, C., Huang, C., Tang, Y., Zhong, F., Kong, X., Wei, X., 2024. Wearable sensor based on covalent organic framework humidity films for long-term monitoring of tomato physiology under abiotic stress. ACS Nano 18 (48), 33105–33118. https://doi.org/10.1021/acsnano.4c09916.

Janová, J., Kubásek, J., Grams, T.E.E., Zeisler-Diehl, V., Schreiber, L., Šantrůček, J., 2024. Effect of light-induced changes in leaf anatomy on intercellular and cellular components of mesophyll resistance for CO_2 in Fagus sylvatica. Plant Biol. 26 (5), 842–854. https://doi.org/10.1111/plb.13655.

Jia, Y., Chen, L., Kang, L., Fu, X., Zheng, S., Wu, Y., Wu, T., Cai, R., Wan, X., Wang, P., Yin, X., Pan, C., 2024. Nano-selenium and glutathione enhance cucumber resistance to botrytis cinerea by promoting jasmonic acid-mediated cucurbitacin biosynthesis. ACS Nano 18 (31), 20576–20590. https://doi.org/10.1021/acsnano.4c05827.

Jin, K., Chen, G., Yang, Y., Zhang, Z., Lu, T., 2023. Strategies for manipulating Rubisco and creating photorespiratory bypass to boost C3 photosynthesis: prospects on modern crop improvement. Plant Cell Environ. 46 (2), 363–378. https://doi.org/10.1111/pce.14500.

Kadhila, T., Amesho, K.T.T., Shangdiar, S., Shihomeka, S.P., Shopati, A.K., Edoun, E.I., 2024. Sustainable bioeconomy for next-generation industrial ecology. In: Garg, V.K., Kataria, N. (Eds.), Bioeconomy for Sustainability. Springer Nature Singapore, Singapore, pp. 293–317. https://doi.org/10.1007/978-981-97-1837-5_13.

Kaur, H., Kaur, H., Kaur, H., Srivastava, S., 2023. The beneficial roles of trace and ultratrace elements in plants. Plant Growth Regul. 100 (2), 219–236. https://doi.org/10.1007/s10725-022-00837-6.

Kermeur, N., Pédrot, M., Cabello-Hurtado, F., 2023. Iron availability and homeostasis in plants: a review of responses, adaptive mechanisms, and signaling. Methods Mol. Biol. 2642, 49–81. https://doi.org/10.1007/978-1-0716-3044-0_3.

Kizhisseri, M.I., Sakr, M., Maraqa, M., Mohamed, M.M., 2025. A comparative bench scale study of oxygen transfer dynamics using micro-nano bubbles and conventional aeration in water treatment systems. Heliyon 11 (4). https://doi.org/10.1016/j.heliyon.2025.e41687.

Komkhum, T., Sema, T., Rehman, Z.U., In-na, P., 2025. Carbon dioxide removal from triethanolamine solution using living microalgae-loofah biocomposites. Sci. Rep. 15 (1). https://doi.org/10.1038/s41598-025-90855-x.

Kumar, P., Choudhury, D., Dey, S.R., Raj, A., 2025. Nanofertilizers/nanopesticides for the national economy. In: Kumar, P., Dubey, R.C. (Eds.), Nanofertilizers for Sustainable Agriculture. Springer, Cham. https://doi.org/10.1007/978-3-031-78649-5_17.

Kurihara, J., Nagata, T., Tomiyama, H., 2023. Rice yield prediction in different growth environments using unmanned aerial vehicle-based hyperspectral imaging. Remote Sens. 15 (8), 2004. https://doi.org/10.3390/rs15082004.

Lakhiar, I.A., Yan, H., Syed, T.N., Zhang, C., Shaikh, S.A., Rakibuzzaman, M., Vistro, R.B., 2025. Soilless agricultural systems: opportunities, challenges, and applications for enhancing horticultural resilience to climate change and urbanization. Horticulturae 11 (6). https://doi.org/10.3390/horticulturae11060568.

Lauria, G., Ceccanti, C., Lo Piccolo, E., El Horri, H., Guidi, L., Lawson, T., Landi, M., 2024. "Metabolight": how light spectra shape plant growth, development and metabolism. Physiol. Plantarum 176 (6). https://doi.org/10.1111/ppl.14587.

Li, M., Li, H., Zhou, Z., Chen, Y., Wang, Y., Li, T., Fu, Q., 2024a. Spatial simulation and optimization of cropping structure under climate and land use change conditions considering synergistic economic benefits and carbon reduction in crop growth processes. Earths Future 12 (10). https://doi.org/10.1029/2024EF004684.

Li, P., Xia, Y., Song, K., Liu, D., 2024b. The impact of nanomaterials on photosynthesis and antioxidant mechanisms in gramineae plants: research progress and future prospects. Plants 13 (7), 984. https://doi.org/10.3390/plants13070984.

Li, Z., Lanster, D.L., Badran, A.H., 2025. Synthetic approaches to enhance biological carbon capture. Curr. Opin. Biotechnol. 95. https://doi.org/10.1016/j.copbio.2025.103350.

Madhupriyaa, D., Baskar, M., Sherene Jenita Rajammal, T., Kuppusamy, S., Rathika, S., Umamaheswari, T., Sriramachandrasekran, M.V., Mohanapragash, A.G., 2024. Efficacy of chelated micronutrients in plant nutrition. Commun. Soil Sci. Plant Anal. 55 (22), 3609–3637. https://doi.org/10.1080/00103624.2024.2397019.

Marcelino, K.R., Ling, L., Wongkiew, S., Nhan, H.T., Surendra, K.C., Shitanaka, T., Lu, H., Khanal, S.K., 2023. Nanobubble technology applications in environmental and agricultural systems: opportunities and challenges. Crit. Rev. Environ. Sci. Technol. 53 (14), 1378–1403. https://doi.org/10.1080/10643389.2022.2136931.

Meloni, M., Gurrieri, L., Fermani, S., Velie, L., Sparla, F., Crozet, P., Henri, J., Zaffagnini, M., 2023. Ribulose-1,5-bisphosphate regeneration in the Calvin-Benson-Bassham cycle: focus on the last three enzymatic steps that allow the formation of Rubisco substrate. Front. Plant Sci. 14. https://doi.org/10.3389/fpls.2023.1130430.

Nakachew, K., Yigermal, H., Assefa, F., Gelaye, Y., Ali, S., 2024. Review on enhancing the efficiency of fertilizer utilization: strategies for optimal nutrient management. Open Agric. 9 (1). https://doi.org/10.1515/opag-2022-0356.

Nazeer, A., Ahmad, F., Verma, N., Ahmad, S., 2023. Nanofertilizers in agriculture. In: Targeted Delivery of Nanopesticides and Nanofertilizers in Sustainable Agricultural Farming. Springer Nature Switzerland, Cham, pp. 87–98. https://doi.org/10.1007/978-3-031-41333-9_5.

Pudhuvai, B., Koul, B., Das, R., Shah, M.P., 2025. Nanofertilizers (NFs) for resurgence in nutrient use efficiency (NUE): a sustainable agricultural strategy. Curr. Pollut. Rep. 11 (1). https://doi.org/10.1007/s40726-024-00331-9.

Sahoo, R., Samanta, D., Sow, S., Ranjan, S., Nath, D., Sadhu, S., Kumar, N., Rana, L., Kumar, A., Roy, D.K., Dharminder, R.K., Raigar, B.L., Das, S., Panday, R., 2025a. Nexus between photosynthesis and radiation use efficiency towards achieving sustainability in the era of climate change: an overview. Discov. Environ. 3 (1). https://doi.org/10.1007/s44274-025-00252-4.

Sahoo, S.K., Dash, G.K., Pati, S., Sahoo, D., Lenka, B., Nayak, L., Kumar, A., Swain, P., Guhey, A., 2025b. Physiological and Biochemical Mechanisms in Underground Vegetable Crops for Growth and Development Abiotic Stress in Underground Vegetables: Response and Mitigation. Elsevier, pp. 39–53. https://doi.org/10.1016/B978-0-443-23961-8.00003-6.

Sahu, V., Patel, A.K., Ahmad, F., 2025. The effect of nanoparticles on nitrogen fixation by trees. In: Nanomaterials in Agroforestry Systems. Springer Nature Singapore, Singapore, pp. 143–170. https://doi.org/10.1007/978-981-96-1337-3_7.

Shah, W.H., Saleem, S., Mushtaq, N.U., Rasool, A., Tahir, I., Rehman, R.U., 2022. C4 and CAM plants with better resilience to environmental stresses. In: Photosynthesis and Respiratory Cycles During Environmental Stress Response in Plants. Apple Academic Press, pp. 163–191. https://doi.org/10.1201/9781003315162.

Shanker, A.K., Gunnapaneni, D., Bhanu, D., Vanaja, M., Lakshmi, N.J., Yadav, S.K., Prabhakar, M., Singh, V.K., 2022. Elevated CO_2 and water stress in combination in plants: brothers in arms or partners in crime? Biology 11 (9). https://doi.org/10.3390/biology11091330.

Sharma, N., Chaudhary, A., Sharma, N., Shukla, S., Joshi, N. C., Vishwakarma, K., Singh, P.K., Mishra, A., 2025. Plant Nutrient Requirements and Nutrient Homeostasis in Plants Soil Health and Nutrition Management. CABI International, pp. 1–20. https://doi.org/10.1079/9781800624597.0001.

Sidhu, A.K., Sharma, M., Bhickchand Agrawal, S., Bhavsar, P., Samota, M.K., 2024. Nanomaterial strategies for enhancing plant resilience in the face of temperature stress. CABI Agric. Biosci. 5 (1). https://doi.org/10.1186/s43170-024-00255-w.

Singh, V.P., Siddiqui, M.H., Graciano, C., Ullah, A., Sadia, S., Ahmed, Z., Murtaza, G., Ismoilov, K., Zhang, Z., 2023. Regulation of Metabolites by Nutrients in Plants Plant Ionomics: Sensing, Signaling, and Regulation. Wiley, pp. 1–18. https://doi.org/10.1002/9781119803041.ch1.

Singh, M., Goswami, S.P., Ranjitha, G., Sachan, P., Kumar Sahu, D., Beese, S., Pandey, S.K., 2024. Nanotech for fertilizers and nutrients-improving nutrient use efficiency with nano-enabled fertilizers. J. Exp. Agric. Int. 46 (5), 220–247. https://doi.org/10.9734/jeai/2024/v46i52372.

Smith, E.N., Van Aalst, M., Weber, A.P.M., Ebenhöh, O., Heinemann, M., 2025. Alternatives to photorespiration: a system-level analysis reveals mechanisms of enhanced plant productivity. Sci. Adv. 11 (13). https://doi.org/10.1126/sciadv.adt9287.

Thangavelu, R.M., da Silva, W.L., Zuverza-Mena, N., Dimkpa, C.O., White, J.C., 2024. Nano-sized metal oxide fertilizers for sustainable agriculture: balancing benefits, risks, and risk management strategies. Nanoscale 16 (43), 19998–20026. https://doi.org/10.1039/d4nr01354a.

Traiphop, S., Temprom, L., Thopan, P., Seemakram, W., Boonlue, S., Krongsuk, S., 2025. Optimizing CO_2 micro-nanobubble formulations with non-ionic surfactants for improving plant growth and photosynthesis efficiency. Surf. Interfaces 60, 106067. https://doi.org/10.1016/j.surfin.2025.106067.

Tripathi, S., Tiwari, K., Mahra, S., Victoria, J., Rana, S., Tripathi, D.K., Sharma, S., 2024. Nanoparticles and root traits: mineral nutrition, stress tolerance and interaction with rhizosphere microbiota. Planta 260 (2). https://doi.org/10.1007/s00425-024-04409-y.

Wang, X.Q., Zeng, Z.L., Li, Y.Y., Huang, W., 2024. Photoinhibition and photosynthetic regulation in fluctuating light under compound stresses of drought and heat. Physiol. Plantarum 176 (3). https://doi.org/10.1111/ppl.14406.

Yadav, A., Yadav, K., Abd-Elsalam, K.A., 2023. Nanofertilizers: types, delivery and advantages in agricultural sustainability. Agrochemicals 2 (2), 296–336. https://doi.org/10.3390/agrochemicals2020019.

Zhang, Y., Fan, Y., Lv, X., Zeng, X., Zhang, Q., Wang, P., 2023. Deficiency in NDH-cyclic electron transport retards heat acclimation of photosynthesis in tobacco over day and night shift. Front. Plant Sci. 14. https://doi.org/10.3389/fpls.2023.1267191.

Zhang, H., Zheng, T., Wang, Y., Li, T., Chi, Q., 2024. Multifaceted impacts of nanoparticles on plant nutrient absorption and soil microbial communities. Front. Plant Sci. 15. https://doi.org/10.3389/fpls.2024.1497006.

Zhu, T., Ning, P., Liu, Y., Liu, M., Yang, J., Wang, Z., Li, M., 2025. Knowledge of microalgal Rubiscos helps to improve photosynthetic efficiency of crops. Planta 261 (4). https://doi.org/10.1007/s00425-025-04645-w.

CHAPTER

8

Photocatalytic nanomaterials for enhanced light utilization and energy conversion

Prasann Kumar[1], *Amit Raj*[1] *and Padmanabh Dwivedi*[2]

[1]Department of Agronomy, School of Agriculture, Lovely Professional University, Phagwara, Punjab, India; [2]Department of Plant Physiology, Institute of Agricultural Sciences, Banaras Hindu University, Varanasi, Uttar Pradesh, India

1. Introduction

The study of photocatalytic nanomaterials has become one of the most vibrant fields in materials science and renewable energy research because of their ability to convert solar energy into chemical energy through redox reactions. As global energy demand grows and fossil fuel reserves decline, the disparity in solar utilization underscores the inefficiency of current photovoltaic and photocatalytic systems in harnessing the entire solar spectrum. To fill this gap, nanostructured photocatalysts leverage quantum confinement, surface plasmon resonance (SPR), and hot-electron injection, as well as bandgap engineering, to enhance light absorption, charge separation, and surface reactivity (Jiang et al., 2023). The history of photocatalysis in bulk materials has been transformed by the development of nanoscale structures, including nanoparticles, nanorods, nanosheets, and heterojunction nanocomposites, which have fundamentally altered the kinetics of photoinduced electron-hole pair generation and migration. At the nanoscale, quantum-mechanical effects govern light-matter interactions, modify the electronic density of states, and enhance the effective surface-area-to-volume ratio. On a mechanistic level, a photon of sufficient energy tends to affect a semiconductor nanoparticle such as TiO_2, ZnO, or g-C_3N_4 when it strikes an electron in the valence band, and one of the electrons moves to the conduction band, leaving behind a hole that can undergo oxidation reactions (Fan et al., 2025). Phasmonic coupling is used to increase the efficiency of this process, in which noble metal nanoparticles (Au, Ag, or Pt) generate localized electromagnetic fields that enhance photon capture and promote photoinduced charge transfer (Kumar et al., 2022). At the same time, defect engineering, via oxygen vacancies, dopant

Nanotechnology in Photosynthesis
https://doi.org/10.1016/B978-0-443-44376-3.00008-2

insertion, or lattice deformation, creates shallow trap states that extend charge-carrier lifetimes and reduce recombination.

Another paradigm shift in design is the heterojunction design mechanism, in which semiconductors with opposite band structures are stacked to generate internal electric fields that drive directional charge migration. Such examples include type-II, Z-scheme, and S-scheme heterojunctions, each with distinct pathways for charge transport and recombination control (Wu et al., 2026). Additionally, the formation of Schottky junctions between metals and semiconductors serves as a charge sink, thereby enhancing electron extraction and stabilizing photocatalyst performance under prolonged irradiation. This further control of photon pathways, achieved by integrating photonic crystal structures and plasmonic waveguides into nanomaterials, enables selective spectral harvesting and minimizes scattering. Moreover, the photoelectrochemical (PEC) mechanism underscores the importance of aligning the redox potentials of the photocatalyst and the electrolyte. The water-oxidation or hydrogen-evolution activity of surface-anchored co-catalysts such as MoS2, RuO2, and NiFe-LDH enhances the kinetics of multi-electron transfer reactions (Huo et al., 2025). The advent of methods of atomic layer deposition (ALD) and solvothermal synthesis of nanomaterials permits the accurate manipulation of morphology, crystallinity, and interfacial bonding of nanomaterials-crucial determinants of photon absorption cross-sections and interface chemisorption strength. Phonon-plasmon coupling and non-radiative energy transfer are also essential processes in nanophotocatalysis, as they can enhance carrier mobility and mitigate thermal degradation. Photo-Fenton, surface hydroxylation, and the generation of superoxide radicals ashore (O_2) are the synergistic processes that effectively oxidize organic contaminants and split water molecules (Feng et al., 2024). Hybrid organic-inorganic systems, such as perovskite nanostructures, have the potential to enable novel energy band structures, extending visible-light absorption to over 700 nm via dielectric confinement and band-edge modulation.

In addition, computational modeling and density functional theory (DFT) simulations have revealed new reaction patterns and optimal orbital hybridizations in nanomaterials by predicting dopant patterns, energetics, and surface reactivity. An exploration of 2D materials such as MXenes, black phosphorus, and MoS2 nanosheets reveals tunable bandgaps and ultrafast charge dynamics that enhance catalytic turnover rates (Shirvani et al., 2023). The reaction selectivity and adsorption-desorption equilibrium of photocatalytic conversions at nanosurfaces are determined by the energy-entropy compensation mechanism, making thermodynamic control of catalytic reversibility. Altogether, photocatalytic nanomaterials operate through a network of mechanisms: photon excitation, electron-hole transport, interfacial charge-recombination control, and catalysis of surface reactions, tailored by nanoscale architecture, compositional heterogeneity, and photophysical coupling (Hayat et al., 2023). Their critical contribution to the global solar energy use dilemma and to the development of sustainable avenues for energy, environmental, and industrial change is based on this multi-mechanistic orchestration.

2. Structural design principles of photocatalysts

Photophysical efficiency, charge-carrier mobility, and surface catalytic reactivity are dictated by the structural design of photocatalysts. Structural tuning processes, including quantum confinement, bandgap modulation, morphological anisotropy, and heterointerface engineering, are currently the basis for achieving high photon-to-chemical conversion efficiencies, as photocatalysis is no longer a discovery field but rather a rational design field dominated by algorithms. The atom-level structure, surface geometry, and defect structure determine the electron excitation, transport dynamics, and catalytic turnover of nanostructured photocatalysts.

2.1 Quantum confinement and bandgap tuning

Quantum confinement effects. This is due to the reduction of semiconductor particles to dimensions comparable to or smaller than their exciton Bohr radius, typically less than 10 nm. Within this regime, the bulk-phase electronic bands of the continuous mode are discontinuously divided into a quantized optical absorption spectrum, leading to an increased bandgap. Quantum well, quantum wire, and quantum dot structures are mechanically manipulated to achieve control over the electron-hole overlap integral and the density of electronic states (Huebener, 2025). Dimensionality reduction enables localization of the wavefunction and quantized excitonic transitions, thereby enhancing control over photoresponse. For example, the conduction band minima in TiO2 and CdS quantum dots shift to higher potentials, thereby increasing the reducing capacity of hydrogen evolution reactions. Previously, advanced band-structure engineering mechanisms have been used to tune the bandgap energy, including anion/cation replacement, the introduction of defect states, and lattice strain, to make materials responsive to visible light (Huang et al., 2023). The integration of plasmonic nanostructures (e.g., Au or Ag nanoparticles) further enables hot-electron injection via plasmons. It enhances the localized electromagnetic field, extending the absorption edge and increasing quantum efficiency under sunlight.

2.2 Surface area and active site exposure

Photocatalysts are intrinsically associated with their respective surface area and concentration of available active sites. At the nanoscale, adsorption dynamics, charge transfer, and photon-adsorbate coupling are enhanced with increasing surface curvature. Defect engineering, porosity modulation, and atomic-scale exfoliation are strategies that create reactive facets and edge sites that inhibit electron-hole recombination. For example, TiO2 has distinct oxidation and reduction propensities on two aspects, (001) and (101), which exhibit different surface energies and atomic coordination (Yan et al., 2025b). Surface reconstruction processes, including anion vacancy migration, lattice rearrangement, and dynamic hydroxylation, drive self-passivation of defects and the formation of reactive intermediates. Also, light-matter interface interactions are further improved by surface plasmon amplification and polarization of charge by adsorption. Hierarchically porous structures, including mesoporous, microporous, and macroporous components, are used to promote reactant diffusion to the catalyst and to scatter photons, thereby enhancing the residence time of light within the catalyst matrix (Fig. 8.1).

2.3 Heterostructures for charge separation

The effective photocatalysis is based on reducing the electron-hole recombination, and this can be achieved by building heterostructures where the interfacial electric fields are utilized to direct the movement of charges. Type-II heterojunctions facilitate electron transfer from the higher to the lower conduction band. In contrast, the Z-scheme and S-scheme designs maintain redox potentials by differentiating charge-carrier transport via band-gap sorting. Interfacial polarization layers in core-shell

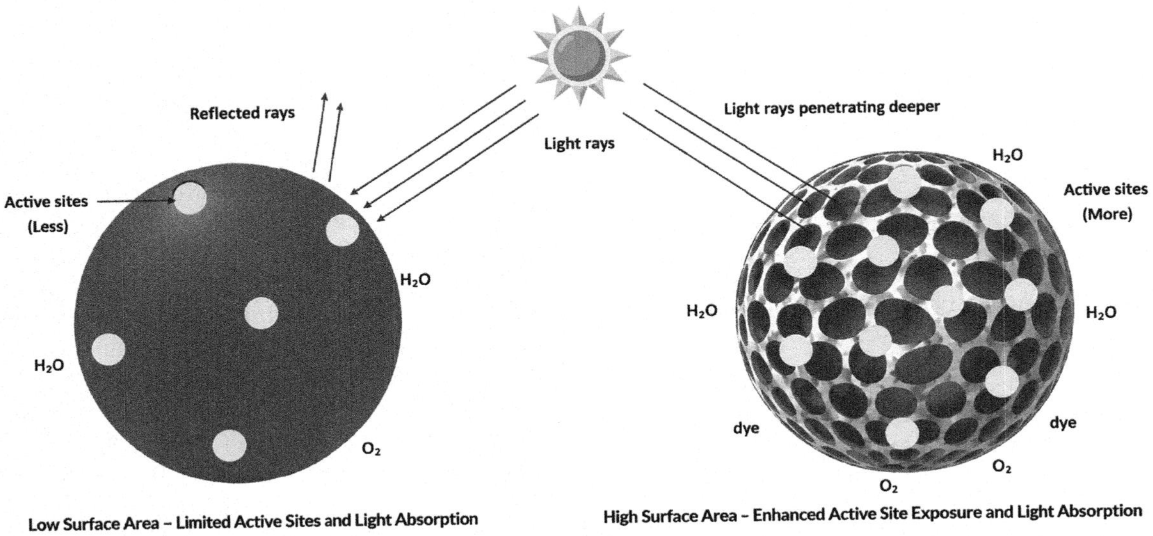

FIGURE 8.1 Enhanced surface area, defect engineering, and hierarchical porosity improve active site exposure, charge transfer efficiency, and photon utilization, significantly boosting photocatalytic performance.

nanostructures act as charge rectifiers, stabilizing photogenerated electrons and holes in spatially separated regions. Additionally, the formation of Schottky barriers between semiconductors and noble metals creates an electron sink that delocalizes charge, thereby increasing photostability (Qin et al., 2024). New homodimeric heterostructure systems, including dual Z-schemes and cascade band arrangements, have been constructed to match photon capture to multielectron transfer reactions. Interlayer coupling and charge-tunneling features, introduced through van der Waals heterointerfaces such as MoS2/g-C3N4, increase the rate of photocarrier separation without compromising optical transparency (Table 8.1).

2.4 Role of morphology (nanorods, sheets, dots)

The morphological control is a critical tool to adjust optical path lengths, charge mobility, and kinetics on the surface reaction. One-dimensional nanorods and nanowires enable charge transport along their longitudinal axis, reducing recombination losses by making carrier diffusion anisotropic. Nanosheets, for example, graphene and layered metal oxides, exhibit ultrathin surface-to-volume ratios that facilitate photon trapping and delocalize electrons. The small size of zero-dimensional quantum dots, combined with discrete quantized energy levels, leads to size-dependent emission, multiphoton absorption, and suppression of Auger recombination, thereby increasing photoluminescence and catalytic turnover (Li et al., 2024). Hierarchy. Three-dimensional morphologies are hierarchical, with flower-like, dendritic, and hollow nanostructures. Three-dimensional hierarchical morphologies are multiple-scattering morphologies and light-path elongation morphologies, which enhance optical absorption. It is developed through several sophisticated morphogenesis processes, including oriented attachment, self-assembly, and template-assisted growth, that selectively form tuneable architectures with aspect ratios

TABLE 8.1 Structural strategies enhancing photocatalytic performance.

Aspect	Key feature	Mechanistic role	Photocatalytic benefit	
Heterostructure formation	Type-II heterojunction	Band alignment enables directional electron transfer	Reduces electron–hole recombination	Qin et al. (2024)
	Z-scheme system	Maintains high redox ability while separating charges	Enhances oxidation–reduction efficiency	
	S-scheme interface	Band bending enables spatial charge separation	Improves charge spatial selectivity	
Core–Shell nanostructures	Interfacial polarization layer	Acts as a charge rectifier at interfaces	Stabilizes photogenerated carriers	Li et al. (2024)
Metal semiconductor junction	Schottky barrier formation	Noble metal acts as an electron sink	Enhances photo-stability and conductivity	Yan et al. (2025a)
Advanced heterostructures	Dual Z-scheme cascade	Multi-step charge separation pathways	Matches complex multi-electron reactions	Huang et al. (2023)
1D morphology	Nanorods/Nanowires	Anisotropic carrier diffusion along the axis	Minimizes recombination losses	Huebener (2025)
2D morphology	Nanosheets (e.g., MoS_2, g-C_3N_4)	Ultrathin geometry supports electron delocalization	Promotes high surface redox contacts	Qin et al. (2024)
0D morphology	Quantum dots	Quantum confinement alters electronic states	Enables size-tuned optical absorption	
Quantum dots	Suppression of auger recombination	Limits non-radiative losses	Improves photoluminescence and turnover	
3D hierarchical structures	Flower-like or dendritic forms	Multiple light scattering and long paths	Enhances photon absorption capacity	Li et al. (2024)
Hollow nanostructures	Diffusion-controlled voids	Increases reactive surface exposure	Improves mass transport and reactant accessibility	Yan et al. (2025b)
Growth mechanisms	Oriented attachment/Self-assembly	Controls facet exposure and crystal flow	Customizes active site density	Huang et al. (2023)
Lattice engineering	Strain and mismatch tuning	Regulates surface energy and charge mobility	Enhances catalytic stability under stress	Huebener (2025)
Integrated design strategy	Morphology + heterostructure + quantum effects	Synergistic optimization of all charge steps	Enables next-generation photocatalyst efficiency	Qin et al. (2024)

and facet exposure. Additionally, strain modulation, lattice-mismatch engineering, and dynamic self-healing morphologies are among the adaptive strategies that stabilize photocatalysts under photothermal stress. The combination of quantum confinement, surface exposure, heterostructure design, and morphological anisotropy provides an integrated platform for optimizing light harvesting, charge-carrier kinetics, and reaction selectivity, thereby forming the structural basis of next-generation photocatalytic nanomaterials (Table 8.1).

3. Charge carrier dynamics and photonic enhancement

The mechanisms of charge carrier generation, migration, recombination suppression, as well as photon-matter interaction, under the dominion of nanomaterials control the performance of photocatalytic nanomaterials that determine the overall quantum efficiency. Engineering the dynamics of charge carriers is essential at the atomic and nanoscale to realize efficient redox reactions in solar energy conversion, pollutant degradation, and photoelectrochemical reactions. The interaction among band-edge alignment, the distribution of surface electric fields, and photonic localization determines the kinetic mechanisms of photon-made reactions.

3.1 Mechanisms of electron hole pair generation

Photon absorption and electron excitation of the electrons in the valence band (VB) to the conduction band (CB) initiates the photocatalytic activity to form an electron-hole pair (exciton) (Tuama et al., 2024). This occurs in response to photoinduced interband transitions, in which the energy of the incident photon (hv) is larger than the bandgap (e.g.) of the semiconductor. At the nanoscale, quantum confinement enhances exciton binding energy, thereby increasing charge separation. The Franck-Condon rule governs the fastest electron transitions to excited states on the femtosecond timescale, whereas phonon-assisted processes govern carrier cooling (Emelianov et al., 2024). Interfaces can increase the density of localized states by forming surface exciton complexes, thereby enabling high absorption cross-sections. Hot-electron injection and charge transfer via plasmons enhance the concentration of active electrons in redox reactions within plasmonic-semiconductor composites. Additionally, photoelectric field-assisted tunneling, multiphoton absorption, and modulation of band-bending processes contribute to broad-spectrum excitation across the visible and near-infrared.

3.2 Strategies to suppress recombination

The inhibition of electron-hole recombination is a key to the enhancement of photocatalytic quantum yield. Recombination can be radiative, non-radiative, or trap-assisted, and typically occurs on a timescale of nanoseconds. Advanced materials are engineered with internal electric fields, heterojunctions, and surface-defect modulation to dissociate charge carriers spatially. Type-II heterostructures support vectorial electron flow. In contrast, Z-scheme and S-scheme systems enable photocatalytic reactions that replicate natural photosynthesis by maintaining high redox potentials between the two semiconductors (Hou et al., 2024). The formation of oxygen vacancies, anion substitution, and defect dipoles facilitates electron localization, thereby minimizing recombination centers. Moreover, ferroelectric polarization processes introduce internal electric fields that drive electrons and holes across domains in opposite directions.

COFs and MOFs have demonstrated outstanding capability to entrap carriers by using p-p conjugation and delocalization of charge pathways (Shahmirzaee and Nagai, 2024). In addition, core-shell structures can create barriers that guide electrons toward the core and holes toward the shell, thereby reducing the recombination kinetics. Through layers of surface passivation, either via thin oxide coatings or self-assembled monolayers, charge trapping is diminished, and catalytic interfaces are stabilized to operate over extended periods.

3.3 Plasmonic resonance and light trapping

Noble metal nanoparticles are used in plasmonic enhancement mechanisms to focus the electromagnetic fields around semiconductor surfaces to enhance the absorption of light and the transport of charges. Localized Surface Plasmon Resonance (LSPR) arises from the coherent oscillation of conduction electrons in metallic nanostructures upon excitation by specific frequencies of light (Mcoyi et al., 2025). The effect generates a near-field enhancement that boosts the photoexcitation efficiency in the neighboring semiconductor. Landau damping and plasmon decay channels that generate hot electrons inject charge into the CB on short timescales, thereby extending photoreactivity into the visible and infrared spectra. Plasmonic antenna-reactors have been developed that combine light-trapping and catalytic functions within a single nanostructure, thereby capturing and enhancing the directional capture and lifetime of photons. Bimetallic nanocomposites (e.g., Au-Ag, Ag-Pt) can be used to obtain hybrid plasmon coupling, and dielectric-plasmonic multilayers can be used to achieve pathlength amplification using Mie scattering and photonic localization (Ganesh et al., 2024). Fabry-Perot interference effects and resonant-cavity structures are further combined to maximize light-matter overlap, ensuring that energy conversion is as high as possible (Fig. 8.2).

3.4 Photonic crystals and nanostructured interfaces

Photonic crystals and fabricated nanostructured interfaces provide a new perspective on the control of photon flow and carrier kinetics. Photonic bandgap engineering enables selective reflection and transmission of light at specific wavelengths and traps photons within the catalytic medium. The propagation of light waves in standing-wave fields occurs via Bragg scattering and Bloch-mode propagation, which increase light absorption by increasing the optical path length (Sandik et al., 2025). The synergistic coupling of light confinement and charge separation is achieved by combining 1D, 2D, and 3D photonic architectures with plasmonic layers. Nanostructured dielectric metasurfaces also regulate the local density of optical states, which facilitates anisotropic scattering and spin-orbit coupling of photons. Such structures can be optimized by exploiting refractive-index gradients, nanoslit resonators, and grating-coupled waveguide modes to increase photon utilization efficiency. Responsive (temperature- or electric-field-driven) dynamic photonic lattices enable adaptive light steering and self-regulating optical feedback. Furthermore, the effects of interfacial polarization of charges, plasmon-exciton coupling, and coherent multiple scattering are synergetically increased in photocatalytic reactions (Liang et al., 2023). Due to the presence of Janus-type nanointerfaces, in which asymmetric electric potential landscapes are combined, unidirectional charge migration is induced, thereby improving reaction kinetics. Hybrid plasmon-polariton interactions between photonic nanostructures and 2D heterointerfaces broaden the range of photoactivity across the solar spectrum. All these photonic and electronic design schemes provide a

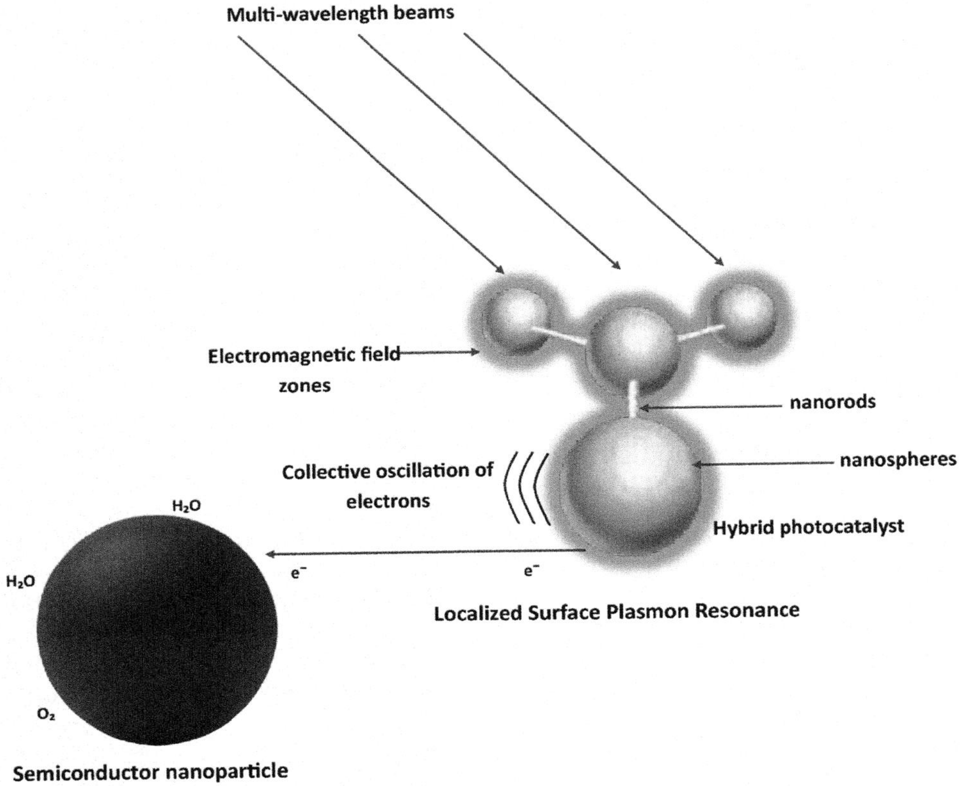

FIGURE 8.2 Plasmonic nanostructures enhance light absorption, hot-electron generation, and photon confinement, significantly improving charge transfer efficiency and broad-spectrum photocatalytic activity.

coherent approach to maximizing solar energy by detailed control of carrier generation, transport, and lifetime.

4. Bandgap engineering and material modification

Photocatalytic nanomaterials can be tuned to efficiently harvest solar energy across the ultraviolet (UV), visible, and near-infrared (NIR) spectra. The efficiency of photocatalytic processes in semiconductors primarily depends on their bandgap energy, which determines their absorption threshold, redox potential, and chain-transfer interactions. As a way of overcoming the intrinsic constraints of wide-bandgap semiconductors such as TiO2 and ZnO, scholars have come up with various solutions such as doping, co-doping, alloying, Z-schemes, and hybrid semiconductor-metal structures. Such methods alter the atomic, electronic, and interfacial properties to enhance carrier separation, light absorption, and structural stability under operating conditions.

4.1 Doping and co-doping strategies

One of the most effective ways to adjust band structure and to control the charge carrier density in semiconductors is doping. Cationic doping (e.g., Fe^{3+}, Cr^{3+}, Cu^{2+}) adds impurity energy levels between the conduction and valence bands to increase visible-light absorption by reducing the effective bandgap (Saleem et al., 2024). Anionic doping (e.g., N, S, C), which alters the valence band by hybridizing with O 2p orbitals, extends the absorption edge and increases the redox ability (Jia et al., 2024). Mechanically, such dopants produce localized trap states, transient charge reservoirs that increase carrier lifetime and reduce recombination. Co-doping (e.g., N-F, C-S) is synergistic, with the effect of compensating the charge and minimizing the strain on the lattice. Additionally, doping via defects, especially oxygen vacancies, affects surface reactivity by modulating local electronic potentials. Recent advances in atomic layer deposition (ALD) and ion-beam implantation have enabled precise nanoscale placement of dopants, thereby improved charge-distribution uniformity and enhancing interactions between light and matter.

4.2 Alloyed and composite photocatalysts

Another potent bandgap tuning mechanism by electronic coupling of dissimilar atoms is through alloying. One way to achieve more efficient use of photons is through metal-semiconductor nanoalloys such as $Zn_{1-x}Cd_xS$ or Cu–Zn–In–S (CZIS), which provide electronic states that are sufficiently intermediate to enable photon use while retaining structural coherence. Such a lattice strain field, induced by the introduction of several cations, can alter orbital hybridization and carrier mobility (Hou et al., 2023). Simultaneously, composite photocatalysts, synthesized from multiple semiconductors, form heterojunction interfaces that facilitate directional charge separation. Type-II heterojunctions allow electrons to move to lower-energy bands and maintain the capacity of holes to oxidize. In contrast, S-scheme heterostructures maintain a high redox potential comparable to that of natural photosynthetic systems (Bilal et al., 2025). A radial electric field required to move the electrons out of the core to the shell is created by the core-shell structure, which can be TiO_2@CdS or ZnO@Ag_2O, and reduction of recombination is achieved. Furthermore, localized surface plasmon resonance (LSPR) is employed in plasmon-assisted alloys to promote photoabsorption and the activation of inert bonds under visible-light irradiation.

4.3 Z-scheme photocatalytic systems

The Z-scheme system, which was inspired by natural photosynthesis, is an innovative way of attaining efficient separation of charge, without compromising redox strength. In this system, the positions of the semiconductors are staggered and linked via an electron mediator (e.g., Au nanoparticles or reduced graphene oxide). When illuminated, electrons in one semiconductor recombine with holes in the other, leaving the remaining electrons and holes available for independent redox reactions. Direct Z-scheme systems do not require mediators because they rely on band-alignment-driven charge-recombination zones. The built-in electrostatic field at the interface facilitates the vectorial flow of charge, thereby enhancing carrier utilization efficiency (Yue et al., 2022). Dual Z-schemes and step-scheme (S-scheme) photocatalysts also improve light absorption, with cascaded band annealing providing broad spectral coverage. Photoinduced injection of hot electrons may enhance the quantum yield by incorporating plasmonic metals such as Ag or Au. The mechanisms governing charge-flow directionality are controlled by either dynamic interfacial polarization or Fermi-level equilibration and provide the maximum possible photoreactivity under fluctuating light conditions (Fig. 8.3).

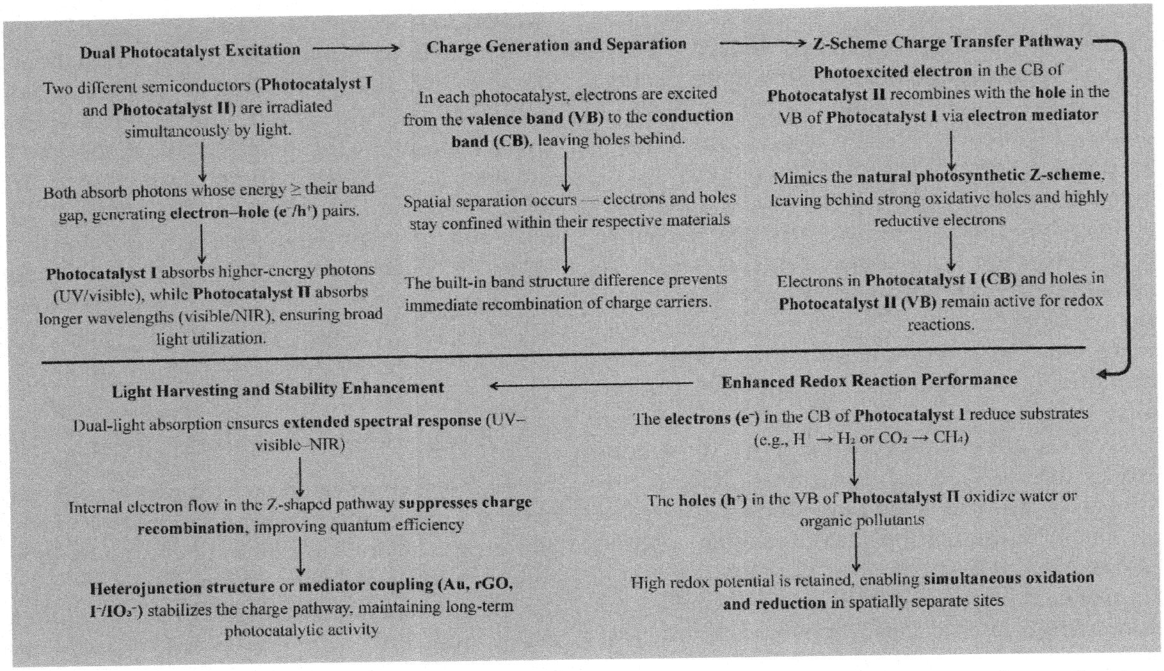

FIGURE 8.3 Z-scheme photocatalytic systems maintain strong redox potentials while promoting efficient directional charge separation through interfacial recombination pathways, enhancing solar-driven catalytic performance.

4.4 Hybrid semiconductor–metal frameworks

Multifunctional interfaces Multifunctional interfaces have been achieved by using Hybrid semiconductor-metal frameworks (SMFs), which are based on the electronic facilitation of semiconductors and optical and catalytic capabilities of metals. The most important operating principle is Schottky junction formation, in which the semiconductor donates electrons to the metal until Fermi-level equilibrium is reached, thereby creating an internal electric field that separates charge. Noble metals (Au, Ag, Pt) serve as electron sinks and slow down the possibility of recombination as well as increase the selectivity of the catalyst. Also, bimetallic (e.g., Pt-Ni, Au-Pd) can be tuned to the d-bands, to maximize the surface adsorption energies of reaction intermediates (Krishna et al., 2025). These hybrids are stabilized by covalent and electrostatic interactions, which reduce the energy barriers to carrier transport. Molecular diffusion and efficient charge transport are achieved by integration with metal-organic frameworks (MOFs) or covalent organic frameworks (COFs). Plasmon-exciton coupling, quantum tunneling, and photoinduced charge polarization contribute to increasing the absorption cross-section and thereby catalytic turnover (Yan et al., 2025b). Altogether, doping, alloying, and Z-structure and hybrid platforms represent a versatile and flexible bandgap engineering approach to optimizing light harvesting, charge dynamics, and photocatalytic stability, and thus to optimizing solar-to-chemical energy conversion technologies of the forthcoming generation.

5. Applications in energy conversion

The use of photocatalytic nanomaterials for energy conversion is among the most promising pathways toward carbon-neutral technologies. They can harness solar photons and perform complex redox reactions to produce hydrogen and fuels, as well as value-added carbon compounds, in a sustainable manner. These processes are based on photon absorption, charge separation, catalytic transfer, and molecular activation at nanostructured interfaces optimized through advanced material engineering.

5.1 Water splitting and hydrogen evolution

Photocatalytic water splitting is a paradigm reaction in which the light energy is directly transformed into chemical energy. This is achieved by the reduction of protons to hydrogen by photogenerated electrons and the oxidation of water to oxygen by holes. More advanced semiconductors, such as TiO2, g-C3N4, ZnIn2S4, and doped perovskites, enable efficient band alignment for this redox pair. The deposition of cocatalysts such as Pt, Ru, Ni2P, or MoS2 nanoparticles forms Schottky junctions that reduce overpotential and enhance charge-transfer kinetics. Directional carrier flow is generated by dual cocatalyst systems, in which electrons move to the reduction sites and holes to the oxidation sites, thereby increasing the spatial separation between them. Photo-induced Fermi-level equilibration, surface plasmon resonance (SPR)-based hot-electron injection, and lattice-distortion-based charge polarization are other mechanisms that enhance hydrogen production (Zhang et al., 2024). Heterojunctions (Z-scheme and S-scheme) and defect-mediated electron relay channels are introduced to increase charge lifetimes. In contrast, proton-coupled electron transfer (PCET) and water dissociation channels are used to evolve hydrogen in broad-spectrum GFPs selectively.

5.2 CO_2 photoreduction pathways

Photocatalytic CO_2 reduction converts atmospheric carbon dioxide into methane, methanol, and formic acid fuels. It resembles natural carbon fixation, except that engineered semiconductor-cocatalyst systems are employed. Mechanistically, adsorbed CO_2 molecules on the surface are subject to multi-electron-proton transfer through intermediate products such as CO_2^-, COOH, or *CHO. This specificity is regulated by surface work function, CO_2 adsorption energy, and the availability of oxygen vacancies or dual active centers. Mixed metal-organic frameworks (MOFs), graphitic carbon nitride, and plasmonic nanoparticles can synergistically enhance light absorption, charge confinement, and CO_2 activation in hybrid photocatalysts (Zhang et al., 2025). Co-catalytic promoters, such as Cu, Ag, or Re complexes, have been shown to control the reduction pathway via sequential hydrogenation steps. In addition, photoinduced surface polarization, interfacial reorientation of dipoles, and band-edge bending enhance charge mobility. Substantially improved turnover frequency with single-atom catalysts anchored on semiconductor supports (e.g., Cu-TiO2, Fe-C3N4) has been observed to result from atomic coordination and quantum confinement of electrons, which affect electron density and reaction selectivity.

5.3 Photocatalytic biofuel synthesis

The photocatalytic transformation of biomass derivatives to liquid fuels is a new direction in the study of renewable energy. The mechanism involves oxidative and reductive coupling of biomass intermediates, including glycerol, ethanol, and 5-hydroxymethylfurfural (HMF), under visible-light irradiation. Semiconductor-metal composites are used to catalyze dehydrogenation, decarbonylation, and C-C coupling using photogenerated carriers (Zhang et al., 2023). Plasmonic enhancement, band-edge modulation, and photothermal coupling facilitate the reaction process by accelerating the activation of molecules and co-catalysts, such as Pd, Ni, or Au, which activate specific reactions toward alkanes or alcohols. The combination of semiconductor nanorods, 2D sheets, and quantum dots improves reactant adsorption and photonic confinement. Localized surface field amplification and quantum tunneling between catalytic clusters, which are more efficient mechanistically, enhance energy efficiency (Chen et al., 2023). Reactive sites in oxygenated intermediates are created by defect engineering, particularly by generating sulfur and nitrogen vacancies, thereby making the process highly selective and limiting carbon loss (Fig. 8.4).

5.4 Integration with artificial photosynthesis

Artificial photosynthesis is the final combination of photocatalytic systems to reproduce the process of natural conversion of energy by sunlight—this system couples water oxidation and CO_2 reduction in architectured photoelectrochemical cells or heterostructured nanoreactors. The innovations encompassed in modern mechanistic innovation include tandem photoelectrode systems, redox mediator coupling, and charge transfer across multi-junction interfaces. The photon management, in the form of light-harvesting antenna structures and photonic

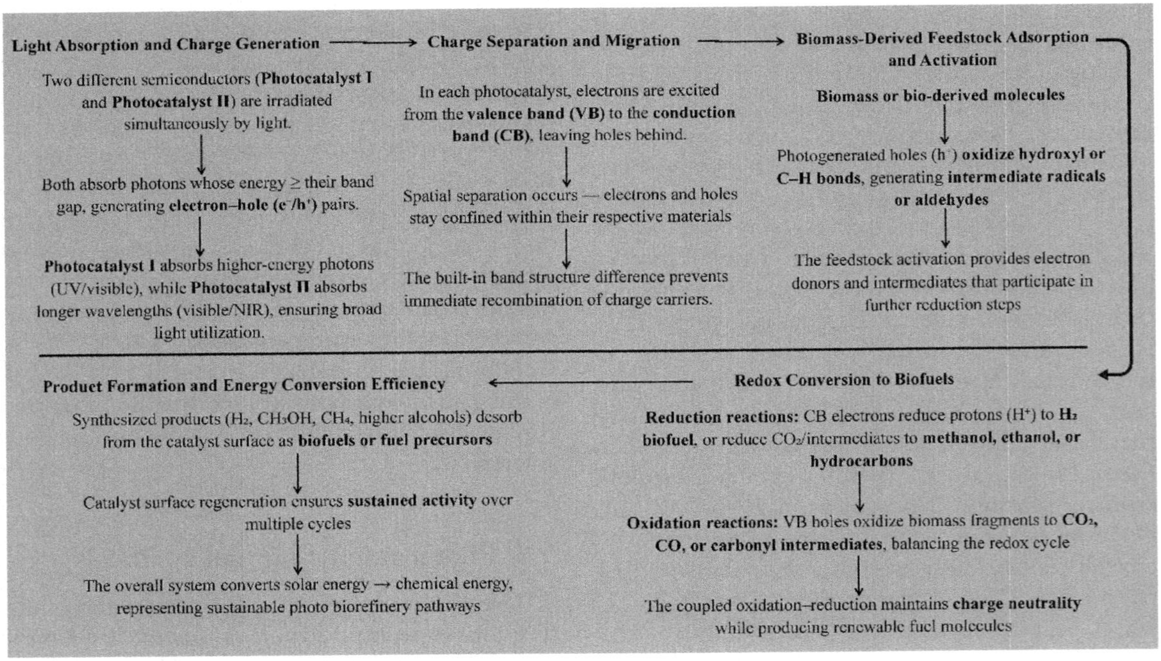

FIGURE 8.4 Photocatalytic biofuel synthesis converts biomass-derived intermediates into liquid fuels through light-driven redox processes, enhanced by plasmonic coupling, defect engineering, and selective co-catalyst activation.

crystal scaffolds, is facilitated, and the directionality of charge transport is maintained by nanoarray electrodes (Ruan et al., 2024). Ion flow is controlled by proton-exchange membranes and catalytic redox shuttles, which assist in reaction compartmentalization. Site-specific activation via ligand-field tuning and orbital hybridization is enabled by molecular catalysts attached to semiconductors, including cobalt porphyrins, ruthenium complexes, and polyoxometalates. Recent advances in spin-selective charge transport, dynamic exciton migration, and dielectric-field modulation have increased the total conversion efficiency. The integration of photocatalytic nanomaterials into energy-conversion processes depends on complex charge dynamics, interfacial structures, and quantum-scale processes that link photon capture to molecular transformation. These systems offer scalable and efficient solar-to-fuel technologies, enabled by cooperative processes such as plasmonic resonance, quantum confinement, defect engineering, and catalytic coupling.

6. Stability and scalability challenges

The activity of photocatalytic nanomaterials is not just determined by the activity of the material but also by the stability, durability, and scalability of the material under natural conditions of operation. Photocorrosion, surface deterioration, and structural wear are among the factors that limit the lifetime of photoactivity and charge dynamics, despite significant progress in this field. Industrial irrelevance can only be achieved by incorporating mechanistic stability plans, scalable synthesis pathways, and system-level flexibility into energy conversion platforms.

6.1 Photocorrosion resistance strategies

Self-oxidation or reduction of the catalyst during illumination causes photocorrosion, which is still a significant drawback in semiconductor materials, including CdS, ZnO, and $BiVO_4$. Mechanistically, photocorrosion occurs when holes generated by photogeneration oxidize the semiconductor lattice, or when electrons generated by the reduction of semiconductor surfaces reduce the metal center. In response to this, passivation coatings, including TiO2, Al2O3, or graphene, are formed to create protective heterointerfaces that deflect carriers to sites of recombination (Shen and Wang, 2024) radial electric field. Core-shell architectures apply radial electric fields to suppress surface redox reactions without slowing photocharge migration. The other stabilizing mechanism is the addition of surface sacrificial agents or redox mediators, such as Na2S/Na2SO3, which are preferentially oxidized, thereby protecting the catalyst structure. Stabilization of the crystal lattice is achieved by atomic doping with elements (Mo, W, or N) that enhance the crystal lattice by modulating the defect-states and enhancing charge delocalization. Mechanistic solutions such as oxygen-vacancy passivation, electron relay via conductive carbon scaffolds, and self-healing via dynamic surface reconstructions further increase operating lifetimes.

6.2 Durability under variable illumination

Under natural conditions of the solar system, the dynamic stress of photocatalysts is placed under varying light intensity, temperature, and humidity. The processes that control the durability are related to photon flux modulation, thermal expansion dislocation, and exciton recombination equilibrium. Light-dark cycling with periodicity induces surface charge, which can cause structural distortion and interfacial delamination (Kim et al., 2022). Adaptive band-edge tuning can self-regulate the light-harvesting state via dynamic doping or photochromic intermediates to mitigate these effects. Mechanisms Photonic stress relaxation mechanisms Photonic stress relaxation mechanisms, including thermally-driven lattice rejoindance, phonon-assisted energy loss, and so on, are used to preserve structural coherence (Oudich et al., 2023). By attaching flexible substrates, such as a polymeric matrix or carbon aerogels, thermal fatigue can be reduced, and conformal coverage of oxynitride or perovskite layers can enhance spectral stability. High-tech processes, such as spin-orbit coupling control, photoinduced surface reoxidation, and plasmonic damping modulation, provide steady photonic absorption and charge transport despite fluctuating illumination.

6.3 Economic feasibility of large-scale synthesis

Photocatalytic nanomaterials are scalable because they require the ability to undergo cost-effective synthesis without affecting performance. Mechanistically, large-scale production is hindered by non-uniform particle growth, energy-intensive calcination, and loss of nanoscale properties during scale-up. Mechanochemical control of nucleation and crystallinity, using sol-gel, hydrothermal, and microwave-assisted methods of green synthesis, requires lower energy input (Naik et al., 2024), dynamic control of supersaturation, and laminar transport of nanoparticles. Nanoparticle formation can be reproducibly achieved using continuous-flow microreactors and aerosol-assisted pyrolysis. Hierarchical structures can be assembled via self-assembly via electrostatic and van der Waals forces, without complex post-processing (Cai et al., 2024). Photocatalytic electrodeposition can achieve economically viable doping and co-catalyst deposition by harnessing photon energy for surface modification, reducing reagent waste. The yield and structural homogeneity are ensured by mechanisms such as ligand-directed crystallization, surfactant-induced porosity, or atomic layer deposition (ALD)- based film growth. Further optimization of synthesis kinetics and reduction of cost is achieved by process intensification by in situ monitoring (UV-vis or Raman spectroscopy).

6.4 Integration into energy devices and reactors

Scalability is not restricted to synthesis and effective incorporation into working devices and reactors. The integration process is based on interfacial interactions, charge-transfer alignment, and maximization of the photonic field. The incorporation of photocatalysts into photoelectrochemical (PEC) cells, membrane reactors, and solar concentrator systems requires strict regulation of electron transport layers (ETLs), ion diffusion pathways, and optical confinement structures (Abbas et al., 2023). Conductive scaffolds, such as reduced graphene oxide (rGO) or carbon nanotube (CNT) networks, exemplify mechanisms that enhance interparticle conductivity and reduce electron-hole recombination rates. Hierarchical reactor designs that incorporate microfluidic channels can aid reactant transport in a laminar fashion and focus light via total internal reflection. The magnetic nanocomposites are readily recoverable and reusable as integrated catalysts and are also flexible with respect to changing illumination geometries as thin-film assemblies. Light management is further optimized through surface engineering with anti-reflective coatings and photon-funneling structures. Also, hybrid reactors can be mechanistically coupled with photocatalytic and electrocatalytic modules to generate hydrogen and value-added chemicals in real time and control reaction parameters (Teng et al., 2024). The stability, durability, and scalability of photocatalytic nanomaterials are based on the synergy among mechanistic surface control, energy-efficient synthesis, and intelligent system integration. Self-passivation, adaptive doping, dynamic photonic regulation, and continuous-flow fabrication are strategies that ensure that next-generation photocatalysts are developed beyond prototypical laboratory versions to scalable materials that can be industrially implemented to enable continuous conversion of solar energy (Table 8.2).

7. Future horizons and smart integration

The next wave of photocatalytic nanomaterials will be the point of convergence of artificial intelligence (AI), bio-inspired design, multi-functionality, and large-scale commercial translation. The development of intelligent systems and photocatalysis toward self-optimization represents a new paradigm in solar energy use, as the synthesis process is controlled in the laboratory. The use of innovative integration systems that combine computational prediction, adaptive material interfaces, and hybrid architectures is redefining the design, operation, and implementation of photocatalysts in real-world scenarios.

7.1 Coupling with AI-driven material discovery

The AI-based material discovery is changing the process of developing photocatalysts using predictive modeling, generative design, and high-throughput screening. Random forest regression, deep neural networks (DNNs), and Gaussian process optimization are used to learn from large-scale datasets of bandgap energies, surface energies, and charge-transfer coefficients to determine optimal material compositions (Lamouadene et al., 2025). The reinforcement learning mechanisms also improve the synthesis conditions by matching the reaction temperature, dopant concentration, and crystal morphology to photocatalyst activity. The explainable AI (XAI) model is used to reveal mechanistic correlations between electronic structure and performance, enabling the discovery of designs without trial and error. Self-driving laboratories for true self-fabrication and testing of new photocatalysts in robotics rely on automated synthesis robots but must be based on Bayesian optimization frameworks. It is also true that quantum machine learning (QML) algorithms compute exciton dynamics and spin-orbit coupling at the atomic scale and improve predictions of light absorption and recombination processes.

7.2 Multi-functional nanomaterials for dual roles

Photocatalysts of the future are being considered as multi-functional devices that can carry out two or synergistic functions-such as degrading pollutants and producing hydrogen, reducing CO_2, and producing organic matter. The mechanical process of such materials relies on tandem charge separation and dual-pathway electron transport. For example, Janus-type nanostructures with spatially defined reduction and oxidation domains can facilitate redox reactions without cross-interference. Heterostructured polymer-semiconductor-metal triads are employed to enhance localized surface plasmon resonance (LSPR) and Schottky barriers, thereby improving photoelectric conversion and selectivity (Khanam and Rout, 2022). Additionally, magneto-photocatalytic and piezo-photocatalytic hybrids can convert mechanical vibrations and magnetic fields into alternative driving forces for charge separation, thereby revealing multiphysical coupling processes. Another functional synergy that enhances the functionality of plasmonic nanoparticles is photothermal-assisted photocatalysis, in which reaction kinetics are accelerated by heating the nanoparticles with sunlight.

TABLE 8.2 Stability, durability, and scalability strategies for photocatalytic nanomaterials.

Challenge	Mechanistic strategy	Underlying principle	Resulting benefit	References
Photocorrosion	Surface passivation coatings (TiO_2, Al_2O_3, graphene)	Protective interfacial barriers redirect charge carriers	Prevents catalyst surface degradation	Shen and Wang (2024)
	Core–Shell radial electric field structures	Radial fields discourage surface redox reactions	Enhances long-term photostability	Li et al. (2022)
	Sacrificial agents/Redox mediators (Na_2S/Na_2SO_3)	Preferential oxidation prevents lattice attack	Maintains catalyst integrity during reaction	Kim et al. (2022)
	Atomic doping (Mo, W, N)	Tunes defect states and strengthens the crystal lattice	Improves charge delocalization & stabilization	Oudich et al. (2023).
	Oxygen vacancy passivation	Reduces reactive defect centers	Prevents self-oxidation under illumination	Naik et al. (2024)
	Electron relay carbon scaffolds	Conductive support enhances charge transport	Minimizes recombination and structural wear	Cai et al. (2024)
Durability (variable illumination)	Adaptive band-edge tuning	Dynamic doping adjusts absorption under flux changes	Stabilizes photocatalytic efficiency in sunlight variations	Abbas et al. (2023)
Durability	Photochromic or self-regulating states	Light-responsive modulation of electronic structure	Reduces structural strain from cycling	
	Phonon/Heat-assisted stress relaxation	Thermal dissipation and lattice recovery	Preserves crystal coherence under heat load	
	Flexible polymeric or aerogel supports	Mechanical compliance reduces thermal fatigue	Maintains catalyst adhesion and stability	
	Perovskite/Oxynitride surface reinforcement	Enhances spectral and environmental stability	Resist humidity and long-term photodegradation	
Scalability	Sol–Gel/Hydrothermal/Microwave synthesis	Low-energy reaction control and nucleation precision	Enables cost-effective large-scale production	
	Continuous-flow microreactors	Uniform nanoparticle growth via controlled fluid dynamics	Improves reproducibility and particle homogeneity	
	Surfactant/Template-assisted porosity engineering	Tailored pore architecture via guided self-assembly	Enhances surface area and catalytic accessibility	
Device integration	Conductive scaffolds (rGO, CNT networks)	Promotes interparticle electron transport	Reduces recombination and boosts device efficiency	

7.3 Bio-inspired photocatalytic frameworks

Nature offers the design of high-efficiency light harvesting in systems of photosystem I and II. The bio-inspired photocatalytic structures replicate these structures by combining organic chromophores, peptide structures, and biomimetic assemblies of cofactors—nanostructured mechanisms. Mechanistically, funneling of energy by pigment-protein complexes and the exciton transfer by the Forster resonance energy transfer (FRET) principles are replicated. Artificial leaf systems use redox-active proteins and enzyme-like cofactors immobilized on semiconductor surfaces to enable energy transduction at the molecular level. Hydrogen-bonded networks are employed to control charge flow in hybrid bio-organic photocatalysts, including chlorophyll-functionalized TiO2 and cytochrome-modified photo-carbon nitrides (Li et al., 2025). Recent innovations also investigate DNA-templated nanowire assemblies, which are biologically inspired helical arrangements for electron transfer. These biospirations create new opportunities for sustainable, low-toxicity, and high-selectivity catalytic designs.

7.4 Roadmap toward commercial adoption

To commercialize photocatalytic nanomaterials, one needs to address the gap between performance and practicality using smart integration approaches. The list of mechanical innovations includes scalable roll-to-roll coating of thin-film photocatalysts, modular reactor integration, and predictive maintenance using digital twins. Internet of Things (IoT) network coupling enables real-time monitoring of photocatalytic performance and the environment. Additive manufacturing (3D printing) enables the fabrication of reactor designs with internal light channels to optimize photon utilization. Photovoltaic-photocatalytic systems are hybrid systems that allow sustained solar-to-chemical conversion at industrial scales. Self-regenerative surface repair, plasma-assisted cleaning, and electrochemical reactivation mechanisms help extend the operational period. Circular material-flow-based economic models and decentralized microreactors also ensure sustainable deployment.

8. Conclusion

Photocatalytic nanomaterials establish new paradigms for light-based energy conversion through the integration of nanoscale engineering with photonic control. Their distinct properties, such as quantum confinement effects, increased surface reactivity, and tunable electronic structures, enable them to harvest photons and separate charge more efficiently than conventional photocatalysts. Morphology-controlled synthesis, doping, and heterojunction engineering of these materials allow bandgap modulation, thereby maximizing light absorption in the visible and near-infrared spectra. The advantage is that they can inhibit electron-hole recombination, a significant drawback of bulk photocatalysis. Plasmonic nanostructures and photonic crystal designs enhance the intensity of incident light at catalytic interfaces, thereby increasing the efficiency of photoactivation under natural sunlight. This has led to extraordinary advances in applications such as hydrogen generation via water splitting, CO_2-to-solar energy conversion, and pollutant treatment, and aligns with global concerns about renewable energy and green sustainability. Although these developments have been made, challenges remain associated with material stability, scalability, and integration in real-world energy systems. Photocorrosion, high production costs, and performance variability under varying irradiation are barriers to their wider use. Addressing these requires the convergence of nanoscience, computational modeling, and innovative material design. New directions to accelerate breakthroughs include AI-assisted discovery, bio-inspired frameworks, and hybrid, multifunctional catalysts. It is not just incremental improvements to photocatalytic nanomaterials, but the creation of robust, cost-effective, and multifunctional platforms that integrate with artificial photosynthesis, innovative reactors, and decentralized energy systems, that will shape the future of photocatalytic nanomaterials. These materials have the potential to revolutionize our world by integrating key concepts of nanoscale science with translational engineering to advance a sustainable, low-carbon energy economy powered by solar innovation.

References

Abbas, T., Yahya, H.S.M., Amin, N.A.S., 2023. Insights and progress on photocatalytic and photoelectrocatalytic reactor configurations and materials for CO_2 reduction to solar fuels. Energy Fuels 37 (23), 18330–18368. https://doi.org/10.1021/acs.energyfuels.3c02863.

Bilal, M., Alfaifi, M.Q., Ben Ahmed, S., Abduljawad, M.M., Alrashed, Y.I., Aldurahim, E.S., Alassmy, Y.A., 2025. A review of strategies to switch heterojunction system from type-II to S-scheme for photocatalytic applications. Mater. Sci. Semicond. Process. 186. https://doi.org/10.1016/j.mssp.2024.109051.

Cai, M., Yang, J., Lu, X., Lu, X., 2024. Layer-by-layer self-assembly strategies of atomically thin two-dimensional nanomaterials: principles, methods, and functional applications. ACS Appl. Nano Mater. 7 (24), 27940–27959. https://doi.org/10.1021/acsanm.3c06286.

Chen, L., Ren, J.T., Yuan, Z.Y., 2023. Enabling internal electric fields to enhance energy and environmental catalysis. Adv. Energy Mater. 13 (11). https://doi.org/10.1002/aenm.202203720.

Emelianov, A.V., Pettersson, M., Bobrinetskiy, I.I., 2024. Ultrafast laser processing of 2D materials: novel routes to advanced devices. Adv. Mater. 36 (30). https://doi.org/10.1002/adma.202402907.

Fan, Y., Chang, X., Wang, W., Fan, H., 2025. Graphitic carbon nitride for photocatalytic hydrogen production from water splitting: nano-morphological control and electronic band tailoring. Nanomaterials 15 (1), 45. https://doi.org/10.3390/nano15010045.

Feng, C., Zhang, H., Guo, J., Si-Ying, Y., Luo, M., Zhang, J., Ren, Y., Liu, Y., Zhou, P., He, C.-S., Xiong, Z., Yue, Y., Wu, Y., Lai, B., 2024. Boosted $H2O_2$ utilization and selective hydroxyl radical generation for water decontamination: synergistic roles of dual active sites in $H2O_2$ activation. Water Res. 267, 122453. https://doi.org/10.1016/j.watres.2024.122453.

Ganesh, K.M., Bhaskar, S., Cheerala, V.S.K., Battampara, P., Reddy, R., Neelakantan, S.C., Reddy, N., Ramamurthy, S.S., 2024. Review of gold nanoparticles in surface plasmon-coupled emission technology: effect of shape, hollow nanostructures, nano-assembly, metal–dielectric and heterometallic nanohybrids. Nanomaterials 14 (1). https://doi.org/10.3390/nano14010111.

Hayat, A., Sohail, M., El Jery, A., Al-Zaydi, K.M., Alshammari, K.F., Khan, J., Ali, H., Ajmal, Z., Taha, T. A., Ud Din, I., Altamimi, R., Hussein, M.A., Al-Hadeethi, Y., Orooji, Y., Ansari, M.Z., 2023. Different dimensionalities, morphological advancements and engineering of g-C3N4-based nanomaterials for energy conversion and storage. Chem. Rec. 23 (5). https://doi.org/10.1002/tcr.202200171.

Hou, Z., Cui, C., Li, Y., Gao, Y., Zhu, D., Gu, Y., Zhang, T., 2023. Lattice-strain engineering for heterogeneous electrocatalytic oxygen evolution reaction. Adv. Mater. 35 (39), 2209876. https://doi.org/10.1002/09876.

Hou, H., Bowen, C.R., Yang, D., Yang, W., 2024. Cation-exchange-upgraded nanostructures for photocatalysts. Chem 10 (3), 800–831. https://doi.org/10.1016/j.chempr.2023.11.009.

Huang, X., Sun, L., Liu, X., Ge, M., Zhao, B., Bai, Y., Wang, Y., Han, S., Li, Y., Han, Y., Zhang, C., 2023. Increase and enrichment of active electrons by carbon dots induced to improve TiO_2 photocatalytic hydrogen production activity. Appl. Surf. Sci. 630, 157494. https://doi.org/10.1016/j.apsusc.2023.157494.

Huebener, R.P., 2025. Nanostructures: superlattices, quantum wires, and quantum dots. In: Conductors, Semiconductors, Superconductors: An Introduction to Solid-State Physics. Springer Nature Switzerland, Cham, pp. 193–220. https://doi.org/10.1007/978-3-031-93666-1_11.

Huo, J., Dou, Y., Wu, C., Liu, H., Dou, S., Yuan, D., 2025. Defect engineering of metal-based atomically thin materials for catalyzing small-molecule conversion reactions. Adv. Mater. 37 (7). https://doi.org/10.1002/adma.202416483.

Jia, X.B., Wang, J., Liu, Y.F., Zhu, Y.F., Li, J.Y., Li, Y.J., Chou, S.L., Xiao, Y., 2024. Facilitating layered oxide cathodes based on orbital hybridization for sodium-ion batteries: marvelous air stability, controllable high voltage, and anion redox. Adv. Mater. 36 (15). https://doi.org/10.1002/adma.202307938.

Jiang, W., Low, B.Q.L., Long, R., Low, J., Loh, H., Tang, K.Y., Chai, C.H.T., Zhu, H., Zhu, H., Li, Z., Loh, X.J., Xiong, Y., Ye, E., 2023. Active site engineering on plasmonic nanostructures for efficient photocatalysis. ACS Nano 17 (5), 4193–4229. https://doi.org/10.1021/acsnano.2c12314.

Khanam, S., Rout, S.K., 2022. A photocatalytic hydrolysis and degradation of toxic dyes by using plasmonic metal–semiconductor heterostructures: a review. Chemistry 4 (2), 454–479. https://doi.org/10.3390/chemistry4020034.

Kim, K.G., Park, G.H., Agumba, D.O., Kim, S.Y., 2022. In-situ measurement of drying and electrochemical-cycling-induced mechanical behavior and delamination of polypyrrole electrodes for energy storage applications. Surf. Interfaces 34. https://doi.org/10.1016/j.surfin.2022.102297.

Krishna, L., Thomas, P., Cherian, A.S., Nair, A.K., 2025. Noble Metal-Based Catalysts for Water Splitting Photochemical Splitting of Water: Fundamentals to Applications. Elsevier, pp. 163–192. https://doi.org/10.1016/B978-0-443-29064-0.00010-3.

Kumar, A., Choudhary, P., Kumar, A., Camargo, P.H.C., Krishnan, V., 2022. Recent advances in plasmonic photocatalysis based on TiO_2 and noble metal nanoparticles for energy conversion, environmental remediation, and organic synthesis. Small 18 (1). https://doi.org/10.1002/smll.202101638.

Lamouadene, H., EL Kassaoui, M., El Yadari, M., El Kenz, A., Benyoussef, A., 2025. Exploring modeling techniques for predicting band gaps of Doped-ZnO: a machine learning approach. Chem. Phys. 591, 112603. https://doi.org/10.1016/j.chemphys.2025.112603.

Li, Q., Wu, K., Zhu, H., Yang, Y., He, S., Lian, T., 2024. Charge transfer from quantum-confined 0D, 1D, and 2D nanocrystals. Chem. Rev. 124 (9), 5695–5763. https://doi.org/10.1021/acs.chemrev.3c00742.

Li, J., Zhang, H., Li, Y., Wu, S., Wei, S., Zhang, F., Huang, L., Chen, L., Li, J., Xiao, H., 2025. Asymmetrical intramolecular junction enables redox active sites on biodecorated carbon nitride toward $H2O_2$ photosynthesis. Chem. Eng. J. 519, 165018. https://doi.org/10.1016/j.cej.2025.165018.

Liang, W.Z., Huang, J., Sun, J., Zhang, P., Li, A., 2023. Multiscale modeling and simulation of surface-enhanced spectroscopy and plasmonic photocatalysis. Wiley Interdiscip. Rev. Comput. Mol. Sci. 13 (5). https://doi.org/10.1002/wcms.1665.

Mcoyi, M.P., Mpofu, K.T., Sekhwama, M., Mthunzi-Kufa, P., 2025. Developments in localized surface plasmon resonance. Plasmonics 20 (7), 5481–5520. https://doi.org/10.1007/s11468-024-02620-x.

Naik, R., Nagaswarupa, H.P., Darukesha, B.H.M., Tejashwini, D.M., 2024. Fundamentals of nanomaterial synthesis. In: Advances in Space Radiation Detection: Novel Nanomaterials and Techniques. Springer Nature Switzerland, Cham, pp. 23–36. https://doi.org/10.1007/978-3-031-74551-5_2.

Oudich, M., Gerard, N.J.R.K., Deng, Y., Jing, Y., 2023. Tailoring structure-borne sound through bandgap engineering in phononic crystals and metamaterials: a comprehensive review. Adv. Funct. Mater. 33 (2). https://doi.org/10.1002/adfm.202206309.

Qin, L., Ma, C., Zhang, J., Zhou, T., 2024. Structural motifs in covalent organic frameworks for photocatalysis. Adv. Funct. Mater. (41), 34. https://doi.org/10.1002/adfm.202401562.

Ruan, X., Li, S., Huang, C., Zheng, W., Cui, X., Ravi, S.K., 2024. Catalyzing artificial photosynthesis with TiO_2 heterostructures and hybrids: emerging trends in a classical yet contemporary photocatalyst. Adv. Mater. 36 (17). https://doi.org/10.1002/adma.202305285.

Saleem, S., Jameel, M.H., Alothman, A.A., Mayzan, M.Z.H. B., Yousaf, T., Ahmad, M.R., Ali, A., Zaman, A., 2024. A band gap engineering for the modification in electrical properties of Fe3O4 by Cu2+ doping for electronic and optoelectronic devices applications. J. Sol. Gel Sci. Technol. 109 (2), 471–482. https://doi.org/10.1007/s10971-023-06287-4.

Sandik, G., Feist, J., García-Vidal, F.J., Schwartz, T., 2025. Cavity-enhanced energy transport in molecular systems. Nat. Mater. 24 (3), 344–355. https://doi.org/10.1038/s41563-024-01962-5.

Shahmirzaee, M., Nagai, A., 2024. An appraisal for providing charge transfer (CT) through synthetic porous frameworks for their semiconductor applications. Small 20 (27). https://doi.org/10.1002/smll.202307828.

Shen, S., Wang, S., 2024. Photoelectrochemical Water Splitting Water Photo- and Electro-Catalysis: Mechanisms, Materials, Devices, and Systems. Wiley, pp. 143–206. https://doi.org/10.1002/9783527831005.ch4.

Shirvani, S., Ghashghaee, M., Smith, K.J., 2023. Two-dimensional nanomaterials in thermocatalytic reactions: transition metal dichalcogenides, metal phosphorus trichalcogenides and MXenes. Catal. Rev. 65 (1), 1–51. https://doi.org/10.1080/01614940.2021.1899605.

Teng, J., Li, W., Wei, Z., Hao, D., Jing, L., Liu, Y., Dai, H., Zhu, Y., Ma, T., Deng, J., 2024. Coupling photocatalytic hydrogen production with key oxidation reactions. Angew. Chem. Int. Ed. 63 (50). https://doi.org/10.1002/anie.202416039.

Tuama, A.N., Alzubaidi, L.H., Jameel, M.H., Abass, K.H., Bin Mayzan, M.Z.H., Salman, Z.N., 2024. Impact of electron–hole recombination mechanism on the photocatalytic performance of ZnO in water treatment: a review. J. Sol. Gel Sci. Technol. 110 (3), 792–806. https://doi.org/10.1007/s10971-024-06385-x.

Wu, X., Sayed, M., Wang, G., Yu, W., Zhu, B., 2026. COF-based S-scheme heterojunction photocatalyst. Adv. Mater. 38 (2). https://doi.org/10.1002/adma.202511322.

Yan, Q., Chen, Y., Tang, B., Wu, X., Zhou, H., Wang, H., Li, H., Lu, L., Zhang, H., Yang, S., Xu, C., Ma, T., 2025a. Precise engineering of asymmetric tri-active sites by symbiotic strategy for photocontrolled directional reforming of biomass. Angew. Chem. 137 (26). https://doi.org/10.1002/ange.202505718.

Yan, J., Wei, Z., Takagi, K., Motodate, M., Jiang, Z., Terashima, C., Shangguan, W., 2025b. An overview of dynamic descriptions for nanoscale materials in particulate photocatalytic systems from spatiotemporal perspectives. Nano-Micro Lett. 17 (1). https://doi.org/10.1007/s40820-025-01687-3.

Yue, X., Fan, J., Xiang, Q., 2022. Internal electric field on steering charge migration: modulations, determinations and energy-related applications. Adv. Funct. Mater. 32 (12). https://doi.org/10.1002/adfm.202110258.

Zhang, Q., Gupta, N.K., Rose, M., Gu, X., Menezes, P.W., Chen, Z., 2023. Mechanistic insights into the photocatalytic valorization of lignin models via C−O/C−C cleavage or C−C/C−N coupling. Chem. Catalysis 3 (2). https://doi.org/10.1016/j.checat.2022.11.009.

Zhang, X., Wang, C., Zhang, M., Luo, D., Ye, S., Weng, B., 2024. Surface plasmon resonance-mediated photocatalytic H_2 generation. ChemSusChem 17 (21). https://doi.org/10.1002/cssc.202400513.

Zhang, J., Xia, L., Deng, D., Jia, X., Song, D., Wang, L., Chang, Y., Xie, X., Dou, L., Wang, W., 2025. Metal-–organic framework-based heterojunction materials for photocatalytic CO_2 reduction reaction. Sol. RRL 9 (17). https://doi.org/10.1002/solr.202500424.

CHAPTER

9

Nanotechnology-enabled chloroplast engineering and gene delivery

Nilabja Banerjee[1,2], Anjali Tiwari[3], Joydeep Eshore[1], Vinai Kumar[4], Mohammad Faizan[5], Sajid Fiaz[6] and Dipti Bisarya[1]

[1]School of Agriculture, Lovely Professional University, Phagwara, Punjab, India; [2]Department of Agriculture, Srinath University, Jamshedpur, Jharkhand, India; [3]Department of Agricultural Science, University of Calcutta, Ballygunge, West Bengal, India; [4]Department of Plant Physiology, Govind Ballabh Pant University of Agriculture & Technology, Pantnagar, Uttarakhand, India; [5]Botany Section, School of Sciences, Maulana Azad National Urdu University, Hyderabad, Telangana, India; [6]Institute of Molecular Biology and Biotechnology, The University of Lahore, Lahore, Punjab, Pakistan

1. Introduction

Chloroplast biotechnology holds significant promise for addressing critical global challenges by reducing the cost of renewable biofuel production, enhancing food security through increased crop yields, and improving overall plant productivity. This approach capitalizes on the intrinsic capacity of chloroplasts to efficiently generate biomass and synthesize high-value bioproducts, thereby offering sustainable solutions for both energy and agricultural sectors. By optimizing chloroplast functions, biotechnology can contribute to the development of resilient, high-yielding crops capable of meeting the demands of a growing global population while supporting environmental sustainability goals (Newkirk et al., 2021a).

Chloroplasts are highly attractive targets for bioengineering because of their unique genetic and functional attributes that distinguish them from other cellular compartments. Along with mitochondria and the nucleus, chloroplasts possess their own genome; however, their endosymbiotic origin from cyanobacteria has allowed them to retain prokaryotic-like gene expression machinery and regulatory systems. As central biosynthetic hubs, chloroplasts drive essential metabolic processes including photosynthesis, carbon fixation, starch synthesis, and the production of amino acids, fatty acids, pigments, and key sulfur- and nitrogen-containing compounds. Their capacity for high-level transgene expression, combined with predominantly maternal inheritance, makes chloroplasts particularly appealing for genetic engineering applications. Compared

Nanotechnology in Photosynthesis
https://doi.org/10.1016/B978-0-443-44376-3.00009-4

with nuclear transformation, chloroplast engineering enables precise site-specific integration via homologous recombination, supports polycistronic gene expression, avoids position effects, and reduces unintended gene silencing. Although chloroplast transformation has advanced from early successes in tobacco to more than 20 flowering plant species, its broader adoption remains constrained by species-specific regeneration systems and limited selection strategies (Grevich and Daniell, 2005; Daniell et al., 2002).

The chloroplast genome is circular, polyploid, and encodes genes essential for photosynthesis and metabolic regulation. These characteristics provide several advantages over nuclear genetic modification, including elevated expression of foreign genes, predictable integration sites, and restricted transgene dissemination through pollen (An et al., 2022; Bock, 2015). Consequently, chloroplast engineering has facilitated improvements in photosynthetic efficiency, stress tolerance, nutritional quality, and the production of pharmaceutical and industrial proteins. However, conventional transformation approaches remain largely restricted to model species due to challenges associated with transformation efficiency, tissue culture dependence, and regeneration capacity. Recent developments such as CRISPR-based plastid genome editing and nanoparticle-mediated delivery systems are beginning to address these limitations, offering new opportunities to expand chloroplast biotechnology to diverse crop species (Demirer et al., 2021; Rani et al., 2025; Ramakrishnan et al., 2025).

Targeted chloroplast transformation plays a critical role in enhancing photosynthetic performance by enabling direct genetic modification of the organelle responsible for light harvesting and carbon assimilation. Engineering chloroplast-encoded components of the photosynthetic machinery—including ATP synthase, cytochrome b_6f complex subunits, and ferredoxin–$NADP^+$ reductase—has been shown to improve electron transport efficiency, ATP generation, and carbon fixation, resulting in increased biomass accumulation and growth (Hanson et al., 2013; Nazari et al., 2024). The high copy number of chloroplast genomes further amplifies transgene expression relative to nuclear transformation, thereby enhancing photosynthetic gains (Meyers et al., 2010; Berry et al., 2013). Beyond photosynthesis, chloroplasts serve as major metabolic engineering platforms, enabling coordinated expression of multigene biosynthetic pathways for amino acids, fatty acids, pigments, pharmaceuticals, and biofuel precursors. The inherent biosafety advantage conferred by maternal inheritance further strengthens the appeal of chloroplast-based metabolic engineering strategies (Bock, 2015; An et al., 2022).

Despite these advantages, conventional chloroplast transformation techniques such as particle bombardment, PEG-mediated protoplast transformation, and protoplast fusion present significant limitations. Biolistic transformation, while relatively species-independent, often causes extensive tissue damage, random or multicopy transgene integration, and unstable expression patterns. PEG-mediated protoplast transformation is technically demanding, highly genotype-dependent, and limited by poor regeneration efficiency in most crop species. Protoplast fusion enables somatic hybridization but is frequently associated with chromosomal instability, somaclonal variation, and low regeneration success. Collectively, these approaches rely heavily on tissue culture systems that are time-consuming, costly, and prone to introducing unintended genetic and phenotypic variation, thereby restricting scalability and reproducibility (Liu et al., 2019; Keshavareddy et al., 2018; Mukundan et al., 2025).

In this context, nanotechnology has emerged as a transformative strategy to overcome long-standing bottlenecks in chloroplast engineering.

Nanotechnology-enabled delivery systems utilize nanoscale materials—such as carbon nanotubes, lipid-based nanoparticles, silica nanoparticles, and quantum dots—to facilitate targeted and minimally invasive transport of DNA, RNA, proteins, and genome-editing complexes into chloroplasts (Newkirk et al., 2021a). Unlike conventional microcarriers used in biolistics, nanocarriers possess tunable physicochemical properties that allow efficient penetration of plant cell walls and membranes with reduced cytotoxicity and tissue damage. Surface functionalization with chloroplast-targeting peptides further enhances organelle specificity, enabling precise intracellular trafficking and cargo delivery (Lv et al., 2020).

Nano-enabled approaches also support innovative applications in chloroplast synthetic biology, including multiplexed gene delivery, marker-free transformation, and iterative genome editing using CRISPR–Cas systems. These strategies improve homologous recombination efficiency and expand the capacity for metabolic pathway engineering, photosynthesis optimization, and biopharmaceutical production (Fig. 9.1). In parallel, nanosensor technologies based on graphene and carbon nanotubes enable real-time monitoring of chloroplast metabolites, stress markers, and photosynthetic intermediates, providing dynamic feedback to guide metabolic engineering and ensure physiological stability during genetic interventions (Law et al., 2023; An et al., 2022) (San). Certain nanomaterials further enhance photosynthetic efficiency by improving light harvesting and mitigating oxidative stress, thereby supporting both genetic modification and physiological resilience under environmental challenges (Yadav et al., 2023; Gupta et al., 2023).

While nanotechnology offers substantial advantages, its application in agriculture necessitates careful evaluation of environmental fate, nanoparticle persistence, and potential phytotoxicity. Ongoing research is therefore focused on developing biodegradable and biocompatible nanomaterials that ensure safety for plants, soil ecosystems, and human consumers (Tschiche et al., 2022; Ma et al., 2024). Against this backdrop, this chapter examines how nanotechnology can overcome key limitations of conventional chloroplast genetic engineering by enabling targeted, efficient, and minimally invasive delivery of biomolecules. It highlights advances in nano-enabled transformation strategies, their applications in genome editing and synthetic biology, and the biosafety considerations and future challenges associated with translating chloroplast nanobiotechnology into sustainable agricultural and bioindustrial practices.

2. Chloroplast structure, function, and genetic architecture

Chloroplasts are semiautonomous organelles of endosymbiotic origin that function as the primary sites of photosynthesis and several essential metabolic pathways in plants. In addition to carbon assimilation, chloroplasts play key roles in amino acid, fatty acid, and phytohormone biosynthesis, making them strategic targets for genetic engineering aimed at enhancing crop productivity and stress tolerance (Taiz et al., 2015; Bock, 2015).

2.1 Chloroplast ultrastructure and membrane organization

Chloroplasts are bounded by a double-membrane envelope composed of an outer and inner membrane that regulates molecular exchange with the cytosol. Internally, the thylakoid membrane system differentiates into stacked grana and unstacked stroma lamellae, providing spatial separation of the photosynthetic complexes. Photosystem II is predominantly localized within grana stacks, whereas Photosystem I and ATP synthase are enriched

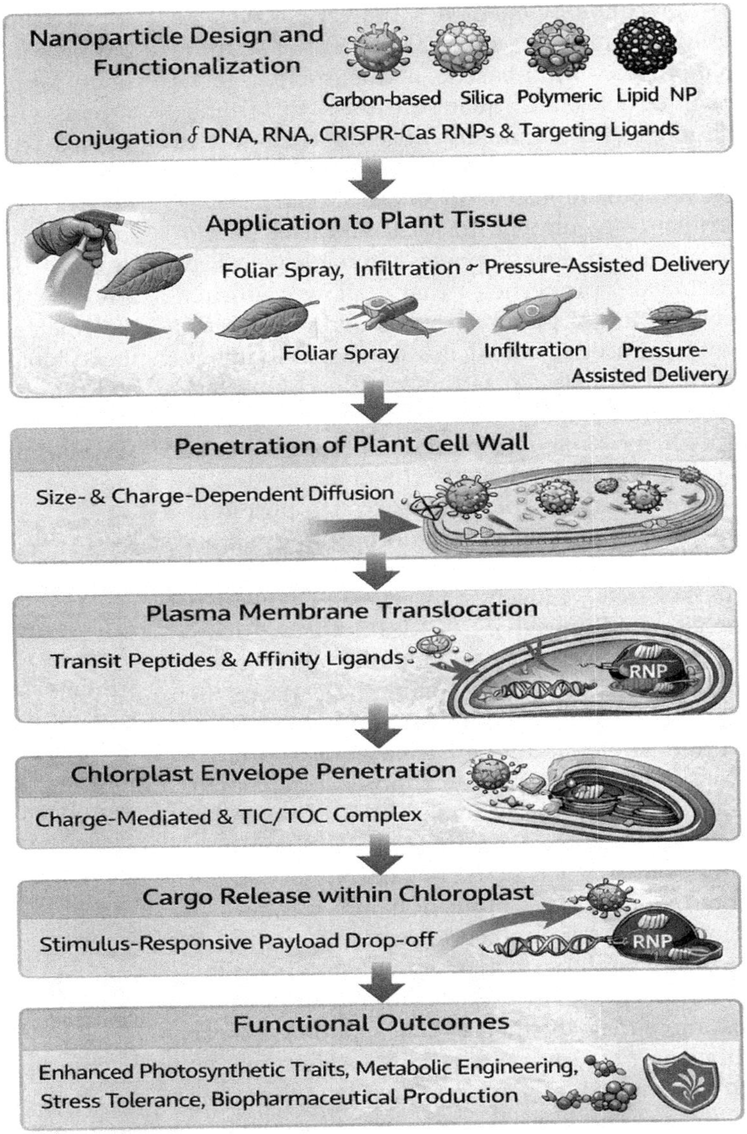

FIGURE 9.1 Schematic workflow of nanotechnology-enabled chloroplast engineering and gene delivery in plants.

in stroma thylakoids, optimizing the efficiency of light-driven electron transport. The stroma contains enzymes of the Calvin–Benson cycle, plastid ribosomes, and the chloroplast genome, supporting both metabolic activity and autonomous gene expression (Alberts et al., 2002; Pogson et al., 2015).

2.2 Chloroplast genome composition, gene expression, and regulation

The chloroplast genome is a circular double-stranded DNA molecule typically 120–160 kb in size, encoding approximately 120–130 genes primarily involved in photosynthesis, transcription, and translation (Daniell et al., 2016; Mathuria et al., 2024). It exhibits a conserved quadripartite structure comprising a large single-copy region, a small single-copy region, and two inverted repeats that enhance genome stability and replication fidelity. Chloroplast gene expression is regulated by a dual transcription system involving a plastid-encoded RNA polymerase (PEP) and a nuclear-encoded RNA polymerase (NEP). PEP primarily transcribes photosynthesis-related genes, while NEP is responsible for housekeeping genes and early plastid development (Pfannschmidt et al., 2015; Tadini et al., 2020). Extensive post-transcriptional regulation—including RNA splicing, editing, and stabilization—further fine-tunes gene expression. Translation occurs on bacterial-type 70S ribosomes within the chloroplast, reflecting the organelle's prokaryotic origin. Coordination between nuclear and plastid genomes is essential, as many chloroplast proteins are nuclear-encoded and imported post-translationally. This nucleo-plastid communication ensures synchronized assembly of multisubunit complexes and adaptive regulation in response to developmental and environmental signals (Jarvis, 2008; Allen et al., 2003).

2.3 Distinction between nuclear and plastid gene expression systems

Chloroplast gene expression differs fundamentally from nuclear gene expression in its organization and regulatory logic. Plastid genes are frequently arranged in operon-like clusters and transcribed as polycistronic mRNAs, reflecting their prokaryotic ancestry. Translation occurs on 70S ribosomes and is often coupled with transcription, allowing rapid protein synthesis within the organelle. In contrast, nuclear genes are transcribed as monocistronic units, undergo extensive RNA processing, and are translated in the cytosol before proteins are imported into chloroplasts via targeting peptides. These differences create unique opportunities for coordinated and high-level transgene expression in plastids (Zoschke and Bock, 2018; Zhang et al., 2022a, 2022b, 2023).

2.4 Advantages of chloroplast genome engineering

Chloroplast genome engineering provides several distinct advantages over nuclear transformation. The high copy number of plastid genomes per cell enables exceptionally strong and stable transgene expression without position effects or epigenetic silencing. Maternal inheritance of chloroplasts in most angiosperms offers intrinsic transgene containment, minimizing the risk of pollen-mediated gene flow. Moreover, the capacity for polycistronic expression allows multiple genes to be introduced and regulated as a single transcriptional unit, facilitating efficient metabolic pathway engineering. Together, these attributes establish chloroplasts as a robust and biosafe platform for plant genetic modification (Maliga, 2004; Daniell et al., 2016).

3. Mechanistic insights into nanomaterial-mediated chloroplast gene delivery

Nanomaterial-mediated chloroplast gene delivery is an emerging strategy that exploits engineered nanoparticles (NPs) to introduce genetic material directly into chloroplasts, overcoming limitations of conventional transformation methods (Bock, 2015). This approach relies on the tunable physicochemical properties of nanomaterials—such as size, surface charge, and functionalization—to bypass the plant cell wall, plasma membrane, and the chloroplast's double-envelope barrier (Newkirk et al., 2021a). Nanocarriers including single-walled carbon nanotubes (SWCNTs), gold nanoparticles, and chitosan-based systems exhibit high biocompatibility and efficient nucleic acid binding, enabling targeted delivery and controlled release within the chloroplast stroma (Demirer et al., 2019). A key mechanism governing chloroplast entry is the Lipid Exchange Envelope Penetration (LEEP) model, whereby positively charged nanoparticles interact electrostatically with negatively charged chloroplast membrane lipids, facilitating translocation into the stroma without membrane disruption (Wong et al., 2016; Kwak et al., 2019). Optimal delivery is generally achieved with nanoparticles in the 10–20 nm size range, which effectively traverse chloroplast membranes and release their genetic cargo in response to stromal conditions such as pH and redox potential (Santana et al., 2020; Newkirk et al., 2021a).

3.1 Barriers to chloroplast transformation

Efficient chloroplast transformation is constrained by a series of hierarchical biological barriers that restrict the delivery and stable integration of exogenous genetic material. These include the plant cell wall, plasma membrane, and the chloroplast double-envelope, each imposing distinct physicochemical limitations on transgene entry. The plant cell wall acts as the initial size-selective barrier, with pore diameters typically limited to 5–20 nm, thereby restricting the penetration of large DNA constructs and nanocarriers (Wang et al., 2025; Newkirk et al., 2021a). Nanoparticle uptake at this stage is strongly influenced by particle size, morphology, and surface charge, as electrostatic interactions with negatively charged wall polysaccharides modulate mobility and retention (Sembada and Lenggoro, 2024; Khan et al., 2019). Following wall traversal, the plasma membrane presents an energetically unfavorable lipid bilayer that limits passive diffusion of hydrophilic macromolecules. Successful internalization relies on endocytosis or membrane-assisted uptake, which can be enhanced through cationic surface modification or peptide functionalization of nanocarriers (Demirer et al., 2019; Kwak et al., 2019).

The chloroplast double-envelope constitutes the most stringent barrier, tightly regulating access to the plastid stroma and genome. Nanoparticles within the optimal size range (~10–20 nm) may exploit lipid exchange envelope penetration (LEEP), while conjugation with chloroplast transit peptides enables receptor-mediated import via the TOC/TIC translocon machinery, significantly improving targeting efficiency (Wong et al., 2016; Santana et al., 2020). Together, these multilayered constraints necessitate precise nanocarrier engineering that integrates size optimization, surface chemistry, and organelle-specific targeting to achieve efficient and reproducible chloroplast transformation (Zhao et al., 2024; Bock, 2015).

3.2 Mechanisms of nanoparticle penetration and translocation to chloroplasts

Nanoparticle (NP) delivery to chloroplasts is governed by coordinated physicochemical interactions and cellular transport processes that enable sequential traversal of the plant cell wall, plasma membrane, cytosol, and chloroplast envelope (Khan et al., 2019). Initial penetration occurs at the plant cell wall, a charged polysaccharide matrix that restricts passage primarily to particles below 5–20 nm, with permeability strongly influenced by NP size, surface charge, and hydrophobicity (Sembada and Lenggoro, 2024). Electrostatic interactions between engineered nanocarriers and negatively charged wall components further modulate NP retention and mobility during this stage. Following cell wall entry, NPs interact with the plasma membrane through charge-mediated mechanisms that facilitate cytoplasmic internalization. Among these, lipid exchange envelope penetration (LEEP) represents a dominant, energy-independent pathway. In this model, cationic NPs associate with anionic phospholipids in the membrane, inducing localized membrane fluidization and transient lipid rearrangements that allow direct translocation without vesicle formation or ATP expenditure (Wong et al., 2016; Newkirk et al., 2021a). Limited endocytosis may also occur depending on NP composition and surface chemistry. The presence of an ionic corona surrounding NPs reduces the energetic barrier to membrane passage and enhances membrane compatibility.

Once within the cytoplasm, intracellular trafficking toward chloroplasts is enhanced by surface functionalization with chloroplast transit peptides, which promote preferential association with the plastid envelope (Santana et al., 2020). The chloroplast double membrane constitutes the final and most selective barrier to NP entry. Penetration of the outer and inner envelope membranes is facilitated by optimal NP size (~10–20 nm), electrostatic compatibility, and the lipid composition of the envelope, which is enriched in galactolipids and sulfolipids. These features support LEEP-mediated translocation and, in some cases, receptor-assisted interactions with the TOC/TIC translocon complexes. Upon successful entry into the stroma, NPs release their genetic cargo—such as plasmid DNA, RNA, or CRISPR–Cas ribonucleoprotein complexes—in response to the chloroplast's redox environment and pH conditions (Fig. 9.2). The delivered cargo may undergo transient expression or stable plastome

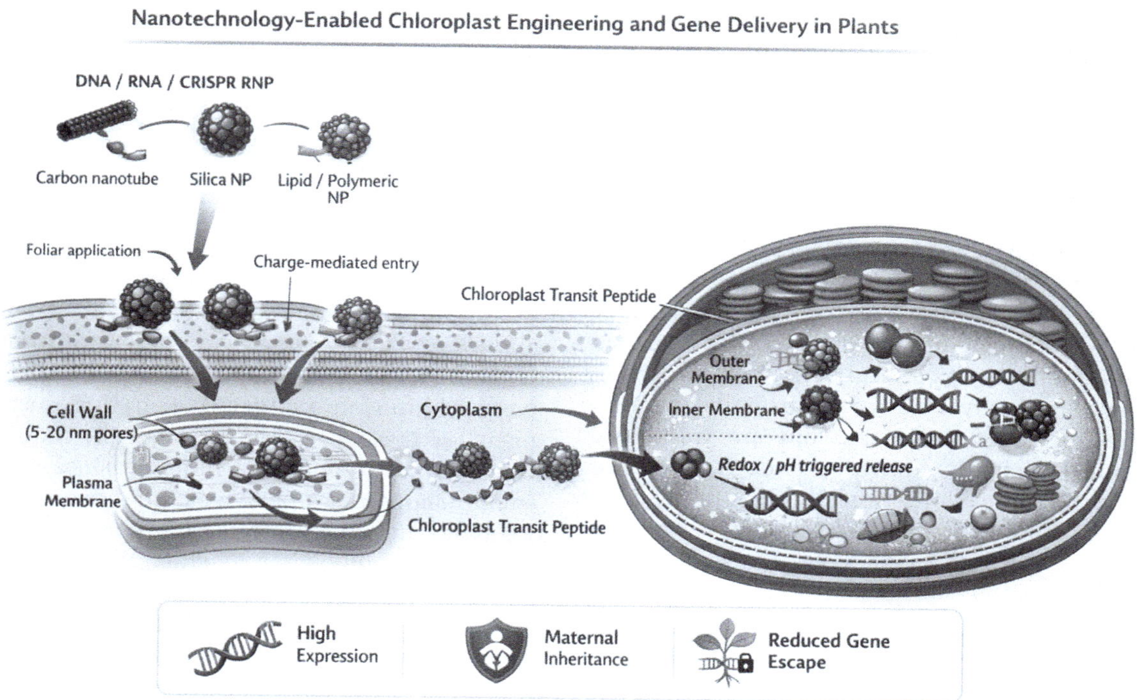

FIGURE 9.2 Mechanisms of nanotechnology-enabled chloroplast engineering illustrating nanoparticle penetration across the cell wall, plasma membrane, cytosol, and chloroplast envelope, culminating in stromal delivery of genetic cargo for transient expression or stable plastome integration.

integration via homologous recombination, enabling high-level transgene expression, polycistronic regulation, and intrinsic gene containment through maternal inheritance. Collectively, these mechanisms underpin the precision, efficiency, and biosafety of nanotechnology-enabled chloroplast genome engineering (Kwak et al., 2019; Rani et al., 2023; Bock, 2015).

3.3 Role of particle size, surface charge, and functionalization in targeting efficiency

The success of nanoparticle-mediated delivery systems heavily depends on the physicochemical properties of the nanocarriers, with particle size, surface charge, and surface functionalization playing pivotal roles in targeting efficiency. Nanoparticle targeting efficiency is critically determined by three interrelated parameters.

3.3.1 *Particle size*

Particle size is a critical factor influencing the efficiency of nanoparticle delivery to chloroplasts. Nanoparticles smaller than approximately 20 nm have enhanced capability to penetrate plant cell walls and chloroplast envelopes due to the limited size exclusion zones of these biological barriers. Studies show that positively charged hydrophilic nanoparticles smaller than 10–20 nm are transported more efficiently in maize and cotton leaves, overcoming cellular porosity constraints (Khan et al., 2019; Newkirk et al., 2021a).

3.3.2 *Surface charge*

Surface charge strongly modulates nanoparticle interaction with negatively charged plant cell surfaces and membranes. Positively charged nanoparticles show superior adhesion and membrane penetration via mechanisms such as Lipid Exchange Envelope Penetration (LEEP), which transiently disrupts lipid bilayers. Neutral or negatively charged particles tend to show less efficient uptake and are more prone to aggregation, diminishing their bioavailability in planta (Wong et al., 2016; Santana et al., 2020).

3.3.3 *Functionalization*

Functionalization with targeting ligands, especially chloroplast transit peptides (cTPs), greatly enhances nanoparticle specificity and efficiency for chloroplast delivery. These peptides mimic natural targeting signals that engage import machinery such as the TIC/TOC complex. Bioconjugation of nanoparticles with peptides derived from the Rubisco small subunit or plastid ribosomal proteins significantly improves chloroplast localization, surpassing delivery based solely on size and charge (Friedman et al., 2013; Santana et al., 2020) (D).

3.4 Endocytosis-independent uptake and stromal localization pathways

Beyond classical endocytosis, nanoparticles can enter plant cells through energy-independent pathways, particularly when engineered to evade membrane recognition. The initial barrier, the plant cell wall, with its size exclusion limit of approximately 5–20 nm, allows smaller nanoparticles to pass through without triggering endocytosis, indicating a passive diffusion or direct penetration mechanism for these particles. Endocytosis requires energy and cellular machinery; hence, very small nanoparticles or those highly engineered to avoid membrane recognition may bypass this process altogether to enter the symplast. This energy-independent uptake is particularly relevant under physiological or stress conditions where endocytotic pathways may be limited. Once past the plasma membrane, the stromal localization of nanoparticles occurs without the classical endocytic vesicle trafficking. Nanoparticles may traverse via plasmodesmata, the intercellular cytoplasmic channels allowing symplastic movement, or directly diffuse through the cytosol aided by their physicochemical properties and chaperone interactions (Santana et al., 2020; Wang et al., 2023). Fluorescent nanoparticle studies have demonstrated effective gene silencing and expression even in the absence of detectable endocytosis, underscoring the importance of non-endocytic pathways in chloroplast delivery (Sembada and Lenggoro, 2024).

3.5 Interaction of nanoparticles with chloroplast DNA, RNA, and protein complexes

Nanoparticles (NPs) have emerged as powerful vectors for targeted delivery of genetic material and bioactive molecules to chloroplasts, enabling novel plant engineering applications. Their effective function relies critically on their interactions with the chloroplast's fundamental biomolecular components: DNA, RNA, and protein complexes. Understanding these interactions is essential for optimizing nanomaterial design and delivery strategies to enhance gene expression, stability, and overall transformation efficiency (Kwak et al., 2019; Wang et al., 2025).

3.5.1 *Interaction with chloroplast DNA*

Chloroplast DNA (cpDNA) is the genetic blueprint housed within the plastid stroma, governing photosynthesis, and other vital metabolic functions. Nanoparticles interact with cpDNA mainly through electrostatic and π–π stacking interactions, especially when the NP surface is functionalized with positively charged polymers or aromatic moieties that can bind nucleic acids efficiently. This complexation not only stabilizes the DNA against nucleolytic degradation but also facilitates its delivery across chloroplast membranes (Demirer et al., 2019; Giraldo et al., 2014). For instance, single-walled carbon nanotubes (SWCNTs) functionalized with cationic polymers exhibit high DNA loading capacity and promote efficient transgene delivery and expression in chloroplasts. This binding must be stable enough to protect DNA during transit but easily reversible within the stroma to enable gene expression (Kwak et al., 2019).

3.5.2 *Interaction with chloroplast RNA*

Post-delivery, nanoparticles may interact with chloroplast RNA-mRNA, tRNA, rRNA, and regulatory RNAs-impacting transcript stability and protein synthesis (Yong et al., 2024). Nanoparticles are engineered to deliver RNA interference (RNAi) molecules or CRISPR-Cas components targeted to plastid genomes or transcripts, modulating gene expression combinatorial networks in the chloroplast. The reversible binding of nanoparticles to RNA or RNA-protein complexes enables transient regulation without permanent genome alterations. Understanding these RNA interactions allows the design of NPs that control translation efficiency and gene silencing with spatial and temporal precision (Mujtaba et al., 2021; Sembada and Lenggoro, 2024).

3.5.3 *Interaction with protein complexes*

Chloroplast protein complexes, including those involved in photosystems, ribosomal machinery, and membrane import systems (TIC/TOC), are pivotal interaction targets for nanoparticles. Surface modifications on NPs with peptides or ligands that mimic transit sequences improve recognition by the protein import channels, directly facilitating translocation across the chloroplast envelope. Additionally, intracellular protein coronas forming on NPs influence their stability, biodistribution, and functionalization within the chloroplast environment (Paila et al., 2015; Santana et al., 2020; Prakash et al., 2023).

4. Classes of nanomaterials for chloroplast engineering

Nanotechnology has become a powerful platform for chloroplast genome engineering by enabling precise delivery of genetic material and bioactive molecules across multiple cellular and organellar barriers. Diverse nanomaterials have been developed based on composition, structure, and functional behavior, each offering distinct advantages for chloroplast targeting, cargo protection, and controlled release (Table 9.1). These nanomaterials broadly include carbon-based, metal and metal oxide, silica-based, polymeric/lipid-based, and hybrid smart nanocarriers.

4.1 Carbon-based nanomaterials

Carbon-based nanomaterials, including single-walled carbon nanotubes (SWCNTs), graphene derivatives, and carbon dots (CDs), are among the most effective carriers for chloroplast delivery due to their nanoscale dimensions, high surface area, and ease of functionalization (Kwak et al., 2019). SWCNTs are cylindrical nanostructures; interact strongly with nucleic acids via π–π stacking and electrostatic forces, facilitating stable cargo loading and efficient penetration through plant cell walls and chloroplast envelopes, often via the Lipid Exchange Envelope Penetration (LEEP) mechanism (Wong et al., 2016). Graphene oxide and reduced graphene oxide further offer high adsorption capacity, chemical tunability, and intrinsic photoluminescence, enabling simultaneous delivery and intracellular tracking (Bellier et al., 2022). Carbon dots (CDs) are small, quasi-spherical nanoparticles with excellent photoluminescence and biocompatibility. They can be functionalized for specific targeting to chloroplasts, enabling both imaging and delivery of genetic cargoes. CDs provide favorable dispersion and reduced aggregation compared to larger carbon nanostructures, facilitating cellular uptake and intracellular trafficking Despite their promise, careful surface engineering is required to minimize cytotoxicity and oxidative stress (Santana et al., 2020; Díez-Pascual, 2021; Peng et al., 2024).

4.2 Metal and metal oxide nanoparticles

Metal-based nanomaterials such as gold (AuNPs), silver (AgNPs), titanium dioxide (TiO_2), and zinc oxide (ZnO) nanoparticles exhibit multifunctional roles in chloroplast engineering, combining delivery capability with antimicrobial and stress-modulating properties (Yiblet and Sisay, 2024). AuNPs are particularly valued for their chemical inertness, tunable size, and optical traceability, allowing targeted chloroplast localization with minimal toxicity. In contrast, AgNPs and metal oxides generate reactive oxygen species that contribute to antimicrobial defense and stress signaling but require strict dose regulation to avoid phytotoxic effects. Surface functionalization enhances their compatibility and targeting efficiency (Rastogi et al., 2017; Kwak et al., 2019).

4.3 Silica-based and mesoporous nanocarriers

Mesoporous silica nanoparticles (MSNs) are distinguished by their ordered pore structure, high loading capacity, and controlled release behavior, making them highly suitable for chloroplast-targeted gene delivery. Their surface silanol groups allow facile functionalization with chloroplast transit peptides, improving organelle specificity. MSNs protect nucleic acids from degradation and release cargo in response to intracellular stimuli such as pH and ionic changes. However, synthesis complexity and aggregation under physiological conditions remain challenges (Narayan et al., 2018; Huq et al., 2025; Liu et al., 2025).

TABLE 9.1 Nanomaterials in chloroplast engineering and gene delivery: mechanisms of action of advantages and limitations.

Nanomaterial class	Mechanisms of action	Advantages	Limitations	References
Carbon-based nanomaterials (carbon nanotubes, graphene derivatives, carbon dots)	Penetrate plant cell walls and chloroplast envelopes via size exclusion and lipid exchange envelope penetration (LEEP); bind DNA/RNA through π–π stacking and electrostatic interactions; enable passive and active chloroplast targeting	High gene-loading capacity; direct chloroplast entry; non-biological gene delivery; enables in planta imaging	Dose-dependent oxidative stress; potential cytotoxicity	Wong et al. (2016), Kwak et al. (2019), Santana et al. (2020), Díez-Pascual (2021), and Peng et al. (2024)
Metal and metal oxide nanoparticles (Au, Ag, TiO_2, ZnO)	Adsorb genetic cargo on particle surface; transiently increase membrane permeability; generate controlled ROS for stress signaling and antimicrobial defense; enable optical tracking	Dual function in gene delivery and stress tolerance; antimicrobial activity; facile surface chemistry	ROS-induced phytotoxicity at high doses; low biodegradability	Rastogi et al. (2017), Kwak et al. (2019), and Yiblet and Sisay (2024)
Silica-based and mesoporous nanocarriers	Encapsulate nucleic acids within mesopores; enable sustained and controlled release; surface functionalization with chloroplast transit peptides improves organelle specificity	High cargo protection; tunable release kinetics; improved chloroplast targeting	Complex synthesis; possible aggregation	Narayan et al. (2018), Huq et al. (2025), and Liu et al. (2025)
Polymeric nanovectors (PLGA, chitosan, PEI)	Encapsulate DNA/RNA protecting from nuclease degradation; promote endocytosis-mediated uptake; gradual intracellular release	Biodegradable; high biocompatibility; prolonged gene expression	Limited penetration across chloroplast membranes	Campos et al. (2015) and Moniri Javadhesari et al. (2022)
Lipid-based nanovectors (liposomes, solid lipid nanoparticles)	Fuse with plasma and chloroplast membranes; deliver cargo via membrane-mimetic interactions; enhance intracellular trafficking	Low toxicity; efficient membrane fusion; high biocompatibility	Structural instability; limited cargo loading	Jiang et al. (2024) and Rahat et al. (2024)
Hybrid nanocomposites and smart nanocarriers (stimuli-responsive systems)	Combine organic–inorganic features; trigger cargo release in response to pH, redox state, light, or enzymatic signals; enable spatiotemporal control of gene delivery	High targeting precision; reduced off-target effects; CRISPR/Cas compatibility	Structural complexity; high production cost	Newkirk et al. (2021) and Kamaraj et al. (2025)

4.4 Polymeric and lipid-based nanovectors

Polymeric (e.g., PLGA, chitosan, PEI) and lipid-based nanocarriers (liposomes, SLNs) offer high biocompatibility, biodegradability, and flexible cargo encapsulation (Campos et al., 2015). Polymeric nanoparticles provide sustained release and protection of genetic material, while lipid-based systems mimic biological membranes, enhancing fusion and endocytic uptake. Hybrid polymer–lipid systems further improve stability and delivery efficiency by combining structural robustness with membrane compatibility (Rahat et al., 2024). Nevertheless, formulation complexity and stability under plant physiological conditions require careful optimization.

4.5 Hybrid nanocomposites and smart nanocarriers

Hybrid nanocomposites integrate inorganic and organic components to achieve synergistic functionality, while smart nanocarriers incorporate stimuli-responsive elements for controlled, spatiotemporal cargo release These systems respond to internal cues (pH, redox status, enzymes) or external triggers (light, temperature), enabling precise chloroplast targeting and reduced off-target effects (Kamaraj et al., 2025; Newkirk et al., 2021a). Although highly promising for advanced chloroplast engineering and CRISPR delivery, their complexity and long-term biosafety require further investigation.

5. Molecular mechanisms of nanotechnology-enabled gene delivery

The formidable barriers of the plant cell wall and the complex double-membrane envelope of the chloroplast have long been significant hurdles in plant genetic engineering. Nanotechnology has emerged as a revolutionary approach, offering sophisticated solutions to these challenges. The core of this paradigm shift lies in the unique molecular mechanisms that nanomaterials employ to conjugate, protect, deliver, and release genetic cargo with precision. This section delves into the intricate processes that underpin nanomaterial-mediated gene delivery to chloroplasts, from initial complexation to final transgene expression.

5.1 Nanoparticle–nucleic acid conjugation and protection mechanisms

The efficacy of nanomaterial-based delivery begins with the stable and functional conjugation of nucleic acids (DNA, RNA) to the nanocarrier. This process is a deliberate engineering step designed not only for transport but, crucially, for protection. Naked nucleic acids are rapidly degraded by nucleases present in the apoplast and cytoplasm. Nanocarriers overcome this through two primary mechanisms.

5.1.1 Electrostatic complexation

This is the most common strategy, leveraging the strong electrostatic attraction between positively charged (cationic) nanocarriers and the negatively charged phosphate backbone of nucleic acids. Cationic polymers like polyethylenimine (PEI) and chitosan entangle DNA or RNA, condensing them into compact, stable structures known as "polyplexes." Similarly, lipid nanoparticles (LNPs) form "lipoplexes." This condensation is critical as it significantly reduces the hydrodynamic size of the genetic material, facilitating cellular entry, and, most importantly, shields it from enzymatic degradation by sterically hindering nuclease access (Wong et al., 2016).

5.1.2 Covalent conjugation and encapsulation

For enhanced stability and controlled release, nucleic acids can be covalently tethered to nanoparticle surfaces or encapsulated within them. Gold nanoparticles (AuNPs) can be functionalized with thiol-modified DNA strands. Mesoporous silica nanoparticles (MSNs) can load DNA within their porous network, physically protecting it from the external environment. A key advancement is the use of chloroplast-targeting peptides (e.g., the transit peptide of the small subunit of Rubisco) covalently attached to the nanocarrier surface. These peptides act as molecular "ZIP codes," guiding the entire complex directly to the chloroplast after cellular uptake (Santana et al., 2020).

5.2 Controlled release of genetic cargo within chloroplasts

Successful delivery is futile if the genetic cargo is not released in a functional form at the target site. "Controlled release" mechanisms are engineered into nanocarriers to respond to the unique biochemical stimuli within the chloroplast stroma (Santana, 2022).

5.2.1 *Redox-triggered release*

The chloroplast stroma is a highly reducing environment, rich in antioxidants like glutathione (GSH). Nanocarriers designed with disulfide bonds (-S-S-) in their polymer backbone or as linker molecules are stable in the oxidizing extracellular space but rapidly cleave in the reductive stroma, leading to the disintegration of the carrier and the subsequent release of DNA or RNA (Raza et al., 2018; An et al., 2022; Santana et al., 2022).

5.2.2 *pH-triggered release*

While subtle, the pH gradient between the cytoplasm (neutral) and the chloroplast stroma (slightly alkaline, pH ~8.0) can be exploited. Carriers with pH-sensitive bonds, such as acetal or hydrazone linkages, remain stable during transit but hydrolyze in the alkaline environment, triggering cargo release (Trinh and Masuda, 2022).

5.2.3 *Enzyme-triggered release*

The chloroplast contains a unique suite of enzymes. Nanocarriers can be fabricated from biodegradable polymers like PLGA [poly (lactic-co-glycolic acid)] that slowly degrade through hydrolysis, providing a sustained release of the genetic material over time, which can be beneficial for prolonged expression (Karimi et al., 2016; Gupta et al., 2022).

5.3 CRISPR/Cas-based chloroplast genome editing using nanocarriers

Efficient chloroplast genome editing using CRISPR/Cas systems requires the coordinated delivery of a Cas nuclease and guide RNA (gRNA), a task complicated by the molecular size of Cas proteins and the intrinsic instability of RNA. Nanocarriers offer a robust solution by enabling co-delivery, protection, and controlled release of all CRISPR components within the chloroplast stroma (Demirer et al., 2021). Cationic polymer-based nanostructures, such as polyethylenimine (PEI) polyplexes, can compact plasmids encoding Cas nucleases and gRNAs into stable complexes, facilitating intracellular transport and plastid targeting. A more advanced strategy involves the direct delivery of pre-assembled Cas–gRNA ribonucleoprotein (RNP) complexes. These RNPs can be encapsulated within lipid nanoparticles (LNPs) or conjugated to functionalized gold nanoparticles (AuNPs), providing enhanced protection to gRNAs and improving delivery efficiency. Upon stromal release, RNPs act immediately at the target locus and are subsequently degraded, resulting in transient genome-editing activity that minimizes off-target cleavage and eliminates the need for stable transgene integration. This transient yet precise editing framework is particularly advantageous for plastid metabolic engineering, enabling targeted gene knock-outs, knock-ins, and pathway optimization while maintaining genomic stability (Ahmar et al., 2021; Xu et al., 2025a, 2025b).

5.4 Nanoparticle-assisted transgene integration and expression stability

For stable, long-term expression of foreign genes, integration into the chloroplast genome (plastome) is essential. Chloroplast transformation is advantageous because it often occurs via homologous recombination, a precise gene replacement mechanism. Nanoparticles can be co-loaded with the transgene cassette (flanked by chloroplast-specific homology regions) and other factors to enhance recombination efficiency. The high copy number of the plastome (1000–10,000 copies per cell) means that a successful integration event can lead to an enormous accumulation of the transgene. The primary goal is to achieve homoplasmy—a state where all copies of the chloroplast genome carry the transgene, eliminating wild-type genomes. Nanoparticle-mediated delivery, by ensuring a high local concentration of intact, functional DNA directly in the stroma, significantly increases the probability of homologous recombination and the subsequent recovery of homoplasmic lines, which are crucial for stable, high-level transgene expression without gene silencing (Narra et al., 2025).

5.5 Visualization and tracking of nanomaterial-mediated gene delivery

Monitoring nanomaterial-assisted gene delivery in plants relies on advanced imaging and analytical approaches that enable real-time and spatially resolved tracking of genetic cargo (DNA, siRNA, dsRNA) as it traverses structural barriers such as the cell wall and intracellular membranes.

5.5.1 Fluorescent confocal microscopy

This is the workhorse technique. Nanoparticles are tagged with fluorescent dyes (e.g., FITC, Cy5) or quantum dots. Using confocal laser scanning microscopy (CLSM), researchers can track the real-time uptake of particles, their trafficking through the cytoplasm, and their ultimate localization within chloroplasts, distinguished by chlorophyll autofluorescence (red) (Mandal et al., 2015; Kolesnikova et al., 2025).

5.5.2 Transmission electron microscopy (TEM)

TEM offers ultrastructural resolution to directly observe nanoparticles within subcellular compartments, including chloroplasts, vacuoles, and cytosol, confirming organelle-level localization (Zhao et al., 2022).

5.5.3 Mass spectrometry imaging (MSI)

Label-free platforms such as LA-ICP-MS and MALDI-MSI facilitate spatial mapping of nanomaterial distribution and their interactions with endogenous biomolecules across plant tissues.

5.5.4 Super-resolution microscopy (SRM)

SRM surpasses the diffraction limit of light microscopy, enabling nanoscale tracking of dynamic interactions between nanomaterials and intracellular structures.

5.5.5 Correlative light and electron microscopy (CLEM)

CLEM integrates fluorescence-based live-cell imaging with high-resolution electron microscopy, allowing precise correlation between functional delivery and subcellular localization. This dual approach is particularly useful for confirming chloroplast-targeted delivery while preserving spatial context (Verkade and Collinson, 2019; Han, 2021).

6. Current progress and experimental evidence

The theoretical promise of nanotechnology-enabled chloroplast engineering is now being validated by a growing body of compelling experimental evidence. Research over the past decade has transitioned from proof-of-concept studies to demonstrations of high-efficiency transformation and significant transgene expression in diverse model organisms. This section reviews the key experimental milestones, compares the technology to conventional methods, and assesses its performance through standardized efficiency metrics.

6.1 Experimental models demonstrating nanoparticle-mediated plastid transformation

The foundational work in this field has established several robust experimental models. Initial breakthroughs often utilized cell wall-free plant systems (protoplasts) to simplify delivery. However, recent advances have conclusively demonstrated that nanoparticles can effectively penetrate the intact cell walls of whole plants, a critical requirement for practical application. The general experimental workflow involves: (1) synthesizing and functionalizing nanoparticles with chloroplast-targeting peptides and nucleic acids; (2) applying these nanocarriers to plant tissues via vacuum infiltration, leaf injection, or incubation; (3) tracking their uptake and localization via fluorescence microscopy; and (4) assaying for transgene expression and genomic integration. Success in these models has been demonstrated through the expression of reporter genes (e.g., GFP, GUS), antibiotic resistance markers, and, more recently, the editing of endogenous plastid genes via CRISPR/Cas (Rani et al., 2025; Newkirk et al., 2021a).

6.2 Case studies: *Arabidopsis thaliana*, *Nicotiana tabacum*, and *Chlamydomonas reinhardtii*

Three model organisms have been pivotal for advancing nanomaterial-mediated chloroplast delivery.

6.2.1 *Arabidopsis thaliana*

As the premier model for plant genetics, *Arabidopsis* has been instrumental in demonstrating the *in planta* functionality of nanocarriers. A landmark study used chitosan-complexed single-walled carbon nanotubes (SWCNTs) to deliver DNA vectors encoding GFP into *Arabidopsis* leaves. Confocal microscopy confirmed the localization of GFP fluorescence specifically within chloroplasts, providing visual evidence of successful targeting and transgene expression without the aid of external force (Geng et al., 2022; Newkirk, 2023). This work established that nanoparticle design, rather than physical force, is the key to breaching biological barriers.

6.2.2 *Nicotiana tabacum*

Tobacco is the most well-established model for chloroplast transformation due to its ease of transformation and high plastome copy number. It has been the system of choice for demonstrating high-level protein production. For instance, mesoporous silica nanoparticles (MSNs) functionalized with transit peptides have been used to deliver linear DNA cassettes into tobacco chloroplasts, resulting in the stable expression of reporter genes and therapeutic proteins at levels competitive with biolistics (Newkirk et al., 2021a). The success in tobacco provides a direct and favorable comparison to the established gold-standard method.

6.2.3 *Chlamydomonas reinhardtii*

This unicellular green alga, with its single, large chloroplast, is a powerful model for chloroplast biology. Nanocarrier studies in

Chlamydomonas have been crucial for quantifying delivery efficiency. Experiments using lipid nanoparticles (LNPs) to deliver DNA and siRNA have shown high transformation frequencies and effective gene silencing, respectively. The simplicity of the *Chlamydomonas* system allows for high-throughput screening of different nanoparticle formulations, accelerating the optimization of delivery parameters (Newkirk, 2023; Rani et al., 2025; Wang et al., 2023).

6.3 Comparative analysis with conventional gene delivery systems

The advantages of nanotechnology become starkly evident when compared to conventional methods like biolistics and *Agrobacterium*-mediated transformation (Table 9.2).

6.3.1 Biolistics (gene gun)

This method bombards tissues with micron-sized gold or tungsten particles coated with DNA. While successful, it is characterized by

TABLE 9.2 Comparative analysis of different chloroplast transformation strategies.

Feature	Biolistics (particle bombardment)	Agrobacterium-mediated transformation	Nanotechnology-enabled chloroplast engineering
Delivery principle	Physical propulsion of DNA-coated microprojectiles	Biological DNA transfer via *Agrobacterium* T-DNA machinery	Physicochemical targeting using engineered nanocarriers
Chloroplast targeting capability	Moderate (requires plastid-specific vectors)	Very low (strong nuclear bias)	High (size-, charge-, and ligand-engineered specificity)
Genome integration route	Homologous recombination (plastome)	Random nuclear integration	Homologous recombination or transient plastid expression
Equipment requirement	Very high (gene gun, pressurized systems)	Low (standard tissue culture facilities)	Low to moderate (nanomaterial synthesis and formulation)
Cellular and tissue damage	High (mechanical and oxidative injury)	Low	Minimal
Cargo versatility	DNA only (limited RNA delivery)	DNA	DNA, RNA, proteins, CRISPR–Cas RNPs
Control over cargo release	None	None	High (redox-, pH-, or enzyme-responsive release)
Species dependence	Low	High (host-range limitations)	Low to moderate
Transgene containment potential	Moderate	Low (pollen transmission risk)	High (maternal plastid inheritance; transient delivery options)
Scalability and field applicability	Limited	Laboratory-restricted	Potentially high (spray–or infiltration-based delivery)
Overall suitability for chloroplast engineering	Established but invasive	Poor	Highly promising and next-generation

high cost, specialized equipment, and significant cell damage. Furthermore, it often leads to complex integration patterns and the delivery of multiple transgene copies, making it difficult to achieve homoplasmy. In contrast, nanoparticles offer a gentler, more scalable, and cost-effective alternative. They cause minimal tissue damage and, by delivering a more controlled cargo load, can facilitate cleaner integration events (Wang et al., 2019; Lacroix and Citovsky, 2020).

6.3.2 Agrobacterium-*mediated transformation*

This method is highly efficient for nuclear transformation but is notoriously ineffective for stable chloroplast transformation, as *Agrobacterium* primarily transfers T-DNA to the nucleus. Nanoparticles, being non-biological and synthetically programmable, are agnostic to the target organelle. Their targeting is determined by their surface chemistry, allowing for direct and specific delivery to the chloroplast, a feat *Agrobacterium* cannot naturally accomplish (Shruti et al., 2024).

6.4 Efficiency metrics: transformation frequency, transgene tability, and expression levels

The ultimate validation of any transformation technology lies in its quantitative performance.

6.4.1 *Transformation frequency*

This metric measures the number of successful transformation events relative to the number of treated cells or explants. Early nanoparticle studies reported frequencies lower than optimized biolistics. However, recent advances with optimized nanocarriers (e.g., LNPs, peptide-functionalized polymers) have dramatically improved these numbers. In model systems like *Chlamydomonas*, nanoparticle-mediated transformation frequencies have begun to approach and even surpass those of conventional methods for certain cargo types (Rani et al., 2025; Shivashakarappa et al., 2025).

6.4.2 *Transgene stability*

Stability is assessed over multiple generations to ensure the transgene is stably integrated and maintained. Nanoparticle-delivered transgenes have been shown to integrate via homologous recombination into the plastome, leading to stable, Mendelian inheritance in subsequent generations once homoplasmy is achieved. The precision of this process often results in cleaner

genetic outcomes compared to the multi-copy, random insertions common with biolistics (Dietz-Pfeilstetter, 2010; Narra et al., 2025).

6.4.3 *Expression levels*

The high copy number of the chloroplast genome enables extraordinarily high levels of protein expression. Nanoparticle-mediated delivery has successfully achieved recombinant protein yields exceeding 5% of total soluble protein (TSP) in early-generation plants, with the potential to reach over 40% in homoplasmic lines—a figure comparable to the best results from biolistics (Yu et al., 2020). This demonstrates that nanocarriers do not compromise the chloroplast's innate capacity for high-level expression (Daniell et al., 2002; Kwak et al., 2019).

7. Challenges, limitations, and safety considerations

Although nanotechnology offers novel opportunities for chloroplast genome manipulation, its practical deployment is constrained by biological, ecological, and regulatory challenges. Addressing these limitations is essential for the responsible advancement of nano-enabled plant biotechnology.

7.1 Cytotoxicity and nanoparticle-induced oxidative stress

A major limitation of nanoparticle-based delivery systems is their potential to induce oxidative stress in plant cells. Many engineered nanomaterials, particularly metal and metal oxide nanoparticles, stimulate excessive generation of reactive oxygen species (ROS) due to their high surface reactivity (Manke et al., 2013). In chloroplasts, elevated ROS levels can disrupt thylakoid membrane integrity, impair electron transport, and reduce photosynthetic efficiency (Li and Kim, 2022). While low levels of ROS may activate antioxidant defenses, sustained or high nanoparticle exposure can overwhelm protective enzymes such as superoxide dismutase and catalase, leading to growth inhibition and cellular damage (Sharma et al., 2019). Cytotoxic responses are strongly influenced by nanoparticle size, surface charge, coating chemistry, and dosage, emphasizing the need for carefully optimized nanocarrier design (Gowtham et al., 2024; Abegunde et al., 2025).

7.2 Genomic stability and off-target effects

Chloroplast genomes exhibit high structural stability and undergo homologous recombination, making them attractive targets for genetic engineering (Bock, 2015). Nevertheless, nanoparticle-mediated gene delivery may interfere with plastome replication or transcription if nanomaterials persist within the chloroplast stroma (Kwak et al., 2019). In CRISPR/Cas-assisted chloroplast editing, unintended mutations may arise from prolonged nuclease activity or imperfect guide RNA specificity (Ahmar et al., 2021). Although off-target effects are generally lower in plastids than in nuclear genomes, rigorous validation of edited lines is essential. Transient delivery approaches, particularly ribonucleoprotein-based systems, offer improved precision by limiting the duration of genome-editing activity (Newkirk et al., 2021b).

7.3 Biocompatibility, environmental fate, and biosafety

The long-term compatibility of nanomaterials with plant systems and surrounding ecosystems remains insufficiently understood. Non-biodegradable nanoparticles may accumulate in plant tissues or persist in soil, potentially altering microbial communities and nutrient cycling processes (Bellanthudawa et al., 2023; Sathyamoorthy et al., 2025). Nanoparticles released into the rhizosphere can influence beneficial plant–microbe interactions, including mycorrhizal associations, with possible consequences for plant health and ecosystem stability. These concerns highlight the importance of life-cycle assessments, environmental monitoring, and the development of biodegradable nanocarriers for agricultural use (Ahmad et al., 2024; Khan et al., 2022).

7.4 Regulatory and ethical considerations

Current regulatory frameworks inadequately address nanotechnology-based plant genetic engineering, as most biosafety guidelines focus on transgenic organisms rather than nano-enabled delivery platforms. This regulatory gap complicates risk evaluation and approval processes.Ethical considerations include transparency in risk communication, environmental stewardship, and equitable access to emerging technologies. Establishing standardized safety assessment protocols and harmonized international regulations is essential to ensure that nanotechnology applications in plant biotechnology are both scientifically sound and socially acceptable (Kumari et al., 2023a, 2023b).

8. Future prospects and research directions

The successful demonstration of nanotechnology-enabled chloroplast engineering marks not an endpoint, but the dawn of a new paradigm in plant biotechnology. The foundational knowledge and experimental evidence now pave the way for a future where genetic manipulation of chloroplasts is more precise, scalable, and impactful than ever before (Table 9.3). This section outlines the key research trajectories and the transformative potential they hold.

8.1 Designing next-generation nanocarriers with enhanced targeting precision

The current generation of nanocarriers, while effective, represents the first step in a journey toward absolute cellular specificity. Future research will focus on developing "smart" nanocarriers with multi-layered targeting capabilities. This involves moving beyond single peptide ligands to multi-valent systems that recognize a sequence of cellular "addresses." For instance, a nanocarrier could be engineered with one ligand that binds to a specific cell wall component (e.g., in leaf mesophyll cells), a second that facilitates passage through the plasma membrane, and a final, high-affinity chloroplast transit peptide for organellar import. Furthermore, the incorporation of stimuli-responsive "gates" that only open upon encountering the unique redox or pH environment of the chloroplast stroma will ensure zero off-target release in the cytosol, maximizing delivery efficiency and minimizing unintended effects (Demirer and Landry, 2017; Antonelli and Palma, 2025).

8.2 Integration with synthetic biology and AI-guided nano-design

The convergence of nanotechnology with synthetic biology and artificial intelligence (AI) will accelerate the design process from art to science. Synthetic biology provides the tools to create standardized, modular genetic parts (promoters, coding sequences, terminators) optimized for plastid expression. These "plug-and-play" cartridges can be efficiently delivered by nanocarriers to build complex genetic circuits within the chloroplast, enabling sophisticated

TABLE 9.3 Scaling nano-chloroplast engineering from lab to field.

Development phase	Primary challenge	Future research direction
Lab-scale (model systems: *Arabidopsis, Tobacco*)	Proof-of-concept delivery and transgene expression	Optimize delivery efficiency, demonstrate stable and multigene chloroplast engineering
Extended to crops in vitro(e.g., rice, potato)	Species-specific barriers to chloroplast targeting	Identify and apply species-specific chloroplast transit peptides and optimize nanoparticle formulations
Greenhouse scalability	Developing scalable, reproducible application methods	Engineer nanoparticle formulations for foliar sprays, root uptake, or seed treatments; optimize dosage and exposure
Field deployment	Ensuring stability, efficacy, and environmental safety	Conduct large-scale efficacy trials, assess environmental impacts, and support regulatory compliance frameworks

tasks such as environmental sensing and programmable metabolite production. Simultaneously, AI and machine learning are poised to revolutionize nanocarrier design. By training algorithms on large datasets linking nanoparticle parameters (e.g., size, zeta potential, surface chemistry, core composition) to functional outcomes (e.g., delivery efficiency, expression level), it will be possible to computationally predict the optimal nanocarrier for a given plant species and cargo (Kumari et al., 2026). This will drastically reduce the time and cost associated with empirical trial-and-error, ushering in an era of rationally designed, high-performance delivery vectors (An et al., 2022; Ji et al., 2025).

8.3 Toward species-independent and scalable chloroplast transformation

A significant limitation of conventional chloroplast transformation is its confinement to a few amenable species, primarily tobacco. Nanotechnology offers the most promising path to "universal" chloroplast transformation. The core strategy involves identifying and utilizing species-specific targeting peptides. By isolating the transit peptides from key chloroplast proteins in recalcitrant species like maize, soybean, or rice, and conjugating them to nanocarriers, we can potentially trick the chloroplast import machinery of any plant into admitting the genetic cargo. For scalability, the focus will shift from syringe-based infiltration to scalable application methods. Research into nanoparticle formulations compatible with foliar sprays, seed priming, or hydroponic systems is critical. Designing nanocarriers that can penetrate the cuticle and be efficiently taken up by field-grown plants will be the ultimate test for transitioning this technology from the lab to the field (Demirer et al., 2019; Shivashakarappa et al., 2025).

8.4 Development of climate-resilient and high-value crop systems

Nano-enabled chloroplast engineering has significant potential for developing crops resilient to climate-induced stresses such as heat, drought, salinity, and oxidative damage. Targeted plastid manipulation can enhance photosynthetic efficiency, antioxidant capacity, and stress-responsive metabolic pathways (Bock, 2015; Zulfiqar et al., 2024). Beyond stress tolerance, chloroplasts can serve as biofactories for the sustainable production of vaccines, nutraceuticals, industrial enzymes, and specialty metabolites (Tanwar et al., 2023) Nanotechnology-mediated delivery systems will facilitate precise metabolic reprogramming while maintaining genetic containment through maternal inheritance of plastids (Kumari et al., 2023a, 2023b; Akhter Bhat et al., 2025).

8.5 Vision for sustainable agricultural biotechnology through nano-enabled chloroplast engineering

Nano-enabled chloroplast engineering offers a forward-looking framework that closely aligns with the core principles of sustainable agriculture. By enabling precise plastid-level genetic modification, this approach reduces dependence on chemical inputs, enhances resource-use efficiency, and supports environmentally responsible crop improvement strategies. The chloroplast's predominantly maternal inheritance further provides an intrinsic biosafety advantage by limiting transgene dissemination through pollen, addressing long-standing ecological and regulatory concerns. The long-term vision emphasizes the development of safe-by-design nanomaterials—preferably biodegradable systems such as chitosan-, lipid-, or polymer-based carriers—coupled with harmonized regulatory oversight and transparent risk assessment protocols. Compared with conventional biolistic or tissue culture–dependent methods, nano-enabled delivery minimizes cellular damage, lowers energy requirements, and improves transformation accessibility across diverse crop species. Looking ahead, the convergence of nanotechnology, synthetic biology, and artificial intelligence is expected to yield an integrated and highly adaptable toolkit. AI-assisted nano-design may enable the creation of fully biodegradable, species-optimized nanocarriers capable of delivering synthetic gene circuits for climate resilience, metabolic enhancement, or stress tolerance via non-invasive application methods such as foliar sprays (Fig. 9.3). Collectively, these advances position nano-enabled chloroplast engineering as a cornerstone technology for resilient food systems and a sustainable bio-based agricultural economy.

9. Conclusion

Nanotechnology-enabled chloroplast engineering represents a robust and biosafe strategy for next-generation plant genetic improvement.

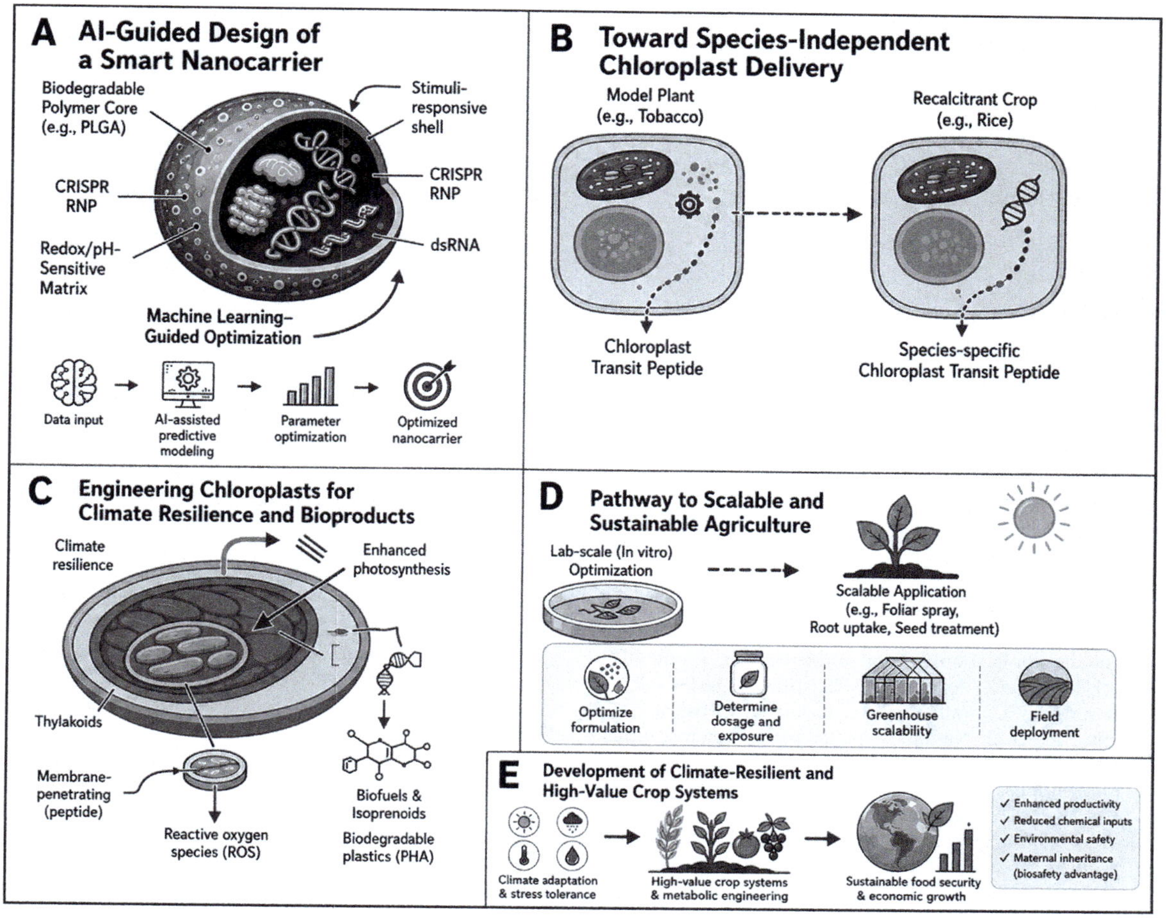

FIGURE 9.3 A multi-scale roadmap for nano-enabled chloroplast engineering.

Advanced nanomaterials capable of traversing cellular and plastid membranes facilitate targeted and controlled delivery of genetic cargo directly into chloroplasts, effectively addressing the limitations of nuclear transformation. The high copy number, transcriptional efficiency, and predominantly maternal inheritance of plastid genomes support stable transgene expression while minimizing risks of gene escape. Integration of nanocarrier engineering with synthetic biology and genome-editing tools further expands opportunities for improving photosynthetic efficiency, stress tolerance, and production of high-value biomolecules. Continued progress will depend on the development of biocompatible, species-independent delivery systems and comprehensive biosafety assessment. Collectively, nano-enabled plastid engineering holds strong potential to underpin sustainable, climate-resilient agricultural biotechnology.

References

Abegunde, S.M., Alaka, M.O., Isaac Awonyemi, O., 2025. Nanomaterial toxicity: a comprehensive review of mechanisms and mitigation strategies. Disc. Hazard. 1 (1). https://doi.org/10.1007/s44475-025-00003-2.

Ahmad, Z., Niyazi, S., Firdoos, A., Wang, C., Manzoor, M. A., Ramakrishnan, M., Upadhyay, A., Ding, Y., 2024. Enhancing plant resilience: nanotech solutions for sustainable agriculture. Heliyon 10 (23). https://doi.org/10.1016/j.heliyon.2024.e40735.

Ahmar, S., Mahmood, T., Fiaz, S., Mora-Poblete, F., Shafique, M.S., Chattha, M.S., Jung, K.H., 2021. Advantage of nanotechnology-based genome editing system and its application in crop improvement. Front. Plant Sci. 12. https://doi.org/10.3389/fpls.2021.663849.

Akhter Bhat, T., Rather, R.A., Bashir, S., Mishra, N.S., John, R., 2025. Nanoparticles in plant systems: omics-based perspectives on stress adaptation and toxicological implications. Plant Nano Biol. 13. https://doi.org/10.1016/j.plana.2025.100181.

Alberts, B., Johnson, A., Lewis, J., Raff, M., Roberts, K., Walter, P., 2002. Chloroplasts and photosynthesis. In: Molecular Biology of the Cell, fourth ed. Garland Science.

Allen, J.F., Raven, J.A., Gray, J.C., Sullivan, J.A., Wang, J.-H., Jerome, C.A., MacLean, D., 2003. Coordination of plastid and nuclear gene expression. Philos. Trans. R. Soc. Lond. Ser. B Biol. Sci. 358 (1429), 135–145. https://doi.org/10.1098/rstb.2002.1180.

An, Y., Wang, Y., Wang, X., Xiao, J., 2022. Development of chloroplast transformation and gene expression regulation technology in land plants. Front. Plant Sci. 13. https://doi.org/10.3389/fpls.2022.1037038.

Antonelli, A., Palma, F., 2025. Nanocarrier-based delivery systems for natural compounds across research stages. Materials 18 (21), 4960. https://doi.org/10.3390/ma18214960.

Bellanthudawa, B.K.A., Nawalage, N.M.S.K., Handapangoda, H.M.A.K., Suvendran, S., Wijayasenarathne, K.A.S.H., Rathnasuriya, M.L.D., Wickramasinghe, P.G.M.U., Aberathna, A.A.A.U., Tennakoon, A., Perera, I.J.J.U.N., 2023. A perspective on biodegradable and non-biodegradable nanoparticles in industrial sectors: applications, challenges, and future prospects. Nanotechnol. Environ. Eng. 8 (4), 975–1013. https://doi.org/10.1007/s41204-023-00344-7.

Bellier, N., Baipaywad, P., Ryu, N., Lee, J.Y., Park, H., 2022. Recent biomedical advancements in graphene oxide- and reduced graphene oxide-based nanocomposite nanocarriers. Biomater. Res. 26 (1). https://doi.org/10.1186/s40824-022-00313-2.

Berry, J.O., Yerramsetty, P., Zielinski, A.M., Mure, C.M., 2013. Photosynthetic gene expression in higher plants. Photosynth. Res. 117 (1–3), 91–120. https://doi.org/10.1007/s11120-013-9880-8.

Bock, R., 2015. Engineering plastid genomes: methods, tools, and applications in basic research and biotechnology. Annu. Rev. Plant Biol. 66 (1), 211–241. https://doi.org/10.1146/annurev-arplant-050213-040212.

Campos, E.V.R., De Oliveira, J.L., Da Silva, C.M.G., Pascoli, M., Pasquoto, T., Lima, R., Abhilash, P.C., Fernandes Fraceto, L., 2015. Polymeric and solid lipid nanoparticles for sustained release of carbendazim and tebuconazole in agricultural applications. Sci. Rep. 5. https://doi.org/10.1038/srep13809.

Daniell, H., Khan, M.S., Allison, L., 2002. Milestones in chloroplast genetic engineering: an environmentally friendly era in biotechnology. Trends Plant Sci. 7 (2), 84–91. https://doi.org/10.1016/S1360-1385(01)02193-8.

Daniell, H., Lin, C.S., Yu, M., Chang, W.J., 2016. Chloroplast genomes: diversity, evolution, and applications in genetic engineering. Genome Biol. 17 (1). https://doi.org/10.1186/s13059-016-1004-2.

Demirer, G.S., Landry, M.P., 2017. Delivering genes to plants. Chem. Eng. Prog. 113 (4).

Demirer, G.S., Zhang, H., Matos, J.L., Goh, N.S., Cunningham, F.J., Sung, Y., Chang, R., Aditham, A.J.,

Chio, L., Cho, M.J., Staskawicz, B., Landry, M.P., 2019. High aspect ratio nanomaterials enable delivery of functional genetic material without DNA integration in mature plants. Nat. Nanotechnol. 14 (5), 456–464. https://doi.org/10.1038/s41565-019-0382-5.

Demirer, G.S., Silva, T.N., Jackson, C.T., Thomas, J.B., Ehrhardt, D.W., Rhee, S.Y., Mortimer, J.C., Landry, M. P., 2021. Nanotechnology to advance CRISPR–Cas genetic engineering of plants. Nat. Nanotechnol. 16 (3), 243–250. https://doi.org/10.1038/s41565-021-00854-y.

Dietz-Pfeilstetter, A., 2010. Stability of transgene expression as a challenge for genetic engineering. Plant Sci. 179 (3), 164–167. https://doi.org/10.1016/j.plantsci.2010.04.015.

Díez-Pascual, A.M., 2021. Carbon-based nanomaterials. Int. J. Mol. Sci. 22 (14), 7726.

Friedman, A., Claypool, S., Liu, R., 2013. The smart targeting of nanoparticles. Curr. Pharm. Des. 19 (35), 6315–6329. https://doi.org/10.2174/13816128113199990375.

Geng, M., Li, L., Ai, M., Jin, J., Hu, D., Song, K., 2022. Recent advances in metal-based nanoparticle-mediated biological effects in Arabidopsis thaliana: a mini review. Materials 15 (13), 4539.

Giraldo, J.P., Landry, M.P., Faltermeier, S.M., McNicholas, T. P., Iverson, N.M., Boghossian, A.A., Reuel, N.F., Hilmer, A.J., Sen, F., Brew, J.A., Strano, M.S., 2014. Plant nanobionics approach to augment photosynthesis and biochemical sensing. Nat. Mater. 13 (4), 400–408.

Gowtham, H.G., Shilpa, N., Singh, S.B., Aiyaz, M., Abhilash, M.R., Nataraj, K., Amruthesh, K.N., Ansari, M.A., Alomary, M.N., Murali, M., 2024. Toxicological effects of nanoparticles in plants: mechanisms involved at morphological, physiological, biochemical and molecular levels. Plant Physiol. Biochem. 210. https://doi.org/10.1016/j.plaphy.2024.108604.

Grevich, J.J., Daniell, H., 2005. Chloroplast genetic engineering: recent advances and future perspectives. Crit. Rev. Plant Sci. 24 (2), 83–107. https://doi.org/10.1080/07352680590935387.

Gupta, A., Tripathy, D.B., Kumar, G., Agarwal, P., Ghosal, A., 2022. Nanopesticides, Nanoherbicides, and Nanofertilizers. CRC Press, Taylor and Francis.

Gupta, A., Tripathy, D.B., Kumar, G., Agarwal, P., Ghosal, A., 2023. Nanopesticides, Nanoherbicides, and Nanofertilizers: Formulations and Applications. CRC Press, pp. 1–179. https://doi.org/10.1201/9781003364429.

Han, S., 2021. Visualisation of Nanoparticle-Cell Interactions by Correlative Microscopy (Doctoral Dissertation, Johannes Gutenberg-Universität Mainz).

Hanson, M.R., Gray, B.N., Ahner, B.A., 2013. Chloroplast transformation for engineering of photosynthesis. J. Exp. Bot. 64 (3), 731–742. https://doi.org/10.1093/jxb/ers325.

Huq, T.B., Anil Kumar Jeeja, P., Dam, S.K., Das, S.S., Vivero-Escoto, J.L., 2025. Recent applications of mesoporous silica nanoparticles in gene therapy. Adv. Healthcare Mater. 14 (26). https://doi.org/10.1002/adhm.202404781.

Jarvis, P., 2008. Targeting of nucleus-encoded proteins to chloroplasts in plants. New Phytol. 179 (2), 257–285.

Ji, J., Li, L., Guo, W., Zhang, J., Yao, Y., Chen, H., Liao, F., Jin, Z., Liu, L., Ouyang, J., Liang, X.J., 2025. Nanoscale synthetic biology with innovative medicinal applications. Fundam. Res. 5 (5), 1889–1910. https://doi.org/10.1016/j.fmre.2024.11.006.

Jiang, Y., Li, W., Wang, Z., Lu, J., 2024. Lipid-based nanotechnology: liposome. Pharmaceutics 16 (1), 34. https://doi.org/10.3390/pharmaceutics16010034.

Kamaraj, I., Kamaraj, S., Shanmugam, G., 2025. Emerging strategies and multifunctional applications of nanomaterials in modern nanomedicine. Med. Nov. Technol. Dev. 100400.

Karimi, M., Ghasemi, A., Sahandi Zangabad, P., Rahighi, R., Moosavi Basri, S.M., Mirshekari, H., Amiri, M., Shafaei Pishabad, Z., Aslani, A., Bozorgomid, M., Ghosh, D., Beyzavi, A., Vaseghi, A., Aref, A.R., Haghani, L., Bahrami, S., Hamblin, M.R., 2016. Smart micro/nanoparticles in stimulus-responsive drug/gene delivery systems. Chem. Soc. Rev. 45 (5), 1457–1501. https://doi.org/10.1039/c5cs00798d.

Keshavareddy, G., Kumar, A.R.V., Ramu, V.S., 2018. Methods of plant transformation-a review. Int. J. Curr. Microbiol. Appl. Sci. 7 (7), 2656–2668.

Khan, M.R., Adam, V., Rizvi, T.F., Zhang, B., Ahamad, F., Jośko, I., Zhu, Y., Yang, M., Mao, C., 2019. Nanoparticle–plant interactions: two-way traffic. Small 15 (37). https://doi.org/10.1002/smll.201901794.

Khan, S.T., Adil, S.F., Shaik, M.R., Alkhathlan, H.Z., Khan, M., Khan, M., 2022. Engineered nanomaterials in soil: their impact on soil microbiome and plant health. Plants 11 (1). https://doi.org/10.3390/plants11010109.

Kolesnikova, O.A., Shikvin, D.A., Antonova, A.O., Iureva, A.M., Komedchikova, E.N., Obozina, A.S., Kachan, V.S., Svetlakova, A.V., Kukushkin, I.D., Shipunova, V.O., 2025. Nanoparticle detection in biology and medicine: a review. Fed. Biosens. 15 (12). https://doi.org/10.3390/bios15120809.

Kumari, A., Rana, V., Yadav, S.K., Kumar, V., 2023a. Nanotechnology as a powerful tool in plant sciences: recent developments, challenges and perspectives. Plant Nano Biol. 5, 100046.

Kumari, R., Suman, K., Karmakar, S., Mishra, V., Lakra, S.G., Saurav, G.K., Mahto, B.K., 2023b. Regulation and safety measures for nanotechnology-based agri-products. Front. Genome Ed. 5. https://doi.org/10.3389/fgeed.2023.1200987.

Kumari, A., Singh, J., Yadav, S.K., 2026. Harnessing nanobiotechnology for sustainable agriculture: from plant nanobionics to AI-driven precision farming. Biosci. Nanotechnol. 2 (1), 3.

Kwak, S.Y., Lew, T.T.S., Sweeney, C.J., Koman, V.B., Wong, M.H., Bohmert-Tatarev, K., Snell, K.D., Seo, J.S., Chua, N.H., Strano, M.S., 2019. Chloroplast-selective gene delivery and expression in planta using chitosan-complexed single-walled carbon nanotube carriers. Nat. Nanotechnol. 14 (5), 447–455. https://doi.org/10.1038/s41565-019-0375-4.

Lacroix, B., Citovsky, V., 2020. Biolistic approach for transient gene expression studies in plants. In: Biolistic DNA Delivery in Plants: Methods and Protocols, pp. 125–139.

Law, S.S.Y., Miyamoto, T., Numata, K., 2023. Organelle-targeted gene delivery in plants by nanomaterials. Chem. Commun. 59 (47), 7166–7181. https://doi.org/10.1039/d3cc00962a.

Li, M., Kim, C., 2022. Chloroplast ROS and stress signaling. Plant Commun. 3 (1). https://doi.org/10.1016/j.xplc.2021.100264.

Liu, J., Nannas, N.J., Fu, F.F., Shi, J., Aspinwall, B., Parrott, W.A., Dawe, R.K., 2019. Genome-scale sequence disruption following biolistic transformation in rice and maize. Plant Cell 31 (2), 368–383. https://doi.org/10.1105/tpc.18.00613.

Liu, S., Zheng, Y., Pan, L., Wang, W., Li, Y., Liu, Z., Zhang, X., 2025. Nanodelivery of nucleic acids for plant genetic engineering. Disc. Nano 20 (1). https://doi.org/10.1186/s11671-025-04207-9.

Lv, Z., Jiang, R., Chen, J., Chen, W., 2020. Nanoparticle-mediated gene transformation strategies for plant genetic engineering. Plant J. 104 (4), 880–891. https://doi.org/10.1111/tpj.14973.

Ma, X., Tian, Y., Yang, R., Wang, H., Allahou, L.W., Chang, J., Williams, G., Knowles, J.C., Poma, A., 2024. Nanotechnology in healthcare, and its safety and environmental risks. J. Nanobiotechnol. 22 (1). https://doi.org/10.1186/s12951-024-02901-x.

Maliga, P., 2004. Plastid transformation in higher plants. Annu. Rev. Plant Biol. 55 (1), 289–313. https://doi.org/10.1146/annurev.arplant.55.031903.141633.

Mandal, S., Zhou, Y., Shibata, A., Destache, C.J., 2015. Confocal fluorescence microscopy: an ultra-sensitive tool used to evaluate intracellular antiretroviral nano-drug delivery in HeLa cells. AIP Adv. 5 (8). https://doi.org/10.1063/1.4926584.

Manke, A., Wang, L., Rojanasakul, Y., 2013. Mechanisms of nanoparticle-induced oxidative stress and toxicity. BioMed Res. Int. 2013 (1), 942916.

Mathuria, A., Chaudhary, A., Mehak, Sharma, H., Singla, S., Verma, C., Saini, A., Mani, I., 2024. Chloroplast Genomics and Their Uses in Crop Improvement Advances in Genomics: Methods and Applications. Springer Nature, pp. 331–356. https://doi.org/10.1007/978-981-97-3169-5_15.

Meyers, B., Zaltsman, A., Lacroix, B., Kozlovsky, S.V., Krichevsky, A., 2010. Nuclear and plastid genetic engineering of plants: comparison of opportunities and challenges. Biotechnol. Adv. 28 (6), 747–756. https://doi.org/10.1016/j.biotechadv.2010.05.022.

Moniri Javadhesari, S., Jabraili, M., Koohi, M., 2022. A review on the application of nanoparticles for targeted gene delivery. Adv. Ceram. Prog. 8 (1), 44–55.

Mujtaba, M., Wang, D., Carvalho, L.B., Oliveira, J.L., Espirito Santo Pereira, A.D., Sharif, R., Jogaiah, S., Paidi, M.K., Wang, L., Ali, Q., Fraceto, L.F., 2021. Nanocarrier-mediated delivery of miRNA, RNAi, and CRISPR-Cas for plant protection: current trends and future directions. ACS Agric. Sci. Technol. 1 (5), 417–435. https://doi.org/10.1021/acsagscitech.1c00146.

Mukundan, N.S., Satyamoorthy, K., Babu, V.S., 2025. Advancing plant protoplasts: innovative techniques and future prospects. Plant Biotechnol. Rep. 19 (6), 659–678. https://doi.org/10.1007/s11816-025-00957-1.

Narayan, R., Nayak, U.Y., Raichur, A.M., Garg, S., 2018. Mesoporous silica nanoparticles: a comprehensive review on synthesis and recent advances. Pharmaceutics 10 (3). https://doi.org/10.3390/pharmaceutics10030118.

Narra, M., Nakazato, I., Polley, B., Arimura, S.I., Woronuk, G.N., Bhowmik, P.K., 2025. Recent trends and advances in chloroplast engineering and transformation methods. Front. Plant Sci. 16. https://doi.org/10.3389/fpls.2025.1526578.

Nazari, M., Kordrostami, M., Ghasemi-Soloklui, A.A., Eaton-Rye, J.J., Pashkovskiy, P., Kuznetsov, V., Allakhverdiev, S.I., 2024. Enhancing photosynthesis and plant productivity through genetic modification. Cells 13 (16). https://doi.org/10.3390/cells13161319.

Newkirk, G.M., 2023. Nanotechnology Approaches for Arabidopsis and Chlamydomonas Chloroplast Bioengineering (Doctoral Dissertation, University of California, Riverside).

Newkirk, G.M., de Allende, P., Jinkerson, R.E., Giraldo, J.P., 2021. Nanotechnology approaches for chloroplast biotechnology advancements. Front. Plant Sci. 12. https://doi.org/10.3389/fpls.2021.691295.

Paila, Y.D., Richardson, L.G.L., Schnell, D.J., 2015. New insights into the mechanism of chloroplast protein import and its integration with protein quality control, organelle biogenesis and development. J. Mol. Biol. 427 (5), 1038–1060. https://doi.org/10.1016/j.jmb.2014.08.016.

Peng, W., Li, P., Zia, M., Bhuiyan, S.A., Liu, Y., Chen, D., Tudi, M., Gao, Y., Yan, X., Jia, Y., Li, Q., 2024. A comparative review on carbon nanomaterials and their impact on plant growth with the lens of green chemistry principles. Glob. Chall. 8 (11). https://doi.org/10.1002/gch2.202400008.

Pfannschmidt, T., Blanvillain, R., Merendino, L., Courtois, F., Chevalier, F., Liebers, M., Grübler, B., Hommel, E., Lerbs-Mache, S., 2015. Plastid RNA polymerases: orchestration of enzymes with different evolutionary origins controls chloroplast biogenesis during the plant life cycle. J. Exp. Bot. 66 (22), 6957–6973. https://doi.org/10.1093/jxb/erv415.

Pogson, B.J., Ganguly, D., Albrecht-Borth, V., 2015. Insights into chloroplast biogenesis and development. Biochim. Biophys. Acta Bioenerg. 1847 (9), 1017–1024. https://doi.org/10.1016/j.bbabio.2015.02.003.

Prakash, S., Gupta, R., Deswal, R., 2023. Nanoparticle protein corona: understanding NP biomolecule interactions for safe and informed nanotechnological applications including stress alleviation in plants. J. Plant Growth Regul. 42 (10), 6377–6396. https://doi.org/10.1007/s00344-022-10857-1.

Rahat, I., Yadav, P., Singhal, A., Fareed, M., Purushothaman, J.R., Aslam, M., Balaji, R., Patil-Shinde, S., Rizwanullah, M., 2024. Polymer lipid hybrid nanoparticles for phytochemical delivery: challenges, progress, and future prospects. Beilstein J. Nanotechnol. 15, 1473–1497. https://doi.org/10.3762/BJNANO.15.118.

Ramakrishnan, M., Kaul, R., Sharma, A., Ahmad, Z., Vijayakanth, V., Keerthana, K., Gao, Z., Zhou, M., Wei, Q., 2025. CRISPR RNP-mediated transgene-free genome editing in plants: advances, challenges and future directions for tree species. Plant Cell Environ. https://doi.org/10.1111/pce.70176.

Rani, S., Kumari, N., Sharma, V., 2023. Uptake, translocation, transformation and physiological effects of nanoparticles in plants. Arch. Agron Soil Sci. 69 (9), 1579–1599. https://doi.org/10.1080/03650340.2022.2103549.

Rani, N., Kumari, K., Hooda, V., 2025. The role of nanoparticles in transforming plant genetic engineering: advancements, challenges and future prospects. Funct. Integr. Genomics 25 (1). https://doi.org/10.1007/s10142-025-01528-x.

Rastogi, A., Zivcak, M., Sytar, O., Kalaji, H.M., He, X., Mbarki, S., Brestic, M., 2017. Impact of metal and metal oxide nanoparticles on plant: a critical review. Front. Chem. 5. https://doi.org/10.3389/fchem.2017.00078.

Raza, A., Hayat, U., Rasheed, T., Bilal, M., Iqbal, H.M.N., 2018. Redox-responsive nano-carriers as tumor-targeted drug delivery systems. Eur. J. Med. Chem. 157, 705–715. https://doi.org/10.1016/j.ejmech.2018.08.034.

Santana, I., Wu, H., Hu, P., Giraldo, J.P., 2020. Targeted delivery of nanomaterials with chemical cargoes in plants enabled by a biorecognition motif. Nat. Commun. 11 (1). https://doi.org/10.1038/s41467-020-15731-w.

Santana, I., 2022. Impact of Targeted Nanomaterials for Chloroplast Bioengineering on Arabidopsis thaliana. Doctoral Dissertation, University of California, Riverside.

Santana, I., Jeon, S.J., Kim, H.I., Islam, M.R., Castillo, C., Garcia, G.F.H., Newkirk, G.M., Giraldo, J.P., 2022. Targeted carbon nanostructures for chemical and gene delivery to plant chloroplasts. ACS Nano 16 (8), 12156–12173. https://doi.org/10.1021/acsnano.2c02714.

Sathyamoorthy, S., Venugopal, A., Lenin, M.J., Tirkey, S., Sevanan, M., 2025. Nanomaterials and Microbial Compatibility: Synergistic and Antagonistic Mechanisms Nano-Bioremediation for Wastewater Treatment. Wiley, pp. 35–72. https://doi.org/10.1002/9781394271641.ch2.

Sembada, A.A., Lenggoro, I.W., 2024. Transport of nanoparticles into plants and their detection methods. Nanomaterials 14 (2) , p.131. https://doi.org/10.3390/nano14020131.

Sharma, S., Singh, V.K., Kumar, A., Mallubhotla, S., 2019. Effect of nanoparticles on oxidative damage and antioxidant defense system. In: Plants Molecular Plant Abiotic Stress: Biology and Biotechnology. Wiley, pp. 315–333. https://doi.org/10.1002/9781119463665.ch17.

Shivashakarappa, K., Marriboina, S., Dumenyo, K., Taheri, A., Yadegari, Z., 2025. Nanoparticle-mediated gene delivery techniques in plant systems. Front. Nanotechnol. 7. https://doi.org/10.3389/fnano.2025.1516180.

Shruti, A., Tripathi, Shukla, S., 2024. Methods of genetic transformation: major emphasis to crop plants. J. Microbiol. Biotechnol. Food Sci. 13 (4). https://doi.org/10.55251/jmbfs.10276.

Tadini, L., Jeran, N., Peracchio, C., Masiero, S., Colombo, M., Pesaresi, P., 2020. The plastid transcription machinery and its coordination with the expression of nuclear genome: plastid-encoded polymerase, nuclear-encoded polymerase and the genomes uncoupled 1-mediated retrograde communication. Phil. Trans. R. Soc. B. Biol. Sci. 375 (1801), 20190399. https://doi.org/10.1098/rstb.2019.0399.

Taiz, L., Zeiger, E., Møller, I.M., Murphy, A., 2015. Plant Physiology and Development, sixth ed. Sinauer Associates, Sunderland.

Tanwar, N., Arya, S.S., Rookes, J.E., Cahill, D.M., Lenka, S. K., Bansal, K.C., 2023. Prospects of chloroplast metabolic engineering for developing nutrient-dense food crops. Crit. Rev. Biotechnol. 43 (7), 1001–1018. https://doi.org/10.1080/07388551.2022.2092717.

Trinh, M.D.L., Masuda, S., 2022. Chloroplast pH homeostasis for the regulation of photosynthesis. Front. Plant Sci. 13. https://doi.org/10.3389/fpls.2022.919896.

Tschiche, H.R., Bierkandt, F.S., Creutzenberg, O., Fessard, V., Franz, R., Giese, B., Greiner, R., Haas, K.H.,

Haase, A., Hartwig, A., Hund-Rinke, K., Iden, P., Kromer, C., Loeschner, K., Mutz, D., Rakow, A., Rasmussen, K., Rauscher, H., Richter, H., Schoon, J., Schmid, O., Som, C., Günter, G.E., Westerhoff, P., Wohlleben, W., Luch, A., Laux, P., 2022. Environmental considerations and current status of grouping and regulation of engineered nanomaterials. Environ. Nanotechnol. Monit. Manag. 18. https://doi.org/10.1016/j.enmm.2022.100707.

Verkade, P., Collinson, L., 2019. Correlative Imaging: Focusing on the Future Correlative Imaging: Focusing on the Future. Wiley, pp. 1–231. https://doi.org/10.1002/9781119086420.

Wang, J.W., Grandio, E.G., Newkirk, G.M., Demirer, G.S., Butrus, S., Giraldo, J.P., Landry, M.P., 2019. Nanoparticle-mediated genetic engineering of plants. Mol. Plant 12 (8), 1037–1040. https://doi.org/10.1016/j.molp.2019.06.010.

Wang, T., Li, J., Hu, R., Shentu, X., Ye, Z., Yu, X., Sun, K., 2025. Nanoparticle-mediated nucleic acid delivery systems in plant biotechnology: recent advances and emerging challenges. Plants 14 (23), 3649. https://doi.org/10.3390/plants14233649.

Wang, X., Xie, H., Wang, P., Yin, H., 2023. Nanoparticles in plants: uptake, transport and physiological activity in leaf and root. Materials 16 (8), 3097.

Wong, M.H., Misra, R.P., Giraldo, J.P., Kwak, S.Y., Son, Y., Landry, M.P., Swan, J.W., Blankschtein, D., Strano, M. S., 2016. Lipid exchange envelope penetration (LEEP) of nanoparticles for plant engineering: a universal localization mechanism. Nano Lett. 16 (2), 1161–1172. https://doi.org/10.1021/acs.nanolett.5b04467.

Xu, J., Cui, Y., Wang, J., Liu, Z., Zhao, C., Liu, Z., Niu, N., Chen, L., Zhao, X., Fu, Y., 2025a. Advances of nano-photofertilizers in plant research. Nanoscale 17 (27), 16153–16192. https://doi.org/10.1039/d5nr00906e.

Xu, P., Zhao, C., Li, S., Li, S., Li, A., Zhao, J., Ma, A., Wang, Q., Guo, D., Zhou, J., Feng, S., 2025b. Gene editing tools promote the development of chloroplast gene engineering. Curr. Plant Biol. 44, 100540. https://doi.org/10.1016/j.cpb.2025.100540.

Yadav, A., Yadav, K., Ahmad, R., Abd-Elsalam, K.A., 2023. Emerging frontiers in nanotechnology for precision agriculture: advancements, hurdles and prospects. Agrochemicals 2 (2), 220–256. https://doi.org/10.3390/agrochemicals2020016.

Yiblet, Y., Sisay, M., 2024. Metal oxide nanoparticles as a promising method to reduce biotic stress in plant cell wall: a review. Heliyon 10 (19). https://doi.org/10.1016/j.heliyon.2024.e37939.

Yong, J., Wu, M., Carroll, B.J., Xu, Z.P., Zhang, R., 2024. Enhancing plant biotechnology by nanoparticle delivery of nucleic acids. Trend. Genet. 40 (4), 352–363.

Yu, Y., Yu, P.C., Chang, W.J., Yu, K., Lin, C.S., 2020. Plastid transformation: how does it work? Can it be applied to crops? What can it offer? Int. J. Mol. Sci. 21 (14), 4854.

Zhang, H., Goh, N.S., Wang, J., Pinals, R.L., Demirer, G.S., 2022a. Nanoparticle cellular internalization is not required for RNA delivery to mature plant leaves. Nat. Nanotechnol. 17 (2), 197–205.

Zhang, C., Zhang, P., Ren, H., Jia, P., Ji, J., Cao, L., Yang, P., Li, Y., Liu, J., Li, Z., You, M., Duan, X., Hu, J., Xu, F., 2022b. Synthetic biology-powered biosensors based on CRISPR/Cas mediated cascade signal amplification for precise RNA detection. Chem. Eng. J. 446, 136864. https://doi.org/10.1016/j.cej.2022.136864.

Zhang, Y., Tian, L., Lu, C., 2023. Chloroplast gene expression: recent advances and perspectives. Plant Commun. 4 (5). https://doi.org/10.1016/j.xplc.2023.100611.

Zhao, B., Luo, Z., Zhang, H., Zhang, H., 2022. Imaging tools for plant nanobiotechnology. Front. Genome Ed. 4. https://doi.org/10.3389/fgeed.2022.1029944.

Zhao, Y., Thenarianto, C., Sevencan, C., Rajappa, S., Shen, D., Puangpathumanond, S., Yao, X., Lew, T.T.S., 2024. Rational nanoparticle design for efficient biomolecule delivery in plant genetic engineering. Nanoscale 16 (46), 21264–21278. https://doi.org/10.1039/d4nr03760j.

Zoschke, R., Bock, R., 2018. Chloroplast translation: structural and functional organization, operational control, and regulation. Plant Cell 30 (4), 745–770. https://doi.org/10.1105/tpc.18.00016.

Zulfiqar, H.F., Afroze, B., Shakoor, S., Bhutta, M.S., Ahmed, M., Hassan, S., Batool, F., Rashid, B., 2024. Nanoparticles in agriculture: enhancing crop resilience and productivity against abiotic stresses. In: Abiotic Stress in Crop Plants-Ecophysiological Responses and Molecular Approaches. IntechOpen.

CHAPTER

10

Nanotechnology driven optimization of Photosystem I and Photosystem II functions

Shoaib Ahmad[1,a], *Adiba Khan Sehrish*[1,a] *and Hafiza Sadia*[2]

[1]State Key Laboratory of Water Pollution Control and Green Resource Recycling, School of the Environment, Nanjing University, Nanjing, Jiangsu, China; [2]State Key Laboratory of Pharmaceutical Biotechnology, NJU Advanced Institute for Life Sciences, School of Life Sciences, Nanjing University, Nanjing, China

1. Introduction

All life on Earth is directly or indirectly dependent upon solar energy. The green engine that drives life on Earth converts sunlight into chemical energy by photosynthesis. Photosynthesis literally means "synthesis with light," which is a term with rather broad implications (Kouřil et al., 2020). Photosynthesis is a biological process in which plants, algae, and many microbes capture sunlight and turn it into chemical energy. Indeed, as Blankenship pointed out, photosynthesis may be described as the process by which an organism collects and stores light. The stored energy is used to power energy-intensive cellular operations. Photosynthesis is the most major reaction on the planet, producing more than 10 billion tons of dry biomass/year (Blankenship, 2021).

Plant leaves are the primary places of photosynthesis, and chloroplasts are usually present in a 100 on average in any single mesophyll cell (Woodson, 2016). They bear significance in capturing light energy and driving its conversion into chemical energy; in essence, they are the major providers of carbon and energy in higher plants. A chloroplast is a huge and complicated organelle that structurally has a double membrane, known as the chloroplast envelope, with internal membranes called thylakoids (Kirchhoff, 2013). The thylakoid system exists in two parts: the grana, consisting of the stacked multiple thylakoid disks, which appear as cylindrical formations, and the stroma lamellae, which are unstacked areas of the membrane that connect adjacent grana (Dekker and Boekema, 2005).

[a] These authors have contributed equally to this work.

Nanotechnology in Photosynthesis
https://doi.org/10.1016/B978-0-443-44376-3.00010-0

Crucially, this process is responsible for the release and maintenance of atmospheric oxygen necessary for aerobic respiration (Eckardt et al., 2024). Beyond sustenance, photosynthesis greatly contributes to regulating the global climate as a major carbon sink, where immense amounts of CO_2 are sequestered with the mitigation of its greenhouse effect (Bachle, 2021). The continuing efficiency of this complex system is now a major focus of research, and ways to enhance the photosynthetic capacity for increasing global crop yields and further answering the challenges of food safety and climate modification are being studied (Croce et al., 2024).

The photochemical conversion of energy in photosynthesis involves the important role of photosystems. They are high molecular weight complexes consisting of many protein subunits, pigments, and cofactors embedded in the cytoplasmic membrane or special membrane systems of bacteria and chloroplasts of eukaryotic cells that have their origin from bacteria. These complexes capture light energy and are promoted to an excited state, initiating a stepwise electron transport process-charge separation (Li et al., 2024). The reductants produced are passed down the transport chain with proton translocation across the membrane to ultimately form NADPH and ATP. Energy carriers like these support a wide variety of metabolic reactions, which include fixation of carbon dioxide and nitrogen gas (Cardona et al., 2012).

There are two principal photosystems, photosystem I and Photosystem, in higher plants and cyanobacteria. Both are large, multisubunit complicated embedded within the thylakoid membrane system (Table 10.1). Within the thylakoid membranes, Photosystem I is primarily located in the stroma lamellae, while Photosystem II exhibits a high concentration within the stacked regions known as the grana (Fan et al., 2015). PSI itself is an essential membrane-bound complex that transfers light-energized electrons from plastocyanin or cytochrome 6 on the lumenal face to either ferredoxin or flavodoxin on the stromal side. The PSI complex contains 12 unique subunits in cyanobacteria and 13 in higher plants, whose genes are encoded as PsaA through PsaX (Golbeck, 2018). In cyanobacteria, it exists in a trimeric arrangement. Numerous cofactors embedded in PSI support light harvesting, charge separation, and photoprotection against photodamage (Hladík et al., 2020).

TABLE 10.1 Comparative overview of PSI and PSII.

Features	Photosystem II (PSII)	Photosystem I (PSI)
Reaction center pigment	P680	P700
Primary function	Oxidation of water and release of O_2	Reduction of $NADP^+$ to NADPH
Location	Thylakoid membrane (grana region)	Thylakoid membrane (stroma lamellae)
Major cofactors	Mn_4CaO_5, cluster, plastoquinone	Fe-S cluster, ferredoxin
Key proteins	D1, D2, CP43, CP47	PsaA, PsaB, PsaC
Sensitivity	High light sensitive, prone to photoinhibition	More stable, but limited red light absorption
Electron acceptor	Plastoquinone	NADP+
Electron donor	Water	Plastocyanin
Limiting factor	Damage to D1 protein, ROS accumulation	Back electron transfer losses

Photosystem II is an important multisubunit complex embedded in the thylakoid membranes of higher plants. It carries out one of the most energetically demanding reactions known in biological systems, including the oxidation of water to release molecular oxygen and produce reducing equivalents. This process is driven by solar energy and represents a vital link in the support of nearly all life forms on Earth. PSII accomplishes this by splitting water; this reaction provides the oxygen needed for respiration, helps protect the ozone layer against ultraviolet light, and supplies the reducing power that ultimately reduces carbon dioxide into the organic substances forming food, biomass, and biofuels (Xu et al., 2020).

The word "nano" is derived from the Greek word for "dwarf." Nanostructured materials are materials with dimensions usually between 1 and 1000 nm, and their shapes and configurations can be engineered into many different forms (Singh et al., 2017). Nanotechnology has brought revolution in various sciences, with food science and technology being among them. It created a broad avenue of opportunities at the molecular level. Over time, it became a revolution in the food industry, and revealed excellent uses in improving flavors, colors, textures, and stability in food products, bioavailability of nutraceuticals and other bioactive agents, antimicrobial agents, and improved packaging material with improved mechanical strength and shelf life. Besides these, nanosensors have also been developed to detect and monitor the potential health risks that are used to guarantee food safety and traceability. To this end, continuous studies have continuously increased the direct application of nanotechnology in the food systems, thereby providing researchers with a new step to analyze and control nanomaterial (He and Hwang, 2016).

Nanotechnology has opened new lines in plant biology, where the capacity for chloroplast molecular biology and genetics has been greatly improved. Some of these versatile tools include biomolecule and chemical delivery systems designed for specific purposes, nanosensors to monitor cellular processes, and nanotherapeutics with precise targeting and guarded release capabilities (Avellan et al., 2019). The magnitude of nanomaterials generally ranges from 1 to 100 nm and may take various shapes, aspect ratios, surface charges, and chemical compositions. Such nanoparticles can be synthesized from different biocompatible materials like silica, gold, carbon, and polymers. Functionalization or loading of nanoparticles with biomolecules by changing their surface properties, charge, and size allows the targeted delivery of cargo into plant cells and organelles, including chloroplasts (Hu et al., 2020). Moreover, the inclusion of biorecognition coatings increases the specificity and efficiency in plants even more (Santana et al., 2020).

For example, SWCNTs are functionalized in different ways to develop targeted delivery into plant cells. Wrapping with single-stranded DNA allows delivery of genetic material into chloroplasts, while modification with polyethylenimine imparts a positive surface charge to the nanomaterial for efficient binding and transport of plasmid DNA into the nuclei of mature terrestrial plants (Demirer et al., 2021). Chitosan-coated SWCNTs have also been used for delivering plasmid DNA into the mature plant chloroplast. Fluorescent nanoparticles, such as carbon nanotubes and quantum dots, have been developed for their applications in the advanced investigation of plant-signaling pathways, stress, and interaction with the environment (Giraldo et al., 2019).

Considering the immense prospect of nanotechnology in tuning biological systems, integration of nanomaterials with photosynthetic components represents the frontier in optimization of light energy conversion efficiency. Photosystem I and Photosystem II, representing the fundamental photochemical complexes of photosynthesis, have provided unique molecular interfaces for nanoengineering. However, despite extensive research on the structural and functional aspects of these complexes, the mechanistic basis of improvement in electron transfer dynamics, photostability, and energy conversion facilitated by nanomaterials remains inadequately explored. Thus, the present chapter attempts to outline the principles underlying nanotechnology-mediated modulation of PSI and PSII with the aim of assessing the role of various nanomaterials in enhancing their photochemical efficiency and discussing their applications pertaining to biohybrid solar energy systems. This chapter will critically synthesize recent advances with emphasis on highlighting innovative strategies toward bridging plant photobiology with nanoscience for sustainable and efficient use of light energy.

2. Structural and functional overview of photosystems

2.1 Architecture of Photosystem II

Photosystem II is an important multi-subunit pigment-protein complex anchored within thylakoid membranes of plants, algae, and cyanobacteria. As a key player in photosynthesis, it catalyzes specifically the light-dependent process, dissociating water into oxygen, protons, and electrons (Xue et al., 2015). This photochemical step is of great importance since it supplies electrons to drive further chemical energy production and releases the oxygen of the atmosphere that is so crucial for life in the aerobic regime. PSII exists as a dimer, with each monomer comprising about 27–28 distinct subunits organized into two main parts: the core complex and the light-harvesting antenna. In the core complex, key proteins house the reaction center P680, necessary for electron transport, along with D1 and D2 subunits. CP47 and CP43 also reside in the core and bind chlorophyll a molecules to form the inner light-harvesting antenna. Besides these, several smaller subunits are present whose functions are still a subject of active investigation. The highly intricate pigment arrangement of the PSII core in cyanobacteria, which includes 35 chlorophyll molecules, 2 pheophytins, and 12 beta carotene molecules, provides the basis for highly effective light capture and subsequent electron transfer (Kim et al., 2020).

In higher plants, the water-splitting apparatus, also called the Oxygen Evolving Complex (OEC), is located on the lumen side of the thylakoid membrane. The complex contains, in addition to the intrinsic subunits, three extrinsic subunits-OEC33, OEC23, and OEC17-encoded by genes psbO, psbP, and psbQ, respectively. These proteins play crucial roles in maintaining the stability of the manganese cluster, which is absolutely required for efficient oxygen evolution. Although the importance of these peripheral subunits is well established, their exact spatial position and specific regulatory role remain, to a large extent, still unknown. The peripheral antenna system has a dual role: it efficiently provides the captured light energy to the reaction center for photosynthesis and serves at the same time as a machinery for the safe dissipation of excessive energy. Under stressful conditions, such as high light, the excess light energy is harmlessly dissipated as heat, a process known as nonphotochemical quenching, NPQ. The antenna system is composed of proteins encoded by the Lhcb multigene family, which coordinate various pigments, chlorophyll a, chlorophyll b, and xanthophylls in a specific ratio. The main light-harvesting complex, LHCII, is heterotrimeric and encoded by genes Lhcb1–3; the monomeric complexes CP29 (Lhcb4), CP26 (Lhcb5), and CP24 (Lhcb6) also make a significant contribution to both efficient energy transfer and photoprotective regulation (Kouřil et al., 2020).

The complex supramolecular organization of Photosystem II-Light-Harvesting Complex II supercomplexes has been investigated by single-particle analysis combined with Transmission Electron Microscopy. This technique is applied to heterogeneous samples, which are either directly isolated by gently solubilizing the thylakoid membranes or enriched through rapid purification methods that select high-molecular-weight complexes. The precise positioning of the major core subunits was confirmed by the examination of plant membranes that had been solubilized under conditions that avoid detachment of the antenna complexes, and the results agreed with earlier cross-linking experiments and single-particle EM data (Sheng et al., 2019). In *Arabidopsis thaliana*, the supercomplex of the largest size contains a dimeric core (C2), two LHCII trimers (S-trimers), which are tightly bound to CP43 and CP26, and two other trimers (M-trimers), which are moderately bound to CP29 and CP24. This complex is denoted as the C2S2M2 supercomplex. Nosek et al. (2017), by cryo-EM, reconstructed a smaller supercomplex at 17 Å resolution with one trimer per reaction center and lacking the CP24 subunit, denoted as C2S2.

2.2 Architecture of Photosystem I

Photosystem I is a multicomponent complex containing several proteins and pigments that drives the electron transport phase in oxygenic photosynthesis. It receives electrons whose origin is from the water-splitting activity of PSII, delivered by either plastocyanin or cytochrome 6, and it provides electrons to ferredoxin. In photosynthetic eukaryotes, PSI contains a central core complex flanked by peripheral light-harvesting complexes known as LHCI. These LHCI proteins absorb light and channel energy into the core containing the P700 reaction center and core antenna proteins. Although the core subunit composition of PSI is extremely conserved across higher plants, the number, protein composition, and pigment content of LHCI vary greatly among different species, reflecting the evolutionary and regulatory strategies involved in allowing organisms to adapt to a heterogeneous light environment (Shevela et al., 2023).

Cryptophytes are unicellular, biflagellated algae that represent an important group of primary producers in various aquatic environments. In contrast with the red algae, which possess extrinsic phycobilisome antennae and transmembrane light-harvesting complexes (LHCs) while lacking chlorophyll c, and the diatoms, which contain Chl a/c-based LHCs but no phycobiliproteins, the cryptophytes have both phycobiliproteins and LHCs, which include chlorophyll a/c_2 and alloxanthins. These alloxanthin Chl a/c-binding proteins (ACPIs) extend the range of absorbable light and represent their evolutionary role in red-lineage plastids. Very recently, structural studies on PSI-LHCI supercomplexes have been reported in red algae and diatoms as PSI-LHCR and PSI-FCPI (fucoxanthin-chlorophyll a/c-binding proteins), respectively (Nagao et al., 2020).

2.3 Interdependence of PSI and PSII in the Z-scheme

The Z-scheme describes sequential, light-powered charge separation taking place within PSII and PSI, along with the exact electron transport pathway that links these two complexes. Light energy input to PSII results in the generation of an electron "hole." In turn, this highly oxidizing "hole" enables the splitting (oxidation) of two water molecules, liberating molecular oxygen and four protons. The excited electron is then transferred along the electron transport chain to PSI (Fig. 10.1), where it gains sufficient potential energy to reduce $NADP^+$ to NADPH (Brettel and Leibl, 2001).

Depending on the mediation method employed, there are two major structural designs for Z-schemes. First, there is an approach

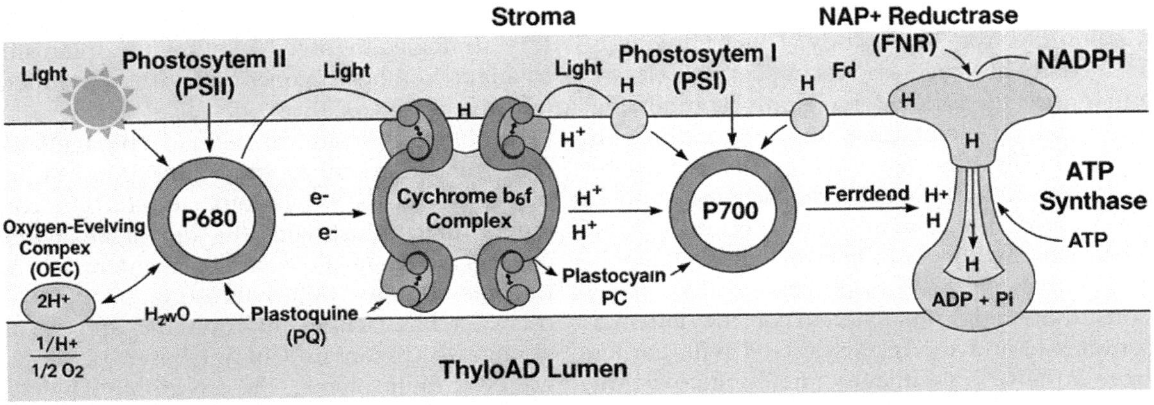

FIGURE 10.1 Schematic overview of Photosystem I and II within the thylakoid membrane.

based on the use of a redox couple consisting of ion pairs, which acts as a mediator between the reaction centers and keeps them physically separated. This arrangement may be subject to undesirable reverse reactions in which electrons and holes are lost, and is possibly problematic in terms of functioning over an extensive pH range. Second, there is an "all-solid-state" Z-scheme in which a conductive material is used to convey electrons directly between the centers, physically connecting them (Zhou et al., 2014).

2.4 Factors affecting photosystem efficiency

The efficiency of Photosystems I and Photosystems II is controlled by a subtle balance of environmental and internal factors (from the plant itself) in accordance with Blackman's Law of Limiting Factors. According to this law, the rate of any process, such as photosynthesis, is controlled by that factor which is closest to its minimum. PSII is usually the bottleneck to overall photosynthetic efficiency since it catalyzes the critical water-splitting reaction and is highly prone to damage (Zivcak et al., 2014).

2.4.1 *Light and photoprotection*

Light is the essential energy input, but its quantity and quality must be precisely regulated to maintain photosystem integrity:

- **Low Light Limitation:** The photon absorption rate is the limiting factor under low light. Photosystems operate below their capacity. Efficiency is generally high, whereas the overall ETR is low (Xu et al., 2023).
- **Excess Light and Damage:** When light intensity is too high, photosystems become overexcited, leading to a phenomenon called photoinhibition. Due to excess energy, it enhances the creation of destructive Reactive Oxygen Species (ROS) that mainly target and damage the D1 protein in the PSII reaction center (Yamane et al., 2015). This damage reduces the maximum quantum efficiency of PSII (Fv/Fm), one of the key indicators of the health status of photosystems. The photosynthetic apparatus continuously tries to repair PSII by synthesizing new D1 protein. When the rate of photodamage surpasses the repair rate, its efficiency falls (Falcioni et al., 2024).

Plants counteract photoinhibition by employing photoprotective mechanisms, of which NonPhotochemical Quenching-NPQ-is the most important. NPQ is a fast, variable process in which excess light energy that has been absorbed is dissipated in the form of harmless heat. The major integral of NPQ, termed energy-dependent quenching (qE), is provoked in the case of high-light-induced accumulation of a proton gradient (ΔpH) across the membrane of thylakoids. This gradient induces conformational changes in the LHCII complex of PSII and triggers the xanthophyll cycle to finally drain excessive energy away from the reaction centers before it can cause damage (Slattery and Ort, 2021). An efficient NPQ mechanism is crucial for maintaining a high effective quantum yield of PSII under fluctuating light conditions (Sharma et al., 2014).

2.4.2 *Temperature stress*

Temperature affects all the stages of photosynthesis by modifying enzyme kinetics and the stability of proteins:

- **Optimum Range:** Enzymes for photosynthesis, particularly RuBisCO in the Calvin cycle, show optimum range. Therefore, between 20 and 30°C, temperature increase increases enzymatic reaction rates and ETR (Falcioni et al., 2024)).
- **High Temperature Damage:** PSII represents the most thermolabile part of the whole photosynthetic apparatus. High temperatures cause the structural destabilization and inactivation of the OEC responsible for water splitting and accelerate the irreversible degradation of the D1 protein. The impact of high heat and high light is synergistically damaging to PSII efficiency (Lattanzio, 2020).

2.4.3 *Water and CO_2 limitations*

The availability of critical raw materials strongly affects the efficiency of photosystems since it determines the demand for ATP and NADPH. Drought forces stomata to close, preventing loss of water, but also reducing further entry of CO_2 into the leaf, which becomes a major limitation for the Calvin cycle. The limitation of the carbon dioxide CO_2 fixation rate, normally by environmental stress such as drought, high temperature, or stomatal closure, creates a major bottleneck in photosynthesis. Because the CO_2 fixation cycle, or the Calvin-Benson-Bassham cycle, is slowed, the consumption of the two important energy carriers, ATP and NADPH, decreases considerably (Pagnani et al., 2024). On the other hand, under low ambient CO_2, RuBisCO can carry out limited carboxylation. Thus, the consumption rate of ATP and NADPH is lower; the electron transport rate becomes low and consequently constrains the photosystems' efficiency (Zhang et al., 2020).

2.4.4 *Internal and nutritional factors*

The structural organization and biochemical composition of a plant determine the upper limits of photosystem performance. The size and efficiency of the light-harvesting complexes (LHCs) depend on the abundance of chlorophyll and accessory pigments. Lower levels of pigments reduce the capacity of the plant to absorb light and effectively couple energy to the reaction centers. Furthermore, the availability of some important nutrients determines the optimal functioning of photosystems: iron (Fe) is needed for cytochrome and ferredoxin synthesis, both highly important electron carriers in the photosynthetic transport chain of electrons, while nitrogen (N) is an important constituent of photosynthetic proteins, including chlorophyll and the D1 reaction center protein.

Deficiency in these nutrients impairs the proper assembly and stability of photosystems, reducing the total ETR (Sharma et al., 2023).

3. Nanotechnology: Concepts and applications in biological systems

Nanotechnology is an important interdisciplinary research field dealing with the design, preparation, and application of materials and devices in which at least one dimension falls in the nanoscale, that is, between 1 and 100 nm. This range is particularly important because materials assume unique physical, chemical, and biological properties at this scale. The property changes are a function of many factors, such as increased surface-area-to-volume ratio, quantum effects, and special surface reactivity, which may be quite different from their bulk materials. A combination of these NMs with living matter constitutes the basics of nanobiotechnology and opened revolutionary avenues for the manipulation and understanding of biological systems (Anjum et al., 2021).

In the nanoscale, materials exhibit enhanced reactivity and selectivity owing to their extremely high surface area. This may relate to changes in size or shape on a very minute scale (e.g., spherical, rodlike, porous) for NPs, coupled with dramatic changes in electronic, magnetic, and optical properties tunable for biological interactions like targeted drug delivery or biosensing, as reported in the literature. Semiconductor nanocrystals, or so-called quantum dots (QDs), demonstrate this quite perfectly. Their emission color is size-dependent, thereby making them highly tunable fluorescent probes for cellular imaging. Large surface areas provide ample sites for functionalization by attachment of specific biological molecules on the NP surface, such as antibodies, peptides, and DNA. Functionalization lends the NP high affinity and specificity for biological targets, a cardinal principle of targeted therapy and diagnostics (Zaman and Manghwar, 2024).

Several classes of nanomaterials have particular relevance for biological applications (Table 10.2):

I. **Metallic Nanoparticles:** AuNPs and AgNPs are widely used because of their unique optical properties, such as plasmon resonance, and antibacterial properties of AgNPs. AuNPs are used as key factors in

TABLE 10.2 Classification of nanomaterials used for photosystem optimization.

Nanomaterial type	Key function	Mode of action	Example
Metal nanoparticles	Light capture enhancement	Localized surface plasmon resonance (LSPR) amplifies light absorption	Au, Ag, Pt
Metal oxides	Antioxidant, redox mediator	Scavenging ROS, stabilizing PSII	TiO_2, CeO_2. ZnO, Fe_2O_3
Quantum dots (QDs)	Photon harvesting, electron donation	Electron injection into reaction centers	CdS, CdSe, ZnS
Carbon-based nanomaterials	Electron transport, structural support	Conductive pathways for PSI electron flow	CNTs, graphene oxide
Polymeric nanoparticles	Delivery and stabilization	Encapsulation of PSI/PSII complexes	PLAG, PEG chitosan
Silica nanostructures	Structural immobilization	Encapsulation for stability under stress	SiO_2 nanoparticles

the diagnosis and photothermal therapy of tumors (Hashmi et al., 2024).

II. **Carbon-Based Nanomaterials:** Carbon Nanotubes (CNTs) and Graphene Oxide (GO) have excellent mechanical strength and electron transport capability. They are widely investigated as vectors of drugs/genes and components in advanced biosensors (Anjum et al., 2021).

III. **Polymeric Nanoparticles:** These are mostly biodegradable and biocompatible, making them ideal for drug encapsulation in controlled release systems. The size and degradation rate can be precisely controlled, offering fine-tuned pharmacokinetics (Sim and Wong, 2021).

IV. **Liposomes and Micelles:** These are self-assembling organic nanostructures of lipids or amphiphilic block copolymers that will automatically encapsulate hydrophobic drugs, protecting them until they can reach their biological site (Mitchell et al., 2021).

3.1 Applications in plant nanobionics and photosynthesis

The incorporation of nanotechnology in plant biology, aka Plant Nanobionics, is a rapidly growing field, particularly in developing ways to improve photosynthesis, very relevant to the main chapter title. Nanomaterials are being designed to integrate directly with the photosynthetic apparatus (Photosystem I and II, PSI and PSII) to increase the efficiency of light harvesting and electron transfer (Mony et al., 2022).

- **Nano-Fertilizers and Biostimulants:** Conventional fertilizers are inefficiently absorbed. Nano-fertilizers, such as ZnO NPs or Si nanocomplexes, have a higher surface area for absorption and a slow-release mechanism that improves nutrient uptake and increases chlorophyll content, which will directly translate into an efficient LHC and, accordingly, into subsequent electron transport. As shown, nano-Si foliar applications may improve PSII efficiency and electron transport rates in different crops (Liu et al., 2024a).
- **Light Harvesting Augmentation:** Some nanoparticles, such as carbon dots (CDs) or specific metal oxides like Titanium Dioxide (TiO_2), can serve as light-harvesting antennae. They absorb a greater number of photons than the natural pigments from a broader spectrum and transfer energy to the chlorophyll molecules in the LHC of both PSI and PSII, thus enhancing photochemistry in a substantial way and improving the quantum efficiency of photosynthesis (Mukarram et al., 2023).
- **Stress Mitigation:** Nanomaterials act like a shield for photosynthetic machinery against various environmental stresses. For instance, some NPs enhance the activity of antioxidant enzymes in plants. This, in turn, reduces oxidative stress due to overproduction of ROS, generally observed under high light, drought, or heavy metal exposure. Thus, the thylakoid membrane and PSII complex maintain their structural integrity (Gautam et al., 2025). More specifically, iron-based nanoparticles have been found to enhance chlorophyll synthesis and overall photosynthetic output in staple crops such as rice and wheat (Tripathi et al., 2023).

4. Role of nanotechnology in modulating Photosystem II functions

Application of nanotechnology represents a frontier approach to significantly enhancing the stability, efficiency, and robustness of PSII, the light-driven water-splitting engine in photosynthesis. In new ways, taking advantage of the unique physical, chemical, and quantum properties of nanomaterials, it has been possible to develop novel strategies aiming at overcoming the natural restrictions imposed by light, heat, and oxidative stress on PSII function in native plants and in artificial biohybrid systems.

4.1 Enhancement of light absorption and energy conversion

Nanomaterials serve as highly efficient light-harvesting elements, which surmount the spectral mismatches and tiny absorption cross-sections of endogenous chlorophyll molecules to ensure maximum utilization of the solar photons by PSII. Among plasmonic NPs, Gold and Silver introduce advanced technologies of light harvesting, based on the LSPR effect: when the incident light is in phase with the resonant frequency of NP free electrons, a strong, highly localized electromagnetic field is generated close to the NP surface (Li et al., 2024).

- **LSPR-Induced Enhancement:** Upon incorporating such plasmonic NPs into the photosynthetic machinery—for example, immobilization in proximity to PSII or LHCII complexes—the photon capture rate of the neighboring chlorophyll pigments increases with the significantly enhanced localized field. This increases the cross-sectional area of effective light absorption for the whole PSII complex, especially increasing absorption in the green-yellow region, where chlorophyll absorption is normally poor, thereby making light available to the P680 reaction center (Joudeh and Linke, 2022).
- **Electron-Hole Separation:** In addition to photoinitiated charge transport, the plasmon resonance can also promote faster electron-hole separation in biohybrid systems where PSII is coupled to metallic nanocrystals, increasing the overall efficiency of the charge transport process and boosting the measured photocurrent response in photovoltaic applications (Fig. 10.2). Bimetallic Au–Ag NPs offer advantages by combining the chemical stability of Au with the high scattering efficiency of Ag, enabling tunability of the LSPR effect for optimized spectral coverage (Mony et al., 2022).

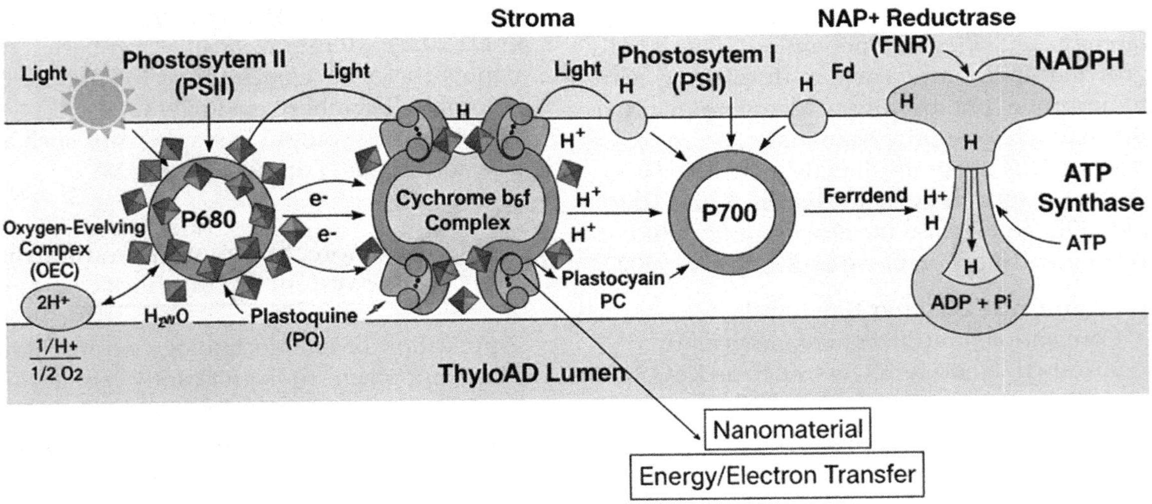

FIGURE 10.2 Mechanism of nanomaterial interaction with PSII.

4.2 Stabilization of the oxygen evolving complex

The Mn_4CaO_5 cluster of OEC thus represents the most sensitive part of PSII. Nanotechnology provides a solution through the creation of both protective environments and biomimetic functional analogs, which serve to enhance the structural stability of the OEC and improve its catalytic turnover rate. It catalyzes the four-electron oxidation of water through a cycle of five intermediate states S0–S4 (Chen et al., 2022). Its function is highly dependent on the stability of the Mn and Ca ions.

- **Nanostructured Immobilization:** Immobilization of intact PSII complexes has been done in an oriented fashion on nanoporous scaffolds and self-assembled monolayers on electrodes. This precise, ordered attachment to a solid-state surface gives a rigid, protective microenvironment that strongly enhances thermal and chemical stability for the embedded PSII and protects the fragile OEC from denaturation or decomposition. This specific orientation provides optimal electron transfer to an external circuit in bio-photoelectrochemical cells (Bae et al., 2024).
- **Biomimetic MnO_x Catalysts:** Nanostructured manganese oxide and calcium-manganese oxide clusters are researched. These inorganic nanoparticles represent functional and structural models of the oxygen-evolving complex (Wang et al., 2022). Using nanotechnology to encapsulate such catalysts within crystalline, stable, nanoporous arrays bypasses the issue of the destructive aggregation or disproportionation of the active site, conferring robustness with a high-water oxidation turnover frequency-key strategy garnered from the naturally sequestered OEC in the protein matrix (Liu et al., 2024b)

Beyond mere stabilization, nanostructured supports are used to create fully synthetic OEC analogs. These synthetic nanostructured scaffolds, typically based on graphene oxide, carbon nanotubes, or engineered Ir or Co oxides, enable the precise tuning of geometry and electronic properties in active sites. The high surface area and ordered nanostructure improve the accessibility of water molecules and optimize the kinetics of charge separation, thus enabling the development of stable, high-performance artificial catalysts for water splitting (Cheng et al., 2023)

4.3 Mitigation of photoinhibition and oxidative stress

Nanomaterials are utilized by taking advantage of their exclusive redox chemistry to serve as superior and reusable ROS scavengers, directly preventing the oxidative damage that limits PSII lifetime and efficiency. Photoinhibition is intrinsically related to uncontrolled production of ROS-primarily singlet oxygen, superoxide, and hydroxyl radicals (Yotsova et al., 2025).

- **Cerium Oxide Nanoparticles:** Due to their mixed valence state, nanoceria are considered excellent self-regenerating antioxidants. These may catalyze the decomposition of ROS. Hence, they effectively mimic the dual actions of antioxidant enzymes such as Superoxide Dismutase, and catalase. CeO_2 NPs combined with photosynthetic microalgae absorb harmful UV radiation and scavenge H_2O_2 to guard the thylakoid membranes successfully and maintain the PSII activity for a longer period (Alabdallah et al., 2024).
- **Titanium Dioxide Nanoparticles:** In specific compositions and concentrations, TiO_2 NPs could act protectively by triggering the

plant's inherent antioxidant defense system, for instance, it increases the activity of SOD and peroxidases. It is this induced defense that decreases the internal load of oxidative stress on the chloroplasts, thereby protecting PSII from indirect damage (Singh et al., 2024).

4.4 Case studies and experimental advances

Practical demonstrations confirm that nanotechnology has the potential to directly modulate and improve PSII functions:

- **Biohybrid Water Oxidation Systems:** The development of oriented PSII complexes on nanostructured ITO electrodes is highly important. Covalent linkage in such a way ensures that the electron acceptor side of PSII is oriented to an optimal position such that there can be an efficient, direct extraction of electrons for use at the bio-photoanode. These photoanodes were considerably more stable and generated high photocurrents, proving the most robust artificial water-splitting system based on PSII to date (Jing et al., 2024).
- **Targeted Nutrient Delivery:** Iron-based nanoparticles prove to be more bioavailable for plants. Since iron acts as an important cofactor for cytochromes and ferredoxin, which are key components in the ETC between PSII and PSI, nano-iron delivery has been successfully used to enhance chlorophyll synthesis and improve the general electron transport rate, especially in cereals under stress (Zhou et al., 2025).
- **Carbon Dot Enhanced Photosynthesis:** Application of specific types of Carbon Dots on plants and algae demonstrated impressive results, including a 20% increase in light-energy conversion efficiency with further improvements in biomass and fruit quality. This was directly due to the CDs acting as the electron shuttle, proving that this technology of using nanomaterials for tuning photosynthetic electron flow is viable (Chen et al., 2025).

5. Nanotechnology in the optimization of Photosystem I functions

PSI is an enormous pigment-protein complex driving the final, critical step of linear electron flow by reducing $NADP^+$ to NADPH using light energy. In the frame of bioelectronics and sustainable energy, the optimization of PSI means maximization of stability, light capture efficiency, and the ability of electron transfer to nonnative, external acceptors, such as electrodes or chemical catalysts. Nanotechnology provides the molecular-scale tools necessary to achieve this enhancement, allowing PSI to become a robust and powerful component for artificial photosynthetic systems (Cherepanov et al., 2023).

5.1 Integration of nanomaterials with PSI complex

The central strategy for optimizing PSI includes the development of hybrid systems in which the biological machinery is directly coupled to highly conductive nanomaterials. This is done with the aim of creating fast, high-yield pathways for electron flow. Hybridizing PSI with various nanoparticles boosts both light harvesting and electron extraction efficiencies (Du et al., 2021). Gold (Au), Silver (Ag) nanoparticles: Because of LSPR, the use of nanoparticles enhances the light absorption of PSI complexes. The local electromagnetic field created due to the LSPR effect enlarges the effective cross-section for capturing light and correspondingly increases the exciton generation rate at the P700 reaction center (Kathpalia and Verma, 2021). Semiconductor Quantum Dots: QDs present themselves as highly efficient artificial antennae. Engineered to absorb a wide range of the solar spectrum, they transfer energy to the light-harvesting complex of photosystem I, LHCI, through FRET. As such, this increases the flow of excitons to the P700 primary donor. This integration has been shown to increase the quantum yield in this system by up to 30% in model systems (Voloshin et al., 2022).

Highly conductive nanostructured scaffolds are required to allow the maximum transfer of electrons from the acceptor side Fe–S clusters FX FAF B of PSI to an external circuit. Carbon Nanotubes (CNTs) and Graphene Oxide (GO): The large surface area and excellent conductivity of CNTs and GO provide an ideal interface for facilitating efficient interaction (Singh et al., 2024). Upon immobilization of PSI complexes onto the electrodes made of CNTs or reduced GO (rGO), their high surface area allows for a maximum loading density, while the respective conductivity facilitates DET of electrons from the PSI complex to the electrode (Paul, 2022). This will avoid the use of soluble mediators, reduce electron loss and recombination, and maximize the current density output in biophotoelectrochemical cells (Kanjer et al., 2023). Metal Oxide Scaffolds: In most cases, nanoporous metal oxides like TiO_2 or ZnO serve as highly stable, high-surface-area platforms that support PSI immobilization. These biomaterials provide mechanical stability and sometimes facilitate electron transport, especially when the PSI complex is chemically oriented to face the electrode surface correctly (Gobbato et al., 2023).

5.2 Enhancement of photocurrent generation

The aim of many PSI-nanotechnology hybrid systems is the development of robust, efficient biohybrid photovoltaic devices able to convert solar energy into electrical current. Nanotechnology plays a crucial role in transforming the intrinsically fluid and fragile PSI complex into a fixed, long-lasting part of a solar cell. Researchers use techniques like self-assembled monolayers on nano-electrodes to enforce the correct orientation of the PSI complex. The complex needs to be oriented in such a way that its electron donor side faces an electron-supplying electrode-or PSII-while its electron acceptor side needs to face an electron-collecting electrode. Such exact nano-architecture is achieved through specific chemical linkers, for example, thiol-or pyridine-functionalized SAMs, which attach to known amino acid residues on the PSI surface to ensure maximum electron collection and higher External Quantum Efficiency. Record Current Densities: A combination of GO or CNT scaffolds with oriented PSI has achieved record-setting power densities for biosolar cells, reaching competitive values with conventional inorganic photovoltaic devices under simulated sunlight (Jing et al., 2024).

5.3 Electron transport and photoconversion efficiency improvements

Nanomaterials are designed to optimize every step along the path that electrons travel, from light absorption to final collection. The CDs as electron shuttles can act as effective electron shuttles or mediators between the biological electron chain (e.g., ferredoxin) and the external electrode in an electrochemical setting. This shuttle mechanism is much faster and more stable compared to traditional soluble chemical mediators, resulting in lowered resistance and increasing the general Photoconversion Efficiency of the device. Nanostructured environments-for example, porous films-quickly sweep away the generated electrons from the P700 hole, thereby inhibiting charge recombination of the primary loss pathway in all solar cells. This quick charge separation is essential for sustaining a high photocurrent (Zhou et al., 2025).

5.4 PSI stabilization under stress conditions

PSI is stable, yet this protein, like PSII, will denature under high heat, extreme pH changes, and continuous high-intensity light. Nanotechnology offers multiple layers of protection. Probably the most direct method of stabilization is the isolation and protection of the PSI protein within a mechanical shield. It is possible to embed PSI complexes in robust, transparent, biocompatible matrices such as polymer hydrogels, silica gels, or liposomal nano capsules (Liu et al., 2024a). This would ensure the three-dimensional structure of this protein and protect it from thermal and pH fluctuations, greatly extending the functional lifetime of immobilized PSI from only several hours in solution to weeks, which is crucial for practical applications. The nano-size of the capsule ensures a high surface-to-volume ratio that will allow nutrient/electron exchange while protecting the nano-capsule (Vijayaraghavareddy et al., 2022).

5.5 Case study 1: Plasmon-enhanced photocurrent in biohybrid solar cells

This case study showcases nanotechnology for modulating Photosystem I function for bioelectronic applications: in biohybrid solar cells, for example. The principal types of nanomaterials used are noble metal nanoparticles, such as silver (Ag) or gold (Au) nanocrystals, which show Localized Surface Plasmon Resonance upon light incidence. By immobilizing PSI near these metallic NPs on a conductive electrode, the strong electromagnetic field of the plasmons enhances the local light intensity around the PSI complex and thus increases light absorption. Critically, this plasmonic effect accelerates the charge separation in PSI and promotes faster electron transfer of light-excited electrons from the complex to the electrode at an increased rate. This leads to a significant increase in photocurrent generation, with reports of up to 6.5-fold enhancements (Aghaei et al., 2018).

The second case study describes the indirect modulation of photosystem I activity through the use of N-CDs, enhancing the overall efficiency of the photosynthetic electron transport chain in living plants. Highly fluorescent and electronically active nanoparticles applied via foliar spray permeate the plant and act artificially as an electron shuttle within the thylakoid membrane. They integrate specifically into the plastoquinone pool, which links Photosystem II with Photosystem I. Excited N-CDs are able to donate electrons to PQ molecules, thereby accelerating P700 rereduction, the reaction center of PSI. Supplementing electron flow, the N-CDs efficiently alleviate the electron transport bottleneck between PSII and PSI, thus increasing the overall operational efficiency of PSI. This approach to modulating photosynthesis, demonstrated to enhance photosynthetic efficiency by more than 20% in crops such as apples, represents a noninvasive nanotechnology approach that can be used to optimize the natural Z-scheme for improved agricultural yield and plant vitality (Zhao et al., 2024).

6. Nano-enabled artificial photosynthesis systems

AP systems, conceptually similar to the key steps of natural photosynthesis, are supposed to capture solar energy and store it in chemical form as a fuel, such as hydrogen (H_2) or hydrocarbons. Nanotechnology has grown to be an indispensable core of modern research into AP, with its engineered materials enabling superior light harvesting, highly efficient charge separation, and very active catalytic sites that were difficult or impossible with either natural systems or synthetic bulk materials. The most recent work, including reports from 2023 to 2024, has continued to focus on two general architecture strategies: nanomaterial biohybrids that incorporate biological components and advanced fully artificial PEC photoelectrochemical cells reliant on synthetic nanomaterials only (Jeevanandham et al., 2025).

In general, NBHs, often termed biophotoelectrochemical cells (BPECs), use nanomaterials to stabilize fragile photosynthetic proteins, such as Photosystem I and II, and electronically wire them to electrodes-a concept currently drawing increased attention in the field of sustainable energy applications (Ruan et al., 2024). Recent work underlines the importance of highly conductive carbon nanostructures, such as Carbon Nanotubes and Graphene Oxide, serving as 3D scaffolds. These nanostructures feature a massively increased surface area for high-density protein loading, while at the same time allowing for DET from the protein complex to the electrode, minimizing energy-wasting recombination pathways (Shlosberg, 2024).

Moreover, incorporating functional nanostructures, including metal oxides and conductive polymers, enables optimization of thylakoid immobilization and greatly enhances both photocurrent and power output in biofuel cells (Sarode and Slaughter, 2025). Apart from wiring, functionalized gold nanoparticles have been used to enhance PSII photocurrent by showing a dual role: harvesting excitation energy and transferring it to PSII, while improving electron transfer to the anode, illustrating the sophistication of control given by nanotechnology at the bio-interface. This synergistic interaction, where nanomaterials enhance light absorption, charge separation, and conversion, constitutes the key to recent progress in NBH-based AP systems for CO_2 fixation and H_2 production (Kim et al., 2025).

Their aim in the design of Fully Artificial Photoelectrochemical Cells remains the construction of highly optimized synthetic catalysts and semiconductor materials at the nanoscale, which can carry out both the light reaction and the "dark" fuel-forming reaction. Photocatalytic water splitting for H_2 generation continues to see important advances through the development of complex heterogeneous photocatalysts (Sohail et al., 2024). Of key importance in materials such as $SrTiO_3$ and a range of transition metal oxides is the use of nanoscale dimensions to control properties and maximize surface area-to-volume ratio, although careful engineering is required to avoid enhanced charge recombination at this length scale. Recent efforts are emphasizing the use of earth-abundant transition metal nanomaterial-based electrocatalysts such as nickel, iron, and cobalt oxides, phosphides, and nitrides for the Oxygen Evolution Reaction and the Hydrogen Evolution Reaction, where structural engineering has been driving enhanced catalytic activity and stability (Maduraiveeran, 2024).

7. Experimental and analytical approaches

Nanophotosynthesis is an interdisciplinary field whereby rigorous suites of experimental and analytical methods in material synthesis and functional understanding are evidenced. Wet chemistry is the dominating synthesis technique for the growth of metal oxides, semiconductor nanoparticles (TiO_2, CeO_2 Quantum Dots), while enabling very precise control over morphology and size (Arora et al., 2024). Characterization techniques to confirm material properties include TEM to confirm nanoscale dimensions and elucidate a very important physical interface to which the photosystem proteins are attached at the surface of the nanoparticle and XRD to confirm crystal structure and phase purity (Reddy et al., 2025).

ChlF spectroscopy often forms the basis for monitoring the function of nano-integrated photosystems. These spectroscopic measurements represent a fast and noninvasive measurement of PSII quantum yield and its stress response, particularly when interacting with the stabilizing nano-antioxidants (Zhao et al., 2024). Imaging and localization are required to understand the functional consequences of nanomaterial integration into the living system: while TEM is still the gold standard for visualizing nanoparticles inside chloroplasts or near the thylakoids, Confocal Microscopy and Fluorescence Lifetime Imaging (FLIM) confirm in real-time the mechanism of energy transfer, a FRET, between fluorescent nanomaterials, like Carbon Dots or QDs, and the photosystems (Shoyhet et al., 2021). Finally, Computational Modeling, including MD simulations, yields atomic-level detail regarding protein-nanoparticle binding and stability, while DFT is able to predict the electronic alignment necessary for efficient charge transfer across the nano-bio interface. It continues to play a critical complementary role, both experimental and computational, toward the design of optimized nano-biohybrid devices (Jávorfi, 2022).

Quantifying PSI activity demands special techniques focused on electron flow dynamics. Flash-induced absorption spectroscopy of the reduction and reoxidation of P700, the reaction center of PSI, has been used to monitor its activities. Monitoring the kinetics of reduction of P700 (oxidized PSI) can serve as a direct measure of the impact a nanomaterial-a Carbon Dot in this case, serving as an artificial electron shuttle-will have on the flow of electrons from the electron transport chain into PSI. For ex vivo biohybrid systems, photocurrent density is the main measure. Improved photocurrent, usually through Plasmon-Enhanced Photocurrent upon coupling PSI with metal nanostructures, is the definitive measure of optimization in solar energy conversion (Aghaei et al., 2018).

One main experimental variable is NP concentration. As stated by many papers, NP interaction usually demonstrates a dual nature: while low concentrations may play a role of biostimulant, enhancing photosynthesis and antioxidant mechanisms, excessive concentrations induce oxidative stress, impairing growth. Besides, materials are engineered to exploit particular photo-physical properties; as an example, plasmonic nanomaterials (like Au nanorods) have been designed with precise spacer layers in order to optimize the distance between the NP and the photosynthetic pigments and get the maximum localized electric field enhancement, boosting light capture both for sensing and energy applications (Optimizing nanostructured platform., 2025).

Successful optimization in vivo, especially within agricultural environments, requires design for effective transportation across complex biological barriers. Current research involves techniques such as the foliar spraying of colloidal NP solutions, dependent upon the capability of the particles to penetrate leaf surfaces through stomata or pores. In addition, nanomaterials intended for use in plant optimization include iron-based and titanium-containing NPs, which continue to be under active investigation due to their potential to greatly improve chlorophyll synthesis and plant growth, a phenomenon that indirectly increases photosystem capacity (Li et al., 2024).

8. Challenges, limitations, and safety considerations

The route to commercialization of nano-enabled photosynthetic technologies has been heavily hindered by serious technical obstacles and critical safety concerns. In the case of BPECs, the primary technical bottleneck involves the relatively short operating lifetime of the photosystem proteins PSI and PSII due to their highly susceptible photodegradation, heat, and electrochemical denaturation. In this regard, continuous investigations of highly stabilizing nano-encapsulation techniques are being pursued with a view to realizing the multi-year durability levels needed for practical solar energy applications (Li et al., 2024).

The reliance on expensive noble metal nanocatalysts for efficient water splitting introduces severe economic barriers toward developing stable, high-performance catalysts using earth-abundant materials. Besides performance, the unique physical and chemical properties of ENMs introduce significant safety and environmental risks. Many studies have shown that the high reactivity of certain metal oxide nanoparticles, including ZnO, can cause oxidative stress and cytotoxicity in beneficial microorganisms. In higher concentrations, nanoparticles can even impair cellular functions in plants (Santos et al., 2025). Critically, there is a significant lack of understanding about the long-term environmental fate and transport of these nanoparticles when used in agriculture as nanofertilizers. The possible effects of nanomaterial bioaccumulation in food chains remain one of the major concerns. Most importantly, the lack of globally comparable and harmonized regulatory frameworks addressing the size-dependent risks of nanomaterials is acting as a brake on responsible innovation and causing erosion of consumer confidence in the nanotechnology industry (Kumari et al., 2023).

9. Future perspectives and research directions

Future nano-enabled photosynthesis will not only intelligently design and integrate systems to overcome present limitations but will also be extended to plant nanobionics and niche market applications (Table 10.3). The key research directions are toward developing highly stable, self-repairing nano-biohybrids that can mimic the dynamic repair mechanisms of natural photosystems, possibly by incorporating genetically engineered proteins or reversible, responsive nanoscaffolds to overcome degradation issues and achieve commercially viable lifespans. Meanwhile, fully artificial systems should move away from noble metal catalysts to Earth-abundant, high-entropy alloy (HEA), or single-atom nano-catalysts that are highly active yet cost-effective for water splitting and CO_2 reduction (Huang et al., 2025).

A significant emerging field is plant nanobionics, whereby a bio-inspired approach involves the use of nanomaterials, such as iron-based or carbon-based nanoparticles, to enhance

TABLE 10.3 Future nano-strategies for photosynthetic optimization.

Strategy	Target	Mechanism	Expected outcomes
Synthetic nano-chloroplasts	Artificial systems	Reconstituted PSI–PSII units	Scalable artificial photosynthesis
AI-driven nano-design	Reaction centers	Predictive optimization of nanointeractions	Precision enhancement
Plasmonic nanostructures	Antenna complex	Broaden the absorption spectrum	Higher quantum efficiency
Biodegradable nanomaterials	Whole chloroplast	Minimize environmental risk	Sustainable application in crops
Smart nanocarriers	PSI and PSII	Delivery of cofactors or antioxidants	Enhanced tolerance and energy yield

the plant's native photosynthetic machinery by increasing chlorophyll content, mitigating oxidative stress, or even improving CO_2 uptake efficiency by regulating stomatal function, hence promising enormous scope for enhancing crop yield and climate resilience. In another direction, recent advances in wearable and implantable bioelectronic devices are driving the development of highly efficient, flexible biohybrid solar cells for biomedical applications, hence constituting an important technology diversification beyond utility-scale energy. In general, scalable, sustainable manufacturing techniques for all nanophotosynthetic components and the establishment of rigorous toxicological and regulatory frameworks to guarantee the safe environmental introduction of these transformational materials will be required in the future.

10. Conclusion

Herein, nanophotosynthesis merges nanotechnology with photosynthesis to completely redesign biological and artificial solar energy conversion into a paradigm-shifting sustainability science. This chapter established that the basic improvements are rooted in proper control of charge carrier dynamics at nano-bio interfaces. Functionalization innovations in sp2 carbon nanomaterials, such as graphene CNTs, allow for unprecedented efficiency in Direct Electron Transfer (DET) and, hence, maximize photocurrent output by overcoming their natural electron mediation barrier. Advanced encapsulation into mesoporous nanoscaffolds stabilizes fragile PSI and PSII against denaturation. It is desirable that future research shift to design-oriented intelligent prediction, by taking into account such targets as incorporation of pan-spectral QDs, construction of a single-atom or high-entropy nanocatalyst for the selective and cost-effective CO_2 reduction, among others. At the same time, plant nanobionics shapes high-impact potential in engineering climate resilient superphotosynthesis catalysts inside through internal targeted delivery of nanoparticles, CeO_2, and Fe-nanofertilizers into plants to boost their native ROS scavenging abilities and photosynthetic performances. Scaling of this science necessitates advancement in low-cost synthetic methodologies and robust mechanism-based nanotoxicological and regulatory frameworks. Nanophotosynthesis is positioned well to underpin essential, scalable, and environmentally benign solutions to global energy and climate problems as a function of mastering angstrom to nanometer scale energy flow.

References

Aghaei, N., Seyedin, H., Sanaeinasab, H., 2018. Strategies for disaster risk reduction education: a systematic review. J. Educ. Health Promot. 7 (1), 98.

Alabdallah, N.M., Alluqmani, S.M., Almarri, H.M., Al-Zahrani, A.A., 2024. Physical, chemical, and biological routes of synthetic titanium dioxide nanoparticles and their crucial role in temperature stress tolerance in plants. Heliyon 10 (4), e09405. https://doi.org/10.1016/j.heliyon.2024.e09405.

Anjum, S., Ishaque, S., Fatima, H., Farooq, W., Hano, C., Abbasi, B.H., Anjum, I., 2021. Emerging applications of nanotechnology in healthcare systems: grand challenges and perspectives. Pharmaceuticals 14 (8), 707. https://doi.org/10.3390/ph14080707.

Arora, R., Roy, T., Adak, P., 2024. A review of the impact of nanoparticles on environmental processes. In: BIO Web of Conferences, 86. EDP Sciences, 01001. https://doi.org/10.1051/bioconf/20248601001.

Avellan, A., Yun, J., Zhang, Y., Spielman-Sun, E., Unrine, J.M., Thieme, J., et al., 2019. Nanoparticle size and coating chemistry control foliar uptake pathways, translocation, and leaf-to-rhizosphere transport in wheat. ACS Nano 13 (5), 5291–5305. https://doi.org/10.1021/acsnano.8b09781.

Bachle, S.R., 2021. Anatomical Constraints on Grass Physiological Responses Depend on Water Availability. Kansas State University.

Bae, S., Kim, M., Jo, N., Kim, K.M., Lee, C., Kwon, T.H., Nam, Y.S., Ryu, J., 2024. Amine-rich hydrogels for molecular nanoarchitectonics of photosystem II and inverse opal TiO_2 toward solar water oxidation. ACS Appl.

Mater. Interfaces 16 (13), 16086–16095. https://doi.org/10.1021/acsami.4c12345.

Blankenship, R.E., 2021. Molecular Mechanisms of Photosynthesis. John Wiley and Sons.

Brettel, K., Leibl, W., 2001. Electron transfer in photosystem I. Biochim. Biophys. Acta Bioenerg. 1507 (1–3), 100–114. https://doi.org/10.1016/S0005-2728(00)00248-9.

Cardona, T., Sedoud, A., Cox, N., Rutherford, A.W., 2012. Charge separation in photosystem II: a comparative and evolutionary overview. Biochim. Biophys. Acta Bioenerg. 1817 (1), 26–43. https://doi.org/10.1016/j.bbabio.2011.07.005.

Chen, Y., Xu, B., Yao, R., Chen, C., Zhang, C., 2022. Mimicking the oxygen-evolving center in photosynthesis. Front. Plant Sci. 13, 929532. https://doi.org/10.3389/fpls.2022.929532.

Chen, F., Yang, N., Huang, X., Lin, J., Zhang, H., Zhang, X., Liu, Y., Li, W., Lei, B., 2025. Carbon dots alleviate photoinhibition and enhance photosynthesis in *Chlorella pyrenoidosa*. Chem. Eng. J. 507, 160429. https://doi.org/10.1016/j.cej.2025.160429.

Cheng, B., Yang, Z., Chen, F., Yue, L., Cao, X., Li, J., Qian, H. L., Yan, X.P., Wang, C., Wang, Z., 2023. Biomass-derived carbon dots with light conversion and nutrient provisioning capabilities facilitate plant photosynthesis. Sci. Total Environ. 901, 165973. https://doi.org/10.1016/j.scitotenv.2023.165973.

Cherepanov, D.A., Milanovsky, G.E., Nadtochenko, V.A., Semenov, A.Y., 2023. Energy barrier of photoinduced charge separation in the reaction centers of photosystems I and II. Russ. J. Phys. Chem. B 17 (3), 594–599. https://doi.org/10.1134/S1990793123030070.

Croce, R., Carmo-Silva, E., Cho, Y.B., Ermakova, M., Harbinson, J., Lawson, T., Zhu, X.G., 2024. Perspectives on improving photosynthesis to increase crop yield. Plant Cell 36 (10), 3944–3973.

Dekker, J.P., Boekema, E.J., 2005. Supramolecular organization of thylakoid membrane proteins in green plants. Biochim. Biophys. Acta 1706 (1–2), 12–39. https://doi.org/10.1016/j.bbabio.2004.09.009.

Demirer, G.S., Silva, T.N., Jackson, C.T., Thomas, J.B., Ehrhardt, D.W., Rhee, S.Y., et al., 2021. Nanotechnology to advance CRISPR-Cas genetic engineering of plants. Nat. Nanotechnol. 16 (3), 243–250. https://doi.org/10.1038/s41565-021-00854.

Du, X., Liu, L., Ye, J., 2021. Plasmonic metal nanoparticles for artificial photosynthesis: advancements, mechanisms, and perspectives. Sol. RRL 5 (12), 2100611. https://doi.org/10.1002/solr.202100611.

Eckardt, N.A., Allahverdiyeva, Y., Alvarez, C.E., Büchel, C., Burlacot, A., Cardona, T., Yano, J., 2024. Lighting the way: compelling open questions in photosynthesis research. Plant Cell 36 (10), 3914–3943.

Falcioni, R., Capalbo, A., Varrà, C., D'Agostino, N., Sabatino, L., Gualtieri, E., 2024. Decreased photosynthetic efficiency in *Nicotine tabacum* L. under transient heat stress. Plants 13 (2), 273. https://doi.org/10.3390/plants13020273.

Fan, M., Li, M., Liu, Z., Cao, P., Pan, X., Zhang, H., et al., 2015. Crystal structures of the PsbS protein essential for photoprotection in plants. Nat. Struct. Mol. Biol. 22 (8), 729–735. https://doi.org/10.1038/nsmb.3068.

Gautam, K., Singh, H., Sinha, A.K., 2025. Nanotechnology in plant nanobionics: mechanisms, applications, and future perspectives. Adv. Biol. 9 (4), 2400589. https://doi.org/10.1002/adbi.202400589.

Giraldo, J.P., Wu, H., Newkirk, G.M., Kruss, S., 2019. Nanobiotechnology approaches for engineering smart plant sensors. Nat. Nanotechnol. 14 (6), 541–553. https://doi.org/10.1038/s41565-019-0470-6.

Gobbato, T., Volpato, G.A., Sartorel, A., Bonchio, M., 2023. A breath of sunshine: oxygenic photosynthesis by functional molecular architectures. Chem. Sci. 14 (44), 12402–12429. https://doi.org/10.1039/d3sc04132a.

Golbeck, J.H., 2018. Advances in Photosynthesis: The Molecular Biology of Cyanobacteria, vol. 1. Kluwer Academic Publishers, pp. 319–360.

Hashmi, K., Gupta, S., Mishra, P., Khan, T., Joshi, S., 2024. The vital role of nanoparticles in enhancing plant growth and development. Eng. Proc. 67 (1), 48. https://doi.org/10.3390/engproc67010048.

He, X., Hwang, H.M., 2016. Nanotechnology in food science: functionality, applicability, and safety assessment. J. Food Drug Anal. 24 (4), 671–681. https://doi.org/10.1016/j.jfda.2016.04.003.

Hladík, J., Posada, C., Stuchlík, Z., 2020. Radial instability of trapping polytropic spheres. Int. J. Modern Phys. D 29 (5), 2050030.

Hu, P., An, J., Faulkner, M.M., Wu, H., Li, Z., Tian, X., et al., 2020. Nanoparticle charge and size control foliar delivery efficiency to plant cells and organelles. ACS Nano 14 (6), 7970–7986. https://doi.org/10.1021/acsnano.9b09178.

Huang, X., Zhang, K., Li, Z., Zhang, J., Zhai, X., Zhang, N., Du, L., Qin, Z., 2025. Exploring the integration of anthocyanins with functional materials in smart food packaging: from stabilization to application. Foods 14 (16), 2896. https://doi.org/10.3390/foods14162896.

Jávorfi, T. (2022). Advances and applications in structural approaches.

Jeevanandham, S., Ramasundaram, S., Vijay, N., Oh, T.H., Selvan, S.T., 2025. Recent progress in designing nanomaterial biohybrids for artificial photosynthesis. Nanomaterials 15 (10), 730. https://doi.org/10.3390/nano1510730.

Jing, X., Liu, Y., Liu, X., Zhang, Y., Wang, G., Yang, F., Zhang, Y., Chang, D., Zhang, Z.L., You, C.X., Zhang, S., 2024. Enhanced photosynthetic efficiency by nitrogen-doped carbon dots via plastoquinone-involved electron transfer in apple. Hortic. Res. 11 (3), uhae016. https://doi.org/10.1093/hr/uh2024016.

Joudeh, N., Linke, D., 2022. Nanoparticle classification, physicochemical properties, characterization, and applications: a comprehensive review for biologists. J. Nanobiotechnol. 20 (1), 262. https://doi.org/10.1186/s12951-022-01511-3.

Kanjer, L., Vaz, M., Wilmotte, A., Bosak, S., 2023. Blue-green travelers: cultivation of cyanobacteria associated with sea turtles. In: Cyano2023 – 8th Early Career Researcher Symposium on Cyanobacteria.

Kathpalia, R., Verma, A.K., 2021. Bio-inspired nanoparticles for artificial photosynthesis. Mater. Today Proc. 45, 3825–3832. https://doi.org/10.1016/j.matpr.2020.12.548.

Kim, A., Watanabe, C.D.P., Duffy, A.V., Ruban, J., Minagawa, J., 2020. Multimeric and monomeric photosystem II supercomplexes represent structural adaptations to low- and high-light conditions. J. Biol. Chem. 295 (43), 14537–14545. https://doi.org/10.1074/jbc.RA120.015219.

Kim, J., Yun, J., Sharipov, M., Moon, J., Ryu, W., 2025. Enhancing the performance of 3D printed bio-photoelectrochemical cells through multi-objective Bayesian optimisation. Virtual Phys. Prototyp. 20 (1), e2449565. https://doi.org/10.1080/17452759.2025.2449565.

Kirchhoff, H., 2013. Architectural switches in plant thylakoid membranes. Photosynth. Res. 116 (4), 481–487. https://doi.org/10.1007/s11120-013-9843-0.

Kouřil, L., Nosek, M., Opatíková, R., Arshad, D.A., Semchonok, I., Chamrád, R., Lenobel, E., Boekema, E.J., Ilík, P., 2020. Unique organization of photosystem II supercomplexes and megacomplexes in Norway spruce. Plant J. 104 (1), 215–225. https://doi.org/10.1111/tpj.14612.

Kumari, R., Suman, K., Karmakar, S., Mishra, V., Lakra, S.G., Saurav, G.K., Mahto, B.K., 2023. Regulation and safety measures for nanotechnology-based agri-products. Front. Genome Ed. 5, 1200987. https://doi.org/10.3389/fgeed.2023.1200987.

Lattanzio, V., 2020. Adjustment of photosynthetic activity to drought and fluctuating light in wheat. Physiol. Plantarum 169 (1), 16–29. https://doi.org/10.1111/ppl.13063.

Li, P., Xia, Y., Song, K., Liu, D., 2024. The impact of nanomaterials on photosynthesis and antioxidant mechanisms in gramineae plants: research progress and future prospects. Plants 13 (7), 984. https://doi.org/10.3390/plants13070984.

Liu, X., Nadeem, M., Rui, Y., 2024a. Effects of nanofertilizers on the mechanism of photosynthetic efficiency in plants: a review. Phyton 93 (12), 3197. https://doi.org/10.32604/phyton.2024.3197.

Liu, S., Han, J., Ma, X., Zhu, X., Qu, H., Xin, G., Huang, X., 2024b. Repeated release of cerium oxide nanoparticles altered algal responses: growth, photosynthesis, and photosynthetic gene expression. Eco-Environ. Health 3 (3), 290–299.

Maduraiveeran, G., 2024. Transition metal nanomaterial-based electrocatalysts for water and CO_2 electrolysis: preparation, catalytic activity, and prospects. Front. Energy Res. 12, 1433103. https://doi.org/10.3389/fenrg.2024.1433103.

Mitchell, M.J., Billingsley, M.M., Haley, R.M., Wechsler, M.E., Peppas, N.A., Langer, R., 2021. Engineering precision nanoparticles for drug delivery. Nat. Rev. Drug Discov. 20 (2), 101–124. https://doi.org/10.1038/s41573-020-00096-2.

Mony, C., Kaur, P., Rookes, J.E., Callahan, D.L., Eswaran, S.V., Yang, W., Manna, P.K., 2022. Nanomaterials for enhancing photosynthesis: interaction with plant photosystems and scope of nanobionics in agriculture. Environ. Sci. Nano 9 (10), 3659–3683. https://doi.org/10.1039/d2en00359g.

Mukarram, M., Khan, M.M.A., Kurjak, D., Lux, A., Corpas, F.J., 2023. Silicon nanoparticles (SiNPs) restore photosynthesis and essential oil content by upgrading enzymatic antioxidant metabolism in lemongrass (*Cymbopogon flexuosus*) under salt stress. Front. Plant Sci. 14, 1116769. https://doi.org/10.3389/fpls.2023.1116769.

Nagao, R., Kato, K., Ifuku, K., Suzuki, T., Kumazawa, M., Uchiyama, I., Kashino, Y., Dohmae, N., Akimoto, S., Shen, J.R., et al., 2020. Structural basis for assembly and function of a diatom photosystem I-light-harvesting supercomplex. Nat. Commun. 11 (1), 2481. https://doi.org/10.1038/s41467-020-16285-0.

Nosek, D., Semchonok, E.J., Boekema, E.J., Ilík, R., Kouřil, L., 2017. Structural variability of plant photosystem II megacomplexes in thylakoid membranes. Plant J. 89 (1), 104–111. https://doi.org/10.1111/tpj.13358.

Pagnani, G., Santini, E., Caldarelli, G., Nassi, O., Vitti, A., 2024. Effects of nanofertilizers on the mechanism of photosynthetic efficiency in plants: a review. Phyton 93 (12), 59193. https://doi.org/10.32604/phyton.2024.059193.

Paul, N., 2022. Using Photosynthetic Proteins for Energy Storage and Sensory Devices. Doctoral dissertation. National University of Singapore, Singapore. https://doi.org/10.25443/nus.2022.001.

Reddy, L.J., Ghosh, K., Panwar, A., 2025. Transcriptomic approaches to examine the nanomaterials' impact on the physiological and molecular responses of plants. In: Nanobiostimulants in Innovative Agriculture. Springer

Nature, Singapore, pp. 169–201. https://doi.org/10.1007/978-981-99-9987-2_9.

Ruan, X., Li, S., Huang, C., Zheng, W., Cui, X., Ravi, S.K., 2024. Catalyzing artificial photosynthesis with TiO_2 heterostructures and hybrids: emerging trends in a classical yet contemporary photocatalyst. Adv. Mater. 36 (17), 2305285. https://doi.org/10.1002/adma.202305285.

Santana, I., Wu, H., Hu, P., Giraldo, J.P., 2020. Targeted delivery of nanomaterials with chemical cargoes in plants enabled by a biorecognition motif. Nat. Commun. 11, 2045. https://doi.org/10.1038/s41467-020-15731.

Sarode, A., Slaughter, G., 2025. Nanomaterial-enabled enhancements in thylakoid-based biofuel cells. Nanomaterials 15 (14), 1092. https://doi.org/10.3390/nano15141092.

Sharma, D., Saini, R., Tiwari, M., 2014. The Effect of Drought-Induced Changes in Chlorophyll Fluorescence and Photosynthetic Pigments in *Triticum aestivum* L. Taylor & Francis Online.

Sharma, A., Kumari, R., Singh, R., 2023. The vital role of nanoparticles in enhancing plant growth and development. Plants 12 (23), 4056. https://doi.org/10.3390/plants12234056.

Sheng, A., Watanabe, A.J., Li, E., Kim, C.H., Song, K., Murata, D.F., Song, J., Minagawa, Z.F., Liu, Z.F., 2019. Structural insight into light harvesting for photosystem II in green algae. Nat. Plants 5 (11), 1320–1330. https://doi.org/10.1038/s41477-019-0531-7.

Shevela, D., Kern, J.F., Govindjee, G., Messinger, J., 2023. Solar energy conversion by photosystem II: principles and structures. Photosynth. Res. 156 (3), 279–307.

Shlosberg, Y. (2024). Photoelectricity generation from beet juice in a bio-electrochemical cell.

Shoyhet, H., Pavlopoulos, N.G., Amirav, L., Adir, N., 2021. Functionalized gold-nanoparticles enhance photosystem II driven photocurrent in a hybrid nano-bio-photoelectrochemical cell. J. Mater. Chem. A 9 (32), 17231–17241. https://doi.org/10.1039/d1ta04601a.

Sim, S., Wong, N.K., 2021. Nanotechnology and its use in imaging and drug delivery. Biomed. Rep. 14 (5), 42. https://doi.org/10.3892/br.2021.1420.

Singh, T., Shukla, S., Kumar, P., Wahla, V., Bajpai, V.K., Rather, I.A., 2017. Application of nanotechnology in food science: perception and overview. Front. Microbiol. 8, 1501. https://doi.org/10.3389/fmicb.2017.01501.

Singh, A., Rajput, V.D., Agrawal, S., Ghazaryan, K., Minkina, T., Al Tawaha, A.R.M., Chauhan, A., Mandzhieva, S.S., Singh, R.K., Papadakis, M., Alexiou, A., 2024. Nanoparticles mediated salt stress resilience: a holistic exploration of physiological, biochemical, and nano-omics approaches. Rev. Environ. Contam. Toxicol. 262 (1), 19. https://doi.org/10.1007/s41742-024-00456-7.

Slattery, R.A., Ort, D.R., 2021. Photosynthesis | Photosynthetic efficiency improvement. In: Encyclopedia of Environmental Science. Elsevier Inc.

Sohail, M., Rauf, S., Irfan, M., Hayat, A., Alghamdi, M.M., El-Zahhar, A.A., Ghernaout, D., Al-Hadeethi, Y., Lv, W., 2024. Recent developments, advances and strategies in heterogeneous photocatalysts for water splitting. Nanoscale Adv. 6 (5), 1286–1330. https://doi.org/10.1039/d3na00745k.

Tripathi, S., Mahra, S., Tiwari, K., Rana, S., Tripathi, D.K., Sharma, S., Sahi, S., 2023. Recent advances and perspectives of nanomaterials in agricultural management and associated environmental risk: a review. Nanomaterials 13 (10), 1604. https://doi.org/10.3390/nano13101604.

Vijayaraghavareddy, P., Lekshmy, S.V., Struik, P.C., Makarla, U., Yin, X., Sreeman, S., 2022. Production and scavenging of reactive oxygen species confer to differential sensitivity of rice and wheat to drought stress. Crop Environ. 1 (1), 15–23.

Voloshin, R.A., Shumilova, S.M., Zadneprovskaya, E.V., Zharmukhamedov, S.K., Alwasel, S., Hou, H.J.M., Allakhverdiev, S.I., 2022. Photosystem II in biophotovoltaic devices. Photosynthetica 60 (1), 121–130. https://doi.org/10.32615/ps.2022.010.

Wang, C., O'Hagan, M.P., Willner, B., Willner, I., 2022. Bioinspired artificial photosynthetic systems. Chem. Eur J. 28 (9), e202103595. https://doi.org/10.1002/chem.202103595.

Woodson, J.D., 2016. Chloroplast quality control – balancing energy production and stress. New Phytol. 212 (1), 36–41. https://doi.org/10.1111/nph.14134.

Xu, C., Pi, X., Huang, Y., Han, G., Chen, X., Qin, X., Huang, G., Zhao, S., Yang, Y., Kuang, T., et al., 2020. Structural basis for energy transfer in a huge diatom PSI-FCPI supercomplex. Nat. Commun. 11 (1), 5081. https://doi.org/10.1038/s41467-020-18813-2.

Xu, Z., Zhou, G., Shimizu, H., 2023. The Interactive Effects of Drought and Heat Stress on Photosynthetic Efficiency, Biomass Production and Yield of Soybean (*Glycine max* L.). NIH.

Xue, R., Tokutsu, S., Bergner, S.V., Scholz, M., Minagawa, J., Hippler, M., 2015. Photosystem II subunit R is required for efficient binding of light-harvesting complex stress-related protein 3 to photosystem II–light-harvesting supercomplexes in *Chlamydomonas reinhardtii*. Plant Physiol. 167 (4), 1566–1578. https://doi.org/10.1104/pp.114.255347.

Yamane, Y., Kashino, Y., Koike, H., 2015. Molecular mechanisms of photoinhibition in higher plants. Plant Cell Physiol. 56 (2), 232–242. https://doi.org/10.1093/pcp/pcu156.

Yotsova, E., Stefanov, M., Rashkov, G., Dobrikova, A., Apostolova, E., 2025. Mitigating salinity stress in pea plants with titanium oxide nanoparticles. Int. J. Plant Biol. 16 (1), 34. https://doi.org/10.53765/ijpb.2025.16.1.34.

Zaman, W., Manghwar, H., 2024. Nanoparticles and their biological applications: recent advances in 2022–2023. Microorganisms 12 (3), 489. https://doi.org/10.3390/microorganisms12030489.

Zhang, X., Li, X., Shi, Y., Wei, S., 2020. Response of photosynthetic performance to drought duration and re-watering in maize (*Zea mays* L.). Plants 9 (8), 1056. https://doi.org/10.3390/plants9081056.

Zhao, S., Li, C., Wu, C., Hu, J., Zhang, Z., Lei, B., Li, W., Hu, C., Liu, Y., Zheng, M., 2024. Effects of multifunctional cerium-doped carbon dots on photosynthetic capacity and nutritional quality of lettuce. Environ. Sci. Nano 11 (7), 3137–3149. https://doi.org/10.1039/d4en00321f.

Zhou, P., Yu, J., Jaroniec, M., 2014. Adv. Mater. 26, 4920–4930. https://doi.org/10.1002/adma.201401237.

Zhou, X., El-Sappah, A.H., Khaskhoussi, A., Huang, Q., Atif, A.M., Elhamid, M.A.A., Ihtisham, M., El-Maati, M. F.A., Soaud, S.A., Tahri, W., 2025. Nanoparticles: a promising tool against environmental stress in plants. Front. Plant Sci. 15, 1509047. https://doi.org/10.3389/fpls.2025.1509047.

Zivcak, M., Brestic, M., Kunderlikova, K., Sytar, O., Zivcakova, I., 2014. Photosynthetic responses and production of reactive oxygen species of *Triticum aestivum* L. exposed to high-temperature and high-light stress. Photosynthetica 52 (1), 59–74. https://doi.org/10.1007/s11099-014-0004-8.

CHAPTER

11

Nanoparticle assisted nutrient delivery for enhancing plant photosynthetic performance

Yusheng Li[1], *Saddam Hussain*[2], *Mohammad Nauman Khan*[2], *Zhiyong Ai*[3] *and Mohammad Faizan*[4]

[1]School of Breeding and Multiplication (Sanya Institute of Breeding and Multiplication), Hainan University, Sanya, China; [2]Laboratory of Plant Molecular Genetics & Crop Gene Editing, School of Life Sciences, Linyi University, Linyi, China; [3]National Center of Technology Innovation for Saline-Alkali Tolerant Rice in Sanya, Sanya, China; [4]Botany Section, School of Sciences, Maulana Azad National Urdu University, Hyderabad, Telangana, India

1. Introduction

1.1 Importance of plant photosynthesis

Photosynthesis underpins Earth's life systems by converting solar energy into chemical energy, driving global biogeochemical cycles. This process comprises two essential stages: the light reactions on the thylakoid membrane and the Calvin–Benson cycle within the chloroplast stroma (Johnson, 2016). Through these mechanisms, plants efficiently fix atmospheric carbon dioxide (CO_2) into carbohydrates, providing the metabolic foundation for growth and the global carbon cycle. Yet, intensifying climate change poses unprecedented threats to this critical process. Rising temperatures, drought frequency, and precipitation anomalies, compounded by elevated atmospheric CO_2, impose complex physiological stresses on crops (Heyduk et al., 2025). With global food demand projected to rise by 70% by 2050, transcending the limitations of traditional agricultural models to sustain photosynthetic activity, particularly in staple crops, has become a central scientific challenge (Huang, 2014; van Dijk et al., 2021).

1.2 Conventional nutrient delivery systems and their limitations

Traditional chemical fertilizers, such as urea and diammonium phosphate, have long supported modern agriculture but are increasingly recognized for their profound inefficiencies

Nanotechnology in Photosynthesis
https://doi.org/10.1016/B978-0-443-44376-3.00011-2

and environmental costs (Mehta et al., 2024). Seasonal utilization rates for nitrogen and phosphorus hover between 30%–50% and 15%–30%, respectively. The substantial fraction of unabsorbed nutrients is either immobilized in soil, forming insoluble compounds with iron/aluminum ions in acidic soils or calcium ions in alkaline conditions, or lost via leaching and volatilization (Bender et al., 2023). Consequently, nutrient runoff drives eutrophication in aquatic ecosystems (Rahman et al., 2025), while gaseous losses (e.g., N_2O) significantly amplify global warming potentials (Tu et al., 2026). Fundamentally, the "burst release" kinetics of conventional fertilizers fail to match the dynamic, stage-specific nutrient demands of crops (Dai et al., 2024). This misalignment is particularly critical for organelle-level requirements, such as the surge in magnesium and iron needed for chloroplast synthesis during rapid growth, thereby constraining the development of optimal photosynthetic machinery (Newkirk et al., 2021; Santana et al., 2022).

1.3 Nanotechnology as a tool for better nutrients delivery

To overcome the bottlenecks in traditional agricultural nutrient management, nanotechnology offers a promising strategy (Guimarães et al., 2018; Khan et al., 2025). Nanoparticles (NPs; 1–100 nm) possess unique physicochemical properties that reshape nutrient delivery dynamics (Klaic et al., 2018; Wu et al., 2025). Unlike traditional ionic uptake dependent on specific transporters, NPs can penetrate physical barriers such as the cell wall (pore diameter approx. 5–20 nm) and enter vascular systems via apoplastic or symplastic pathways for long-distance transport (Liu et al., 2022). Foliar application further exploits stomatal entry and cuticular permeation for direct delivery to mesophyll cells (Husted et al., 2023). Through precise structural design and surface functionalization, nano-fertilizers can achieve "smart" controlled release (Shen et al., 2023). For instance, biodegradable polymer coatings (e.g., chitosan) can synchronize nutrient release with crop growth cycles, ensuring an "on-demand" supply that optimizes bioavailability while minimizing environmental loss (Leonardi et al., 2021).

2. Fundamentals of photosynthetic activity and nutrient demand

2.1 Overview of photosynthesis

Photosynthesis is the fundamental biological process that converts solar energy into stable chemical energy, centered on the coordinated operation of two functional modules: the light reactions and the Calvin cycle. However, the efficiency and stability of this energy conversion chain are not inherently constant. They are highly dependent on a precise redox balance and energy allocation mechanism mediated by the photosynthetic electron transport chain (ETC) (Liu et al., 2023a; Lu et al., 2025). At the molecular level, the light reactions convert light energy into ATP and NADPH on the thylakoid membrane, while the Calvin cycle utilizes this active chemical energy to fix CO_2 within the chloroplast stroma. The ETC, connecting these two processes, serves not only as a conduit for linear electron flow but also as a critical regulatory hub determining the energy yield ratio (ATP/NADPH) and safeguarding against photooxidative damage (Santos-Merino et al., 2021; Shomali et al., 2024). Understanding the elemental foundation of the key molecules, such as chlorophyll, Rubisco, and cytochromes, namely plant nutrients, is a prerequisite for elucidating photosynthetic regulatory mechanisms and exploring their improvement potential (Wang et al., 2025; Hu et al., 2026).

2.2 Key nutrients influencing photosynthesis

The normal functioning of photosynthesis relies on a complex molecular machinery precisely constructed and maintained by multiple mineral nutrients. Nutrient deficiencies become the primary abiotic stress factor limiting photosynthetic performance by directly disrupting the structural integrity or catalytic functions of these molecular machines (Shah et al., 2024).

During light energy capture and conversion, multiple nutrients synergistically ensure efficient photon energy absorption and transfer. Magnesium (Mg) serves as the central ion in the chlorophyll porphyrin ring, directly determining the spectral properties and stability of pigment molecules. Mg deficiency accelerates chlorophyll degradation, significantly reducing the maximum photochemical efficiency of Photosystem II (Croce and van Amerongen, 2020; dos Santos et al., 2023). Nitrogen (N), as a fundamental constituent of chlorophyll molecules and light-harvesting complex proteins, leads to systemic reductions in chlorophyll content and key photosynthetic protein abundance when deficient, thereby undermining light energy absorption and carbon assimilation capacity at the source (Zhang et al., 2022). The integrity and efficient operation of the electron transport chain critically depend on adequate iron (Fe) and manganese (Mn) supply. Fe serves as the core component of the heme cofactor in the cytochrome b_6f complex and the iron-sulfur clusters in photosystem I. Fe deficiency directly impedes electron transfer to ferredoxin, disrupting electron flow and exacerbating reactive oxygen species accumulation (Nam et al., 2021; Therby-Vale et al., 2022). Mn forms the core of the manganese cluster unique to the oxygen-evolving complex in photosystem II, driving water splitting and electron supply. Mn deficiency leads to loss of oxygen evolution function, causing the entire ETC blocked (Schmidt et al., 2016). Recent proteomics studies indicate that manganese deficiency also disrupts the accumulation of core subunits in Photosystem II. Even with upregulation of related repair proteins, maintaining the dynamic equilibrium of photosynthetic machinery remains challenging (Li et al., 2024). During carbon fixation and metabolism, nutrients like nitrogen, phosphorus, potassium, and zinc play critical roles. N is a major component of the Rubisco enzyme, directly determining the maximum potential for carbon dioxide fixation (Chai et al., 2024; Suganami et al., 2020). P serves not only as a structural component of ATP and NADPH but also regulates the transport of Calvin cycle intermediates (Kayoumu et al., 2023). K influences intercellular CO_2 concentration by modulating stomatal opening and closing, while acting as an activator for multiple enzymes. It thus synergistically impacts carbon fixation rates through both substrate supply and enzyme activity regulation (Hou et al., 2019; Sarwar et al., 2024). Zinc (Zn) serves as an essential cofactor for carbonic anhydrase. Its deficiency reduces the efficiency of converting carbon dioxide into bicarbonate, thereby indirectly limiting Rubisco's substrate availability (Rumeau et al., 2004; Zuo et al., 2017).

2.3 Relationship between nutrient deficiencies and reduced photosynthetic performance

Nutrient deficiencies negatively impact photosynthetic performance not only at the direct molecular level (e.g., reduced enzyme activity, structural damage) but may also be amplified through indirect physiological and metabolic pathways (Lu et al., 2025). For instance, deficiencies in macronutrients such as N, P, and K can ultimately lead to a significant decline in net photosynthetic rate (P_n) by affecting photosynthetic electron transport efficiency, carbon fixation rates, and energy allocation (Hu et al., 2025; Liu et al., 2025b). Nutrient deficiencies may trigger photooxidative damage (e.g., reactive oxygen species accumulation), structural reorganization of photosynthetic machinery (e.g., reduced PSII repair efficiency), and metabolic pathway reprogramming (e.g., alterations in carbohydrate metabolism) (Xu et al., 2025). These complex processes collectively constitute the dynamic equilibrium between "vulnerability" and "adaptability" in maintaining photosynthetic function under abiotic stress.

2.4 Physiological markers of nutrient-mediated photosynthetic efficiency

To systematically evaluate the effects of nutrients on photosynthesis and their regulatory mechanisms, researchers typically employ a series of interrelated physiological indicators. Together, these indicators form a comprehensive phenotypic system reflecting photosynthetic functional status and stress responses (Liu et al., 2025a).

2.4.1 *Chlorophyll content and its fluorescence parameters*

Chlorophyll content forms the basis of light energy absorption capacity, while fluorescence parameters such as Fv/Fm, ΦPSII, and NPQ dynamically indicate the integrity of photosystem II, actual photochemical efficiency, and non-radiative energy dissipation capacity. These parameters are particularly useful for revealing the damage and adaptive regulation of photosynthetic mechanisms caused by nutrient deficiencies such as magnesium and iron (Han et al., 2022; Wang et al., 2022; Guidi et al., 2019).

2.4.2 *Gas exchange parameters*

Net photosynthetic rate, stomatal conductance, and intercellular CO_2 concentration, collectively characterize the overall efficiency and limiting factors of carbon assimilation. Net photosynthetic rate directly reflects CO_2 fixation capacity, stomatal conductance indicates stomatal regulation of CO_2 supply, and intercellular CO_2 concentration helps distinguish between biochemical and stomatal limitations in photosynthetic metabolism. This clarifies the regulatory roles of nutrients like nitrogen and phosphorus in carbon fixation pathways (Stinziano et al., 2017; Yang et al., 2023; Salter et al., 2018).

2.4.3 *Key enzyme activities and antioxidant systems*

Rubisco and carbonic anhydrase activities directly correlate with the biochemical potential for carbon fixation and CO_2 transport. Meanwhile, antioxidant enzyme activities, such as superoxide dismutase and catalase, characterize the scavenging capacity of reactive oxygen species during photosynthetic electron transport. These activities reflect the redox state and metabolic adaptation strategies of photosynthetic machinery under nutrient stress (Li et al., 2025c; Lu et al., 2022).

3. Nanoparticles in agriculture: A brief overview

3.1 Types of nanoparticles used in agriculture

Agricultural nanoparticles are categorized based on their core chemical composition. These categories exhibit highly specialized functional potential in crop nutrient supply and photosynthetic physiological regulation due to inherent differences in structure and physicochemical properties (Khan et al., 2025).

Metal-based nanoparticles, such as zinc oxide (ZnO), magnetite (Fe_3O_4), and copper oxide (CuO) nanoparticles, offer the core advantage of slow-release trace metals (Chen et al., 2020; Purkayastha et al., 2025). Through high surface reactivity, they dissolve gradually within plant tissues or soil environments, continuously releasing essential metal ions (e.g., Zn^{2+}, Fe^{3+}, Cu^{2+}) to precisely meet crop requirements. Additionally, these nanoparticles possess reactive oxygen species (ROS) scavenging capabilities (Zeng et al., 2024). They quench excess ROS generated during photosynthesis under stress conditions (e.g., intense light, drought), effectively shielding the photosystem II (PSII) reaction center from oxidative damage and maintaining the stability and efficiency of photosynthetic machinery (Faizan et al., 2021; Tighe-Neira et al., 2018).

Unlike metal-based nanoparticles that directly supply trace elements, polymeric nanoparticles (e.g., chitosan, polylactic acid) and porous silica nanoparticles (e.g., mesoporous silica (MSNs)) primarily function as nutrient carriers (Bekkam and Thiyagarajan, 2024; Nandini et al., 2025). Biopolymers like chitosan can encapsulate bulk nutrients such as urea, delaying their release to reduce losses from leaching or volatilization while preventing the inhibitory effects of excessive quick-release nitrogen on photosynthetic systems (Demeke et al., 2025; Thirumurugan et al., 2025). Porous silica, leveraging its high specific surface area, ordered pore structure, and easily functionalize surface, enables the loading and on-demand release of various nutrients (e.g., Mg^{2+}, Fe^{2+}), helping maintain stomatal function and CO_2 supply under stress conditions (Prasad et al., 2023; Sun et al., 2016).

Carbon-based nanomaterials (e.g., carbon nanotubes, graphene quantum dots) primarily exert effects by improving the rhizosphere microenvironment (Ge et al., 2018; You et al., 2023). Carbon-based materials efficiently adsorb nutrient ions in soil, enhancing nutrient retention rates and creating favorable conditions for root absorption (Joshi et al., 2018; Lahiani et al., 2018). They also boost electron transport efficiency between mitochondria and chloroplasts within cells, thereby optimizing nutrient utilization and carbon fixation at the energy conversion level (Huo et al., 2025).

In practical applications, these distinct nanoparticle types are often integrated through composite modification strategies to achieve synergistic nutrient delivery effects, driving agricultural inputs toward smarter, more efficient solutions (Fig. 11.1).

3.2 Nano-fertilizers versus conventional fertilizers: Comparative efficiency

Based on the core mechanism of "targeted delivery-controlled release," nano-fertilizers demonstrate significant advantages over traditional fertilizers in nutrient utilization efficiency, environmental impact, and crop physiological regulation (Kumar et al., 2023). Compared to traditional fertilizers characterized by rapid nutrient release and high loss rates, nano-fertilizers achieve dynamic matching of release rates with crop demand by encapsulating nutrients within nano-carriers or directly preparing them in nano-form, thereby enhancing nutrient utilization efficiency (Ranjan et al., 2025; Pudhuvai et al., 2025). This enhanced efficiency also substantially reduces environmental emissions and application rates. The slow-release properties, combined with the carrier's adsorption of nutrients, minimize nutrient migration into the environment (Lu et al., 2024; Zhu et al., 2024). Studies indicate that chitosan-coated urea nanoparticles can reduce ammonia volatilization and nitrate leaching in paddy fields (Easwaran et al., 2025). Nano-fertilizers also enable more direct pathways for photosynthetic enhancement (Xiao et al., 2021). Through foliar application, rationally designed nanoparticles can overcome physiological barriers to deliver nutrients precisely to chloroplasts (Li et al., 2025a). For instance, foliar application of nano-selenium directly participates in chlorophyll synthesis and activates ATP synthase, increasing chlorophyll content and net

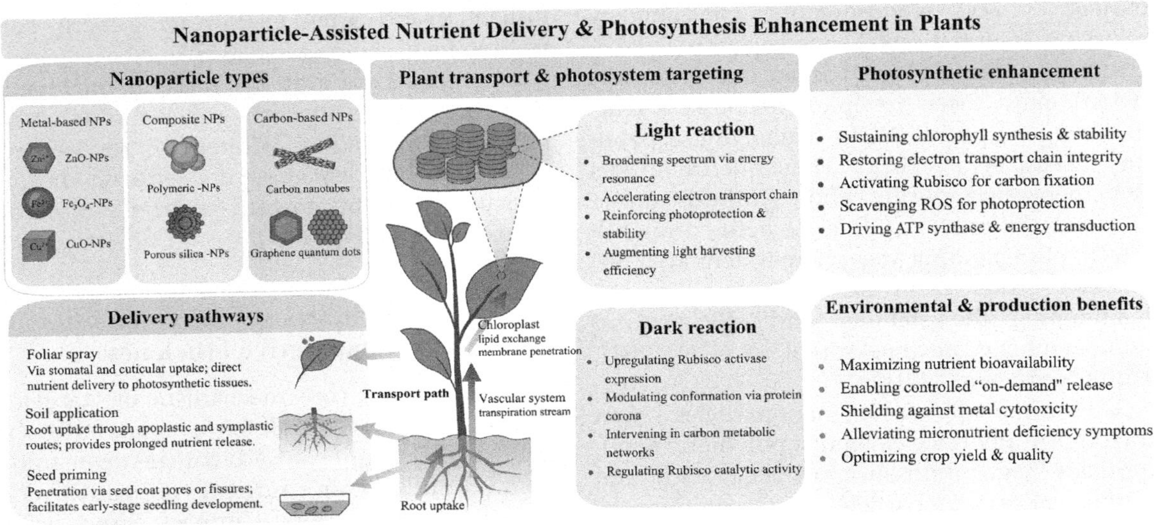

FIGURE 11.1 Nanoparticle-assisted nutrient delivery and photosynthesis enhancement in plants.

photosynthetic rate, ultimately improving fruit quality (Liu et al., 2023b; Wang et al., 2025). This "point-to-point" delivery method achieves efficient coupling from nutrient uptake to enhanced photosynthetic performance, highlighting the significant capability of nanofertilizers in sustainable agricultural production.

3.3 Properties of NPs relevant to nutrient delivery

The exceptional performance of nanoparticles in nutrient delivery stems from the synergistic interaction of four key factors: size, surface charge, surface modification, and controlled-release properties. These characteristics collectively regulate the entire process of nutrient "absorption-transport-utilization" within plants, serving as the cornerstone for achieving efficient and precise delivery (Bansal et al., 2025; Sardoiwala et al., 2018). Particle size is the primary factor determining whether a nanoparticle can effectively penetrate plant tissue barriers. Typically, particles smaller than the pores in cell walls (approximately 50 nm) are readily absorbed by roots and transported within the plant. Particles that are too small (<10 nm) exhibit strong penetration but tend to migrate rapidly with transpiration streams, making it difficult to retain them effectively at target sites and potentially leading to higher biotoxicity. Conversely, particles that are too large (>100 nm) are easily retained by root surfaces or cortex, hindering their entry into the vascular system for long-distance transport. Surface charge modulates interactions between particles and biological interfaces or environmental media via electrostatic forces (Chen et al., 2024; Sharma et al., 2025). Since plant cell walls carry a net negative charge under physiological conditions, positively charged nanoparticles enhance binding to cell walls through electrostatic adsorption, promoting uptake. Conversely, negatively charged particles reduce aggregation with soil colloids, improving migration in soil and rhizosphere accessibility (Juárez-Maldonado et al., 2019; Zhu et al., 2012). Surface modification serves as a core strategy to enhance particle stability, dispersibility, and targeted functionality (Abhinayaa et al., 2019; Hosseinifard et al., 2024; Dev et al., 2018). For instance, grafting hydrophilic polymers like polyethylene glycol (PEG) onto particle surfaces creates steric hindrance layers that effectively inhibit agglomeration. Introducing specific ligands enables directed delivery to targeted tissues or organelles (Hu et al., 2022; Selli et al., 2019). Controlled release properties of nanoparticles are achieved through precise structural design, with the core objective being to synchronize nutrient release kinetics with crop growth stages, particularly the critical photosynthetic phase, precisely matching nutrient demands. This effectively avoids the supply imbalances common in traditional fertilization, such as excessive early-stage supply and insufficient late-stage availability (Helal et al., 2023; Mallik et al., 2025).

4. Mechanisms of nanoparticle-assisted nutrient delivery

Nanoparticle-assisted nutrient delivery achieves full-chain regulation through "precision application-efficient absorption-targeted transport-on-demand release," overcoming the spatial and temporal limitations of traditional nutrient delivery systems. This enables precise matching with the dynamic nutrient demands of photosynthesis (Fig. 11.1).

4.1 Foliar versus soil versus seed priming application routes

The application route of nanoparticles as the starting point of the delivery process directly determines the initial interface for nutrient entry into plants, the absorption rate, and subsequent transport pathways. Foliar application, soil application, and seed incorporation exhibit significant differences in their sites of action, transport barriers, and applicable scenarios. Their selection requires comprehensive consideration of crop type, growth stage, and environmental conditions.

4.1.1 Foliar application

Foliar application delivers nutrients directly to photosynthetic organs, with absorption primarily occurring through stomatal and cuticular pathways. Particles smaller than 50 nm can enter through open stomata, bypassing the cuticular barrier; alternatively, they may dissolve into cuticular waxes or penetrate nanoscale pores to reach epidermal cells (Freire et al., 2024; Hong et al., 2021; Li et al., 2018). This method offers rapid response, with increased chlorophyll content detectable within 24–48 h after spraying magnesium oxide nanoparticles. It effectively prevents soil fixation of elements like phosphorus and iron, making it suitable for rapid nutrient correction during mid-to-late growth stages (Li et al., 2024). However, foliar application is susceptible to rainwash and exhibits low retention rates on varieties with dense trichomes, making uniform coverage challenging.

4.1.2 Soil application

Soil application represents the primary pathway for nanoparticle uptake into plants, involving both apoplastic and symplastic routes (Asgari et al., 2018; Djanaguiraman et al., 2024). Particles may diffuse through intercellular spaces or enter root hairs via endocytosis. However, they are blocked by the Casparian strip in the endodermis and require specific transporters to enter cells before being transported to the xylem. This method enables sustained nutrient supply and enhances nutrient availability in soil (Rashid et al., 2023; Yan et al., 2023). However, challenges include soil adsorption, root barriers, and high costs associated with covering entire root zones.

4.1.3 Seed priming application

Seed-triggered delivery introduces nanoparticles into seeds through seed coat pores or germination cracks (Fonseca et al., 2024; Khan et al., 2023). Nutrients released in the endosperm or cotyledons are directly utilized by the radicle and plumule. This method significantly enhances photosynthetic capacity during the seedling stage. For instance, corn seedlings treated with seed priming exhibit markedly increased chlorophyll and soluble protein content. It effectively avoids nutrient competition between seedlings and soil microorganisms and is simple to implement, facilitating mechanization (Neto et al., 2020; Del Buono et al., 2021) (Esper). However, its total nutrient capacity is limited, typically sufficient only for early seedling growth, necessitating supplemental fertilization later.

4.2 Uptake and translocation of NPs within plant tissues

The absorption and transport of nanoparticles within plants constitute a cascade process spanning multiple biological barriers (Tripathi et al., 2017), including cellular uptake, tissue transport, and organelle targeting. Cell entry depends on particle size and surface properties: smaller particles enter via passive diffusion, while larger particles rely on endocytosis. Subsequently, particles move short distances within tissues via the symplastic or apoplastic pathways, entering the vascular system through the symplastic pathway at the root endodermis. Upon reaching the xylem, particles are transported upward over long distances via the transpiration stream. Ultimately, particles with higher surface charge can penetrate the chloroplast bilayer membrane through a "lipid exchange membrane penetration" mechanism, enabling precise nutrient delivery to photosynthetic sites (Gogos et al., 2016; Raliya et al., 2016; Fares et al., 2024).

4.3 Controlled-release and sustained supply of nutrients aligned with photosynthesis

The one-time nutrient release pattern of traditional fertilizers struggles to match plants' fluctuating nutrient demands across growth stages and circadian rhythms (Eddarai et al., 2024). Smart controlled-release systems based on nanotechnology can achieve precise, on-demand nutrient release through environmentally responsive designs (Singh et al., 2022; Zain et al., 2024). For instance, leveraging the dynamic pH changes in the rhizosphere, nutrients can be loaded into polymeric nanogels containing responsive functional groups such as carboxyl or amino groups. When roots secrete organic acids, causing a local pH drop, the polymer structure undergoes protonation changes, triggering rapid nutrient release that directly responds to the plant's physiological needs (Conway and Keller, 2016; Li et al., 2019; Zahoor et al., 2025). Additionally, light-responsive nanosystems utilize photosensitive molecules like azobenzene and spiropyran as regulatory "switches." Under specific wavelengths of light, these molecules undergo conformational changes that alter the carrier structure and release internal nutrients. This synchronizes nutrient supply with the light intensity rhythm of photosynthesis, achieving smarter spatiotemporal matching to plant demands (Gong et al., 2024; Xu et al., 2017).

4.4 Interaction with photosynthetic apparatus

Nanoparticles not only serve as nutrient carriers; certain specific materials can also directly interact with photosynthetic apparatus (Saeedi et al., 2024; Zsiros et al., 2019). For example, materials such as carbon nanotubes and quantum dots possess unique optical properties that enable them to absorb ultraviolet or near-infrared light. Through energy resonance transfer, they deliver captured light energy to chlorophyll, thereby broadening plants' utilization range of light energy (Khan et al., 2024). Certain metal oxide nanoparticles like cerium dioxide and titanium dioxide can function as auxiliary components in the electron transport chain. They facilitate electron transfer from photosystem II to photosystem I or scavenge excess reactive oxygen species, thereby maintaining smooth electron transport chain operation and enhancing photoprotection (Abbas et al., 2020; Hamed et al., 2024; Zhuang et al., 2025).

In carbon fixation, nanoparticles can also modulate Rubisco enzyme activity (Castillo et al., 2025). Studies indicate that titanium dioxide nanoparticles upregulate the gene expression and protein levels of Rubisco activase in spinach, indirectly promoting Rubisco activation. A more direct regulatory mechanism involves positively charged nanoparticles entering chloroplasts, potentially forming electrostatic "protein crowns" with negatively charged Rubisco proteins. This interaction alters Rubisco's conformation and catalytic activity, either enhancing or inhibiting enzyme function depending on the nanoparticles' physicochemical properties (Li et al., 2023a; Weng et al., 2024). These findings demonstrate nanoparticles' potential as biological modulators capable of directly intervening in photosynthetic metabolic networks.

5. Nanoparticles and enhanced photosynthetic performance

As an emerging precision agriculture technology carrier, nanoparticles demonstrate significant potential in enhancing crop photosynthetic performance (Kaur et al., 2024). This section systematically analyzes how several nanomaterials, represented by zinc oxide (ZnO), iron (Fe), magnesium (Mg), and EDTA-iron nano-chelates, enhance plant photosynthetic efficiency by precisely delivering core nutrients. This occurs at molecular, physiological, and field application levels, improving light energy capture, electron transport, and carbon assimilation to ultimately boost crop yield and quality.

5.1 ZnO NPs: Enhancing chlorophyll synthesis and enzyme activation

Zn is an essential micronutrient for plant growth and development. As a key component or activator of over 300 enzymes within plants, it plays an irreplaceable regulatory role in multiple critical stages of photosynthesis. It serves as the core cofactor for carbonic anhydrase (CA) and participates in chlorophyll synthesis and the regulation of Rubisco enzyme (Liu et al., 2025a; Mellis et al., 2024). However, zinc in soil is easily immobilized, resulting in low bioavailability. Foliar application of ZnO NPs effectively circumvents complex soil environments. Leveraging their controlled ion release kinetics, ZnO NPs continuously and slowly release Zn^{2+} after entering the leaf intercellular spaces or through stomata (Elshoky et al., 2021). This simultaneously enhances local CO_2 bioavailability and photosynthetic pigment stability, maintaining the functional integrity of Photosystem II (Akhondi et al., 2025; Tryfon et al., 2023).

5.2 Fe NPs: Improving electron transport and reducing chlorosis

Fe serves as the important element of the photosynthetic ETC, forming the core of cytochrome b_6f complex ($Cytb_6f$) and multiple iron-sulfur clusters (Fe–S clusters) within photosystem I (PSI) (Tikhonov, 2024). Fe deficiency ranks among the primary factors globally limiting plant productivity, directly causing impaired photosynthetic electron transport and leaf chlorosis. Fe_3O_4 NPs, as an efficient iron supplement, dissolve upon plant uptake in acidic organelles (e.g., vacuoles) or through root exudates, releasing Fe^{2+}/Fe^{3+} ions. These ions enter chloroplasts and directly participate in the resynthesis and assembly of critical iron-sulfur clusters in the PSI core subunit. These iron-sulfur clusters serve as critical relay stations for electron transfer from ferredoxin (Fd) to $NADP^+$ reductase (FNR), ultimately generating NADPH (Feng et al., 2022). Their structural repair directly unblocks the entire photosynthetic electron transport chain from PSII to PSI, significantly enhancing ETR (Zhang et al., 2025). Supplemented iron ions serve as cofactors or activators for crucial antioxidant enzymes like catalase (CAT) and superoxide dismutase (SOD). This substantially enhances the cell's capacity to scavenge ROS (Gama et al., 2025), mitigating oxidative stress damage to photosynthetic machinery and thereby safeguarding the functional integrity of PSII (Elbasuney et al., 2022; Sebastian et al., 2017).

5.3 Mg NPs: Role in chlorophyll core and ATP synthesis

Mg occupies the central position within the chlorophyll molecule, serving as the core for light energy capture. Simultaneously, it acts as an essential activator for ATP synthase and multiple key enzymes in the Calvin cycle, such as Rubisco. The availability of Mg directly determines the efficiency of light energy absorption and energy conversion (Suliman et al., 2025). Mg^{2+} ions slowly released from MgO NPs on leaf surfaces enter chloroplasts, where they bind to chlorophyll precursor porphyrin IX, ensuring consistent chlorophyll synthesis rates. This enables plants to maintain higher chlorophyll concentrations per unit leaf area (Ali et al., 2024). Mg^{2+} serves as the key messenger linking the light-driven proton motive force across thylakoid membranes to ATP synthesis in the stroma. It binds to and activates the active site of the β-subunit of ATP synthase, significantly promoting the condensation reaction of ADP with inorganic phosphate (Pi) to form ATP. This accelerates carbohydrate synthesis and facilitates its transport and accumulation (Chitrashalini and Kumar, 2025).

5.4 Nano-chelates for micronutrient availability

Although Fe_3O_4 NPs demonstrate remarkable efficacy in Fe supplementation, the released free Fe ions remain susceptible to rapid oxidation and precipitation in certain soil environments (e.g., high-pH calcareous soils), leading to reduced bioavailability. Excess free Fe ions may generate highly oxidative hydroxyl radicals (•OH) via the Fenton reaction, inducing cytotoxicity (Kobayashi et al., 2019). The integration of the potent metal chelator ethylenediaminetetraacetic acid (EDTA) with Fe nanoparticles to create organic-inorganic hybrid nanomaterials represents a cutting-edge direction in nanofertilizer design (Rui et al., 2016). EDTA's strong chelating action shields Fe ions during transport, preventing precipitation reactions with soil phosphates, hydroxides, and other ions, thereby maintaining their bioavailability as they traverse cell walls and membranes (Haydar et al., 2022). Near target organelles (e.g., chloroplasts), changes in the local environment (e.g., pH, specific enzymes) trigger the dissociation of EDTA from iron, enabling "on-demand release" of Fe ions. This precise release mechanism ensures Fe ions do not become excessively free before being utilized by the photosynthetic system. This approach maximizes their positive effects while minimizing oxidative side effects caused by free Fe ions (Avrahami et al., 2026).

These cases collectively demonstrate that nanoparticles achieve precise matching between nutrient supply and photosynthetic demand across temporal, spatial, and physiological dimensions through the synergistic integration of structural design, controlled release, and targeted delivery. Consequently, they exhibit significant advantages in enhancing photosynthetic efficiency, optimizing yield and quality, and reducing environmental emissions.

6. Integration with stress mitigation

Against the backdrop of climate change, abiotic stresses such as drought, salinity, high temperatures, and heavy metal pollution severely inhibit crop photosynthesis, limiting food production. These stresses cause oxidative damage by disrupting root nutrient uptake, damaging photosynthetic organs, and inducing reactive oxygen species accumulation (Li et al., 2025a). Nanoparticle-assisted nutrient delivery can specifically alleviate the suppression of the photosynthetic system by different stresses through a synergistic mechanism of nutrient supply and enhanced stress tolerance (Table 11.1). Its core lies in property adaptation, precisely regulating nutrient release and physiological responses based on the type of stress (Zhao et al., 2020, 2023).

6.1 Role of NP-mediated nutrient delivery under abiotic stresses

6.1.1 Drought stress

Under drought stress, insufficient soil moisture leads to increased plant transpiration. To reduce water loss, guard cells undergo dehydration and shrinkage, causing stomatal closure and decreased stomatal conductance (Song et al., 2026). This not only restricts intercellular CO_2 supply and impairs photosynthesis but also triggers reactive oxygen species (ROS) accumulation, exacerbating oxidative damage (Cao et al., 2024). Research indicates that certain nanomaterials can simultaneously regulate stomatal behavior and scavenge ROS, thereby alleviating drought stress (Ahmed et al., 2021). For instance, functional carbon nanodots (FCNs) mitigate reductions in tissue water content and water use efficiency by enhancing root vitality and increasing levels of osmolytes (soluble sugars and proline). This helps regulate enzymatic and non-enzymatic antioxidant systems, scavenge reactive oxygen species (ROS), and inhibit lipid peroxidation. Consequently, plant cell membrane structure and function are protected under stress conditions, thereby sustaining photosynthesis and nutrient uptake (Chen et al., 2022). Meanwhile, polyacrylic acid-coated manganese oxide (Mn_3O_4) nanoparticles (PMO) exhibit enhanced photosynthetic performance under drought stress by catalyzing ROS scavenging and regulating stomatal aperture, thereby increasing cotton fresh weight (Li et al., 2025b).

6.1.2 Salt stress

Salt stress primarily inhibits plant growth through ionic toxicity and osmotic stress. High concentrations of Na^+ in soil, once absorbed by roots, disrupt intracellular ionic equilibrium (Li et al., 2023b). It competitively inhibits K^+ uptake and transport, affecting K^+-dependent physiological processes (such as photosynthetic electron transport), leading to reduced electron transfer rates (Li et al., 2025a). Simultaneously, Na^+-induced osmotic stress diminishes root water uptake capacity, causing leaf dehydration and desiccation damage to photosynthetic organs. Salts inhibit the uptake of trace elements like Fe^{2+} and Zn^{2+}, leading to nutrient imbalances (Atta et al., 2023). Nanomaterials can alleviate salt stress through multiple pathways. For example, treating rice with 150 μg/mL magnesium-doped carbon dots (Mg-CDs) promotes Na^+-K^+ ion balance, enhancing photosynthetic system stability and antioxidant levels (Liu et al., 2024). Polyacrylic acid-coated nano-cerium oxide (PNC) increases K^+ content in cotton leaves under salt stress, while the

TABLE 11.1 Nanomaterials as modulators of plant physiological responses to abiotic stresses.

Abiotic stress	Nanomaterials	Mode of action	Observed plant response	Reference
Drought	Functional carbon nanodots (FCNs)	Water retention and antioxidant enzymes activities↑	Root vitality, photosynthesis and nutrient uptake↑	Chen et al. (2022)
	Manganese oxide nanoparticles (Mn_3O_4-NPs)	ROS scavenging and regulation of stomatal aperture↑	Photosynthetic performance and cotton fresh weight↑	Li et al. (2025b)
	Iron oxide nanoparticles (IO-NPs)	Antioxidant enzyme contents and photosynthesis efficiency↑	Nutrient acquisition, plant height and biomass in rice↑	Ahmed et al. (2021)
	Graphene oxide (GO)	Drought-related hormones and genes expression↑	Root parameters↑, oxidative stress damage in soybeans↓	Zhao et al. (2022)
Salinity	Magnesium-doped carbon dots (Mg-CDs)	Photosynthetic system and antioxidant activity↑	Plant biomass and chlorophyll content↑, O_2^- contents↓	Liu et al. (2024)
	Polyacrylic acid coated nanoceria (PNC)	Mesophyll conductance to CO_2 and K^+ content in leaves↑	Chloroplast surface area and net photosynthetic rate↑	Yang et al. (2024)
	Oxidized chitosan–selenocystamine hydrochloride engineered nanomaterials (Cs-Se NMs)	Transcription of oxidative stress genes↑, the abundance and variety of rhizobacteria↑	Activity of antioxidant enzymes↑, MDA levels↓	Fang et al. (2024)
	Tannic acid-iron nanomaterials (TA-Fe NMs)	Reactive oxygen species scavenging capabilities↑	Rice seed germination, root and shoot growth↑	Cheng (2025)
Heat	Selenium-iron (Se-Fe) nanocomposite	Plant's intrinsic antioxidant defense system↑	Grain yield and protein content↑	Abasi et al. (2025)
	Selenium nanoparticles (Se-NPs)	Chelate toxic ions, detoxifying enzymes↑	Pollen germination percentage and seed yield in sorghum↑	Djanaguiraman et al. (2018)
	Zinc oxide nanoparticles (ZnO-NPs)	Production of reactive oxygen species↓, chlorophyll contents and gas exchange attributes↑	Seeds per pod, pods per plant and grain yield in mungbean↑	Kareem et al. (2022)
Heavy metal	Graphite carbon nitride nanosheets (gC_3N_4 NS)	Root cell wall-cadmium retention↑, migration to the shoot↓	Alleviation of cadmium toxicity and photosynthesis↑	Xu et al. (2024)
	Ferric oxide and zinc oxide nanoparticles (Fe_3O_4 and ZnO-NPs)	Fe_3O_4 and ZnO-NPs in roots/leaves↑, metabolites, including proline, 6-hydroxynicotinic acid and farrerol↑	Plant height, root length, shoot and root fresh weight under cadmium toxicity in tobacco↑	Zou et al. (2022)
	Molybdenum disulfide nanomaterials (MoS_2 NMs)	Adsorb heavy metal ions, the risk of the transfer of toxic heavy metals↓	Bioaccumulation of heavy metals in roots↓	Song et al. (2019)

enlarged chloroplast surface area also boosts CO_2 uptake (Yang et al., 2024).

6.1.3 *Heat stress*

High-temperature stress severely damages the photosynthetic system of plants. On one hand, elevated temperatures significantly accelerate the degradation rate of the D1 protein in photosystem II (PSII) reaction centers while simultaneously inhibiting the activity of D1 repair-related enzymes (Chen et al., 2017). This leads to a reduction in the number of functional PSII reaction centers and impaired system function (Zha et al., 2016). On the other hand, high temperatures cause denaturation of Rubisco activase, resulting in decreased carboxylase activity of Rubisco and a significant reduction in carbon fixation efficiency (Carmo-Silva et al., 2012; Wijewardene et al., 2021). Research indicates that silicon nanoparticles (Si NPs) stimulate the production of heat shock proteins (HSPs), which protect intracellular proteins and membranes from denaturing under high temperatures (Zaman et al., 2025). Spraying sorghum leaves with 10 mg L^{-1} nano-selenium under high-temperature stress enhances the antioxidant defense system by boosting antioxidant enzyme activity, thereby improving pollen germination rates and yield (Djanaguiraman et al., 2018).

6.1.4 *Heavy metal stress*

Heavy metal ions (such as Cd^{2+} and Pb^{2+}) absorbed by roots can be transported via the xylem to leaves, where they accumulate in chloroplasts, causing multiple damages to photosynthetic machinery (Panda et al., 2025). Key mechanisms include disrupting the phospholipid bilayer structure of thylakoid membranes and inhibiting the activity of the cytochrome b_6/f complex in the electron transport chain (ETC) and the iron-sulfur clusters in photosystem I (PSI), leading to impaired electron transfer (Hu et al., 2023; Song et al., 2025). Nanomaterials can mitigate heavy metal toxicity and restore photosynthetic function through multiple pathways (Xu et al., 2024). Research indicates that nanomaterials can reduce heavy metal uptake in crops, particularly MoS_2 NMs which decreased the bioaccumulation of heavy metals (As, Cd, Cr, Cu, Ni, Al, Ti, and Pb) by 36.4%–60.6% (Song et al., 2019). Graphite carbon nitride nanosheets (g-C_3N_4 NSs) enhance photosynthetic efficiency by increasing pectin content, inhibiting cadmium translocation to shoot tissues through enhanced pectin methylesterase (PME) activity, and protecting chloroplast structure (Xu et al., 2024).

6.2 NP-assisted modulation of antioxidant systems to sustain photosynthesis

Non-biotic stress triggers excessive accumulation of reactive oxygen species (ROS, such as O_2^-, H_2O_2, and ·OH) within plants, constituting a key common pathway causing damage to photosynthetic organs (Choudhury et al., 2017). To address this challenge, nanoparticles can effectively maintain mitochondrial redox homeostasis through antioxidant and ROS scavenging activities, thereby protecting photosynthetic function integrity and enhancing plant stress resistance (Guo et al., 2023; Fallah et al., 2024). Once introduced into plant cells, these NPs function as direct ROS scavengers or modulate antioxidant defense systems. They influence signaling ions like calcium, signaling molecules such as nitric oxide (NO) and plant hormones, and regulatory factors including transcription factors (TFs), thereby modulating gene expression and antioxidant enzyme production (Zia-ur-Rehman et al., 2023). Studies indicate that boron nanoparticles effectively stimulate phenolic accumulation in pea plants under drought stress, significantly enhancing CAT, APX, and GPX activities. This accelerates O_2^- scavenging, effectively reduces H_2O_2 levels, and minimizes the accumulation of membrane lipid peroxidation products such as malondialdehyde (MDA) (Sutulienė et al., 2023). Beyond activating enzymatic systems, nanoparticles also promote the synthesis of non-enzymatic antioxidants by regulating metabolic pathways, achieving dual functions of ROS scavenging and osmotic regulation. Examples include ascorbic acid (AsA) and glutathione (GSH), which further enhance plants' tolerance to oxidative stress and photosynthetic capacity (Singh et al., 2025).

6.3 Synergistic effects with plant growth regulators and bio-stimulants

Plant growth regulators (PGRs) and biostimulants, as traditional agricultural inputs, can regulate crop growth, development, and stress responses to some extent. However, they are generally limited by chemical instability, susceptibility to environmental degradation, poor targeting, and low utilization efficiency (Campos et al., 2023; Handoko and Lin, 2025). Nanoparticles not only serve as highly efficient and stable carriers enabling precise, controlled release of active substances, but their unique physicochemical properties can also directly or indirectly regulate plant physiological processes, amplifying improvements in crop stress resistance and photosynthetic performance (Pérez-de-Luque, 2017; Xia et al., 2021).

6.3.1 *Nano-auxin synergy*

Auxins (such as indole-3-acetic acid, IAA) serve as key hormones regulating plant root development and stomatal behavior. Their efficient utilization is crucial for enhancing water and nutrient uptake. However, exogenous IAA readily degrades and inactivates in soil environments. Utilizing functionalized nanoparticles to protect and deliver IAA enables significant improvements in controlled release and biostability, further enhancing auxin activity (Butova et al., 2023). Lignin nanoparticles encapsulating IAA (LNP–IAA) effectively modulate xylem development in cherry tomatoes and wheat, enabling controlled hormone action without inducing phytotoxic effects. Compared to free hormone application, nanoparticle encapsulation preserves IAA's bioactivity while achieving sustained release and enhanced physiological performance (Faleiro et al., 2026). Research also indicates that gold nanoparticles (AuNPs) stabilized with IAA and indole-3-butyric acid (IBA) significantly promote plant growth under stress conditions, demonstrating potential as safe and efficient growth promoters (Sahil et al., 2025).

6.3.2 *Nano-humic acid synergy*

Humic acid (HA) is a naturally occurring macromolecular organic compound rich in carboxyl and phenolic hydroxyl groups. It binds to soil clay particles via hydrogen bonds and van der Waals forces, promoting soil aggregate formation and enhancing soil porosity, making it an excellent soil conditioner and biostimulant (Canellas et al., 2020). Chelating it with metal-based nanoparticles (e.g., ZnO NPs) significantly reduces NPs agglomeration in soil solutions through electrostatic repulsion and steric hindrance effects, thereby improving their dispersion and mobility (Rahale et al., 2021). Studies indicate that foliar application of iron (III) oxide nanoparticles (IONPs) stabilized by humic substances significantly increases wheat leaf iron content by 70%, demonstrating that humic substances act as surfactants to synergistically enhance iron absorption from goethite nanoparticles (Zimbovskaya et al., 2020). HA modification may alter nanoparticle environmental behavior and biotoxicity. Its long-term impacts on soil microbial communities and heavy metal migration risks require careful evaluation through systematic field trials—a prerequisite for ensuring the sustainability of this technology's agricultural applications (Bondarenko et al., 2020; Hasnain et al., 2023).

7. Conclusion

Nanoparticles, with their unique physicochemical properties, theoretically enable intelligent controlled release and targeted delivery of nutrients, offering new possibilities for overcoming the bottlenecks in photosynthetic efficiency and nutrient utilization in traditional agriculture. However, the practical application of this technology still faces multiple challenges. Current research primarily focuses on short-term effects, and reported yield increases mostly originate from laboratory or small-scale field trials, leaving long-term ecological risks unclear. Furthermore, the production costs of nanomaterials are significantly higher than those of conventional fertilizers, and there is a lack of large-scale, multi-season field validation and systematic cost-benefit analysis. Simultaneously, scaling up laboratory-scale synthesis processes to industrial levels requires solving a series of engineering challenges, including batch consistency and energy consumption control. Only by overcoming these critical obstacles can nanotechnology truly become a vital force supporting the sustainable development of future agriculture.

References

Abasi, Raja, Sohail, Mashwani, Ahmad, Mehmood, 2025. Efficacy of bio-inspired SeFe nanocomposite to improve protein content and antioxidant mechanism in wheat under heat stress. Int. J. Biol. Macromol. 331.

Abbas, Q., Liu, G., Yousaf, B., Ali, M.U., Ullah, H., Mujtaba Munir, M.A., Ahmed, R., Rehman, A., 2020. Biochar-assisted transformation of engineered-cerium oxide nanoparticles: effect on wheat growth, photosynthetic traits and cerium accumulation. Ecotoxicol. Environ. Saf. 187. https://doi.org/10.1016/j.ecoenv.2019.109845.

Abhinayaa, R., Jeevitha, G., Mangalaraj, D., Ponpandian, N., Meena, P., 2019. Toxic influence of pristine and surfactant modified halloysite nanotubes on phytopathogenic bacteria. Appl. Clay Sci. 174, 57–68. https://doi.org/10.1016/j.clay.2019.03.022.

Ahmed, T., Noman, M., Manzoor, N., Shahid, M., Abdullah, M., Ali, L., Wang, G., Hashem, A., Al-Arjani, A.B.F., Alqarawi, A.A., Abd_Allah, E.F., Li, B., 2021. Nanoparticle-based amelioration of drought stress and cadmium toxicity in rice via triggering the stress responsive genetic mechanisms and nutrient acquisition. Ecotoxicol. Environ. Saf. 209. https://doi.org/10.1016/j.ecoenv.2020.111829.

Akhondi, M., Amiri, M.S., Beyk-Khormizi, A., Mousavi-Kouhi, S.M., Gautam, R.K., Mottaghipisheh, J., Badrolnojoum, M., Taghavizadeh Yazdi, M.E., 2025. Zinc oxide nanoparticles as a way to improve physiology, photosynthesis, and anti-oxidative systems in Salvia leriifolia. Plant Nano Biology 14. https://doi.org/10.1016/j.plana.2025.100211.

Ali, S., Ulhassan, Z., Shahbaz, H., Kaleem, Z., Yousaf, M.A., Ali, S., Sheteiwy, M.S., Waseem, M., Ali, S., Zhou, W., 2024. Application of magnesium oxide nanoparticles as a novel sustainable approach to enhance crop tolerance to abiotic and biotic stresses. Environ. Sci. Nano 11 (8), 3250–3267. https://doi.org/10.1039/d4en00417e.

Asgari, F., Majd, A., Jonoubi, P., Najafi, F., 2018. Effects of silicon nanoparticles on molecular, chemical, structural and ultrastructural characteristics of oat (Avena sativa L.). Plant Physiol. Biochem. 127, 152–160. https://doi.org/10.1016/j.plaphy.2018.03.021.

Atta, K., Mondal, S., Gorai, S., Singh, A.P., Kumari, A., Ghosh, T., Roy, A., Hembram, S., Gaikwad, D.J., Mondal, S., Bhattacharya, S., Jha, U.C., Jespersen, D., 2023. Impacts of salinity stress on crop plants: improving salt tolerance through genetic and molecular dissection. Front. Plant Sci. 14. https://doi.org/10.3389/fpls.2023.1241736.

Avrahami, G., Yecheskel, Y., Balous, H., Ben Mordechay, E., Richter, S., Zucker, I., 2026. Iron-oxide nanoparticle release from jellyfish-based hydrogels for agricultural fertilization. Environ. Sci. Nano 13 (1), 377–388. https://doi.org/10.1039/d5en00505a.

Bansal, M., Jyoti, N., Bharti, A., Deepika, Sharma, S., Thakur, M., Chauhan, V., Kumar, R., 2025. Green synthesis of nanomaterials used as nano-fertilizer for sustainability in crop production: a overview on recent advancements and future perspectives. Plant Nano Biology 11, 100143. https://doi.org/10.1016/j.plana.2025.100143.

Bekkam, R., Thiyagarajan, C., 2024. Foliar application of rice husk derived nanosilica for mitigating drought stress in hybrid maize by boosting antioxidant defense system. J. Hazard. Mater. Adv. 16, 100482. https://doi.org/10.1016/j.hazadv.2024.100482.

Bender, S.F., Schulz, S., Martínez-Cuesta, R., Laughlin, R.J., Kublik, S., Pfeiffer-Zakharova, K., Vestergaard, G., Hartman, K., Parladé, E., Römbke, J., Watson, C.J., Schloter, M., van der Heijden, M.G.A., 2023. Simplification of soil biota communities impairs nutrient recycling and enhances above- and belowground nitrogen losses. New Phytol. 240 (5), 2020–2034. https://doi.org/10.1111/nph.19252.

Bondarenko, L., Kahru, A., Terekhova, V., Dzhardimalieva, G., Uchanov, P., Kydralieva, K., 2020. Effects of humic acids on the ecotoxicity of Fe_3O_4 nanoparticles and Fe-ions: impact of oxidation and aging. Nanomaterials 10 (10), 2011. https://doi.org/10.3390/nano10102011.

Del Buono, D., Luzi, F., Puglia, D., 2021. Lignin nanoparticles: a promising tool to improve maize physiological, biochemical, and chemical traits. Nanomaterials 11 (4), 846. https://doi.org/10.3390/nano11040846.

Butova, V.V., Bauer, T.V., Polyakov, V.A., Minkina, T.M., 2023. Advances in nanoparticle and organic formulations for prolonged controlled release of auxins. Plant Physiol. Biochem. 201. https://doi.org/10.1016/j.plaphy.2023.107808.

Campos, E.V.R., Pereira, A.d.E.S., Aleksieienko, I., do Carmo, G.C., Gohari, G., Santaella, C., Fraceto, L.F., Oliveira, H.C., 2023. Encapsulated plant growth regulators and associative microorganisms: nature-based solutions to mitigate the effects of climate change on plants. Plant Sci. 331. https://doi.org/10.1016/j.plantsci.2023.111688.

Canellas, L.P., Canellas, N.O.A., Luiz Eduardo, L.E.S., Olivares, F.L., Piccolo, A., 2020. Plant chemical priming by humic acids. Chem. Biol. Technol. Agric. 7 (1). https://doi.org/10.1186/s40538-020-00178-4.

Cao, Y., Yang, W., Ma, J., Cheng, Z., Zhang, X., Liu, X., Wu, X., Zhang, J., 2024. An integrated framework for drought stress in plants. Int. J. Mol. Sci. 25 (17), 9347. https://doi.org/10.3390/ijms25179347.

Carmo-Silva, A.E., Gore, M.A., Andrade-Sanchez, P., French, A.N., Hunsaker, D.J., Salvucci, M.E., 2012. Decreased CO_2 availability and inactivation of Rubisco limit photosynthesis in cotton plants under heat and drought stress in the field. Environ. Exp. Bot. 83, 1–11. https://doi.org/10.1016/j.envexpbot.2012.04.001.

Castillo, C., Jeon, S.J., Hoang, K.N.L., Alford, C., Svendahl, E., Deng, C., Wang, Y., Wang, Y., Wei, X., Hernandez, R., White, J.C., Wheeler, K.E., Murphy, C.J., Giraldo, J.P., 2025. In vivo transformations of positively charged nanoparticles alter the formation and function of RuBisCO photosynthetic protein corona. Nat. Nanotechnol. 20 (8), 1152–1162. https://doi.org/10.1038/s41565-025-01944-x.

Chai, H., Gao, L., Zhao, C., Liu, X., Jiang, D., Dai, T., Tian, Z., 2024. Low nitrogen priming enhances Rubisco activation and allocation of nitrogen to the photosynthetic apparatus as an adaptation to nitrogen-deficit stress in wheat seedling. J. Plant Physiol. 303, 154337. https://doi.org/10.1016/j.jplph.2024.154337.

Chen, Y.E., Zhang, C.M., Su, Y.Q., Ma, J., Zhang, Z.W., Yuan, M., Zhang, H.Y., Yuan, S., 2017. Responses of photosystem II and antioxidative systems to high light and high temperature co-stress in wheat. Environ. Exp. Bot. 135, 45–55. https://doi.org/10.1016/j.envexpbot.2016.12.001.

Chen, Y., Lu, J., Liu, M., Li, H., Sun, Q., Nechitaylo, G.S., Bogoslovskaya, O.A., Olkhovskaya, I.P., Glushchenko, N.N., 2020. Tomato response to metal nanoparticles introduction into the nutrient medium. IET Nanobiotechnol. 14 (5), 382–388. https://doi.org/10.1049/iet-nbt.2019.0183.

Chen, Q., Cao, X., Nie, X., Li, Y., Liang, T., Ci, L., 2022. Alleviation role of functional carbon nanodots for tomato growth and soil environment under drought stress. J. Hazard Mater. 423 (Pt B), 127260. https://doi.org/10.1016/j.jhazmat.2021.127260.

Chen, Y., Jiang, S., Wang, Y., Zhang, F., Wang, L., Zhao, L., Liu, S., Tan, J., Persson, S., Sun, B., Chen, J., Blennow, A., 2024. Small-sized starch nanoparticles for efficient penetration of plant cells. Chem. Commun. 60 (95), 14113–14116. https://doi.org/10.1039/d4cc05493h.

Cheng, X., 2025. Tannic acid-iron nanomaterial enhances rice growth and antioxidant defense under salt stress. Front. Plant Sci. 16. https://doi.org/10.3389/fpls.2025.1565234.

Chitrashalini, S., Kumar, S.V., 2025. Harnessing the potential of plant-mediated magnesium oxide nanoparticles: a sustainable approach towards crop management and soil health. Pedosphere. https://doi.org/10.1016/j.pedsph.2025.11.013.

Choudhury, F.K., Rivero, R.M., Blumwald, E., Mittler, R., 2017. Reactive oxygen species, abiotic stress and stress combination. Plant J. 90 (5), 856–867. https://doi.org/10.1111/tpj.13299.

Conway, J.R., Keller, A.A., 2016. Gravity-driven transport of three engineered nanomaterials in unsaturated soils and their effects on soil pH and nutrient release. Water Res. 98, 250–260. https://doi.org/10.1016/j.watres.2016.04.021.

Croce, R., van Amerongen, H., 2020. Light harvesting in oxygenic photosynthesis: structural biology meets spectroscopy. Science 369 (6506). https://doi.org/10.1126/science.aay2058.

Dai, Z., Huang, Z., Hu, D., Naz, M., Afzal, M.R., Raza, M.A., Benavides-Mendoza, A., Tariq, M., Qi, S., Du, D., 2024. Role of nanofertilization in plant nutrition under abiotic stress conditions. Chemosphere 366. https://doi.org/10.1016/j.chemosphere.2024.143496.

Demeke, E.D., Benti, N.E., Terefe, M.G., Anbessa, T.T., Mengistu, W.M., Mekonnen, Y.S., 2025. A comprehensive review on nano-fertilizers: preparation, development, utilization, and prospects for sustainable agriculture in Ethiopia. Nanoscale Adv. 7 (8), 2131–2144. https://doi.org/10.1039/d4na01068j.

Dev, A., Srivastava, A.K., Karmakar, S., 2018. Nanomaterial toxicity for plants. Environ. Chem. Lett. 16 (1), 85–100. https://doi.org/10.1007/s10311-017-0667-6.

van Dijk, M., Morley, T., Rau, M.L., Saghai, Y., 2021. A meta-analysis of projected global food demand and population at risk of hunger for the period 2010–2050. Nat. Food 2 (7), 494–501. https://doi.org/10.1038/s43016-021-00322-9.

Djanaguiraman, M., Belliraj, N., Bossmann, S.H., Prasad, P. V.V., 2018. High-temperature stress alleviation by selenium nanoparticle treatment in grain sorghum. ACS Omega 3 (3), 2479–2491. https://doi.org/10.1021/acsomega.7b01934.

Djanaguiraman, M., Anbazhagan, V., Dhankher, O.P., Prasad, P.V.V., 2024. Uptake, translocation, toxicity, and impact of nanoparticles on plant physiological processes. Plants 13 (22). https://doi.org/10.3390/plants13223137.

Easwaran, C., Christopher, S.R., Ramasamy, R., Solomon, R. V., Kumar, M., Mohan, P., Marimuthu, R., Koothan, V., Nallusamy, S., 2025. Design and characterization of a chitosan-based nanohybrid fertilizer: molecular insights and nutrient release kinetics. J. Inorg. Organomet. Polym. Mater. 35 (11), 8919–8936. https://doi.org/10.1007/s10904-025-03836-y.

Eddarai, E.M., El Mouzahim, M., Ragaoui, B., Eladaoui, S., Bourd, Y., Bellaouchou, A., Boussen, R., 2024. Review of current trends in chitosan based controlled and slow-release fertilizer: from green chemistry to circular economy. Int. J. Biol. Macromol. 278. https://doi.org/10.1016/j.ijbiomac.2024.134982.

Elbasuney, S., El-Sayyad, G.S., Attia, M.S., Abdelaziz, A.M., 2022. Ferric oxide colloid: towards green nano-fertilizer for tomato plant with enhanced vegetative growth and immune response against fusarium wilt disease. J. Inorg. Organomet. Polym. Mater. 32 (11), 4270–4283. https://doi.org/10.1007/s10904-022-02442-6.

Elshoky, H.A., Yotsova, E., Farghali, M.A., Farroh, K.Y., El-Sayed, K., Elzorkany, H.E., Rashkov, G., Dobrikova, A., Borisova, P., Stefanov, M., Ali, M.A., Apostolova, E., 2021. Impact of foliar spray of zinc oxide nanoparticles on the photosynthesis of Pisum sativum L. under salt stress. Plant Physiol. Biochem. 167, 607–618. https://doi.org/10.1016/j.plaphy.2021.08.039.

Faizan, M., Faraz, A., Mir, A.R., Hayat, S., 2021. Role of zinc oxide nanoparticles in countering negative effects generated by cadmium in Lycopersicon esculentum. J. Plant Growth Regul. 40 (1), 101–115. https://doi.org/10.1007/s00344-019-10059-2.

Faleiro, R., Pace, M.R., Tessmer, M.A., Pereira, A.do E.S., Fraceto, L.F., Mayer, J.L.S., 2026. Smart delivery of auxin: lignin nanoparticles promote efficient and safer vascular development in crop species. Plant Nano Biol. 15, 100226. https://doi.org/10.1016/j.plana.2025.100226.

Fallah, S., Yusefi-Tanha, E., Peralta-Videa, J.R., 2024. Interaction of nanoparticles and reactive oxygen species and their impact on macromolecules and plant production. Plant Nano Biol. 10. https://doi.org/10.1016/j.plana.2024.100105.

Fang, J., Peng, Y., Zheng, L., He, C., Peng, S., Huang, Y., Wang, L., Liu, H., Feng, G., 2024. Chitosan-se engineered nanomaterial mitigates salt stress in plants by scavenging reactive oxygen species. J. Agric. Food Chem. 72 (1), 176–188. https://doi.org/10.1021/acs.jafc.3c06185.

Fares, A., Mahdy, A., Ahmed, G., 2024. Unraveling the mysteries of silver nanoparticles: synthesis, characterization, antimicrobial effects and uptake translocation in plant—a review. Planta 260 (1). https://doi.org/10.1007/s00425-024-04439-6.

Feng, Y., Kreslavski, V.D., Shmarev, A.N., Ivanov, A.A., Zharmukhamedov, S.K., Kosobryukhov, A., Yu, M., Allakhverdiev, S.I., Shabala, S., 2022. Effects of iron oxide nanoparticles (Fe_3O_4) on growth, photosynthesis, antioxidant activity and distribution of mineral elements in wheat (Triticum aestivum) plants. Plants 11 (14). https://doi.org/10.3390/plants11141894.

Fonseca, J.D.d.S., Wojciechowska, E., Kulesza, J., Barros, B. S., 2024. Carbon nanomaterials in seed priming: current possibilities. ACS Omega 9 (45), 44891–44906. https://doi.org/10.1021/acsomega.4c07230.

Freire, B.M., Lange, C.N., Cavalcanti, Y.T., Seabra, A.B., Batista, B.L., 2024. Ionomic profile of rice seedlings after foliar application of selenium nanoparticles. Toxics 12 (7). https://doi.org/10.3390/toxics12070482.

Gama, J.P.S., Campos, F.G., Riccardi, C.d.S., Boaro, C.S.F., 2025. Iron oxide nanoparticles for photosynthetic recovery in iron-deficient 'Micro-Tom' tomato plants. Environments—MDPI 12 (10). https://doi.org/10.3390/environments12100346.

Ge, Y., Shen, C., Wang, Y., Sun, Y.Q., Schimel, J.P., Gardea-Torresdey, J.L., Holden, P.A., 2018. Carbonaceous nanomaterials have higher effects on soybean rhizosphere prokaryotic communities during the reproductive growth phase than during vegetative growth. Environ. Sci. Technol. 52 (11), 6636–6646. https://doi.org/10.1021/acs.est.8b00937.

Gogos, A., Moll, J., Klingenfuss, F., Heijden, M., Irin, F., Green, M.J., Zenobi, R., Bucheli, T.D., 2016. Vertical transport and plant uptake of nanoparticles in a soil mesocosm experiment. J. Nanobiotechnol. 14 (1). https://doi.org/10.1186/s12951-016-0191-z.

Gong, J.Y., Zhou, X.L., Faraj, Y., Zou, L.B., Zhou, C.H., Xie, R., Wang, W., Liu, Z., Pan, D.W., Ju, X.J., Chu, L.

Y., 2024. Coumarin-based light-responsive composite nanochannel membranes for precise controlled release of pesticides. Adv. Funct. Mater. 34 (21). https://doi.org/10.1002/adfm.202314642.

Guidi, L., Lo Piccolo, E., Landi, M., 2019. Chlorophyll fluorescence, photoinhibition and abiotic stress: does it make any difference the fact to be a C3 or C4 species? Front. Plant Sci. 10. https://doi.org/10.3389/fpls.2019.00174.

Guimarães, G.G.F., Klaic, R., Giroto, A.S., Majaron, V.F., Avansi, W., Farinas, C.S., Ribeiro, C., 2018. Smart fertilization based on sulfur-phosphate composites: synergy among materials in a structure with multiple fertilization roles. ACS Sustain. Chem. Eng. 6 (9), 12187–12196. https://doi.org/10.1021/acssuschemeng.8b02511.

Guo, Z., Chen, Q., Liang, T., Zhou, B., Huang, S., Cao, X., Wang, X., Ding, Z., Tu, J., 2023. Functionalized carbon nano-enabled plant ROS signal engineering for growth/defense balance. Nano Today 53, 102045. https://doi.org/10.1016/j.nantod.2023.102045.

Hamed, S.M., El Tablawy, N.H., Mohamed, M.Y.A., Alammari, B.S., AbdElgawad, H., 2024. Accumulation and nano-ecotoxicological impact of cerium oxide nanoparticles on cyanobacteria: understanding photosynthesis, detoxification, and antioxidant responses. J. Environ. Chem. Eng. 12 (2). https://doi.org/10.1016/j.jece.2024.112134.

Han, J., Gu, L., Wen, J., Sun, Y., 2022. Inference of photosynthetic capacity parameters from chlorophyll a fluorescence is affected by redox state of PSII reaction centers. Plant Cell Environ. 45 (4), 1298–1314. https://doi.org/10.1111/pce.14271.

Handoko, R.N.S., Lin, S.Y., 2025. Integrating plant growth regulators and biostimulants to enhance resilient and sustainable raspberry and blackberry production. Sci. Hortic. 350. https://doi.org/10.1016/j.scienta.2025.114296.

Hasnain, M., Munir, N., Abideen, Z., Zulfiqar, F., Koyro, H. W., El-Naggar, A., Caçador, I., Duarte, B., Rinklebe, J., Yong, J.W.H., 2023. Biochar-plant interaction and detoxification strategies under abiotic stresses for achieving agricultural resilience: a critical review. Ecotoxicol. Environ. Saf. 249. https://doi.org/10.1016/j.ecoenv.2022.114408.

Haydar, M.S., Ghosh, S., Mandal, P., 2022. Application of iron oxide nanoparticles as micronutrient fertilizer in mulberry propagation. J. Plant Growth Regul. 41 (4), 1726–1746. https://doi.org/10.1007/s00344-021-10413-3.

Helal, M.I.D., El-Mogy, M.M., Khater, H.A., Fathy, M.A., Ibrahim, F.E., Li, Y.C., Tong, Z., Abdelgawad, K.F., 2023. A controlled-release nanofertilizer improves tomato growth and minimizes nitrogen consumption. Plants 12 (10). https://doi.org/10.3390/plants12101978.

Heyduk, K., Sciulla, S., Hennessy, B., Czymmek, M., McAssey, E.V., Kane, C., Kim, G.Y., Sogunle, I., Heublein, L., Sriram, D., MacNeill, B., Hren, M.T., Esque, T.C., Yoder, J.B., McKain, M.R., Smith, C.I., DeFalco, L.A., 2025. Cryptic CAM photosynthesis in Joshua tree (Yucca brevifolia, Y. jaegeriana). New Phytol. 248 (5), 2583–2598. https://doi.org/10.1111/nph.70437.

Hong, J., Wang, C., Wagner, D.C., Gardea-Torresdey, J.L., He, F., Rico, C.M., 2021. Foliar application of nanoparticles: mechanisms of absorption, transfer, and multiple impacts. Environ. Sci. Nano 8 (5), 1196–1210. https://doi.org/10.1039/d0en01129k.

Hosseinifard, M., Jurga, N., Brandmeier, J.C., Farka, Z., Hlaváček, A., Gorris, H.H., Grzyb, T., Ekner-Grzyb, A., 2024. Influence of surface modification and size of lanthanide-doped upconverting nanoparticles on wheat seedlings. Chemosphere 347. https://doi.org/10.1016/j.chemosphere.2023.140629.

Hou, W., Tränkner, M., Lu, J., Yan, J., Huang, S., Ren, T., Cong, R., Li, X., 2019. Interactive effects of nitrogen and potassium on photosynthesis and photosynthetic nitrogen allocation of rice leaves. BMC Plant Biol. 19 (1). https://doi.org/10.1186/s12870-019-1894-8.

Hu, Y., Zhang, Q., Garcia-Rojas, D., Ling, V., Masterson, C. M., Bi, Y., Xiao, Z., Guo, X., Villanova, J., Dunn, J., Colvin, V.L., 2022. Increasing the antioxidant capacity of ceria nanoparticles with catechol-grafted poly (ethylene glycol). J. Mater. Chem. B 10 (48), 10042–10053. https://doi.org/10.1039/d2tb00779g.

Hu, Z., Zhao, C., Qi, L., Feng, Y., Zhang, X., Lu, Y., Ying, R., Yin, A., Ji, W., 2023. Heavy metals can affect plant morphology and limit plant growth and photosynthesis processes. Agronomy 13 (10), 2601. https://doi.org/10.3390/agronomy13102601.

Hu, X., Zhang, L., Fu, H., Ma, X., Liu, J., Zhang, C., Duan, B., Guan, H., Wang, P., Zhou, Q., 2025. Effects of phosphorus deficiency on lipid composition and photosynthesis process in Zygophyllum xanthoxylum. BMC Plant Biol. 25 (1). https://doi.org/10.1186/s12870-025-07193-3.

Hu, J., Xiong, J., Zhao, Z., Wang, X., 2026. Carbon dots promote plant growth via coordinated regulation of nutrient uptake and photosynthesis: evidence from multivariate modeling. Bioresour. Technol. 440, 133515. https://doi.org/10.1016/j.biortech.2025.133515.

Huang, J.k., 2014. Climate change and agriculture: impact and adaptation. J. Integr. Agric. 13 (4), 657–659. https://doi.org/10.1016/S2095-3119(14)60752-8.

Huo, X., He, M., Qiao, J., Zhao, J., Yang, B., 2025. Regulatory effects of nano-carbon on poplar growth and rhizosphere soil organic carbon accumulation. Environ. Res. 278, 121628. https://doi.org/10.1016/j.envres.2025.121628.

Husted, S., Minutello, F., Pinna, A., Tougaard, S.L., Møs, P., Kopittke, P.M., 2023. What is missing to advance foliar fertilization using nanotechnology? Trends Plant Sci. 28 (1), 90–105. https://doi.org/10.1016/j.tplants.2022.08.017.

Johnson, M.P., 2016. Photosynthesis. Essays Biochem. 60 (3), 255–273. https://doi.org/10.1042/ebc20160016.

Joshi, A., Kaur, S., Dharamvir, K., Nayyar, H., Verma, G., 2018. Multi-walled carbon nanotubes applied through seed-priming influence early germination, root hair, growth and yield of bread wheat (Triticum aestivum L.). J. Sci. Food Agric. 98 (8), 3148–3160. https://doi.org/10.1002/jsfa.8818.

Juárez-Maldonado, A., Ortega-Ortíz, H., Morales-Díaz, A.B., González-Morales, S., Morelos-Moreno, Á., Cabrera-De la Fuente, M., Sandoval-Rangel, A., Cadenas-Pliego, G., Benavides-Mendoza, A., 2019. Nanoparticles and nanomaterials as plant biostimulants. Int. J. Mol. Sci. 20 (1). https://doi.org/10.3390/ijms20010162.

Kareem, H.A., Saleem, M.F., Saleem, S., Rather, S.A., Wani, S.H., Siddiqui, M.H., Alamri, S., Kumar, R., Gaikwad, N.B., Guo, Z., Niu, J., Wang, Q., 2022. Zinc oxide nanoparticles interplay with physiological and biochemical attributes in terminal heat stress alleviation in mungbean (Vigna radiata L.). Front. Plant Sci. 13. https://doi.org/10.3389/fpls.2022.842349.

Kaur, S., Garg, T., Joshi, A., Awasthi, A., Kumar, V., Kumar, A., 2024. Potential effects of metal oxide nanoparticles on leguminous plants: practical implications and future perspectives. Sci. Hortic. 331, 113146. https://doi.org/10.1016/j.scienta.2024.113146.

Kayoumu, M., Iqbal, A., Muhammad, N., Li, X., Li, L., Wang, X., Gui, H., Qi, Q., Ruan, S., Guo, R., Zhang, X., Song, M., Dong, Q., 2023. Phosphorus availability affects the photosynthesis and antioxidant system of contrasting low-P-tolerant cotton genotypes. Antioxidants 12 (2), 466. https://doi.org/10.3390/antiox12020466.

Khan, M.N., Fu, C., Li, J., Tao, Y., Li, Y., Hu, J., Chen, L., Khan, Z., Wu, H., Li, Z., 2023. Seed nanopriming: how do nanomaterials improve seed tolerance to salinity and drought? Chemosphere 310. https://doi.org/10.1016/j.chemosphere.2022.136911.

Khan, M.N., Ali, W., Grillo, R., Wu, H., Nie, L., Khan, M.N., Ali, W., Grillo, R., Wu, H., Nie, L., 2024. Carbon dots and single-walled carbon nanotubes enhances maize shading stress tolerance. bioRxiv, 2024.2007.2011.603111.

Khan, M.N., Li, Y., Mu, Y., Sultan, H., Baloch, A., Din, I., Fu, C., Li, J., Khan, Z., Kumar, S., Wu, H., Grillo, R., Nie, L., 2025. Recent advances in nano-enabled plant salt tolerance: methods of application, risk assessment, opportunities and future prospects. J. Integr. Agric. 24 (5), 1611–1630. https://doi.org/10.1016/j.jia.2024.05.028.

Klaic, R., Giroto, A.S., Guimarães, G.G.F., Plotegher, F., Ribeiro, C., Zangirolami, T.C., Farinas, C.S., 2018. Nanocomposite of starch-phosphate rock bioactivated for environmentally-friendly fertilization. Miner. Eng. 128, 230–237. https://doi.org/10.1016/j.mineng.2018.09.002.

Kobayashi, T., Nozoye, T., Nishizawa, N.K., 2019. Iron transport and its regulation in plants. Free Radic. Biol. Med. 133, 11–20. https://doi.org/10.1016/j.freeradbiomed.2018.10.439.

Kumar, N., Samota, S.R., Venkatesh, K., Tripathi, S.C., 2023. Global trends in use of nano-fertilizers for crop production: advantages and constraints – a review. Soil Tillage Res. 228. https://doi.org/10.1016/j.still.2023.105645.

Lahiani, M.H., Nima, Z.A., Villagarcia, H., Biris, A.S., Khodakovskaya, M.V., 2018. Assessment of effects of the long-term exposure of agricultural crops to carbon nanotubes. J. Agric. Food Chem. 66 (26), 6654–6662. https://doi.org/10.1021/acs.jafc.7b01863.

Leonardi, M., Caruso, G.M., Carroccio, S.C., Boninelli, S., Curcuruto, G., Zimbone, M., Allegra, M., Torrisi, B., Ferlito, F., Miritello, M., 2021. Smart nanocomposites of chitosan/alginate nanoparticles loaded with copper oxide as alternative nanofertilizers. Environ. Sci. Nano 8 (1), 174–187. https://doi.org/10.1039/d0en00797h.

Li, C., Wang, P., van der Ent, A., Cheng, M., Jiang, H., Lund Read, T., Lombi, E., Tang, C., de Jonge, M.D., Menzies, N. W., et al., 2018. Absorption of foliar-applied Zn in sunflower (Helianthus annuus): importance of the cuticle, stomata and trichomes. Ann. Bot. 123, 57–68.

Li, T., Lu, S., Yan, J., Bai, X., Gao, C., Liu, M., 2019. An environment-friendly fertilizer prepared by layer-by-layer self-assembly for pH-responsive nutrient release. ACS Appl. Mater. Interfaces 11 (11), 10941–10950. https://doi.org/10.1021/acsami.9b01425.

Li, X., Ghanizadeh, H., Han, Z., Wang, Q., Li, F., Qiu, Y., Zhang, Y., Chen, X., Liu, J., Wang, A., 2023a. Metabolic profile and physiological mechanisms underlying the promoting effects of TiO_2NPs on the photosynthesis and growth of tomato. Sci. Hortic. 322, 112394. https://doi.org/10.1016/j.scienta.2023.112394.

Li, Y., Ai, Z., Mu, Y., Zhao, T., Zhang, Y., Li, L., Huang, Z., Nie, L., Khan, M.N., 2023b. Rice yield penalty and quality deterioration is associated with failure of nitrogen uptake from regreening to panicle initiation stage under salinity. Front. Plant Sci. 14. https://doi.org/10.3389/fpls.2023.1120755.

Li, A., You, T., Pang, X., Wang, Y., Tian, L., Li, X., Liu, Z., 2024. Structural basis for an early stage of the photosystem II repair cycle in Chlamydomonas reinhardtii. Nat. Commun. 15 (1). https://doi.org/10.1038/s41467-024-49532-2.

Li, K., Liu, D.N., Li, L.Y., Gao, Y., Gao, W.J., Chen, B.W., Luo, F., Yao, Y., 2025a. Effects of nitrogen application amount on nitrogen distribution and photosynthesis in tea leaves. Front. Plant Sci. 16. https://doi.org/10.3389/fpls.2025.1575317.

Li, Y., Tao, Y., Xu, W., Wu, H., Li, G., Lin, Y., Gu, J., Li, F., Wu, H., Pablo Giraldo, J., Li, Z., 2025b. Mn_3O_4 nanoparticles maintain ROS homeostasis to modulate stomatal aperture to improve cotton drought tolerance. Environ. Sci. Nano 12 (2), 1126–1137. https://doi.org/10.1039/d4en00963k.

Li, Y., Xue, Y., Guan, Z., Wang, Z., Hou, D., Zhao, T., Lu, X., Qi, Y., Yanbo, H., Liu, J., Lin, L., Sultan, H., Guo, X., Ai, Z., He, A., 2025c. Salt stress responses of different rice varieties at panicle initiation: agronomic traits, photosynthesis, and antioxidants. Plants 14 (15), 2278. https://doi.org/10.3390/plants14152278.

Liu, Y., Guo, R., Zhang, S., Sun, Y., Wang, F., 2022. Uptake and translocation of nano/microplastics by rice seedlings: evidence from a hydroponic experiment. J. Hazard Mater. 421, 126700. https://doi.org/10.1016/j.jhazmat.2021.126700.

Liu, X., Qiao, Y., Zhou, W., Dong, W., Gu, L., 2023a. Determinants of photochemical characteristics of the photosynthetic electron transport chain of maize. Front. Plant Sci. 14. https://doi.org/10.3389/fpls.2023.1279963.

Liu, Y., Liu, R., Lin, C., Yu, S., Nie, Y., Zhang, H., Jia-Qi, L., Pan, C., Zhu, W., Diao, J., Zhou, Z., 2023b. Improvement by application of three nanomaterials on flavor quality and physiological and antioxidant properties of tomato and their comparison. Plant Physiol. Biochem. 201, 107834. https://doi.org/10.1016/j.plaphy.2023.107834.

Liu, Y., Liu, D., Han, X., Chen, Z., Li, M., Jiang, L., Zeng, J., 2024. Magnesium-doped carbon quantum dot nanomaterials alleviate salt stress in rice by scavenging reactive oxygen species to increase photosynthesis. ACS Nano 18 (45), 31188–31203. https://doi.org/10.1021/acsnano.4c09001.

Liu, J., Sun, S., Lv, R., Lin, S., Golubev, Y.A., Wang, K., Cao, R., Zeng, Y., 2025a. Zinc-based organophyllosilicate nanosheets as novel carbonic anhydrase-like nanozyme for efficient CO_2 fixation. Sep. Purif. Technol. 354. https://doi.org/10.1016/j.seppur.2024.128979.

Liu, J., Gang, H., Dong, Q., Wang, H., Wang, X., Shao, K., Fu, C., Hong, J., Huo, J., 2025b. Carbon quantum dots from fallen leaves of Lonicera caerulea L.: an innovative plant growth promoter and fruit quality enhancer. Environ. Res. 274, 121350. https://doi.org/10.1016/j.envres.2025.121350.

Lu, X., Croft, H., Chen, J.M., Luo, Y., Ju, W., 2022. Estimating photosynthetic capacity from optimized Rubisco-chlorophyll relationships among vegetation types and under global change. Environ. Res. Lett. 17 (1). https://doi.org/10.1088/1748-9326/ac444d.

Lu, K., Abouzeid, R., Wu, Q., Chen, Q., Liu, S., 2024. Hydrogel nanocomposite based slow-release urea fertilizer: formulation, structure, and release behavior. Giant 18, 100270. https://doi.org/10.1016/j.giant.2024.100270.

Lu, Z., Ren, T., Li, Y., Cakmak, I., Lu, J., 2025. Nutrient limitations on photosynthesis: from individual to combinational stresses. Trends Plant Sci. 30 (8), 872–885. https://doi.org/10.1016/j.tplants.2025.03.006.

Mallik, T., Ghosh, S., Ekka, D., 2025. N-doped graphene oxide nanomaterial: synthesis and application as controlled-release of urea for advancement in modern agriculture. Discover Nano 20 (1). https://doi.org/10.1186/s11671-025-04194-x.

Mehta, S., Thakur, A., Kurbah, I., Chauhan, N., Thakur, R., 2024. Nanotechnology in plant nutrition: ensuring sustainable agriculture through nanofertilizers. J. Plant Nutr. Soil Sci. 187 (5), 589–603. https://doi.org/10.1002/jpln.202300288.

Mellis, E.V., Ramos, L.F., Ferreira, A.J., Andrade, R.P., Teixeira, L.A.J., Otto, R., Ferraz-Almeida, R., 2024. Micronized Zn oxide on carbonic anhydrase activity, health, and yield of ratoon sugarcane under tropical conditions. Sugar Tech 26 (6), 1567–1579. https://doi.org/10.1007/s12355-024-01424-x.

Nam, H.I., Shahzad, Z., Dorone, Y., Clowez, S., Zhao, K., Bouain, N., Cho, H., Rhee, S.Y., Rouached, H., 2021. Interdependent iron and phosphorus availability controls photosynthesis through retrograde signaling. bioRxiv. https://doi.org/10.1101/2021.02.11.430802.

Nandini, T., Sudhalakshmi, C., Sivakumar, K., Parameswari, E., Thangamani, C., 2025. A review-chitosan nanoparticles towards enhancing nutrient use efficiency in crops. Int. J. Biol. Macromol. 306 (Pt 2), 141433. https://doi.org/10.1016/j.ijbiomac.2025.141433.

Neto, M.E., Britt, D.W., Lara, L.M., Cartwright, A., Dos Santos, R.F., Inoue, T.T., Batista, M.A., 2020. Initial development of corn seedlings after seed priming with nanoscale synthetic zinc oxide. Agronomy 10 (2). https://doi.org/10.3390/agronomy10020307.

Newkirk, G.M., de Allende, P., Jinkerson, R.E., Giraldo, J.P., 2021. Nanotechnology approaches for chloroplast biotechnology advancements. Front. Plant Sci. 12. https://doi.org/10.3389/fpls.2021.691295.

Panda, A., Fatnani, D., Parida, A.K., 2025. Uptake, impact, adaptive mechanisms, and phytoremediation of heavy metals by plants: role of transporters in heavy metal sequestration. Plant Physiol. Biochem. 221. https://doi.org/10.1016/j.plaphy.2025.109578.

Pérez-de-Luque, A., 2017. Interaction of nanomaterials with plants: what do we need for real applications in agriculture? Front. Environ. Sci. 5. https://doi.org/10.3389/fenvs.2017.00012.

Prasad, T.N.V.K.V., Swethasree, M., Satisha, G.C., Nirmal Kumar, A.R., Sudhakar, P., Ravindra Reddy, B., Saritha, M., Sabitha, N., Bhaskar Reddy, B.V., Rajasekhar, P., Prasanthi, L., Girish, B.P., Choudhury, S.R., 2023. Nanoparticulate silica internalization and its effect on the growth and yield of groundnut (Arachis hypogaea L.). Environ. Sci. Technol. 57 (14), 5881–5890. https://doi.org/10.1021/acs.est.3c00327.

Pudhuvai, B., Koul, B., Das, R., Shah, M.P., 2025. Nano-fertilizers (NFs) for resurgence in nutrient use efficiency (NUE): a sustainable agricultural strategy. Curr. Pollut. Rep. 11 (1). https://doi.org/10.1007/s40726-024-00331-9.

Purkayastha, K.D., Bhattacharya, S.S., Gogoi, N., 2025. Green synthesized Zn-based mono and bi-metallic nanoparticles for improved micronutrient delivery: a futuristic study in hydroponic crop culture system. Chemosphere 380. https://doi.org/10.1016/j.chemosphere.2025.144427.

Rahale, C.S., Lakshmanan, A., Sumithra, M.G., Kumar, E.R., 2021. Humic acid involved chelation of ZnO nanoparticles for enhancing mineral nutrition in plants. Solid State Commun. 333. https://doi.org/10.1016/j.ssc.2021.114355.

Rahman, M.A., Yeasmin, M., Lu, Y., Islam, M.A., Rahman, M.M., Kormoker, T., Pan, D., 2025. Nutrient and trace metal dynamics driving eutrophication in the Cox's Bazar Coast of the Bay of Bengal, Bangladesh. Mar. Pollut. Bull. 221. https://doi.org/10.1016/j.marpolbul.2025.118563.

Raliya, R., Franke, C., Chavalmane, S., Nair, R., Reed, N., Biswas, P., 2016. Quantitative understanding of nanoparticle uptake in watermelon plants. Front. Plant Sci. 7 (2016). https://doi.org/10.3389/fpls.2016.01288.

Ranjan, S., Sow, S., Sadhu, S., Sahoo, R., Nath, D., Samanta, D., Nazim, M., Kumar, N., Rana, L., 2025. Unlocking higher productivity and nutrient use efficiency in cereals through nano-fertilizers. Crop Design 4 (4), 100119. https://doi.org/10.1016/j.cropd.2025.100119.

Rashid, M.I., Shah, G.A., Sadiq, M., Amin, N.u., Ali, A.M., Ondrasek, G., Shahzad, K., 2023. Nanobiochar and copper oxide nanoparticles mixture synergistically increases soil nutrient availability and improves wheat production. Plants 12 (6). https://doi.org/10.3390/plants12061312.

Rui, M., Ma, C., Hao, Y., Guo, J., Rui, Y., Tang, X., Zhao, Q., Fan, X., Zhang, Z., Hou, T., Zhu, S., 2016. Iron oxide nanoparticles as a potential iron fertilizer for peanut (Arachis hypogaea). Front. Plant Sci. 7 (2016). https://doi.org/10.3389/fpls.2016.00815.

Rumeau, D., Bécuwe-Linka, N., Beyly, A., Carrier, P., Cuiné, S., Genty, B., Medgyesy, P., Horvath, E., Peltier, G., 2004. Increased zinc content in transplastomic tobacco plants expressing a polyhistidine-tagged Rubisco large subunit. Plant Biotechnol. J. 2 (5), 389–399. https://doi.org/10.1111/j.1467-7652.2004.00083.x.

Saeedi, R., Seyedi, A., Esmaeilizadeh, M., Seyedi, N., Morteza Zahedi, S., Malekzadeh, M.R., 2024. Improving the performance of the photosynthetic apparatus of Citrus sinensis with the use of chitosan-selenium nanocomposite (CS + Se NPs) under salinity stress. BMC Plant Biol. 24 (1). https://doi.org/10.1186/s12870-024-05462-1.

Sahil, Chaudhary, A.A., Sorout, S., Yadav, K., Ali, M.A.M., Boufahja, F., Kumar, V., Jain, D., Verma, K.S., 2025. Auxin mediated synthesis of gold nanoparticles: a novel approach to enhance shoot and root growth in pearl millet. Front. Plant Sci. 16. https://doi.org/10.3389/fpls.2025.1621007.

Salter, W.T., Gilbert, M.E., Buckley, T.N., 2018. A multiplexed gas exchange system for increased throughput of photosynthetic capacity measurements. Plant Methods 14 (1). https://doi.org/10.1186/s13007-018-0347-y.

Santana, I., Jeon, S.J., Kim, H.I., Islam, M.R., Castillo, C., Garcia, G.F.H., Newkirk, G.M., Giraldo, J.P., 2022. Targeted carbon nanostructures for chemical and gene delivery to plant chloroplasts. ACS Nano 16 (8), 12156–12173. https://doi.org/10.1021/acsnano.2c02714.

dos Santos, L.A., Batista, B.L., Lobato, A.K.d.S., 2023. 24-Epibrasinolide delays chlorophyll degradation and stimulates the photosynthetic machinery in magnesium-stressed soybean plants. J. Plant Growth Regul. 42 (1), 183–198. https://doi.org/10.1007/s00344-021-10539-4.

Santos-Merino, M., Torrado, A., Davis, G.A., Röttig, A., Bibby, T.S., Kramer, D.M., Ducat, D.C., 2021. Improved photosynthetic capacity and photosystem I oxidation via heterologous metabolism engineering in cyanobacteria. Proc. Natl. Acad. Sci. 118 (11). https://doi.org/10.1073/pnas.2021523118.

Sardoiwala, M.N., Kaundal, B., Choudhury, S.R., 2018. Toxic impact of nanomaterials on microbes, plants and animals. Environ. Chem. Lett. 16 (1), 147–160. https://doi.org/10.1007/s10311-017-0672-9.

Sarwar, N., Wasaya, A., Yasir, T.A., Mubeen, K., Hussain, M., Javaid, M.M., Mehboob, K., El Sabagh, A., 2024. Potassium supplementation mitigates flooding stress by regulating antioxidant enzymes, photosynthetic performance, yield attributes, and lint quality in cotton plants. J. Plant Nutr. 47 (19), 3754–3766. https://doi.org/10.1080/01904167.2024.2384565.

Schmidt, S.B., Jensen, P.E., Husted, S., 2016. Manganese deficiency in plants: the impact on photosystem II. Trends Plant Sci. 21 (7), 622–632. https://doi.org/10.1016/j.tplants.2016.03.001.

Sebastian, A., Nangia, A., Prasad, M.N.V., 2017. Carbon-bound iron oxide nanoparticles prevent calcium-induced iron deficiency in oryza sativa l. J. Agric. Food Chem. 65 (3), 557–564. https://doi.org/10.1021/acs.jafc.6b04634.

Selli, D., Motta, S., Di Valentin, C., 2019. Impact of surface curvature, grafting density and solvent type on the PEGylation of titanium dioxide nanoparticles. J. Colloid Interface Sci. 555, 519–531. https://doi.org/10.1016/j.jcis.2019.07.106.

Shah, I.H., Jinhui, W., Li, X., Hameed, M.K., Manzoor, M.A., Li, P., Zhang, Y., Niu, Q., Chang, L., 2024. Exploring the role of nitrogen and potassium in photosynthesis implications for sugar: accumulation and translocation in horticultural crops. Sci. Hortic. 327. https://doi.org/10.1016/j.scienta.2023.112832.

Sharma, D., Sharma, A., Gautam, H.R., 2025. Green synthesized nanoparticles for disease management in vegetable crops: a review. Plant Nano Biol. 13. https://doi.org/10.1016/j.plana.2025.100179.

Shen, M., Liu, S., Jiang, C., Zhang, T., Chen, W., 2023. Recent advances in stimuli-response mechanisms of nano-enabled controlled-release fertilizers and pesticides. Eco-Environ. Health 2 (3), 161–175. https://doi.org/10.1016/j.eehl.2023.07.005.

Shomali, A., Das, S., Sarraf, M., Johnson, R., Janeeshma, E., Kumar, V., Aliniaeifard, S., Puthur, J.T., Hasanuzzaman, M., 2024. Modulation of plant photosynthetic processes during metal and metalloid stress, and strategies for manipulating photosynthesis-related traits. Plant Physiol. Biochem. 206, 108211. https://doi.org/10.1016/j.plaphy.2023.108211.

Singh, P.M., Tiwari, A., Maity, D., Saha, S., 2022. Recent progress of nanomaterials in sustainable agricultural applications. J. Mater. Sci. 57 (24), 10836–10862. https://doi.org/10.1007/s10853-022-07259-9.

Singh, K.M., Jha, A.B., Dubey, R.S., Sharma, P., 2025. Nanoparticle-mediated mitigation of salt stress-induced oxidative damage in plants: insights into signaling, gene expression, and antioxidant mechanisms. Environ. Sci. Nano 12 (6), 2983–3017. https://doi.org/10.1039/d5en00174a.

Song, C., Ye, F., Zhang, H., Hong, J., Hua, C., Wang, B., Chen, Y., Ji, R., Zhao, L., 2019. Metal(loid) oxides and metal sulfides nanomaterials reduced heavy metals uptake in soil cultivated cucumber plants. Environ. Pollut. 255 (Pt 3), 113354. https://doi.org/10.1016/j.envpol.2019.113354.

Song, J., Sun, Z., Saud, S., Fahad, S., Nawaz, T., 2025. Exploring the deleterious effects of heavy metal cadmium on antioxidant defense and photosynthetic pathways in higher plants. Plant Stress 15, 100716. https://doi.org/10.1016/j.stress.2024.100716.

Song, F., Qi, Y., Huang, J., Guo, Z., Li, Y., Wen, D., 2026. Plant drought stress: physiological, biochemical and molecular mechanisms. Plant Stress 19, 101153. https://doi.org/10.1016/j.stress.2025.101153.

Stinziano, J.R., Morgan, P.B., Lynch, D.J., Saathoff, A.J., McDermitt, D.K., Hanson, D.T., 2017. The rapid A–Ci response: photosynthesis in the phenomic era. Plant Cell Environ. 40 (8), 1256–1262. https://doi.org/10.1111/pce.12911.

Suganami, M., Suzuki, Y., Kondo, E., Nishida, S., Konno, S., Makino, A., 2020. Effects of overproduction of Rubisco activase on Rubisco content in transgenic rice grown at different N levels. Int. J. Mol. Sci. 21 (5), 1626. https://doi.org/10.3390/ijms21051626.

Suliman, A.A., El-Dewiny, C.Y., Soliman, M.K.Y., Salem, S. S., 2025. Investigation of the effects of applying bio-magnesium oxide nanoparticle fertilizer to Moringa oleifera plants on the chemical and vegetative properties of the plants' leaves. Biotechnol. J. 20 (3). https://doi.org/10.1002/biot.202400536.

Sun, D., Hussain, H.I., Yi, Z., Rookes, J.E., Kong, L., Cahill, D.M., 2016. Mesoporous silica nanoparticles enhance seedling growth and photosynthesis in wheat and lupin. Chemosphere 152, 81–91. https://doi.org/10.1016/j.chemosphere.2016.02.096.

Sutulienė, R., Brazaitytė, A., Małek, S., Jasik, M., Samuolienė, G., 2023. Response of oxidative stress and antioxidant system in pea plants exposed to drought and boron nanoparticles. Antioxidants 12 (2), 528. https://doi.org/10.3390/antiox12020528.

Therby-Vale, R., Lacombe, B., Rhee, S.Y., Nussaume, L., Rouached, H., 2022. Mineral nutrient signaling controls photosynthesis: focus on iron deficiency-induced chlorosis. Trends Plant Sci. 27 (5), 502–509. https://doi.org/10.1016/j.tplants.2021.11.005.

Thirumurugan, N.K., Velu, G., Murugaiyan, S., Maduraimuthu, D., Ponnuraj, S., Sharmila, D.J., Subramanian, K.S., 2025. Nano-biofertilizers: utilizing nanopolymers as coating matrix—a comprehensive review. Biofabrication 17 (1). https://doi.org/10.1088/1758-5090/ad94a8.

Tighe-Neira, R., Carmora, E., Recio, G., Nunes-Nesi, A., Reyes-Diaz, M., Alberdi, M., Rengel, Z., Inostroza-Blancheteau, C., 2018. Metallic nanoparticles influence the structure and function of the photosynthetic apparatus in plants. Plant Physiol. Biochem. 130, 408–417. https://doi.org/10.1016/j.plaphy.2018.07.024.

Tikhonov, A.N., 2024. The cytochrome b 6 f complex: plastoquinol oxidation and regulation of electron transport in chloroplasts. Photosynth. Res. 159 (2–3), 203–227. https://doi.org/10.1007/s11120-023-01034-w.

Tripathi, D.K., Shweta, Singh, S., Singh, S., Pandey, R., Singh, V.P., Sharma, N.C., Prasad, S.M., Dubey, N.K., Chauhan, D.K., 2017. An overview on manufactured nanoparticles in plants: uptake, translocation, accumulation and phytotoxicity. Plant Physiol. Biochem. 110, 2–12. https://doi.org/10.1016/j.plaphy.2016.07.030.

Tryfon, P., Sperdouli, I., Adamakis, I.D.S., Mourdikoudis, S., Moustakas, M., Dendrinou-Samara, C., 2023. Impact of coated zinc oxide nanoparticles on photosystem II of tomato plants. Materials 16 (17). https://doi.org/10.3390/ma16175846.

Tu, X., Wang, J., Liu, X., Fallah, N., Uwiragiye, Y., Huang, Y., Cai, Z., Chang, S.X., Cheng, Y., Müller, C., 2026. Effects of warming and elevated CO_2 on gross nitrogen dynamics and N_2O emissions disappeared under combined treatment in waterlogged paddy soils. Agric. Ecosyst. Environ. 396. https://doi.org/10.1016/j.agee.2025.109989.

Wang, G., Zeng, F., Peng, S., Sun, B., Wang, Q., Wang, J., 2022. Effects of reduced chlorophyll content on photosystem functions and photosynthetic electron transport rate in rice leaves. J. Plant Physiol. 272, 153669. https://doi.org/10.1016/j.jplph.2022.153669.

Wang, R., Zhang, M., Chen, T., Shen, W., Dai, J., Zhang, H., Zhang, H., 2025. Enhanced leaf photosynthesis, grain yield, rice quality and aroma characteristics in rice grains (Oryza sativa L.) with foliar application of selenium nanoparticles. Plant Physiol. Biochem. 223, 109812. https://doi.org/10.1016/j.plaphy.2025.109812.

Weng, Y., Bai, X., Kang, M., Huang, Y., Ji, Y., Wang, H., Hua, Z., 2024. Comparative analysis of chemically and green synthesized titanium dioxide nanoparticles for the regulation of photosynthesis in Lactuca sativa L. Environ. Sci. Nano 11 (1), 161–174. https://doi.org/10.1039/d3en00589e.

Wijewardene, I., Shen, G., Zhang, H., 2021. Enhancing crop yield by using Rubisco activase to improve photosynthesis under elevated temperatures. Stress Biol. 1 (1). https://doi.org/10.1007/s44154-021-00002-5.

Wu, C., Shen, X., Feng, W., Li, P., Chen, Y., 2025. Agricultural nanotechnology. Coord. Chem. Rev. 543, 216906. https://doi.org/10.1016/j.ccr.2025.216906.

Xia, X., Shi, B., Wang, L., Liu, Y., Zou, Y., Zhou, Y., Chen, Y., Zheng, M., Zhu, Y., Duan, J., Guo, S., Jang, H.W., Miao, Y., Fan, K., Bai, F., Tao, W., Zhao, Y., Yan, Q., Cheng, G., Liu, H., Jiao, Y., Liu, S., Huang, Y., Ling, D., Kang, W., Xue, X., Cui, D., Huang, Y., Cui, Z., Sun, X., Qian, Z., Gu, Z., Han, G., Yang, Z., Leong, D.T., Wu, A., Liu, G., Qu, X., Shen, Y., Wang, Q., Lowry, G. V., Wang, E., Liang, X.-J., Gardea-Torresdey, J., Chen, G., Parak, W.J., Weiss, P.S., Zhang, L., Stenzel, M.M., Fan, C., Bush, A.I., Zhang, G., Grof, C.P. L., Wang, X., Galbraith, D.W., Tang, B.Z., Offler, C.E., Patrick, J.W., Song, C.-P., 2021. From mouse to mouse-ear cress: nanomaterials as vehicles in plant biotechnology. Exploration 1 (1), 9–20. https://doi.org/10.1002/exp.20210002.

Xiao, D., Jiang, M., Luo, X., Liu, S., Li, J., Chen, Z., Li, S., 2021. Sustainable carbon dot-based AIEgens: promising light-harvesting materials for enhancing photosynthesis. ACS Sustain. Chem. Eng. 9 (11), 4139–4145. https://doi.org/10.1021/acssuschemeng.0c09348.

Xu, X., Bai, B., Wang, H., Suo, Y., 2017. A near-infrared and temperature-responsive pesticide release platform through core-shell polydopamine@ PNIPAm nanocomposites. ACS Appl. Mater. Interfaces 9 (7), 6424–6432. https://doi.org/10.1021/acsami.6b15393.

Xu, K., Guo, Y., Xing, C., Fu, R., Zou, B., Liu, R., Cai, L., Yan, J., Wu, X.L., Cai, M., 2024. Graphitic carbon nitride nanosheets mitigate cadmium toxicity in Glycine max L. by promoting cadmium retention in root and improving photosynthetic performance. J. Environ. Sci. (China) 139, 543–555. https://doi.org/10.1016/j.jes.2023.08.027.

Xu, Y., Zhang, Z., Guo, Q., Feng, H., Qin, F., Zhang, G., Ding, H., 2025. Effects of nitrogen allocation and photosynthetic proteins response in peanut leaves on photosynthesis under conditions of water scarcity and nitrogen deficiency. BMC Plant Biol. 25 (1). https://doi.org/10.1186/s12870-025-07232-z.

Yan, G., Jin, H., Yin, C., Hua, Y., Huang, Q., Zhou, G., Xu, Y., He, Y., Liang, Y., Zhu, Z., 2023. Comparative effects of silicon and silicon nanoparticles on the antioxidant system and cadmium uptake in tomato under cadmium stress. Sci. Total Environ. 904, 166819. https://doi.org/10.1016/j.scitotenv.2023.166819.

Yang, Z., Tian, J., Wang, Z., Feng, K., Ouyang, Z., Zhang, L., Yan, X., 2023. Coupled soil water stress and environmental effects on changing photosynthetic traits in wheat and maize. Agric. Water Manag. 282, 108246. https://doi.org/10.1016/j.agwat.2023.108246.

Yang, Y., Yang, X., Dai, K., He, S., Zhao, W., Wang, S., Zhou, Z., Hu, W., 2024. Nanoceria-induced variations in leaf anatomy and cell wall composition drive the increase in mesophyll conductance of salt-stressed cotton leaves. Plant Physiol. Biochem. 216. https://doi.org/10.1016/j.plaphy.2024.109111.

You, Y., Kerner, P., Shanmugam, S., Khodakovskaya, M.V., 2023. Emerging investigator series: differential effects of carbon nanotubes and graphene on the tomato rhizosphere microbiome. Environ. Sci. Nano 10 (6), 1570–1584. https://doi.org/10.1039/d2en01026g.

Zahoor, Faqir, Y., Dai, W., Lai, Q., Ma, J., Tan, C., 2025. Utilization of chitosan for the controlled-release fertilizer and its impact on plant and soil. Mater. Today Commun. 47, 113213. https://doi.org/10.1016/j.mtcomm.2025.113213.

Zain, M., Ma, H., Rahman, S.U., Nuruzzaman, M., Chaudhary, S., Azeem, I., Mehmood, F., Duan, A., Sun, C., 2024. Nanotechnology in precision agriculture: advancing towards sustainable crop production. Plant Physiol. Biochem. 206, 108244. https://doi.org/10.1016/j.plaphy.2023.108244.

Zaman, W., Ayaz, A., Park, S.J., 2025. Nanomaterials in agriculture: a pathway to enhanced plant growth and abiotic stress resistance. Plants 14 (5), 716. https://doi.org/10.3390/plants14050716.

Zeng, Y., Molnárová, M., Motola, M., 2024. Metallic nanoparticles and photosynthesis organisms: comprehensive review from the ecological perspective. J. Environ. Manag. 358, 120858. https://doi.org/10.1016/j.jenvman.2024.120858.

Zha, Q., Xi, X., Jiang, A., Wang, S., Tian, Y., 2016. Changes in the protective mechanism of photosystem II and molecular regulation in response to high temperature stress in grapevines. Plant Physiol. Biochem. 101, 43–53. https://doi.org/10.1016/j.plaphy.2016.01.024.

Zhang, Z., Cao, B., Chen, Z., Xu, K., 2022. Grafting enhances the photosynthesis and nitrogen absorption of tomato plants under low-nitrogen stress. J. Plant Growth Regul. 41 (4), 1714–1725. https://doi.org/10.1007/s00344-021-10414-2.

Zhang, Y., Li, L., Dai, H., Kong, X., Rahman, M., Zhang, B., Zhang, Z., Zhou, Y., Liu, Q., 2025. Iron oxide nanoparticles (FeO-NPs) mitigate salt stress in peanut seedlings by enhancing photosynthesis, osmoregulation, and antioxidant activity. Plant Physiol. Biochem. 227, 110206. https://doi.org/10.1016/j.plaphy.2025.110206.

Zhao, L., Lu, L., Wang, A., Zhang, H., Huang, M., Wu, H., Xing, B., Wang, Z., Ji, R., 2020. Nano-Biotechnology in agriculture: use of nanomaterials to promote plant growth and stress tolerance. J. Agric. Food Chem. 68 (7), 1935–1947. https://doi.org/10.1021/acs.jafc.9b06615.

Zhao, L., Wang, W., Fu, X., Liu, A., Cao, J., Liu, J., 2022. Graphene oxide, a novel nanomaterial as soil water retention agent, dramatically enhances drought stress tolerance in soybean plants. Front. Plant Sci. 13. https://doi.org/10.3389/fpls.2022.810905.

Zhao, W., Wu, Z., Amde, M., Zhu, G., Wei, Y., Zhou, P., Zhang, Q., Song, M., Tan, Z., Zhang, P., Rui, Y., Lynch, I., 2023. Nanoenabled enhancement of plant tolerance to heat and drought stress on molecular response. J. Agric. Food Chem. 71 (51), 20405–20418. https://doi.org/10.1021/acs.jafc.3c04838.

Zhu, Z.J., Wang, H., Yan, B., Zheng, H., Jiang, Y., Miranda, O.R., Rotello, V.M., Xing, B., Vachet, R.W., 2012. Effect of surface charge on the uptake and distribution of gold nanoparticles in four plant species. Environ. Sci. Technol. 46 (22), 12391–12398. https://doi.org/10.1021/es301977w.

Zhu, Y., Cao, Y., Fu, B., Wang, C., Shu, S., Zhu, P., Wang, D., Xu, H., Zhong, N., Cai, D., 2024. Waste milk humification product can be used as a slow release nano-fertilizer. Nat. Commun. 15 (1). https://doi.org/10.1038/s41467-023-44422-5.

Zhuang, G., Mai, Y., Ren, Y., Zhang, M., Zhang, N., Xiong, Z., Mo, Z., 2025. Titanium dioxide nanoparticles regulate the growth, antioxidant response, and photosynthetic pigments in fragrant rice seedlings under different light and nitrogen levels. J. Plant Growth Regul. 44 (5), 2213–2234. https://doi.org/10.1007/s00344-024-11539-w.

Zia-ur-Rehman, M., Anayatullah, S., Irfan, E., Hussain, S.M., Rizwan, M., Sohail, M.I., Jafir, M., Ahmad, T., Usman, M., Alharby, H.F., 2023. Nanoparticles assisted regulation of oxidative stress and antioxidant enzyme system in plants under salt stress: a review. Chemosphere 314. https://doi.org/10.1016/j.chemosphere.2022.137649.

Zimbovskaya, M.M., Polyakov, A.Y., Volkov, D.S., Kulikova, N.A., Lebedev, V.A., Pankratov, D.A., Konstantinov, A.I., Parfenova, A.M., Zhilkibaev, O.T., Perminova, I.V., 2020. Foliar application of humic-stabilized nanoferrihydrite resulted in an increase in the content of iron in wheat leaves. Agronomy 10 (12). https://doi.org/10.3390/agronomy10121891.

Zou, C., Lu, T., Wang, R., Xu, P., Jing, Y., Wang, R., Xu, J., Wan, J., 2022. Comparative physiological and metabolomic analyses reveal that Fe_3O_4 and ZnO nanoparticles alleviate Cd toxicity in tobacco. J. Nanobiotechnol. 20 (1). https://doi.org/10.1186/s12951-022-01509-3.

Zsiros, O., Nagy, V., Párducz, Á., Nagy, G., Ünnep, R., El-Ramady, H., Prokisch, J., Lisztes-Szabó, Z., Fári, M., Csajbók, J., Tóth, S.Z., Garab, G., Domokos-Szabolcsy, É., 2019. Effects of selenate and red Se-nanoparticles on the photosynthetic apparatus of Nicotiana tabacum. Photosynth. Res. 139 (1–3), 449–460. https://doi.org/10.1007/s11120-018-0599-4.

Zuo, Z., Sun, L., Wang, T., Miao, P., Zhu, X., Liu, S., Song, F., Mao, H., Li, X., 2017. Melatonin improves the photosynthetic carbon assimilation and antioxidant capacity in wheat exposed to nano-ZnO stress. Molecules 22 (10), 1727. https://doi.org/10.3390/molecules22101727.

CHAPTER

12

Photocatalytic nanomaterials for improving photosynthetic light use efficiency

Aditi[1], *Ayushi Kumari*[1], *Subham Preetam*[2] *and Sumira Malik*[1,3,4]

[1]Amity Institute of Biotechnology, Amity University, Ranchi, Jharkhand, India; [2]Department of Chemical Engineering, School of Engineering, Monash University Malaysia, Bandar Sunway, Selangor, Malaysia; [3]University Centre for Research & Development (UCRD), Chandigarh University, Mohali, Punjab, India; [4]School of Basic and Applied Sciences, K. R. Mangalam University, Gurgaon, Haryana, India

1. Introduction

Increased food productivity is required in turn calls for improved food quality control. As a result, nanobiotechnology has provided a viable remedy for this pressing problem, improving plant survival rates and growth yield. The Nano-fertilizers, nano-sensors and nano-pesticides have strong applications in the agricultural sector, which can lead to sustainability since the nanomaterials have been recognized as a safe delivery system and have diverse, distinct potentials (Fig. 12.1). Nanotechnology has proved to be among a number of potential technologies that can enhance the productivity of agriculture with the aid of effective pesticides, herbicides, nano-fertilizers, controlling soil nutrition and the remediation of wastewater (Khan et al., 2022). The abundance of engineered nanomaterials (ENMs) has had real and plausible potential in the agricultural and farm/plant productivity fields. Nevertheless, there appears to be significantly more that is yet to be established to strengthen scientific knowledge in order to generate another green revolution in the coming years, to the good of global food security, due to the presence of unfavorable environmental conditions and rising population in third-world nations. Since agricultural products are necessary to sustain life on planet Earth, recent developments in NFs can be considered an attempt to balance accuracy in agriculture with ecological and economic viability. It should be noted that the global green revolution led to improved noted that the global green revolution led to increased food grain production at the expense of overuse of synthetic fertilizers and pesticides, which had a

Nanotechnology in Photosynthesis
https://doi.org/10.1016/B978-0-443-44376-3.00012-4

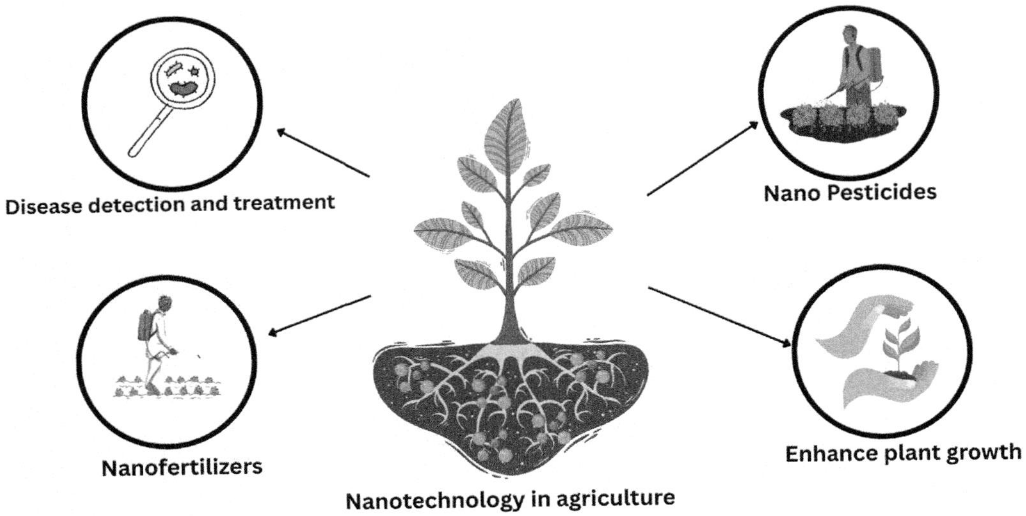

FIGURE 12.1 Application of nanotechnology in agriculture.

devastating effect on our ecosystem. Therefore, both the ills related to farming would be replaced in a friendly manner by the application of NFs and NP that would protect our ecosystem/Agro-climatic areas. The fertilizers that are derived from biological resources could possess a myriad of benefits over the synthetic fertilizers in terms of crop growth, efficiency in the usage of the nutrients and prevention of climate change. Remarkably, NFs have demonstrated significant potential as an innovative solution and can result in in vivo greater Agri-potential through holistic maintenance of crop plants to optimize their metabolic processes, viz., WUE, NUE, LUE, and CUE, coupled with plant performance and plant productivity/carbon economy, which is needed to explore another green/grain revolution to nourish the world with a healthy ecosystem under climate change in the future (Verma et al., 2022). Nanoparticle use centers on safety and sustainability as priorities for these technologies. To assess the risk to human health, ecosystems, and specifically the food chain, further in-depth safety assessments need to be conducted regarding the long-term impacts and risks that nanoparticles have on both humans and ecosystems. In addition, it will be necessary to conduct economic feasibility studies to determine the viability of these technologies across a variety of agricultural settings. All stakeholders must collaborate to use nanoparticles responsibly. Furthermore, ethical development and informed decision-making can only be achieved through open communication between farmers, legislators, and business leaders. Additionally, studies of the long-term effects of nanoparticles on soil health and the ecosystem help build knowledge regarding the environment and its effects over time. Additionally, the use of impact their performance in agriculture. By executing the aforementioned components, we can fully recognize the benefits and opportunities associated with the use of nanoparticles, creating a more prosperous, stable, effective, and sustainable future for agricultural production through responsible development, enhanced collaboration, and research efforts (Atanda et al., 2025).

2. Photocatalytic application of nanomaterials

In drought conditions, the use of nanoparticles helps plants grow and facilitates photosynthesis. Moreover, Au-NPs have been shown to exhibit remarkable anti-oxidative activity, protecting plants from oxidative stress induced by drought conditions. To synthesize Au-NPs, some plant extracts such as green tea, grape seeds, and lemon peel are more efficient. During dry seasons, plants can improve performance and physiology when they come into contact with gold particles—specifically, growth rate and yield increase when green tea extract synthesizes Au-NPs. Au-NPs help plants to take up nutrients efficiently from soil and enhance their photosynthetic pigments that provide the best food resources for plants. During drought stress, when plants are easily affected by microbes, the gold nanoparticles helped fight microbes through their antimicrobial activity, inhibiting infection rates caused by pathogens. The fundamental features of Au-NPs are to combat disease and boost the immune system in plants. The growth and development of plants in contact with Au-NPs enhances the ability to penetrate plant tissues, allowing. Moreover, Au-NPs not only help deliver micronutrients but also enhance seed germination and tissue growth rates. Au-NPs may also promote plant development by regulating gene expression and activating signaling pathways. Beyond influencing gene expression, Au-NPs might also interact with plant membranes and enhance drought tolerance. The exact process of their activity is still unknown. By scavenging oxygen-reactive species, gold nanoparticles help mitigate oxidative damage triggered during drought conditions. There are some safety measures when implanting Au-NPs in plants (Wahab et al., 2023).

2.1 Zinc oxide nanoparticles

The zinc oxide nanoparticles (ZnONPs) are known to have dual properties as anti-microbial agents as well as nutrient delivery systems. Zinc ions, which are necessary for the growth and development of the plants, can be released gradually by ZnONPs. It has also been found that ZnONPs can strongly increase the growth and productivity of crops like wheat and maize since zinc is provided in a constant amount. Besides, ZnONPs have been promising to enhance the photosynthetic rates and biomass of plants. Also, ZnO NPs have the ability to affect the generation of reactive oxygen species (ROS) in target organisms. Being very reactive, these ROS molecules are able to cause damage to important cellular components, including proteins and DNA, which play a role in cell death. In addition to the antimicrobial effects, ZnO NPs enhance plant growth and stress tolerance. They have the ability to serve as vectors of important nutrients such as zinc, thereby assisting the plants in absorbing nutrients, which can increase growth and development. Also, ZnO NPs can regulate plant hormone signaling pathways, which can enhance stress tolerance and enhance plant defense systems (Atanda et al., 2025).

2.2 Silicon dioxide nanomaterials

Silicon, the second most prevalent element in nature after oxygen, is found in soil. Plants require silica to respond to biotic and abiotic stressors. These interactions enhance the amount of water that plants use, their ability to photosynthesize, their structural characteristics, and their stiffness, which keeps leaves from toppling over and prevents disease. To form silica bodies, plants use their vascular systems to carry dissolved silica from the soil water to various cells and tissues. SiO_2 nanoparticles, which are composed of silicon oxide, or silica, are among the most widely used types of nanoparticles. They can come from their natural surroundings or be produced in laboratories. Artificial nanoparticles are amorphous, whereas extracted nanoparticles are mineral silica with a crystalline phase, like quartz. The benefit of creating such NPs is the final product's cleanliness, even though recovered NPs contain metal impurities (Karnwal et al., 2023).

2.3 Silver nanomaterials

Silver nanoparticles (AgNPs) are renowned for their antimicrobial properties, making them valuable tools in crop protection. They are effective against bacteria and fungi, and they suppress the growth of different pathogens, which translates into fewer losses of crops and high yields. As an example, AgNPs have been shown to regulate the development of Pseudomonas syringae, which causes tomato and bean diseases. AgNPs are also promising in the field of nutrient management, besides disease control. Nano fertilizer developed from AgNPs has the ability to release nutrients in step with the uptake by the plant. This novel method reduces nutrition loss in several ways by reducing the leaching, lowering of contamination of soil and groundwater and undesirable chemical reactions in fertilizers. AgNPs can assist in stablising fertilisers from changing into dangerous or ineffective compounds for plants.

Ultimately, this approach helps maintain soil fertility and promotes sustainable agricultural practices. Silver nanoparticles (AgNPs) remain a promising tool in agriculture due to their broad-spectrum antimicrobial properties. However, the precise mechanisms by which AgNPs exert these effects are still under investigation. There is a mechanism that involves the way AgNPs combine with the cell membrane of the target organism results in membrane disruption. This interaction causes structural and functional disruptions, leading to the leakage of cellular contents and, ultimately, cell death. In addition, AgNPs can induce the production of reactive oxygen species (ROS) within target organisms. These highly reactive ROS molecules can disrupt essential cellular components, such as proteins and DNA, thereby contributing to cell death. Additionally, AgNPs may interfere with the electron transport chain, which is essential for energy production in target organisms. This damage can result in energy depletion, ultimately leading to cell death (Atanda et al., 2025).

2.4 Gold nanomaterials

In drought conditions, the use of gold nanoparticles helps the plants to grow and facilitates the process of photosynthesis. Moreover, Au-NPs have been shown to have incredible features in protecting plants from their oxidative reactions caused by drought conditions through anti-oxidative activity. To synthesize Au-NPs, some plant extracts such as green tea, grape seeds, and lemon peel are more efficient to a large extent. During dry seasons, plants can improve performance and physiology when they contact gold particles—specifically, growth rate and yield increase when green tea extract synthesizes Au-NPs. Au-NPs help plants to take up nutrients efficiently from soil and enhance their photosynthetic pigments that provide the best food resources for plants. During drought stress, when plants can easily be affected by microbes, the gold nanoparticles proved to help fight against microbes due to antimicrobial activity by inhibiting infection rates caused by pathogens. The fundamental features of Au-NPs are to combat disease and boost the immune system in plants. The growth and development of plants in contact with Au-NPs enhances the ability to penetrate plant tissues, allowing.

Moreover, Au-NPs not only help deliver micronutrients but also improve the germination of seeds and the growth rate of tissues. Au-NPs may also increase plant development by controlling gene expression and their activating pathways. Above and beyond influencing the expression of a gene, Au-NPs might also cooperate with plant membranes and enhance the ability of drought tolerance. The exact process of their activity is still unknown. For scavenging oxygen-reactive species, gold nanoparticles fight oxidative damage triggered during drought conditions. There are some safety measures when implanting Au-NPs in plants during (Wahab et al., 2023).

3. Interaction of nanomaterials with the plant system

There is continuous research on how plants interact with and absorb nanoparticles. Different types of nanoparticles may interact with plant tissues in different ways and have different absorption mechanisms. The size, shape, and surface chemistry of the particles, as well as the types of plants and environmental conditions, all play major roles in how plants can absorb nanoparticles. Through endocytosis, symplast (living plant parts), or apoplast (non-living plant spaces), NPs in the soil solution can get to plant roots. NPs may also enter plants through the stomata, which are microscopic holes on the surface of leaves. This mechanism is especially significant for airborne nanoparticles. The cuticle, a waxy covering on the plant's aerial parts, can also be a site of nanoparticle uptake, especially for lipid-soluble nanoparticles. This suggests that nanoparticles can benefit from natural plant absorption processes, such as ion channels and transporters, though the specifics may vary depending on the size, charge, and composition of the nanoparticles. Once inside the plant, nanoparticles can travel through the xylem and phloem from the roots to other parts of the plant, but how well this process works depends on the nanoparticles' properties (Thabet and Alqudah, 2024) may depend on the duration of exposure and the metabolic activity of the plant species. Furthermore, a range of biological reactions, such as enhanced nutrient utilization, stress tolerance, and even modified gene expression, may occur in plants that absorb nanoparticles (Wang et al., 2021; Table 12.1).

3.1 Mechanism of nanoparticle uptake by plants

Nanomaterials primarily enter plants through foliar spraying or root uptake, after which they are transported to the target tissues. For root absorption, the apoplastic and symplastic pathways are employed. Nanoparticles enter the circulatory system after penetrating the root tissues, then ascend through the xylem to reach the aerial parts of the plant, such as the leaves and flowers (Fig. 12.2). Foliar spraying is another effective way to deliver nanomaterials, particularly to growth regulators and insecticides. By penetrating the cuticle and stomata, nanomaterials applied to leaves enable them to overcome soil-related obstacles (Zaman et al., 2025).

3.1.1 Root uptake

The small NPs (3–5 nm diameter) have been reported to penetrate plant roots by an osmotic pressure, capillary force, or directly by the root epidermal cells. Large NPs are restricted by the semipermeable epidermal cells of the root cell wall with microscopic holes. Some NPs facilitated their passage to the epidermal cell wall by digging new holes in it. Once through the walls of the cells, NPs can be carried apoplastically via extracellular spaces up to the central vascular cylinder, where the xylem can move unidirectionally. Nonetheless, NPs are

TABLE 12.1 Effect of nanoparticles/nano-fertilizers on crops under extreme conditions (Verma et al., 2022).

S.no.	Crop treated with nanoparticles	Application type	Impacts
1	NPs with tomato (Solanum lycopersicum L.)	Foliar	Increased vitamin C can lead to an increase in Cu accumulation in tomato fruits.
2	NPs with tomato (Solanum lycopersicum L.)	Soil	Better plant growth, development, productivity and fruit condition
3	NPs with barley (Hordeum vulgare L.)	Soil	Increased vegetation performance
4	NPs with barley (Hordeum vulgare L.)	Foliar	Positively improved growth parameters, chlorophyll leaf index, yield and biochemical activities.
5	NPs with wheat (Triticum aestivum L.)	Soil	Lower photosynthetic pigments and seed protein, an increase in activities of antioxidant enzymes.
6	NPs with wheat (Triticum aestivum L.)	Foliar	Increased growth, yield and nutritional condition in comparison to healthy plants
7	NPs with cucumber (Cucumis sativus L.)	Soil	Increased globulin and decreased glutelin.
8	NPs with cucumber (Cucumis sativus L.)	Foliar	Increase in the length of plants, number of leaves, and area growth, biomass, weight and quality of fruits compared to the control plants.
9	NPs with soyabean (Glycine max L.)	Soil	Lessened the toxicity of roots of plants and leaves. Increase gaseous exchange of leaves and enzyme reactions.
10	NPs with soyabean (Glycine max L.)	Foliar	Increased seed-weight-leaf biomass as compared to normal plants.

forced to cross the Casparian strip wall to get into the core vascular cylinder in a symplastic manner. This is done through binding to the endodermal cell membrane carrier proteins by endocytosis, pore formation and transport. The absorption of NPs into the cytoplasm occurs then NPs move between cells via the plasmodesmata. The NPs that they are not able to internalize are collected on the Casparian strip, and the NPs which have already accomplished this, are carried to the shoots and returned to the roots through the phloem. NPs are known to be found in the cytoplasm of cortical cells, the epidermal cell walls and the nuclei of plants. The available NPs that did not settle on the root surface of the soil aggregates could modify the absorption of nutrients. The NPs can enter seeds through direct absorption via parenchymatic intercellular gaps in the coat, after which it diffuses through the cotyledon (Ali et al., 2021).

3.1.2 *Foliar application*

The leaves may be penetrated by the NPs to which the leaves are exposed via the cuticles or the stomata. The principal barrier to the leaf is the cuticle, which excludes NPs that are smaller than 5 nm. NPs bigger than 10 nm penetrate through the stomata and get into the vascular system of the plant through the apoplastic and symplastic routes. The transmission of NPs (between 10 and 50 nm) (symplastic pathway) occurs in the cytoplasm of the following cell. Larger NPs (50–200 nm) are hence moved between the cells through the

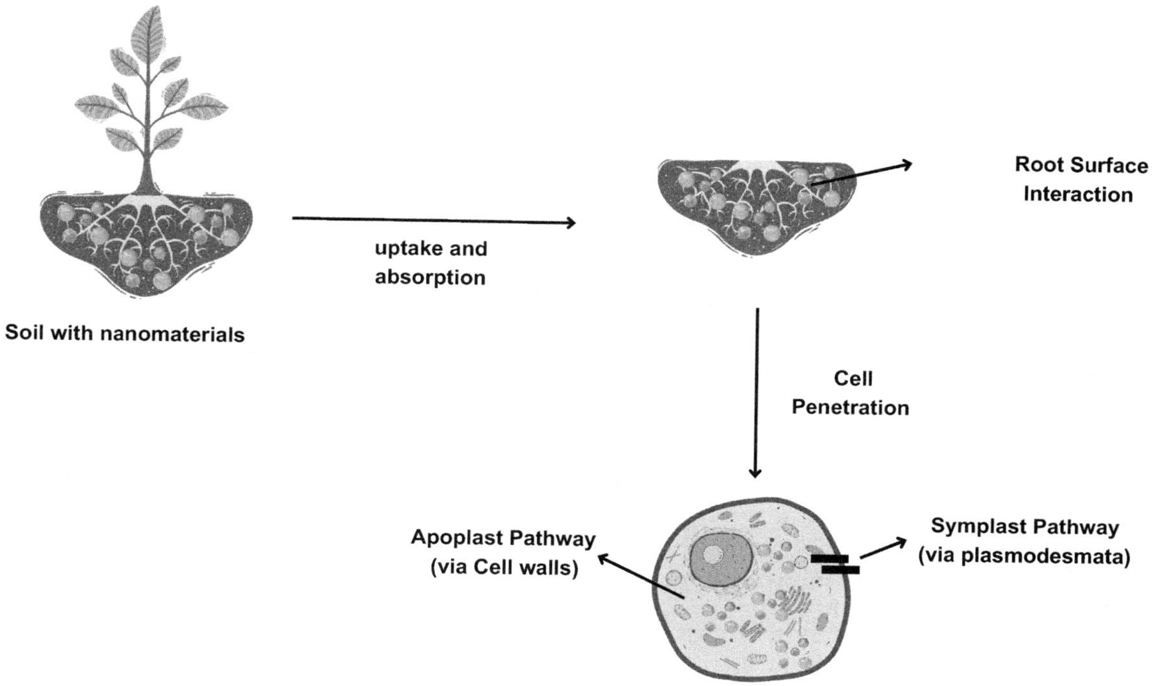

FIGURE 12.2 Nanomaterials interaction with plant roots.

apoplastic route. The sugar flow passes through the phloem sieve tubes in internalized NPs. It is caused by vascular transport through phloem, a strong sink of the sap and therefore NPs may move in either direction and accumulate to different levels in roots, stems, fruits, grains, and young leaves. The term nonselective direction of least resistance is used to refer to the apoplastic passage. It is familiar that the apoplastic way that promotes the movement of soluble components in water and the non-essential metal components. The size, concentration, application methods and climate were the key factors that influenced effective adsorption of NPs following foliar application (Fig. 12.3). The leaf morphology and chemical composition, trichrome presence, and trapping NPs of leaf exudates and waxes are important variables influencing the trapping process of NPs on a leaf surface (Ali et al., 2021).

3.2 Nanoparticles translocation within the plant tissue

Translocation of nanomaterials into plants is then made through the vascular tissues, mainly the xylem and phloem, based on the capabilities of the nanomaterials used in their functionality and location of use. The absorption of NPs into the roots is usually followed by their flow through the xylem by the process of transpiration. Transfer of topically applied NPs, on the other hand, occurs via phloem to non-photosynthetic tissues, including roots and developing fruits. The information regarding the moves and dispersal of nanomaterials is dependent on the size, solubility, and interactions with the plant biopolymers, such as cellulose and lignin. Smaller NPs tend to invade more remote tissues, but larger aggregates tend to be localized in certain locations. Such

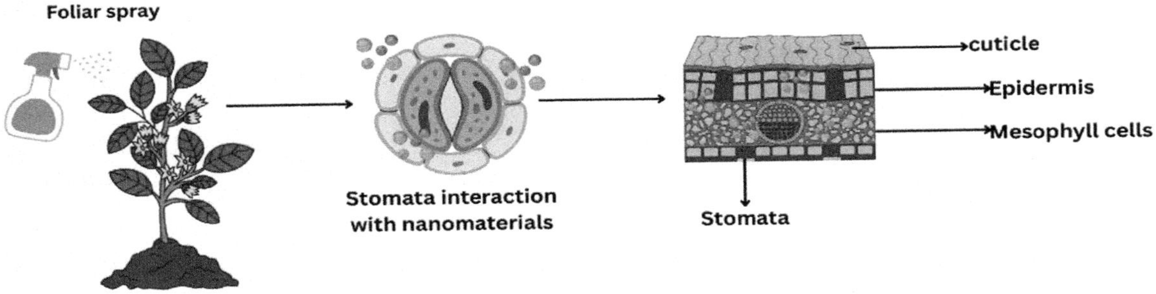

FIGURE 12.3 Nanomaterials foliar interaction.

focused delivery allows the delivery of the nutrients or biocides to the regions where they are the most demanded (Zaman et al., 2025).

4. Nanomaterials in improving plant growth

4.1 Improvement in crop quality

Nanomaterial reduces the level of phytotoxicity of crops and promote seedling development. The optical concentration of an interaction between Zinc oxide nanoparticles (ZnO-NPs) and the pomegranate peel extract, using the seed germination and shoot height (6.5 cm) as outcome parameters, was 2 and 12 ppm, respectively (Shaban et al., 2022). ZnO-NPs have a strong effect in promoting the germination of barley seeds. Green gram and Turkish gram showed significantly increased growth and development in the media where ZnO nanoparticles were used as compared to Bengal gram. Zinc deficiency affects the growth of plants, their chlorophyll, and the concentration of protein. The roots and shoots of maize plants treated with biosynthesized ZnO NPs using Bacillus sp. were found to be longer (Sabir et al., 2020), which demonstrated that plant growth was positively affected by the NPs as compared to the control. Nano-priming may be a useful process to multiply crop production. The authors of Sebastian et al. (2019) also synthesized iron oxide nanoparticles with the help of the extract of phenolic-rich coconut husk and discovered that the NPs had a Cd-stress-reducing and an Fe-boosting effect on rice plants. As a result, biomass, chlorophyll content, and photosynthetic quantum yield of the rice plant were enhanced. Here is the significance of the iron in photosynthesis, respiration, and the amount of chlorophyll. The concentration of higher levels of iron oxide nanoparticles also enhance growth of plants and enhances the enzyme activity. The magnesium level of roots and leaves was higher in the plants that were sprayed with magnesium hydroxide nanoparticles (Mg(OH)2NPs). The biogenic Mg(OH)2NPs enhance the percentage of seed germination. Thus, it demonstrates that seed germination and seedling growth can be enhanced with the help of Mg(OH)2NPs (Sundararajan et al., 2024).

4.2 Playing a role in enhancing the resistance to biotic and abiotic stress

Plant and post-harvest diseases caused by biotic stress necessitate a move away from conventional farming practices and pesticide use. Nanotechnology can reduce the phytotoxicity of some pesticides by increasing their distribution and decreasing their concentration. In order to examine the Fusarium oxysporum plant disease in tomato and pepper plants, for example, reduced graphene oxide nanosheets were subjected to copper oxide nanoparticles to produce nanocomposites. Pesticides can also be substituted by nanotechnology. Indicatively, it has been established that the use of metal oxide nanoparticles (CuO) reduces incidences of tomato bacterial wilt by changing the structure and composition of the rhizospheric bacterium community. Nanomaterials enhance tolerance and resistance to various abiotic stress factors such as drought, salt, heavy metal toxicity, and high temperature fluctuations to enhance plant growth, tolerance and production. Nanomaterials have produced promising materials that are resistant to different biotic and abiotic stressors. Nanomaterials can help in alleviating abiotic stressors by altering antioxidant enzymes and preserving water within plant tissues. It is also a nutrient-enriched agricultural crop due to the close agricultural culture system between the soil and the nanomaterials, which also improves the mechanism and offers an easy way of obtaining the nutrients and protection against overabsorption. Nanomaterials can be used to assist in food production, feed production, and seed production and to contribute to sustainability by increasing stress capacity (Sodhi et al., 2025).

4.3 Improve biomass and yield

Plasmonic nanoparticles, such as gold and silver nanoparticles, are able to enhance the local electromagnetic field intensity of a plant using the surface plasmon resonance (SPR) effect. This makes chlorophyll molecules more vulnerable to the light absorption and hence facilitates photochemical reactions that supply more energy to photosynthesis. ZnO/Ag nanoparticles were transformed into silver-modified (ZnO/Ag) in a green manner using mint extract. The localized surface plasmon resonance (LSPR) of silver enhanced the amount of light absorption and separation of charge carriers to a large degree. Nanomaterials can enhance crop growth, germination and yield, by increasing the accessibility and absorption of nutrients by plants. Mesoporous silica nanoparticles (MSNs) may be taken up by plant roots and delivered above ground tissues. MSNs were absorbed by wheat and lupin and the accumulation of chlorophyll and total leaf protein was promoted thus promoting photosynthesis and growth of the plants. Nepal et al. (2023) claim that biochar nanomaterials (CNMs) at moderate and low concentrations promoted the growth and yield of lettuce and its nutrient uptake. The growth of maize plants in sandy soil was promoted under optimal application of 200 mg/kg of carbon nanoparticles (CNPs). Nanofertilizers enhance the quality of crops, and their potential biotoxicity and pragmatics of the long-term impact on the environment (Chen et al., 2025).

4.4 Nanomaterials in genetic engineering

Food productivity has increased due to climate change, population explosion, genetic modification of plants and introducing desirable genetic features in crops. Plants have cell walls that prevent exogenous chemicals. To overcome this challenge and obtain genetic transformation in plants, many techniques have been locally employed across the globe in transmitting DNA to plant cells, and these methods are based on Agrobacterium transformation or biolistic methods. Plants have developed, but genetic engineering has not been developed, and it has taken time to develop. The number of available plants is relatively small compared to the numerous strategies of complex gene and protein distribution that have been developed to

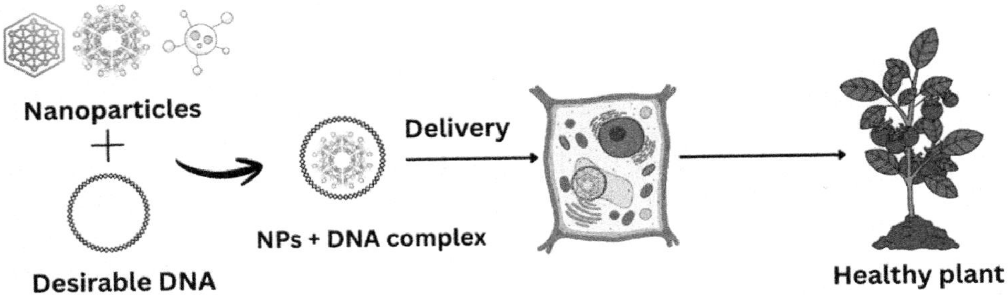

FIGURE 12.4 Nanomaterials have been explored as an alternative to transition gene delivery methods in plant genetic engineering.

work in animal systems. The transfer of genetic delivery and the regeneration of the transformed plant are often the two major processes in the genetic transformation of plants today. The complexity of the latter process and its necessity are greatly affected by the delivery mode and stability of transformation. It is the induction of competent totipotent tissue, the callus-forming tissue culture and selection and progeny segregation that are the three components of the regeneration processes. It has been found that plant cell cultures have been utilized in many innovative studies in plant genetic engineering using nanomaterials (Fig. 12.4). The appearance of nanoparticles makes them appealing materials to use in the transmission of biomolecules because they are used in a broad host range, have highly tailorable physicochemical properties, and can penetrate the plant cell walls without the use of an external force (Ali et al., 2021).

5. Carbon dots for enhancing plant health and plant stress tolerance

Carbon quantum dots (CQDs) are very versatile and recent nanomaterials that have a wide application range. Complex applications of pristine and engineered CQDs in soils and living systems of plants, with special references to soil biochemical processes and various plant development periods. The role of CQDs in increasing the plant tolerance to diverse biotic and abiotic stresses, including nutritional imbalance, salinity, drought, heat, and heavy metal stress, has been given special attention, and research indicates that tolerance to stress was increased, and the seed germination rates improved significantly when seeds were primed with CQDs (Fig. 12.5). Carbon-based nanomaterials (CNMs) are among the nanomaterials that have become very popular because of their peculiar physico-chemical and optical properties. The examples of CNMs are carbon nanotubes (CNTs), carbon quantum dots (CQDs), graphene quantum dots (GQDs), and polymer dots (PDs), etc. (Andleeb et al., 2025).

5.1 Effect of carbon dots on plant growth in both normal and stress conditions

5.1.1 Priming effect on seed germination

The seed priming with pure or doped CQDs offers a new method to promote the early growth of plants. This is highly effective in seed dormancy breaking, synergistic germination enhancement, and the overall health improvement of seedlings, thus indicating the role of CQD in being a powerful growth stimulator. The advantages of CQD priming are not limited to germination, such as promoted plant growth and developmental processes, resistance

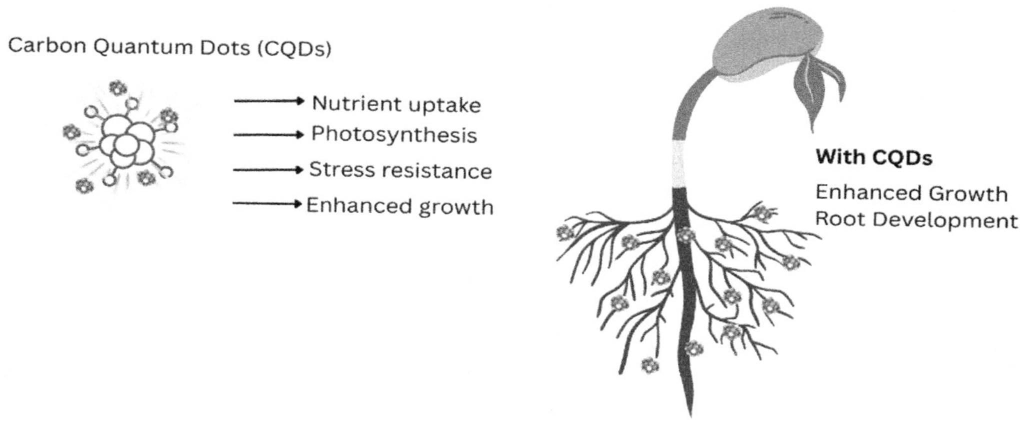

FIGURE 12.5 Carbon quantum dots boosting plant growth (CQDs).

to different stress factors in the environment, premature flowering, and finally improvement of plant growth, development, and productivity (Andleeb et al., 2025).

5.1.2 *Impact of CQDs on soil properties*

Studies regarding the application of nanotechnology, and particularly CQDs, to the soil ecosystem are growing and have colossal potential in affecting the population of microbes, nutrient levels in soils, and pollutant interactions. Nanoscale micronutrients behave highly depending on numerous factors associated with soil such as soil pH, soil organic and inorganic compounds, and soil microbes (bacteria and fungi like arbuscular mycorrhizae). The factors also have the potential to influence CQDs by impacting dissolution, aggregation, and surface properties (such as charge and coats), potentially leading to varied soil and plant outcomes (Andleeb et al., 2025).

5.1.3 *Nutrient significance for CQDs-coated fertilizers*

Applications of CQDs have grown in the last few years to the usage of coating fertilizers to enhance nutrient uptake in plants. CQD-coated fertilizers have a high surface area-to-volume ratio that makes them superior compared to modern and even contemporary fertilizers that is of a polymer nature. These manures enhance resistance to stress, lessen the usage of chemical inputs with high productivity, and enhance the environmental situation in comparison with conventional blends. The CQDs that are coated with fertilizers inhibit the leaching of nutrients between the soil and soil nutrients since they guarantee adequate absorption of nutrients and plant growth, as the nutrients released to the plants are slowed down to match the plant development stages.

Nitrogen-doped CQDs, water-soluble nano fertilizers, increased the amount of chlorophyll, the efficiency of energy conversion to light, the activity of Rubisco, and the activity of electronic transmissions to a significant extent. It was necessary to further conduct research on the nutritional value of these CQD-coated fertilizers even after these positive results, though (Andleeb et al., 2025).

5.1.4 *CQDs' role in modifying soil pH, carbon contents, and nutrient availability*

The introduction of CQDs could alter some of the key characteristics of the soil, such as the ability of soil C to be retained, pH, and nutrient composition. The properties of the soil and the source of the CQDs used highly affect these effects. Previous studies have suggested that CQDs can enhance soil health, nutrient retention and organic C levels, which can promote plant growth and the overall health of the soil. As an example, CQDs may be effective adsorbents. The adsorption ability of Cd-CQDs allows the nutritional media to reduce the concentration of freely dissolved Cd ions in the nutritional medium by 22.5%. CQDs are effective Cd adsorbents that have the maximum adsorption capacity of 2.42 mg g −1, especially in soils that have a pH of 4–7 (Andleeb et al., 2025).

5.1.5 *Impact of CQDs on crop health*

Research has shown that CQDs can be applied as nano fertilizers, improving crop photosynthesis, while also increasing crop yield. Spraying 560 mg L^{-1} of CQD aqueous solution on leaves can increase the yield of dicotyledonous plants, such as soybeans, tomatoes, and eggplants, by 20%. Photosynthesis is the main source of assimilation for crop yield formation, and the level of yield is determined by the accumulation and distribution of photosynthetic products. The continuous spraying of CQDs (50 mg L^{-1}) during the vegetative growth stage of maize can increase carbohydrate accumulation during the reproductive growth stage, resulting in an increase in the 1000-grain weight and a final yield increase of 24.50%. This is because the expression of genes of sucrose transporter (SUT) in leaves improved by 1.61 after CQDs treatment, and the upregulation of SUT expression improved the ability of leaves to transport sucrose in the phloem. Therefore, more carbohydrates are transported from leaves to grains, thereby promoting grain filling and increasing yield. Under drought conditions, spraying nitrogen-doped CQDs on maize leaves promotes the synthesis of carbohydrates by enhancing photosynthesis, resulting in a 30% reduction of maize yield loss; at the same time, the starch, soluble sugar, protein, linoleic acid, and α-linolenic acid contents of grains significantly increased by 7.0%, 9.8%, 49.7%, 10.5%, and 12.3%, respectively, thus improving the grain quality. As the primary nutrient source of crops in contemporary agricultural production, fertilizer directly involves or controls the metabolism and cycling of nutrients in crops, has a close relation with the crop yield and quality, the addition of CQDs as fertilizer enhancers to various types of fertilizers can help with the rapidity of the chemical reaction, nutrient decomposition, the nature of fertilizer delivery, boost the efficiency of fertilizer usage, and the growth and development of crops. Nanocarbon addition to the slow-release fertilizers can enhance the production of rice tillers, raise chlorophyll content in the booting process, enhance the rate of accumulation of dry matter, and increase the number of effective panicles and grains per panicle. The ultimate outcome is higher grain yield and greater efficiency of the nitrogen fertilizer use in rice. That is why CQDs are highly promising toward crop yield and quality enhancement (Li et al., 2023; Table 12.2).

5.2 Application of CQDs for food technology

5.2.1 *Pesticides and herbicides*

Nowadays, CQDs are one of the most efficient and selective fluorescence detectors of hazardous materials. As an example, tetradifon, hexane, atrazine, imidacloprid and diethyl chlorothiophosphate, among other pesticides that have a very low detection limit, can be detected by quenching-enabled photoluminescent CQDs. Moreover, sulfur and nitrogen-doped CQDs were found to detect carbamate insecticides with a high level of sensitivity. These findings indicate that CQDs are flexible and possess outstanding capabilities in the detection of pesticides and herbicides. Modern breeding techniques cannot substitute the use

TABLE 12.2 pCQDs on plant growth of cereal crops (Li et al., 2023).

S. no.	Crop	Treated part of the plant	Effect of CDs on plant	References
1	Rice	Leaf	• Greater shoot length and root dry weight of shoot and root • Growth of greater height of plants and weight of grains by 4.8% and 5.1%. • Growing in height and fresh biomass by 22.34% and 70.60%	Li et al. (2021) Tan et al. (2021) Li et al. (2021)
2	Maize	Leaf	• Enhanced yield and 1000 grains weight by 24.5% and 14.0% • Grew bt 20.9% and 39.6% in height of plants and weight respectively. • Greater photosynthetic parameter • Increased fresh weight of shoots and roots by 62.1% and 50.6%, respectively, and dry weight of shoots and roots by 29.2% and 37.5%, respectively.	Wang et al. (2021) Tan et al. (2021) Milenković et al. (2021) Yang (2022)

of pesticides, herbicides, and integrated agricultural chemicals, which have resulted in massive contamination of soil, water, and biodiversity, and the extinction of biological species. These products pose a threat to the health of organisms (human and animal) in industry, remediation of the environment, and biomedicals because of high residue levels of the products due to their high level of fluorescence, high biocompatibility, and tunable surface properties, electrical, and optical properties. These characteristics have simplified the process of embedding CQDs in active food packaging infrastructure, water treatment systems, real-time food quality measuring tools, as well as state-of-the-art medical therapies (Andleeb et al., 2025).

5.2.2 Monitoring food packaging, quality and product quality

CQDs are excellent functional fillers for active food packaging films due to their antibacterial and antioxidant properties. They are water-soluble and biocompatible, thus making it easy to integrate them into different biopolymer films. These films can stop the transmission of food-borne illnesses. Food rotting can also be detected using CQDs. Most often, CQDs are reacted with certain chemicals or environmental cues to alter their optical properties. These differences may be seen as the alterations in food quality or contamination, and the alteration in the chemical or physical characteristics. Food wrapping sheets made of CQDs have the ability to react to any change in pH and change color since they are a fluorescence-sensitive sheet. This provides an esthetic indicator of the freshness of food. As an example, CQDs made of tan-gerine peel were utilized in pH sensing as fluorescent (Andleeb et al., 2025).

6. Challenges

6.1 Adverse impact on organisms other than the target species

Phyto-synthesized NPs have gained popularity due to their potential to enhance plant growth and stress resistance, and have gained a lot of interest in the agricultural field. Assessment of the risks on non-target organisms in the agroecosystem is also important, as is any other new technology. To use the phyto-synthesized NPs in a responsible manner in the agricultural field, the benefits of using them should be balanced with the necessity to preserve the ecosystem. Their size and special characteristics are the reasons why nanoparticles may behave in different manners concerning the living systems as compared to the larger particles or bulk materials, which raises concerns. Within an agricultural environment, plants can absorb phyto-synthesized NPs and release them on the food chain to other animals, including those that are unintended. This potential raises the need to carry out further research and evaluation on the possible impacts of such nanoparticles on the environment when used in agriculture. Sustainable implementation of this technology ethically needs to know the functionality of these nanoparticles in living systems. Depending on composition, content and length of exposure, these nanoparticles can have a range of consequences on other non-target species. As an example, the total efficiency and sustainability of the agroecosystem can be influenced in case soil microorganisms, which are important in the nutrient circulation and soil health, are impacted negatively.

Equally, exposure to such nanoparticles can cause a direct or indirect adverse impact on beneficial insects that act as pollinators or natural predators of pests. We evaluate these potential impacts so as to make sure that they are used responsibly in the agricultural field and threat to the delicate balance of the environment. In order to make sure that the phyto-synthesized NPs are used safely and responsibly in agriculture, further research is required to fully assess their potential to impact other biological systems other than the unwanted species. There is a need to carry out risk tests and put in thorough consideration of the short-term and long term impacts before engaging in large-scale applications. To find a balance between the need to preserve the delicate ecological balance of the agroecosystem and the need to exploit the benefits of phyto-synthesized NPs to increase the development and tolerance of plants to stress, efforts must be made to reduce the risks. To achieve such a balance, eco-friendly practices should be put in the forefront. By exercising a responsible and informed decision, we will be able to protect the environment and non-target animals and ensure that the application of various phyto-synthesized nanoparticles in agriculture is efficient and can be sustained (Wahab et al., 2023).

6.2 Consequences of metal uptake in plants

6.2.1 Plant health

Toxicity: Excessive absorption of certain metals, such as cadmium (Cd), lead (Pb), and mercury (Hg), can poison plants. By interfering with essential physiological processes like photosynthesis, respiration, and enzyme functioning, these metals cause reduced development, wilting, and even death (Aslam et al., 2023).

Nutritional imbalance: High metal concentrations can disrupt the essential nutrient balance of plants. For example, excess iron (Fe) can hinder the absorption of other essential elements such as manganese (Mn) and zinc (Zn) (Aslam et al., 2023).

6.2.2 Environmental impact

Bioaccumulation: Plants that accumulate heavy metals may contaminate the environment. By seeping into the soil and affecting nearby ecosystems and water bodies, these metals have the potential to contaminate aquatic life and wildlife (Karnwal et al., 2023).

Soil degradation: Metal-contaminated plants can cause the soil to deteriorate, making it unsuitable for farming or other land uses. This could have long-term financial and ecological consequences (Karnwal et al., 2023).

6.2.3 Food chain contamination

Bioavailability: Plants are an essential component of the food chain. When herbivores eat plants tainted with metals, the metals can biomagnify as they ascend the food chain, eventually impacting humans and predators (Karnwal et al., 2023).

Human Exposure: Humans may directly consume heavy metals from polluted crop plants, which poses a serious health concern (Karnwal et al., 2023).

7. Conclusion

Increased food productivity is required due to population growth, which eventually calls for improved food quality control. As a result, nanobiotechnology has provided a viable remedy for this pressing problem that has improved plant survival rates in addition to growth yield. The use of nano-fertilizers, nano sensors, nanopesticides has a potent application in the agricultural sector which can also lead to sustainability since the nanomaterials are recognized as safe means of delivery and have many exclusive possibilities (Khan et al., 2022). Nanotechnology is a useful method for solving problems related to microbial biofertilizers, which can be applied in a variety of situations. These nanoparticles are great options for use in biofertilizer formulations since they have shown positive effects on microorganisms that promote plant growth or the growth of various plant kinds. By incorporating beneficial bacteria-containing nanoparticles into plants, microorganisms in the natural atmosphere may become more resilient and tolerant, enhancing plant growth and agricultural sustainability (Karnwal et al., 2023). When using nanoparticles, sustainability and safety continue to be top priorities. In order to determine the risks of nanoparticles on ecosystem and human health in the long run, particularly through the food chain, extensive safety tests are needed. To ascertain if these technologies are viable in a variety of agricultural contexts, economic feasibility studies are also essential (Atanda et al., 2025). It should be noted that the global green revolution enhanced food grain making at the expense of greater use of synthetic and artificial fertilizers and pesticides, which is hazardous our ecosystem. Therefore, by implementing NFs and NPs to protect our ecosystem and agroclimatic zones, these agricultural problems can be addressed in an environmentally responsible manner (Verma et al., 2022).

References

Ali, S., Mehmood, A., Khan, N., Tan, B., 2021. Uptake, translocation, and consequences of nanomaterials on plant growth and stress adaptation. J. Nanomater. 2021, 1–17. https://doi.org/10.1155/2021/6677616.

Andleeb, N., Zafar, S., Rahim, Z., Iqbal, M.M., Lal, R., Farooq, M.A., 2025. Carbon quantum dots as versatile nanomaterials for improving soil health and plant stress tolerance: a comprehensive review. Planta 262 (2). https://doi.org/10.1007/s00425-025-04758-2.

Aslam, M.M., Okal, E.J., Waseem, M., 2023. Cadmium toxicity impacts plant growth and plant remediation strategies. Plant Growth Regul. 99 (3), 397–412. https://doi.org/10.1007/s10725-022-00917-7.

Atanda, S.A., Shaibu, R.O., Agunbiade, F.O., 2025. Nanoparticles in agriculture: balancing food security and environmental sustainability. Discov. Agric. 3 (1). https://doi.org/10.1007/s44279-025-00159-x.

Chen, X., Wang, S., Lai, H., Deng, L., Zhong, Q., Okoye, C.O., Niu, Q., Jing, Y., Wang, J., Jiang, J., 2025. Nanobiotechnology in soil remediation: use of nanomaterials to promote plant growth and stress tolerance. Nanomaterials 15 (22), 1743. https://doi.org/10.3390/nano15221743.

Karnwal, A., Dohroo, A., Malik, T., Nokhodchi, A., 2023. Unveiling the potential of bioinoculants and nanoparticles in sustainable agriculture for enhanced plant growth and food security. BioMed Res. Int. 2023 (1). https://doi.org/10.1155/2023/6911851.

Khan, F., Pandey, P., Upadhyay, T.K., 2022. Applications of nanotechnology-based agrochemicals in food security and sustainable agriculture: an overview. Agriculture 12 (10), 1672. https://doi.org/10.3390/agriculture12101672.

Li, Y., Pan, X., Xu, X., Wu, Y., Zhuang, J., Zhang, X., Zhang, H., Lei, B., Hu, C., Liu, Y., 2021a. Carbon dots as light converter for plant photosynthesis: augmenting light coverage and quantum yield effect. J. Hazard Mater. 410, 124534. https://doi.org/10.1016/j.jhazmat.2020.124534.

Li, Y., Xu, X., Lei, B., Zhuang, J., Zhang, X., Hu, C., Cui, J., Liu, Y., 2021b. Magnesium-nitrogen co-doped carbon dots enhance plant growth through multifunctional regulation in photosynthesis. Chem. Eng. J. 422, 130114. https://doi.org/10.1016/j.cej.2021.130114.

Li, G., Xu, J., Xu, K., 2023. Physiological functions of carbon dots and their applications in agriculture: a review. Nanomaterials 13 (19), 2684. https://doi.org/10.3390/nano13192684.

Milenković, I., Borišev, M., Zhou, Y., Spasić, S.Z., Leblanc, R. M., Radotić, K., 2021. Photosynthesis enhancement in maize via nontoxic orange carbon dots. J. Agric. Food Chem. 69 (19), 5446–5451. https://doi.org/10.1021/acs.jafc.1c01094.

Nepal, J., Xin, X., Maltais-Landry, G., Wright, A.L., Stoffella, P.J., Ahmad, W., He, Z.L., 2023. Water-dispersible carbon nanomaterials improve lettuce (Latuca sativa) growth and enhance soil biochemical quality at low to medium application rates. Plant Soil 485 (1–2), 569–587. https://doi.org/10.1007/s11104-022-05852-0.

Sabir, S., Zahoor, M.A., Waseem, M., Siddique, M.H., Shafique, M., Imran, M., Muzammil, S., 2020. Dose Response 3, 1.

Sebastian, A., Nangia, A., Prasad, M.N.V., 2019. J. Hazard. Mater. 371, 261.

Shaban, A.S., Owda, M.E., Basuoni, M.M., Mousa, M.A., Radwan, A.A., Saleh, A.K., 2022. Res. Square, PPR483397.

Sodhi, G.K., Wijesekara, T., Kumawat, K.C., Adhikari, P., Joshi, K., Singh, S., Farda, B., Djebaili, R., Sabbi, E., Ramila, F., Sillu, D., Santoyo, G., de los Santos-Villalobos, S., Kumar, A., Pellegrini, M., Mitra, D., 2025. Nanomaterials–plants–microbes interaction: plant growth promotion and stress mitigation. Front. Microbiol. 15. https://doi.org/10.3389/fmicb.2024.1516794.

Sundararajan, N., Habeebsheriff, H.S., Dhanabalan, K., Cong, V.H., Wong, L.S., Rajamani, R., Dhar, B.K., 2024. Mitigating global challenges: harnessing green synthesized nanomaterials for sustainable crop production systems. Glob. Chall. 8 (1). https://doi.org/10.1002/gch2.202300187.

Tan, T.L., Zulkifli, N.A., Zaman, A.S.K., Jusoh, M.b., Yaapar, M.N., Rashid, S.A., 2021. Impact of photoluminescent carbon quantum dots on photosynthesis efficiency of rice and corn crops. Plant Physiol. Biochem. 162, 737–751. https://doi.org/10.1016/j.plaphy.2021.03.031.

Thabet, S.G., Alqudah, A.M., 2024. Unraveling the role of nanoparticles in improving plant resilience under environmental stress condition. Plant Soil 503 (1–2), 313–330. https://doi.org/10.1007/s11104-024-06581-2.

Verma, V., Al-Dossari, M., Singh, J., Rawat, M., Kordy, M., Shaban, M., 2022. A review on green synthesis of TiO_2 NPs: synthesis and applications in photocatalysis and antimicrobial. Polymers 14, 1444.

Verma, K.K., Song, X.P., Joshi, A., Tian, D.D., Rajput, V.D., Singh, M., Arora, J., Minkina, T., Li, Y.R., 2022. Recent trends in nano-fertilizers for sustainable agriculture under climate change for global food security. Nanomaterials 12 (1). https://doi.org/10.3390/nano12010173.

Wahab, A., Batool, F., Muhammad, M., Zaman, W., Mikhlef, R.M., Naeem, M., 2023. Current knowledge, research progress, and future prospects of phyto-synthesized nanoparticles interactions with food crops under induced drought stress. Sustainability 15 (20), 14792. https://doi.org/10.3390/su152014792.

Wang, C., Yang, H., Chen, F., Yue, Le, Wang, Z., Xing, B., 2021. Nitrogen-doped carbon dots increased light conversion and electron supply to improve the corn photosystem and yield. Environ. Sci. Technol. 55 (18), 12317–12325. https://doi.org/10.1021/acs.est.1c01876.

Yang, H., 2022. Mechanism of Nitrogen-Doped Carbon Dots on Photosynthesis and Drought Tolerance of Maize. Master's Thesis. Jiangnan University, Wuxi, China.

Zaman, W., Ayaz, A., Park, S.J., 2025. Nanomaterials in agriculture: a pathway to enhanced plant growth and abiotic stress resistance. Plants 14 (5), 716. https://doi.org/10.3390/plants14050716.

CHAPTER

13

Mesoporous silica-based nanotechnology for chloroplast engineering and targeted delivery of double-stranded RNA and ribonucleoproteins

Rathod Sridhar[1], *M.S. Paarshitha Reddy*[1], *Hanuman Singh Jatav*[2], *K. Aruna*[3], *T. Gohain*[1], *Avinash Pilla*[4] *and Anvesh Sankuratri*[5]

[1]Department of Agronomy, School of Agricultural Sciences, Nagaland University, Medziphema, Nagaland, India; [2]Department of Soil Science and Agricultural Chemistry, Sri Karan Narendra Agriculture University, Jobner, Rajasthan, India; [3]Department of Agronomy, Professor Jayashankar Telangana Agricultural University, Hyderabad, Telangana, India; [4]Department of Agriculture, Koneru Lakshmaiah Education Foundation, Vaddeswaram, Andhra Pradesh, India; [5]Department of Plant pathology, Tennessee State University, Nashville, TN, United States

1. Introduction

Chloroplasts represent multifunctional organelles, indispensable for photosynthetic activity, carbon fixation, amino acids biosynthesis and production of pivotal metabolites, which regulate plant growth and defense mechanisms. Their semi-autonomous genome, which is largely maternal inheritance among major crop species, provides a unique genetically compartment of the plant favorable for high-level transgene expression, metabolic engineering and biosafety through reduced transgene escape (Bock, 2015). As a consequence of these benefits, chloroplast engineering has been explored for the augmentation of photosynthetic efficiency, enhancement of nutrient use efficiency, synthesis of high carbohydrate biomolecules and reinforcement of long lasting disease resistance. Nonetheless, conventional methods for chloroplast transformation, such as biolistic delivery and penetration by polyethylene glycol (PEG), are subject to certain limitations such as being species specific, damaging the tissues, low transformation efficiency for agriculturally important crops, and labor-intensive regeneration procedures (Giraldo et al., 2014; Wang et al., 2016). These constraints have hindered the broad

Nanotechnology in Photosynthesis
https://doi.org/10.1016/B978-0-443-44376-3.00013-6

application of plastid biotechnology to large-scale crop improvement activities and plant-pathology interventions.

At the cellular level, there are many physical and biochemical obstacles to efficient chloroplast-targeted gene delivery. The plant cell wall is a dense matrix of cellulose, hemicellulose, and pectin polymers, which limits the transfer of macromolecules above a certain restricted size (Cunningham et al., 2018). Beneath such a wall, the plasma membrane together with cytosolic enzymes degrade or sequester nucleic acids and protein cargo, and the dual chloroplast envelope, armed with selective TOC-TIC translocon complexes, restricts the spontaneous ingress of exogenous materials (Demirer et al., 2019a,b). In addition, plastid nucleoids, which consist of highly organized DNA- and protein-complexes, pose an additional barrier for stable genome editing and integration (Kwak et al., 2019). These biological constraints then aggregately reduce the effectiveness of the traditional delivery mechanisms, especially in mature tissues and crop species that have highly lignified or structurally complex cell walls.

Nanotechnology has changed this landscape and offered engineered nanomaterials able to move across the plant cell wall, travel through intracellular milieums and deliver biomolecular cargo directly into the chloroplasts with minimum perturbation to the plant cells. Carbon nanotubes, mesoporous silica nanoparticles, lipid-based nanocarriers, DNA nanostructures, polymeric systems and metal-organic frameworks can be engineered in terms of size, charge, shape and surface chemistry for organelle-specific delivery (Demirer et al., 2019a,b; Zhang et al., 2019a,b; Schwartz et al., 2020). These nanocarriers not only protect nucleic acids and proteins from enzymatic degradation, but also should allow species independent delivery, which would overcome long-standing barriers in crop transformation. Crucially, nanomaterials can be designed to release cargo in the stroma of the chloroplasts to allow for the expression of a transgene, the introduction of regulatory RNAs, or the activation of the components of the genome-editing toolbox (CRISPR–Cas).

The combination of nanotechnology and chloroplast engineering has high potential for today's agronomy. Targeted manipulation of pathway processes in chloroplasts by nanoscale delivery can improve photosynthetic efficiency, complement biomass accumulation and increase nutrient-use efficiency, traits of direct relevance for crop yield and crop sustainability. Moreover, chloroplasts also serve as important regulators of plants' response to abiotic stress; nanoparticle-mediated transfer of protective genes or regulatory RNAs provides a fast route for breeding tolerance to drought, salinity, heat and oxidative stress (Newkirk et al.,2021). In plant pathology, chloroplast-targeted expression of antiviral RNAs, antimicrobial peptides and insecticidal molecules is a way to effectuate long-term defense with reduced environmental risk because plastid genomes have limited pollen-mediated gene flow (Keegstra and Cline, 1999). The use of Nanotechnology in this regard has also been furthered by the facts that it enables the use of strategies using RNA interference (RNAi) based methods to simultaneously drive RNA interference through the delivery of dsRNA or siRNA constructs that could be used to knock down essential pathogen or pest genes providing a robust complement to conventional resistance breeding (Wallace and Anderson. 2012).

Collectively, the sum total of nanotechnology, plastid synthetic biology, agronomy, and plant pathology is redefining the success of crop improvement. Nanocarrier based chloroplast engineering provides a cost-effective, species independent platform to overcome the main limitations of traditional transformation techniques allowing for the creation of cultivars with increased stress tolerance, disease resistance and yield potential. This chapter discusses the biological barriers, nanomaterial carriers, delivery systems, agronomic methods, and plant pathology innovations stemming from nanoscale delivery of chloroplasts in gene transfer that provide the necessary perspective on the implications for next generation age biotechnologies.

2. Biological barriers to chloroplast gene delivery

Efficient delivery of genes destined for chloroplasts requires crossing many, hierarchical biological barriers that limit ingress, stability and intracellular trafficking of exogenous biomolecules. These impediments not only modulate the efficacy of plastid-based genetic manipulation but also delineate the practicability of deploying such methodologies in operative agricultural settings, wherein crop tissues display heterogeneity in cell-wall composition, developmental stage and exposure to biotic or abiotic stressors. Consequently, a thorough understanding of such barriers is crucial for the rational design of nanocarriers with applications to agronomics and plant pathology in the (Table 13.1).

2.1 Plant cell wall as the primary physical barrier

The plant cell wall represents the first and hardest barrier to overcome for the gene delivery platforms, due to its complexity consisting of cellulose microfibrils, hemicelluloses, pectins, lignin (mainly in mature tissues) and structural proteins (Carpita and McCann. 2010). Its characteristic pore diameter, normally in the range from 5 to 20 nm, puts a limit on the passive diffusion of macromolecular cargo such as plasmid DNA, proteins and some nanoparticles. In agricultural situations the architectural properties of the cell wall show considerable variation with crop species and tissue type. e.g., grasses have a more lignified and silica-laden cell wall than dicots, and hence exacerbate impediments to delivery in cereals (e.g., rice and wheat). Biotic challenges, like fungal infections or herbivore attacks, for example, can trigger wall fortification and callose deposition and this subsequently reduces permeability (Voigt, 2014). Accordingly, efficacious nanocarrier systems must be calibrated according to disparities in wall rigidity, tissue maturation and species-specific physiological conditions.

2.2 Selectivity and cytosolic degeneration of the plasma membrane

Following translocation through the cell wall, nanocarriers must penetrate the plasma membrane, an extremely selective and stringent barrier that organises the uptake of nutrients, ionic homeostasis and plant defense signaling. Hydrophilic genetic cargoes cannot permeate the lipid bilayer without facilitation, becoming

TABLE 13.1 Biological barriers to chloroplast gene delivery.

Barrier	Description	Agronomy/Pathology relevance
Cell wall	Rigid cellulose–hemicellulose–pectin matrix restricting nanoparticle entry	Highly lignified in cereals; callose deposition during infection reduces permeability
Plasma membrane	Selective lipid bilayer influencing uptake of hydrophilic cargo	Stress alters membrane fluidity; pathogens remodel membrane domains
Chloroplast envelope	Double membrane with TOC–TIC selectivity	Heat/viral stress alters envelope architecture affecting targeting
Plastid nucleoids	Compact DNA–protein complexes limiting genome access	Stress-induced nucleoid reorganization affects editing efficiency

entrapped in endocytic vesicles where they are exposed to degradation (Inoue et al., 2016). In other cultivars exposed to drought, salinity or elevated temperature, the fluidity of the membranes and the lipid composition change substantially, thus the interaction of nanoparticles with membranes is changed. From the point of view of plant pathology, pathogenic invasion can induce fast membrane remodeling and the production of reactive oxygen species (ROS), which lead to the accelerated breakdown of unprotected nucleic acids or proteins. Consequently, nanocarriers will need to be designed with surface chemistries that allow endocytosis-independent cellular entry and which are resistant to cytosolic nucleases, proteases and oxidative agents (Zeng et al., 2017).

2.3 Chloroplast envelope and TOC-TIC import selectivity

Nanocarriers residing inside the cytosol have to pass through the double membrane that constitutes the chloroplast envelope, an outer membrane studded with porin-like channels, and an inner membrane containing the TOC-TIC translocon complexes responsible for the import of nucleus-encoded proteins (Li and Chiu, 2010). These complexes show strict N-terminal transit peptide recognition property and have no intrinsic capacity to facilitate the import of exogenous nucleic acids or nanoparticles. Consequently, this selective barrier represents a significant barrier for incorporation of the engineered photosynthetic pathway, stress responsive genes or pathogen resistance determinants into the chloroplast genome. In tissues infected by pathogens, chloroplasts often experience changes in their structure like swelling, membrane deformation, or changes in lipid composition that may worsen impediments of transport (Bhattacharyya et al., 2015). Addressing these problems requires design of nanomaterials of carefully defined size, surface charge and ligand functionality that can interact constructively with the chloroplast envelope in the (Fig. 13.1).

2.4 Plastid nucleoids and genome availability

Chloroplast DNA is organized into highly compacted nucleoids consisting of DNA-protein complexes associated with the inner envelope or thylakoid membranes (Powikrowska

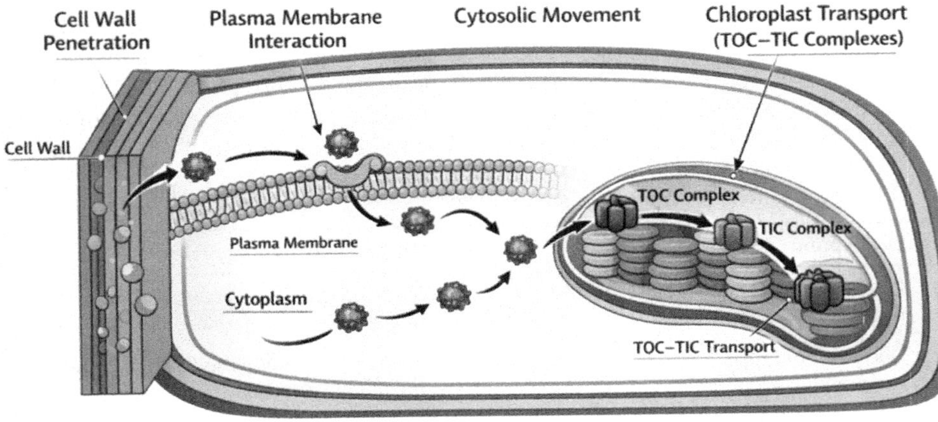

FIGURE 13.1 Nanocarrier uptake pathways in plant cells.

et al., 2014). These structures provide stability of the genome while providing limited access to exogenous genetic material. Genome-editing tools or regulatory RNAs, as well as homologues recombination templates must localize close to these nucleoids in order to be functional. In agronomically important crops like maize, cotton and soybean, nucleoid distribution can be quite different at different stages of development, which will have an impact on transformation results. Under conditions of pathogen attack or abiotic stress, nucleoids can be fragmented or reorganised spatially which can affect accessibility of genome editing reagents. Designing nanocarriers that can be localized precisely inside chloroplasts is thus of utmost importance for stable plastid engineering.

2.5 Tissue specific and developmental constraints in crop species

Chloroplast numbers, structure and metabolic activity vary spectacularly between tissues and developmental stages, affecting the uptake efficiency of genes. Mature mesophyll cells have many suitable chloroplasts for photosynthesis engineering pointing to meristematic cells with lower import capacity (Allorent and Petroutsos. 2017). In crops, highly lignified tissues (massive plant part in cereal crops), thick cuticles (massive plant part in tomatoes) or storage tissues (massive plant part in potato) represent additional obstacles to the diffusion of the nanoparticles. Furthermore, considerations are critical from plant pathology: chloroplasts play central roles in salicylic acid and jasmonic acid signaling pathways and chloroplast morphology and mobility often change during pathogen attack. These physiological variations have to be taken into account when design nanocarriers for better tolerance to stress, resistance to diseases or for enhancing photosynthetic efficiency.

2.6 Environmental and stress induced modulation of the delivery barriers

Agronomic environments subject the crops to varying light, drought, nutrient deprivation and pathogen stress; these various factors work together to change the architecture of the cell. Accumulation of reactive oxygen species, lipid peroxidation and subsequent increase in membrane rigidity under stress can physically impede the translocation of nanoparticles and reduce the efficiency of gene delivery (Wong et al., 2017). Conversely, some stress conditions may increase the permeability of the cell wall or allow flow between the cytosol, and this may provide potential opportunities for improved delivery. Elucidating the effect of environmental and pathological influencing factors on the intracellular trafficking of nanomaterials is crucial to design robust, field deployed chloroplast engineering methodologies.

3. Nanocarrier platforms for chloroplast gene delivery

Nanocarriers have become versatile tools to overcome the biological barriers for the delivery of chloroplast genes in plant systems. Their tunable physicochemical properties, such as size, surface charge, morphology and functionalization through ligands, allow tailoring the loading of cargo with tissue penetration, intracellular trafficking and organelle targeting. Crucially, the nanocarrier systems can be tailored to different agricultural situations, and thus enhance the delivery of gene transfer in recalcitrant cell walls, mature tissues, or stress-induced cellular changes of many species. Within plant pathology, nanocarriers aid in the transport of antiviral RNAs, antimicrobial peptides and genome-editing reagents to the chloroplast, providing the basis for long-term resistance to viruses, fungi and insect pests.

The following subsections describe the main nanomaterial platforms currently explored for chloroplast engineering in the (Fig. 13.2) (Table 13.2).

3.1 Carbon-based nanomaterials

Carbon nanomaterials (single and multi-walled carbon nanotubes S+ and M + SWCNT, graphene quantum dots-carbon nanodots) are one of the most studied families of nanocarriers for plant genetic delivery. Their strong aspect ratios, high mechanical stability and ability to move through the plant cell wall without tissue damage make them especially suitable for agricultural applications (Demirer et al., 2019a,b). SWCNTs are capable of traversing leaf tissue and delivering plasmid DNA/RNA/ribonucleotide protein complexes directly to plant cells by passive energy-independent diffusion. Functionalization using polyethyleneimine (PEI), chitosan or chloroplast targeted peptides improves the electrostatic loading of genetic cargo and improves chloroplast localization (Serag et al., 2011a,b). In agronomy, carbon nanocarriers have also been used to control the photosynthetic efficiency or the nutrient uptake and transport, as well as to introduce a gene providing stress tolerance. Within the discipline of plant pathology, the carbon-based nanoparticles ensure the supply of the dsRNA to the chloroplasts, which will, in turn, trigger antiviral silencing and will be an efficient measure against RNA viruses that replicate in chloroplast-associated membranes.

3.2 Mesoporous silica nanoparticles (MSNs)

Mesoporous silica nanoparticles (MSNs) exhibit homogeneous pore patterns, large surface areas and strong chemical stability that provide ideal conditions for nucleic acid encapsulation, such as DNA and RNA, and

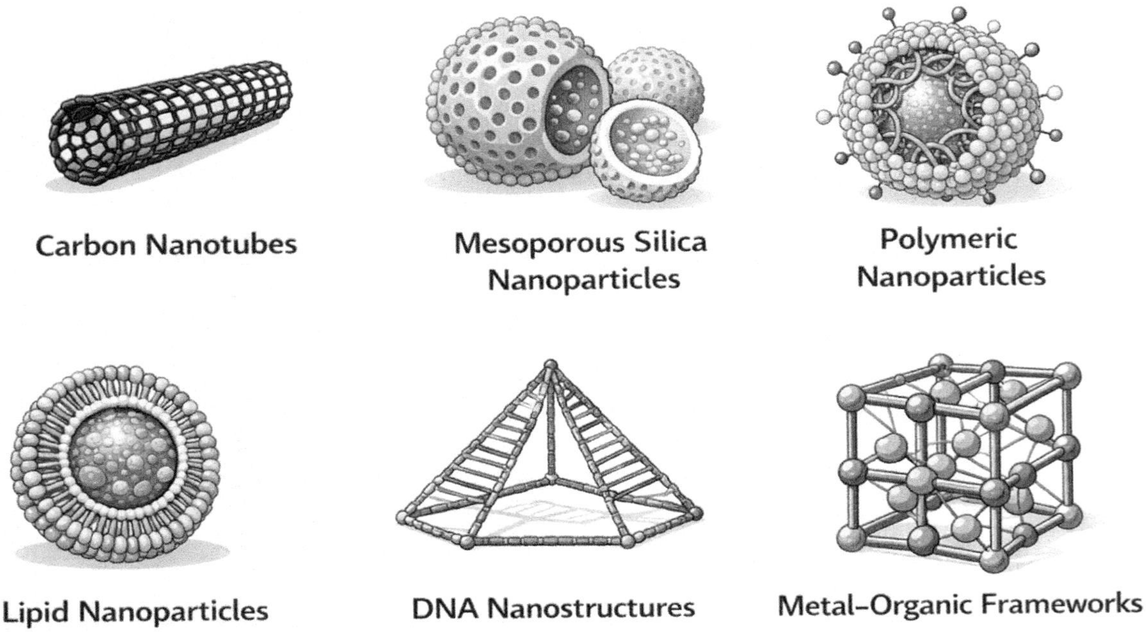

FIGURE 13.2 Major Nanocarrier platforms for chloroplast gene delivery.

TABLE 13.2 Nanocarrier platforms for chloroplast gene delivery.

Nanocarrier type	Key features	Applications	Limitations
Carbon nanotubes	High aspect ratio; excellent penetration; functionalizable surfaces	Gene delivery, RNAi constructs, stress-regulation genes	Possible persistence in soil
Mesoporous silica nanoparticles	High surface area; tunable pores; stable	CRISPR delivery, metabolic pathway engineering	Slow degradation in certain soils
Polymeric nanoparticles	Biodegradable; customizable charge and size	Stress-tolerance genes, RNA delivery	Potential cytotoxicity at high doses
Lipid nanoparticles	Membrane fusion capability; protects nucleic acids	Antiviral RNA delivery; transient expression	Lower UV stability under field conditions
Metal–organic frameworks	pH-responsive release; encapsulates large cargo	CRISPR proteins; immunity enhancement	Metal ion accumulation concerns

proteinaceous cargo for a controlled release (Torney et al., 2007).The tunability of porosity dimensions allows for packaging of large plasmid constructs or for packaging of complex components associated with the delivery of the corresponding Cas protein, while surface silanol functions as versatile functionalization sites to attach targeting peptides, polymer matrices, or protective coatings. In agronomic settings, MSNs have shown efficacy in foliar infiltration, seed coat application, and root uptake, which are therefore useful for field-based delivery modalities. Their biocompatibility and low phytotoxicity are the basis for their use in the management of plant diseases, where MSNs have been used in the transport of antimicrobial proteins or RNA silencing molecules against pathogen genes. Functionalized MSNs have also been used to introduce genes and metabolic pathways for improving photosynthetic performance to chloroplast compartments to yield improvements in major crop species.

3.3 Polymeric nanocarriers

Polymeric nanoparticles, such as poly lactic-co-glycolic acid (PLGA), polyethyleneimine (PEI), poly amido amine (PAMAM) dendrimers and chitosan, offer biodegradability, sustained protection of cargo and tunable surface functionality (Mondal et al., 2021). PEI-based formulations have been found to be effective in compacting DNA or RNA molecules and enhancing chloroplast delivery, as they are positively charged on the surface and have the ability to pass through the plant cellular barriers. PLGA nanoparticles have sustained-release properties, which are favorable for long expression of stress-resilience genes and dendrimers allow multivalent coupling of peptides and ligands for targeted chloroplast interaction. In regard to agronomic applications, polymeric nanocarriers are enabling the delivery of genes for drought tolerance, nitrogen use efficiency and optimization of photosynthetic performance. In plant pathology, antimicrobial nanoparticles, such as chitosan, have been reported to have anti-microbial properties, and can provide a double function by delivering an antiviral or anti-fungal RNAi constructs, while simultaneously priming plant defense pathways.

3.4 Lipid based nanoparticles and liposomes

Lipid nanoparticles (LNPs) and liposomes have high biocompatibility, and are able to encapsulate nucleic acids or proteins in lipids bilayers that are used to protect the cargo from degradation during transportation (Perlatti et al., 2020). Their natural compatibility with plant membranes aids the fusion or lipid exchange which allows the efficient cytosolic release. Tailoring lipid composition, surface charges and peptides enhance chloroplast targeting In crop improvement, LNPs have been exploited to deliver regulatory RNAs and genome-editing reagents for improving photosynthetic efficiency or abiotic stress tolerance. In the field of plant pathology, LNPs can serve as a delivery system for dsRNA or siRNA constructs that target genes involved in fungal virulence, replication of viral pathogens, or insect detoxification and that have the potential to provide species specific and environmentally safe options for chemical pesticides.

3.5 Metal-organic frameworks (MOFs)

Metal–organic frameworks such as zeolitic imidazolate frameworks (ZIFs) have been reported as possible nanocarriers because they are porous, their size is tunable, release behavior is pH sensitive, and they can be used to encapsulate sensitive biomolecules without severe conditions (Sun et al., 2020). Upon deployment, MOFs break down into constituent nutrient ions which help mitigate environmental concerns, and make them suitable for field based applications. Their highly porous crystalline matrix is large enough to accommodate relatively large explanations of the CRISPR-Cas complex, plasmids and proteins, whereas their sensitivity to chloroplast stroma conditions allows efficient cargo discharge. Consequently, the MOFs are especially useful in plant pathology for delivery of antimicrobial peptides or nucleic acids that interfere with pathogen gene expression. In addition to the obvious advantage of more adaptable strategies for engineering the functions of chloroplasts under stress such as heat or oxidative pressure, such a protective encapsulation provided during transit improves the feasibility of these strategies.

3.6 DNA nanostructures

DNA nanostructures, such as DNA origami, tetrahedral DNA frameworks and designer nucleic acid scaffolds, make use of the well-defined rules of base pairing to construct programmable carriers of high structural fidelity (Zhang et al., 2019a,b). Their innate biocompatibility and resistance to degradation by nucleases provided by chemical modification makes these carriers suitable for the delivery of genetic cargo to the chloroplasts. The chloroplast membranes can be selectively interacted with by functionalization with peptides or ligands that have specific interaction specificity. In agricultural applications, such nanostructures have been explored for the delivery of photosynthesis regulators, plastid-e-editing templates and RNA molecules to improve resilience to stress. In plant pathology, DNA-based carriers serve as a platform for delivery of virus-mediated-RNA interference or sequences that initiate targeted silencing of pathogen genes.

3.7 Hybrid and stimulus responsive nanocarrier systems

Hybrid nanocarriers are a blend of various types of materials, lipid - polymer hybrids, silica - polymer or peptide - functionalized carbon nanotubes, which employ synergy of each other to obtain both increased delivery efficiency and a chloroplast specificity (Lam et al., 2021). Stimulus-responsive materials or materials capable of releasing the cargo in response to temperature, pH, light, reactive oxygen species or enzymatic activity can be used to achieve spatial and temporal control of gene expression. Such responsiveness is especially pertinent in situations concerning plant pathology, where pathogen infection causes micro environmental changes that are able to trigger the targeted release of defense genes. In agronomy, stimulus-responsive carriers make it possible to send photosynthetic enhancers or stress tolerance genes only under high light, drought or salinity conditions and thus to optimize the metabolic efficiency and to avoid undesired effects.

4. Mechanisms of nanocarrier uptake and chloroplast localization

The efficiency of chloroplast gene delivery is dependent on how nanocarriers pass through cellular barriers, how they move around intracellular milieua and how they selectively localize within plastids. These processes are controlled by physicochemical properties of the nanomaterials (size, surface charge, hydrophobicity and functionalization) and the physiological status of the plant tissues. In the context of agronomical and plant pathology issues, environmental stresses and pathogenic invaders can also cause membrane fluidity and cytoskeletal dynamics, and chloroplast positioning, which can affect the transit of nanocarriers into the cell. Consequently, a mechanistic understanding of these pathways is imperative for the design of strong, field deployable delivery systems in the (Fig. 13.3) (Table 13.3).

4.1 Penetration of cell wall by passive diffusion and structural interactions

The plant cell wall is the first obstacle that nanocarriers have to overcome. Particles with diameters less than the effective pore threshold of the wall (<20 nm in most tissues) can diffuse

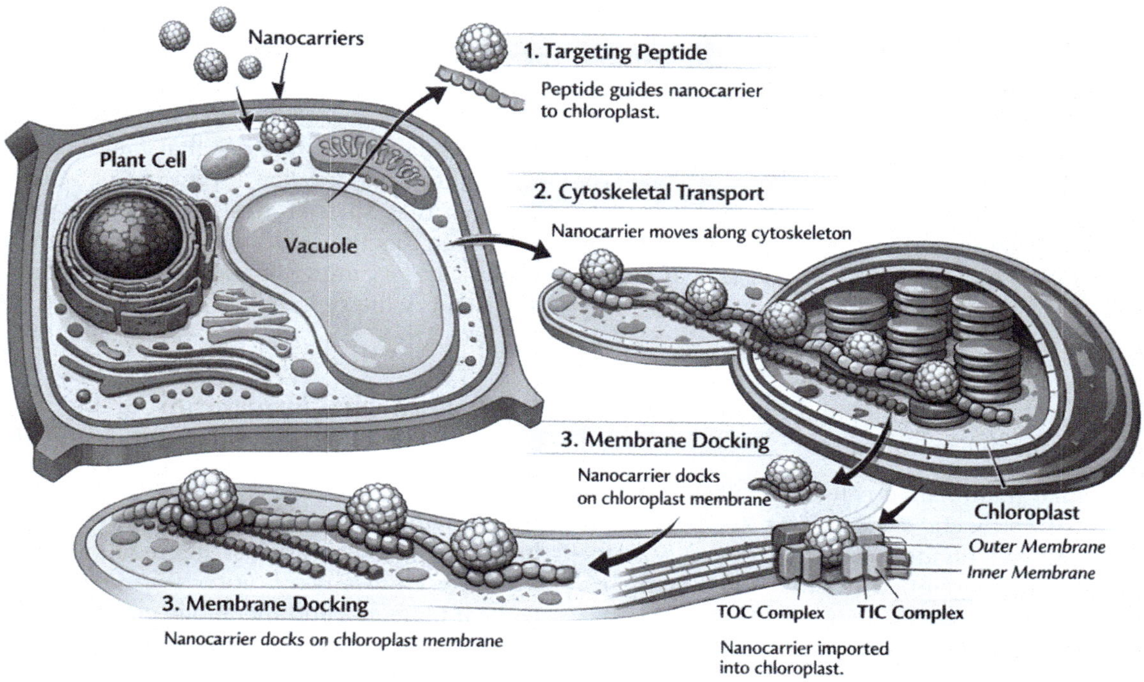

FIGURE 13.3 Mechanisms of chloroplasts targeting by nanocarriers.

TABLE 13.3 Mechanisms of nanocarrier uptake and targeting.

Mechanism	Description	Agronomy/Pathology relevance
Passive diffusion	Nanoparticles pass through naturally occurring wall pores	Reduced in lignified tissues; limited during pathogen-induced wall thickening
Membrane insertion	Direct penetration of lipid bilayer without endocytosis	Stress-modified membranes may alter insertion efficiency
Cytosolic streaming	Actin–myosin driven intracellular movement	Higher in mesophyll tissues of many crops
Peptide-based targeting	CTP or ligand-guided chloroplast localization	Useful for antiviral and immune-modulating gene delivery

passively through networks of the micro fibrils (Wu et al., 2017). High aspect ratio nanomaterials like the carbon nanotube are able to align along microchannels in the wall and slip between bundles of cellulose. In crops such as wheat, maize and sorghum, lignification's and silica deposition increase the rigidity of the walls, thus making the penetration of nanomaterials more difficult; this phenomenon is especially relevant in the engineering of the chloroplast in mature leaf and stem tissues. Pathogen infection often results in callose deposition and oxidative cross-linking which reduces wall permeability. Nanocarriers designed to have a neutral or slightly negative surface charge experience less electrostatic hindrance by constituents of the walls, which enhances penetration efficiency even in the case of modified cell walls (Lew et al., 2018).

4.2 Endocytosis independent movement of the plasma membrane

Unlike mammalian cells, a small number of classical endocytic mechanisms are present in the plant cell, owing to which exogenous materials are often trapped in endosomal compartments where they degrade, impairing the integrity of the delivered genetic cargo. Nanocarriers - especially those with a carbon-based or lipid-based material composition - display the ability to cross the cytosol via endocytosis-independent mechanisms disclosing a direct interaction with the lipid bilayer (Wu et al., 2021a,b). Hydrophobic patches of surface or an amphiphilic coating ease the passage of the carriers into the membrane, inducing transient nanoscopic disruptions, permitting the passage of the cargo without disrupting cell viability. Agronomic environments (i.e., drought, salinity-induced abiotic stresses) modify membrane lipid composition and fluidity and thus modulate the particle-membrane interaction. Pathogen infection also further remodels membrane domains in a manner that may facilitate or hinder nanocarrier permeation. Consequently, the rational design of nanocarriers that retain functional efficacy in changing membrane states is imperative in order to achieve consistent delivery in field grown crops.

4.3 Cytoskeletal inside transport: Electrostatic stability and chemical inside reactions

Following cytosol entry, nanocarriers need structural stability against aggregation and a crowded environment with organelles, cytoskeletal filaments and metabolic enzymes. Electrostatic interactions play an important role in this process: carriers with a positive surface charge prevent encapsulated nucleic acids from being degraded by cytosolic nucleases whereas PEGylated or polymer coated nanocarriers avoid non-specific protein attachment and aggregation (Serag et al., 2011a,b). Cytoskeletal structure: Cytoskeletal, specifically actin filaments, viscosity and the movement of the filaments, dynamic behavior of filaments, causes a change in the mobility of the nanoparticles. In photosynthetically active tissues such as the mesophyll of rice and soybean, increased cytoplasmic streaming increases the directional flow of the nanocarriers to the regions rich in chloroplasts. In the case of pathogen-infected tissues, high levels of reactive oxygen species and defense-related enzymes can degrade the unprotected cargo, making the need for protective coatings to achieve efficient intracellular delivery of an important consideration.

4.4 Targeting of chloroplasts by ligands, peptides and surface modifications

Effective chloroplast targeting can only be achieved through precise interactions between engineered nanocarriers and the plastid envelope. A very common approach is the conjugation of chloroplast transit peptides (CTPs) that mimic endogenous targeting sequences recognized by the TOC-TIC translocons (Lee et al., 2018). CTP-functionalized nanocarriers tend to accumulate into the outer envelope, placing genetic cargo near import sites. Other strategies include.

- Peptide ligands to envelope lipids or proteins
- pH-responsive polymers which are activated in the mildly alkaline stroma of the chloroplast
- Cationic surfaces which increase the electrostatic attraction to the negatively charged plastid membrane

These strategies are particularly useful in agronomic crops, in which the number of chloroplasts and membrane composition differs among plant species. In plant pathology applications, nanocarriers targeted to the chloroplast provide the opportunity to express antiviral RNAs or antimicrobial proteins in the organelle most closely linked to plant defense signaling.

4.5 Trafficking pathways to the cytoskeleton and chloroplasts

Once inside the cytosol, nanoparticles take advantage of passive Brownian diffusion and active mechanisms involving cytoplasmic streaming and actin-myosin-based transport that are used to encounter chloroplasts. The level of cytoplasmic streaming is dramatically increased in photosynthetically active leaves and in C4 crop plants such as maize and sorghum, further increasing the probability of nanocarrier–chloroplast interactions (Oikawa et al., 2011). for example, chloroplasts motility is modulated by environmental perturbation and pathogen invasion, chloroplasts may aggregate close to the infection foci to potentiate defense signaling, a phenomenon which may facilitate selective delivery. Some nanocarriers undergo transient interactions with cytoskeletal filaments suggestive of the existence of a guide trafficking pathway but the exact mechanisms underlying this process are still being studied.

4.6 Intra-chloroplast cargo release and functional deployment

Upon arrival into the chloroplast, the nanocarriers must release their cargo with high efficiency in the stroma of the chloroplast. Some of the release mechanisms used is.

- pH sensitive polymer degradation which is activated by alkaline pH which is characteristic of the chloroplast stroma
- Breaking down of peptide linkages through enzymic cleavage
- Ion triggered disassembly of metal-organic frameworks

- Electrostatic weakening of DNA - polymer complexes under the influence of ionic conditions of stroma
- strand displacement reaction in DNA nanostructure-based carrier (Zhao et al., 2020)

Precise cargo release is imperative for the ability to enable transgene expression, genome editing via the use of nucleases (on the Cas9 level), RNA interference, and protein function. In the domain of agronomy, such targeted release is supporting targeted rewiring of metabolism to improve photosynthetic capacity and nutrient use efficiency. In the plant pathology literature, the surveillance of the virus or pathogen at the organelle where for example, important parts of the immune pathways are, is through the release of antiviral RNAs or antimicrobial molecules from the chloroplasts.

5. Applications of nanotechnology in chloroplast engineering

Nanotechnology based chloroplast engineering represents a flexible mean to the modulation of the photosynthetic processes, the introduction of the stress tolerance traits, the increase of the use of nutrients and the establishment of a lasting resistance to pathogens and pests. Through the accurate transmittance of nucleic acids, proteins, and genome editing renouncements into the chloroplasts, nanocarriers achieve to overcome many constraints of a conventional transformation technique and facilitate the development of desired traits in model and nonmodel cultivars. The following subsections define important areas of application relevant to agronomy, plant physiology and plant pathology in the (Fig. 13.4) (Table 13.4).

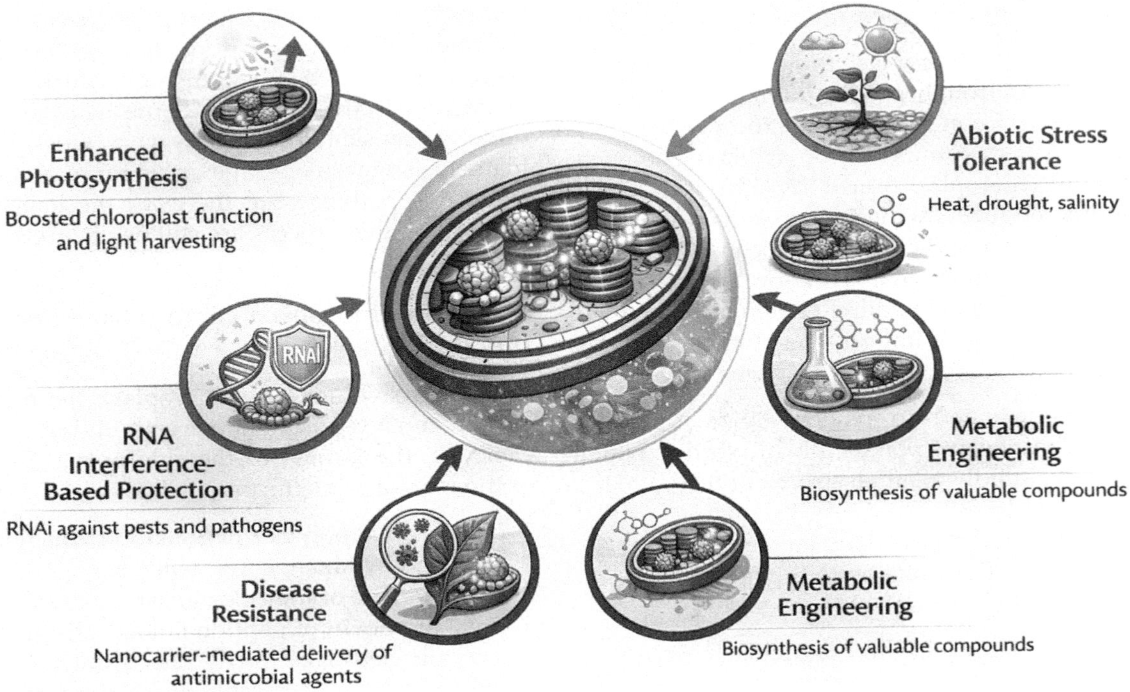

FIGURE 13.4 Applications of nanotechnology-enabled chloroplast engineering.

TABLE 13.4 Applications of nanotechnology-enabled chloroplast engineering.

Application area	Description	Examples
Photosynthesis enhancement	Improved light harvesting and carbon fixation	RuBisCO activators, photorespiration bypass
Abiotic stress tolerance	Genes aiding drought, salinity, heat adaptation	Heat shock proteins, ion homeostasis regulators
Disease resistance	Expression of RNAi and antimicrobial molecules	Antiviral dsRNA, antifungal peptides
Metabolic engineering	Production of high-value compounds	Carotenoids, terpenoids, therapeutic proteins

5.1 Improvement of photosynthetic efficiency

Chloroplasts form a fundamental hub in photosynthetic reactions making them a target for light harvested optimization, electron transport, carbon fixation, and photo protection mechanisms. The use of nanocarrier-based delivery systems is helpful for the accelerated introduction of photosynthetic regulators, synthetic RuBisCO activases and genetic constructs that would enhance the circular flow of electrons in a way that overcomes issues imposed by traditional tissue culture approaches (Nielsen et al., 2016). Transfection of plasmid-based engineered light harvesting complexes or photorespiration bypass pathways can be used to boost the quantum efficiency and enhance biomass accrual in key cereal crops such as rice, wheat and maize. Furthermore, nanoparticles that regulate the redox status of chloroplasts or which enhance the absorption of light, hold the potential to act synergistically with genetic interventions, and to optimize photosynthesis performance under variable field conditions.

5.2 Engineering abiotic stress tolerance

Chloroplasts are central in mediating different aspects of the plant abiotic stress response such as reactive oxygen species detoxification, osmotic regulation and hormone biosynthesis. Application of gene transfer by nanocarriers represents a fast method for the heterologous expression of drought tolerance genes such as transcription factors regulating stomatal conductance, salinity tolerance, such as ion transporter modulators, and heat tolerance, such as heat-shock-proteins located into the plastids (Hasan et al., 2022). The introduction of regulatory RNAs in the chloroplast genomes can fine-tune stress-induced metabolic fluxes and increase photosystem resilience to severe environmental perturbations. Nanocarriers that have been designed for stimulus-responsiveness feature release, for example, pH-sensitive or reactive oxygen species (ROS)-responsive matrices, and thus it is possible to promote stress-triggered gene activation, which enhances dynamic physiological adaptation when in the context of agriculture.

5.3 Nutrient use efficiency and metabolic rewiring

Nanotechnology-empowered chloroplast engineering promotes metabolic optimization for increase of nutrient uptake and assimilation. Delivery of genes involved in nitrogen assimilation pathways (e.g., regulatory elements in the GS-GOGAT cycle), sulfur metabolism, or phosphorus remobilization will be able to enhance nutrient-use efficiency in cereals and legumes (Chen et al., 2019). The introduction of biosynthetic modules for amino acids, vitamins or secondary metabolites with the help of nanocarrier delivery is supporting the establishment of plastid based metabolic factories. Synthetic biology approaches such as compartmentalization of metabolic pathways inside chloroplasts can be enhanced by the provision of precise transient delivery provided by nanoparticles, allowing us to avoid the need for comprehensive transgenic regeneration.

5.4 Chloroplast genome editing by using the CRISPR systems

Historically, genome editing inside the chloroplast has been limited by inefficient delivery of the components of the miRNA producing pathway (CRISPR and Cas) and the absence of simple plastid transformation formats. Nanocarriers offer a versatile platform for the direct delivery of the components of the type II A system (CRISPR-Cas proteins, guide RNAs and donor templates) to the plastids (Safi et al., 2021). Metal-Organic Frameworks (MOFs) (or DNA nanostructures) especially can encapsulate combinations of large protein-RNA complexes, protecting them during their transit through the chloroplast. Editing of chloroplast genomes allows stable modifications of photosynthetic genes, improvement of stress-resistance loci or integration of pathogen-resistance modules. Importantly, editing localized in the plastid reduces the risk of off target effects in the nuclear genome, enhancing biosafety for agricultural deployment.

5.5 RNA interference (RNAi) for plant-pathogen and pests control

Chloroplasts constitute a stable and high expression environment for producing that type of double stranded RNA (dsRNA) or small interfering RNA (siRNA) molecules that are being used in RNA interference. Nanocarriers enable the delivery of dsRNA into chloroplasts, which enable accumulation of the RNA molecules to high levels and determine the fate of key genes of pathogens or insects upon uptake or infection (Koman et al., 2022). For example.

- Delivery of dsRNA against viral replication proteins boosts antiviral immunity.
- Chloroplast expressed dsRNAs directed against fungal virulence genes suppress the development of disease.
- RNAi constructs against insect detoxification enzymes reduce pest tolerance to plant defense metabolites.

Nanocarrier enabled RNAi strategies thus present new species specific and environmentally safe alternatives for conventional agrochemicals.

5.6 Chloroplast engineering plant immunity and resistance to diseases

Chloroplasts serve as central integrators of plant immune signaling processes, generating molecules in salicylic acid (SA) and jasmonic acid (JA) signaling pathways, reactive oxygen species (ROS) outbursts, and defense metabolite production. Nanocarriers can be used to deliver genes or proteins that enhance the chloroplast-mediated immunity such as antimicrobial peptides, antiviral factors and regulators of the defense (Wu et al., 2021a,b). Engineered chloroplasts with pathogen-recognition molecules or ROS modulating enzymes can induce localized as well as systemic resistance Because plastid genomes show little gene flow via the pollen grain, chloroplast-targeted resistance traits minimize the ecological risks associated with the use of nuclear transgenes, an important issue in crop breeding and management of plant pathosystems.

5.7 Metabolic engineering in high value compounds

Chloroplasts are optimal organelles for production of high value metabolites like terpenoids, carotenoids, vitamins and therapeutic proteins. Nanocarrier based delivery allows for compartmentalized pathway installation without requiring stable transformation and thereby enables rapid prototyping of metabolite producing lines (Niu et al., 2019). Transient plastid expression through nanoparticles can be used for the production of pharmaceutical proteins, edible vaccines, or biomaterials of industrial interest. This approach has lower temporal and economic costs when compared with traditional methods of transformations and regenerations.

5.8 Field level and agronomical delivery strategies

Nanocarriers can be applied via foliar sprays, seed coatings, root drenching or nanoparticle infused fertilizers, making them compatible with large scale agricultural production. Seed-delivered nanocarriers can become part of developing plastids and contribute to metabolical improvements in early stages while foliar infiltration can introduce characteristics in the vegetative stage (Beltran et al., 2022). For plant pathology, nanoparticle sprays of topical application that carry an antiviral RNAs or antimicrobial peptides offers a nontransgenic solution for the control of emerging pathogens. These flexible modes of delivery fit the approaches of precision agriculture, which optimize the timing, dosage and spatial distribution of gene delivery interventions.

5.9 Case studies on major crop species

Several studies have demonstrated successful plasmids delivery of chloroplast genes through nanocarriers in economically significant crops, ranging from rice, maize, wheat, soybean, cotton and tomato (Li et al., 2023). Carbon nanotubes and Mesoporous Silica Nanoparticles (MSNs) have been used for high efficiency transient expression in monocots that are traditionally recalcitrant for plastid transformation. In tomato and potato, nanocarriers loaded with viral-interfering RNAs have shown reduced viral load, as well as enhanced resilience. These case studies highlight the generality of chloroplast engineering based on nanotechnology for various agricultural species and production conditions.

6. Challenges, biosafety and regulatory aspects

The significant promise of nanotechnology-enabled chloroplast engineering is balanced by a host of technical, environmental and regulatory complexities that will need to be systematically overcome before large-scale agricultural-level deployment is possible. These challenges include issues both associated with the design of the nanoparticles and their stability inside cells, with ecological consequences, with human safety, as well as with regulatory approval requirements. A thorough knowledge of these constraints is an absolute requirement for the development of responsible, field-compatible gene delivery systems for chloroplasts.

6.1 Nanocarrier design and delivery technical limitations

A major challenge of exhaustion in the implementation of nanocarrier mediated chloroplast engineering has been achieving consistent delivery efficiency in a range of plant species and tissue types. Variations in the composition of the cell wall—for example, increased lignification in cereals, cuticular thickness in horticultural crops and stress-induced callouse deposit—-strongly affect the penetration of nanoparticles

(Mitter et al., 2017). The optimization of nanocarrier size, surface charge, and chemistry for every single crop is a labor intensive process.

Moreover, the intracellular trafficking routes are different between plant species, developmental stages and environmental conditions, which complicates the approach of developing universally applicable nanocarrier systems. Making sure that the cargo is released within the chloroplast stroma and not prematurely dissociated is another important technical challenge.

6.2 Nanoparticle persistence and biodegradation and environmental fate

The environmental fate of nanomaterials is a major biosafety issue around the use of nanomaterials in agriculture. Carbon nanotubes and some metallic nanoparticles have long degradation rates and hence cause concerns about long-term accumulation in the soils, waters and plant tissue (Kalaji et al., 2022). Potential adverse effects on the soil micro-organisms, beneficial insects and aquatic ecosystems must therefore be carefully assessed. In contrast biodegradable formulations, including mesoporous silica nanoparticles, polymeric carriers and metal-organic frameworks tend to follow environmentally compatible degradation pathways. Nevertheless, degradation kinetics may change under specific soil conditions, so that a detailed evaluation of nanoparticle persistence before its deployment in the field is required. Incorporation of materials responsive to the environment and design of nanocarriers that breakdown into innocuous by-products is a critical approach to reducing ecology.

6.3 Biosafety in crops for improvement and plant-pathology usage

Chloroplast engineering presents some of the inherent biosafety advantages of primarily maternal inheritance of plastid genomes in most crops (minimizing gene flow through pollen). This is a very valuable attribute when introducing disease resistance traits, antimicrobial peptides or RNAi constructs directed at pest or pathogen genes (Ghosh et al., 2023). However, there is a need to carefully consider the biosafety profile of the nanocarriers themselves. Phytotoxicity risks such as oxidative stress or membrane destabilization or dysfunction of chloroplasts are dependent on the nanoparticle concentration, coating materials and frequency of application. In the context of plant pathology, off target effects on nonpathogenic or beneficial microbes should be evaluated, particularly in the implementation of RNAi based strategies. Safety evaluations must include both short-term physiological responses as well as long-term ecological interactions.

6.4 Frameworks for regulating nanotechnology in agriculture

Regulation of nanotechnology in agriculture is still in a developing stage all around the world. Nanocarriers used for gene delivery have a unique regulatory niche because they need not always involve permanent genetic changes in heritable progeny: there are many systems that enable temporary expression without the need to perturb the nuclear genome. This creates complexity to the classification under the current genetically modified organism (GMO) regulatory paradigms (van Bezouwen et al., 2017). Regulatory authorities are faced with trying to determine if nanocarrier-enabled chloroplast engineering is in the ambit of GMO, non-GMO, or nanopesticide frameworks. In the European Union, strict environmental and traceability requirements are imposed, while in other jurisdictions, risk based approaches to assessment are taken. Harmonization of regulatory standards, set up of nanoparticle-specific testing protocols and creation of transparent safety evaluation frameworks will be indispensable to successful commercialization.

6.5 Human and food safety issues

For instance, when nanocarriers are used for crops that will be eaten by animals or humans, there might be implications in terms of human health that need to be thoroughly investigated. Residues from persistent nanomaterials may become concentrated in edible tissue, requiring that the absorption, digestion and metabolic fate be rigorously assessed in humans and animals (Fuentes et al., 2016).

Although biodegradable systems address these issues, the safety of system degradation products has been a precondition for approval. For traits that are expressed in chloroplasts, such as vaccines or therapeutic proteins, assessment of food safety is required with regard to dosage, allergenicity, and protein stability. Regulatory endorsement will require extensive risk assessment, toxicological studies and formulation of standardized guidelines describing nanoparticle behavior in food matrices.

6.6 Scalability and field deployment issues

Translating laboratory breakthroughs to field-level implementation is still fraught with challenges that can be attributed to the variability in environmental conditions, crop phenology, and logistics of delivery. Foliar spraying or seed-coating methods should guarantee homogeneous distribution of nanoparticles on a large area and heterogeneous microclimates (Singh et al., 2015). Field conditions, such as rainfall, ultraviolet exposure and temperature fluctuations, can affect the stability of the nanoparticles and their delivery efficacy.

Moreover, integration of nanocarrier based gene delivery to existing agricultural practices requires compatibility with irrigation, fertilization and crop protection regimes. The development of robust, cost-efficient and farmer-friendly application systems will be critical for ensuring the widespread adoption.

6.7 Ethical, social and economic issues

Public perception of nanotechnology in agriculture is highly variable and has a decisive effect on the adoption rates. Concerns about the presence of nanoparticles in food and environmental consequences, as well as unexpected long term effects, often raise skepticism. Ethical considerations involve issues of equitable access to advanced technologies and the risk of economists that socioeconomic disparities will exist between areas that adopt nanotechnology-enable crop improvements and those that do not (Ma et al., 2015). Transparent communication, stakeholder engagement, and participatory decision-making processes are crucial for the promotion of responsible innovation.

7. Future perspectives and new opportunities

Nanotechnology-based chloroplast engineering is likely to play a transformative role in technological improvement of crops in future agriculture, sustainable agriculture and in plant diseases. While significant progress has been made in understanding the mechanism of delivery and engineering of nanocarrier systems, there have been new advancements in materials design, synthetic biology, and plant physiology that are extending choice of applications of this technology. Future studies will focus on improving precision, scalability and environmental compatibility and tackling regulatory challenges and acceptance.

7.1 Precision nanocarriers for species/site specific delivery

Advances in the engineering of nanomaterials will make it possible to create precision nanocarriers tailored for specific crop species or tissue types or developmental phases. By making use of machine learning and predictive modeling, it's potential to design highly flexible physicochemical features of nanocarriers for traversing distinctive cell-wall assorted architectures, equivalent to the distinction between monocots and dicots (Aslani et al., 2015).

Tissue specific delivery of molecules - targeting guard cells, bundle sheath cells in C4 plants, or fruit tissues in horticultural crops - provides possibilities to engineer photosynthesis, nutrient allocation or postharvest traits with unprecedented accuracy. These carrier systems can contain ligand-based recognition, charge-switching behavior or stimulus-responsive behavior that becomes activated only in desired tissues.

7.2 Combining nanotechnology and plastid synthetic biology

As plastid synthetic biology continues to advance, there is a growing need for delivery platforms that are efficient for the rapid prototyping of metabolic pathways, regulatory circuits and gene expression modules. Nanocarriers will allow test becomes possible iterative testing of metabolic designs inside the chloroplasts without the requirement of transgenic regeneration for each construct (Moghaddam et al., 2021). This advancement will lead to the quick development of plastid-wise biosynthetic pathways for terpenoids, pharmaceuticals, nutraceuticals and bio-based polymers. Future systems may include modular DNA nanostructures that are able to assemble multi-genes clusters or scaffold complete pathways within the chloroplasts, to optimize the spatial organization for higher metabolic flux.

7.3 Developments of chloroplast genome editing

As anticipated, the upcoming decade is likely to see dramatic development of the capability to perform chloroplast genome editing. Owing to the development of effective nanocarriers to deliver proteins and RNAs associated with the nucleases and endonucleases to chloroplasts, such as the creation of efficient nucleocarriers to deliver Cas9 and C3CRT into chloroplasts; this feature, coupled with the development of base editors and prime-editing tools to yield durable mutations in chloroplast genomes, will be instrumental in this process. Overcoming existing limitations in plastid editing such as a limited efficiency of homology -directed repair and delivery of exogenous DNA will lead to the ability to accurately edit photosynthetic genes and stress -response regulators and pathogen resistance loci (Bock R and Warzecha H. 2010). Combining ultra-stable nano-carrying particles with programmable release mechanisms will enable spatially controlled editing which may in turn enable organelle-specific gene reprogramming in critical stages of development or under environmental stress.

7.4 Plant immunity enhancement using nanotechnology

Future plant pathology applications are likely to focus on the manipulation of immune-mediated responses by manipulating chloroplasts. Chloroplasts also act as signaling centers for salicylic acid, jasmonic acid, ROS and other defense metabolites; thus, local delivery of regulatory RNAs, antimicrobial peptides, or genome editing reagents can help to fine tune immune reactions with high specificity (Boehm et al., 2016). Emerging strategies include nanoparticle stabilized constructs of dsRNAs that are designed to silence essential genes associated with the pathogen or inhibit viral replication factors. Nanocarriers that carry modules for pathogen recognition could provide rapid and inducible responses upon infection thereby presenting a good alternative to conventional resistant variety breeding.

7.5 Environmentally responsive and smart nanocarrier systems

Stimulus-responsive nanocarriers-which are activated by light or heat, reactive oxygen species (ROS) levels, pH, or enzyme released by pathogens-are very interesting prospects for dynamic regulation of the expression of chloroplast genes (Lim et al., 2021). These "smart" systems are able to release the cargo in response only to drought, heat, salinity or infection and reduce the off target effects, while increasing efficiency of resource use. For example, ROS-responsive MOFs could release antioxidant enzymes at times of oxidative burst or pathogen–enzyme–triggered DNA nanocarriers could respond in a specific way to deliver double–stranded RNA at sites of infection. Such technologies are in support of precision agriculture and can lead to a decrease in the dependency on chemical pesticides.

7.6 Field-deployable ways of delivery and scalable formulations

Transition from proof-of-concept in the laboratory to application at the farm scale requires research focused on delivery methods which are practical, robust, and compatible with mechanized agriculture. One of the advancements can be nanoparticle-coated seeds, foliar sprays with improved adhesion and penetration, microgel-based slow-release systems and drone-mediated application technologies (Pivato et al., 2020). Formulations must be resistant to environmental effects such as UV, rain, temperature and guarantee uniform delivery to large areas of crops. Nanocarriers that complement current fertilizer or irrigation systems will also help with adoption.

7.7 Sustainability and safety and regulatory innovation

Sustainable nanocarriers-shareably made of biodegradable polymers, plant-based or nutrient-releasing matrices-will be fundamental to the compatibility with the ecosystems in the long term (Raliya et al., 2018) Future regulatory frameworks will have to adapt to hybrid technologies that blur the boundaries between genetically modified organisms (GMOs); non-transgenic approaches and nanoparticles-based interventions. Transparent safety assessments, standardized testing protocols and international harmonization of regulation will help to facilitate worldwide deployment. A key area of opportunity, however, is to develop "transient engineering" platforms, where nanocarriers are used to introduce genetic material for a short time to minimize regulatory burdens, whilst allowing rapid trait improvement.

7.8 Intersection of AI, nanotechnology, and plant sciences

The future of chloroplast engineering may be determined by the combination of artificial intelligence (AI), high-throughput phenotyping and computational modeling. AI can help predict nano-bio interactions, optimize the design of nanocontainers and identify the peptides that could target the chloroplasts with high accuracy (Ashraf et al., 2023). Coupled with automated gene-circuits design as well as metabolic modeling, these will allow one to rationally engineer photosynthesis, metabolism and immunity on a systems basis. This coming together will lead to a rapid development of innovation and enable real-time optimization of the gene delivery strategies using nanoparticle based chloroplast.

7.9 Toward climate resilient and disease resistant agriculture

Ultimately, nanotechnology based chloroplast engineering provides a route to building climate resilient crop systems that can endure drought, heat, salinity and future pathogen threats. By allowing the rapid, accuracy and nontransgenic introduction of beneficial characteristics, nanocarrier systems have the potential to greatly reduce breeding cycles and ensure food security. The possibility of synergizing the enhanced efficiency of photosynthesis, efficient metabolism of nutrients, and high-level disease resistance makes chloroplast nanobiotechnology one of the keys to future sustainable agriculture.

8. Conclusion

Nanotechnology-enabled chloroplast engineering is a major breakthrough at the crossroads of plant biotechnology, agronomy, and plant pathology. By overcoming the constant challenges of traditional methods used for the transformation of plastids, nanocarrier systems offer unprecedented levels of precision, versatility, and scalability for the delivery of genetic material directly to chloroplasts. Through the painstaking control of nanoparticle dimensions and surface chemistry, as well as the functionalization of surface ligands, these platforms enable efficient penetration of the cell wall, cytosolic context, and the chloroplast envelope (traditionally limiting gene transfer processes, particularly in mature tissues and agronomically relevant crops).

The ability to engineer chloroplasts with nucleic acids, proteins, and reagents based on RNAi and the so-called "CRISPR" technology opens new possibilities for enhancing the photosynthetic efficiency, the metabolism of nutrients, and tolerance to abiotic stresses such as drought, salinity, and heat. Within the scope of plant pathology, nanocarrier-driven delivery of RNA interference molecules, antimicrobial peptides, and genome editing constructs are promising alternative to chemical pesticides and transgenic methodologies and has the benefit of further biosafety features, with plastid-specific inheritance. These developments are in line with global agricultural goals that focus on climate resilience, sustainable agricultural intensification, and environmentally acceptable plant protection.

However, there are still a number of challenges. Achieving uniform delivery across different crop species, elucidating the long-term fate of nanoparticles in the environment, and farmer-friendly and field-compatible systems is a key prerequisite for widespread deployment of this technology. Equally critical are transparency in regulation, public confidence, and rigorous safety evaluations taking into account the behavior of nanomaterials as well as their biological impact on food and ecosystem health. Emerging trends, for example, stimulus-responsive nanocarriers, artificial intelligence-driven design, and advanced synthetic biology frameworks, have the promise to overcome many of these limitations.

In sum, the trend in chloroplast engineering is tilting increasingly toward integrated nanotechnology-type solutions. As research continues to improve and pragmatic hurdles are overcome, the capacity of nanoscale delivery platforms can be used to revolutionize plant metabolic engineering, fortify plant immunity, and accelerate the development of climate-resilient and disease-resistant cultivars. The blending of nanotechnology and plastid biology offers a unique chance of redefining modern-day agriculture and hence enhance world food security and promote sustainable crop production in the face of increasing environmental uncertainty.

References

Allorent, G., Petroutsos, D., 2017. Chloroplast signaling and plant development under stress. Curr. Opin. Plant Biol. 37, 92–100. https://doi.org/10.1016/j.pbi.2017.03.012.

Ashraf, S., Pelaz, B., del Pino, P., et al., 2023. Concepts, design and perspectives of plant nanobiotechnology. Nat. Nanotechnol. 18, 665–679. https://doi.org/10.1038/s41565-023-01485-y.

Aslani, F., Bagheri, S., Muhd Julkapli, N., Juraimi, A.S., Hashemi, F.S.G., Baghdadi, A., 2015. Effects of engineered nanomaterials on plants growth: an overview. Sci. Rep. 5, 10697. https://doi.org/10.1038/srep10697.

Beltran, J., Ita-Nagy, M., Tsyusko, O., Keller, A.A., 2022. Mesoporous silica nanoparticles in agriculture: mechanisms, applications, and challenges. Agric. Syst. 195, 103314. https://doi.org/10.1016/j.agsy.2021.103314.

Bhattacharyya, D., Gnanasekaran, P., Kumar, R.K., Kushwaha, N.K., Sharma, V.K., Yusuf, M.A., Chakraborty, S., 2015. Chloroplast functions during viral infection: plant defense and pathogen counter-defense. Plant Cell Rep. 34, 987–994. https://doi.org/10.1007/s00299-015-1768-0.

Bock, R., 2015. Engineering plastid genomes: methods, tools, and applications in basic research and biotechnology. Annu. Rev. Plant Biol. 66, 211–241. https://doi.org/10.1146/annurev-arplant-050213-040212.

Bock, R., Warzecha, H., 2010. Solar-powered factories for new vaccines and antibiotics. Trends Biotechnol. 28 (5), 246–252. https://doi.org/10.1016/j.tibtech.2010.02.004.

Boehm, C.R., Pollak, B., Purswani, N., Patron, N., Haseloff, J., 2016. Synthetic biology tools for chloroplast engineering. Plant Physiol. 172 (3), 1309–1322. https://doi.org/10.1104/pp.16.01302.

Carpita, N.C., McCann, M.C., 2010. The plant cell wall. Plant Physiol. 154, 483–486. https://doi.org/10.1104/pp.110.161240.

Chen, L., Sun, T., Song, Y., 2019. Chloroplast membrane interactions with nanomaterials: biophysical insights into targeted delivery. J. Mater. Chem. B 7, 6744–6752. https://doi.org/10.1039/C9TB01124F.

Cunningham, F.J., Goh, N.S., Demirer, G.S., Matos, J.L., Landry, M.P., 2018. Nanoparticle-mediated delivery towards advancing plant genetic engineering. Trends Biotechnol. 36 (9), 882–897. https://doi.org/10.1016/j.tibtech.2018.03.009.

Demirer, G.S., Zhang, H., Matos, J.L., Goh, N.S., Cunningham, F.J., Sung, Y., Chang, R., Aditham, A.J., Chio, L., Cho, M.J., Staskawicz, B., Landry, M.P., 2019a. High-aspect-ratio nanomaterials enable delivery of functional genetic material without DNA integration in mature plants. Nat. Nanotechnol. 14 (5), 456–464. https://doi.org/10.1038/s41565-019-0382-5.

Demirer, G.S., Zhang, H., Matos, J.L., Goh, N.S., Cunningham, F.J., Sung, Y., Chang, R., Aditham, A.J., Chio, L., Cho, M.J., Staskawicz, B., Landry, M.P., 2019b. High-aspect-ratio nanomaterials enable delivery of functional genetic material without DNA integration in mature plants. Nat. Nanotechnol. 14 (5), 456–464. https://doi.org/10.1038/s41565-019-0382-5.

Fuentes, P., Zhou, C., Wang, X., et al., 2016. Encoding novel metabolic pathways in chloroplasts for enhanced plant biosynthesis and productivity. Nat. Commun. 7, 11429. https://doi.org/10.1038/ncomms11429.

Ghosh, S., Sarkar, A., Chattopadhyay, A., Das, S., 2023. Nanoparticle-mediated delivery of dsRNA for effective management of plant viral diseases. Plant Mol. Biol. 111 (1–2), 1–18. https://doi.org/10.1007/s11103-022-01355-7.

Giraldo, J.P., Landry, M.P., Faltermeier, S.M., McNicholas, T.P., Iverson, N.M., Boghossian, A.A., Reuel, N.F., Hilmer, A.J., Sen, F., Brew, J.A., Strano, M.S., 2014. Plant nanobionics approach to augment photosynthesis and biochemical sensing. Nat. Mater. 13 (4), 400–408. https://doi.org/10.1038/nmat3890.

Hasan, M.M., Bhatnagar, I., Grömping, U., Mecklenburg, M., Staiger, D., 2022. Mechanisms of nanomaterial trafficking in plant cells: cytoskeletal and endomembrane interactions. Plant Cell Rep. 41, 1981–1996. https://doi.org/10.1007/s00299-022-02892-3.

Inoue, H., Li, M., Schnell, D.J., 2016. An essential role for chloroplast protein translocation in plant development. Nat. Plants 2, 16085. https://doi.org/10.1038/nplants.2016.85.

Kalaji, H.M., Dabbous, A., Cetner, M.D., Samborska, I.A., Łukasik, I., Brestic, M., 2022. Nanoparticle-mediated modulation of chloroplast performance under drought and salinity stress. Plant Cell Rep. 41, 1643–1658. https://doi.org/10.1007/s00299-022-02866-5.

Keegstra, K., Cline, K., 1999. Protein import and routing in chloroplasts. Nat. Rev. Mol. Cell Biol. 6, 281–291. https://doi.org/10.1038/nrm1613.

Koman, V.B., Lew, T.T.S., Wong, M.H., et al., 2022. Nanoparticle-mediated delivery of functional biomolecules into plant chloroplasts. Sci. Adv. 8 (4), eabn0034. https://doi.org/10.1126/sciadv.abn0034.

Kwak, S.-Y., Lew, T.T.S., Sweeney, C.J., Koman, V.B., Wong, M.H., Bohmert-Tatarev, K., Snell, K.D., Seo, J.-S., Chua, N.-H., Strano, M.S., 2019. Chloroplast-selective gene delivery and expression in planta using chitosan-complexed single-walled carbon nanotube carriers. Nat. Nanotechnol. 14 (5), 447–455. https://doi.org/10.1038/s41565-019-0375-4.

Lam, P.-L., Wong, R.S., Gambari, R., 2021. Polymer–lipid hybrid nanoparticles for nucleic acid delivery:

development and characterization. J. Mol. Liq. 330, 115584. https://doi.org/10.1016/j.molliq.2021.115584.

Lee, D.W., Woo, S., Geem, K.R., Hwang, I., 2018. Chloroplast targeting sequences: structure, function, and mechanism. Front. Plant Sci. 9, 50. https://doi.org/10.3389/fpls.2018.00050.

Lew, T.T.S., Wong, M.H., Kwak, S.Y., Sinclair, R., Koman, V. B., Strano, M.S., 2018. Rational design principles for the transport and subcellular distribution of nanomaterials into plant protoplasts. Small 14 (44), 1802086.

Li, H.-M., Chiu, C.-C., 2010. Protein transport into chloroplasts. Annu. Rev. Plant Biol. 61, 157–180. https://doi.org/10.1146/annurev-arplant-042809-112222.

Li, J., Fan, X., Qi, Y., Zhang, Q., Li, M., Wang, J., Yang, Y., 2023. Chloroplast-targeted CRISPR tools for genome engineering and trait improvement. Trends Plant Sci. 28 (3), 324–339. https://doi.org/10.1016/j.tplants.2022.10.005.

Lim, C.Y., Zhang, Y., Chen, H., Ding, Y., Zhao, Y., 2021. Advanced nanomaterials for precision agriculture: a review of next-generation tools. Adv. Mater. 33 (44), 2103473. https://doi.org/10.1002/adma.202103473.

Ma, C., White, J.C., Dhankher, O.P., Xing, B., 2015. Metal-based nanotoxicity and plant physiology: biological mechanisms and implications. Environ. Sci. Technol. 49 (1), 710–721. https://doi.org/10.1021/es503425a.

Mitter, N., Worrall, E.A., Robinson, K.E., et al., 2017. Clay nanosheet delivery of RNA interference for sustainable crop protection. Nat. Plants 3, 16207. https://doi.org/10.1038/nplants.2016.207.

Moghaddam, P.R., Husain, T., Kharrazi, S., Fassi, M., Taheri, A., 2021. Environmental fate and lifecycle analysis of nanomaterials in agriculture: risks and regulatory perspectives. NanoImpact 24, 100362. https://doi.org/10.1016/j.impact.2021.100362.

Mondal, A., Basu, R., Hasnain, S.E., Ghosh, A., 2021. Polymeric nanocarriers for plant gene delivery. Plant Physiol. Biochem. 162, 67–79. https://doi.org/10.1016/j.plaphy.2021.02.016.

Newkirk, G.M., de Allende, P., Jinkerson, R.E., Giraldo, J.P., 2021. Nanotechnology approaches for chloroplast biotechnology advancements. Front. Plant Sci. 12, 691295. https://doi.org/10.3389/fpls.2021.691295.

Nielsen, A.Z., Ziersen, B., Jensen, K., et al., 2016. Redirecting photosynthetic reducing power toward specific metabolic pathways in chloroplasts. Plant Physiol. 171 (2), 1309–1321. https://doi.org/10.1104/pp.16.00174.

Niu, Y., Jin, G., Zhang, Y., 2019. Nanotechnology strategies to improve plant photosynthesis under environmental stress. Environ. Sci. Nano 6, 3221–3237. https://doi.org/10.1039/C9EN00766G.

Oikawa, K., Yamasato, A., Kong, S.-G., Kasahara, M., Nakai, M., Takahashi, F., Teramoto, H., Tabata, S., Wada, M., 2011. Chloroplast movement mechanisms revisited: actin/motor-dependent and -independent pathways. Nat. Commun. 2, 238. https://doi.org/10.1038/ncomms1236.

Perlatti, B., de Souza Bergo, P.L., Fernandes, J.B., Forim, M. R., 2020. Nanotechnology in agriculture: lipid-based nanoformulations for insect control. J. Agric. Food Chem. 68 (15), 3977–3990. https://doi.org/10.1021/acs.jafc.0c00097.

Pivato, M., Fabris, M., Ravelli, R.B., 2020. Plastid genome editing: current state and future perspectives. Curr. Opin. Biotechnol. 64, 44–51. https://doi.org/10.1016/j.copbio.2019.09.004.

Powikrowska, M., Oetke, S., Jensen, P.E., 2014. Dynamics of plastid nucleoids during chloroplast differentiation. J. Exp. Bot. 65 (9), 2687–2701. https://doi.org/10.1093/jxb/eru145.

Raliya, R., Saharan, V., Dimkpa, C., Biswas, P., 2018. Nanofertilizer for precision and sustainable agriculture: current state and future perspectives. J. Agric. Food Chem. 66 (26), 6480–6490. https://doi.org/10.1021/acs.jafc.8b00691.

Safi, A., Fan, C., Pinals, R.L., Zhang, H., Landry, M.P., 2021. Nanocarrier-assisted delivery enhances chloroplast localization of biomolecules in mature leaves. ACS Nano 15 (6), 10185–10196. https://doi.org/10.1021/acsnano.1c01923.

Schwartz, S.H., Hendrix, B., Hoffer, P., Sanders, R.A., Zheng, W., 2020. Carbon dots for efficient small interfering RNA delivery and gene silencing in plants. Plant Physiol. 184 (2), 647–657. https://doi.org/10.1104/pp.20.00733.

Serag, M.F., Kaji, N., Nakanishi, J., Tokeshi, M., Baba, Y., 2011a. Carbon nanotubes: biomaterial applications. Chem. Soc. Rev. 40 (7), 3964–3974. https://doi.org/10.1039/c0cs00136h.

Serag, M.F., Kaji, N., Niwa, D., Tokeshi, M., Shankla, M., Baba, Y., 2011b. Uptake and intracellular fate of engineered nanomaterials: mechanisms and applications. ACS Nano 5 (8), 6425–6438. https://doi.org/10.1021/nn202025p.

Singh, A., Singh, N.B., Hussain, I., Singh, H., Singh, S.C., 2015. Plant–nanoparticle interactions: an assessment of toxicity and biosafety. Plant Physiol. Biochem. 91, 125–137. https://doi.org/10.1016/j.plaphy.2015.04.014.

Sun, T., Zhang, Y., Zheng, D., Du, Y., Song, Y., Nie, G., 2020. Metal–organic frameworks as carriers for controlled delivery of bioactive molecules in plants. J. Am. Chem. Soc. 142 (52), 21549–21559. https://doi.org/10.1021/jacs.0c10683.

Torney, F., Trewyn, B.G., Lin, V.S.-Y., Wang, K., 2007. Mesoporous silica nanoparticles deliver DNA and chemicals into plants. Nat. Nanotechnol. 2, 295–300. https://doi.org/10.1038/nnano.2007.108.

van Bezouwen, L.S., Caffarri, S., Kale, R.S., Kouřil, R., Thunnissen, A.M.W., Oostergetel, G.T., Boekema, E.J., 2017. Subunit and chlorophyll organization of the plant photosystem II supercomplex. Nat. Plants 3 (7), 1–11.

Voigt, C.A., 2014. Callose-mediated resistance to pathogenic intruders in plant defense-related papillae. Front. Plant Sci. 5, 168. https://doi.org/10.3389/fpls.2014.00168.

Wallace, I.S., Anderson, C.T., 2012. Genetic modulation of plant cell wall structure and function. Plant Sci. 193–194, 61–69. https://doi.org/10.1016/j.plantsci.2012.05.003.

Wang, P., Lombi, E., Zhao, F.J., Kopittke, P.M., 2016. Nanotechnology: a new opportunity in plant sciences. Trends Plant Sci. 21 (8), 699–712. https://doi.org/10.1016/j.tplants.2016.04.005.

Wong, M.H., Giraldo, J.P., Kwak, S.-Y., Koman, V.B., Sinclair, R., Lew, T.T.S., Bisker, G., Liu, P., Strano, M.S., 2017. Nitroaromatic detection and infrared communication from wild-type plants using plant nanobionics. Nat. Mater. 16 (2), 264–272. https://doi.org/10.1038/nmat4771.

Wu, H., Tito, N., Giraldo, J.P., 2017. Anionic cerium oxide nanoparticles protect plant photosynthesis under abiotic stress. Nat. Nanotechnol. 12 (8), 761–765. https://doi.org/10.1038/nnano.2017.99.

Wu, X., Gong, X., Yang, H., et al., 2021a. Nanoparticle-induced cell membrane permeability enhancement: mechanisms and safety assessment. Small 17 (15), 2006231. https://doi.org/10.1002/smll.202006231.

Wu, H., Shabala, L., Azzarello, E., Huang, Y., Pandolfi, C., Su, N., Shabala, S., 2021b. Genome-based engineering of chloroplast function improves photosynthesis and abiotic stress resilience. Plant Physiol. 187 (1), 57–71. https://doi.org/10.1093/plphys/kiab074.

Zeng, Y., Li, B., Jin, Z., 2017. Reactive oxygen species and plant cell death. Plant Signal. Behav. 12 (8), e1305746. https://doi.org/10.1080/15592324.2017.1305746.

Zhang, H., Demirer, G.S., Pinals, R.L., Chang, R., Goh, N.S., Aditham, A.J., Cunningham, F.J., Fan, C., Landry, M.P., 2019a. DNA nanostructures coordinate gene silencing in mature plants. Proc. Nat. Acad. Sci. USA 116 (15), 7543–7548. https://doi.org/10.1073/pnas.1818290116.

Zhang, H., Demirer, G.S., Pinals, R.L., Chang, R., Fan, C., Landry, M.P., 2019b. Engineering DNA nanostructures for delivery of genetic material in plants. ACS Nano 13 (3), 2913–2924. https://doi.org/10.1021/acsnano.8b08728.

Zhao, Q., Chen, W., Qin, W., Li, L., Sun, T., Song, Y., 2020. Stimuli-responsive nanocarriers for precise delivery of genetic cargo in plant cells. ACS Appl. Mater. Interfaces 12 (22), 24899–24910. https://doi.org/10.1021/acsami.0c03429.

CHAPTER

14

Life cycle assessment (LCA) of nano-harvest technologies: Evaluating environmental and economic sustainability

Mohammad Faizan[1], S. Maqbool Ahmed[1], Vishnu D. Rajput[2] and Shamsul Hayat[3]

[1]Botany Section, School of Sciences, Maulana Azad National Urdu University, Hyderabad, Telangana, India; [2]Academy of Biology and Biotechnology, Southern Federal University, Rostov-On-Don, Russia; [3]Department of Botany, Aligarh Muslim University, Aligarh, Uttar Pradesh, India

1. Introduction

Agriculture is undergoing a major technological transformation driven by the need to increase food production under constraints of climate change, land degradation, water scarcity, and rising energy costs (Alam et al., 2025). Modern farming systems face the dual challenge of producing more food while simultaneously reducing environmental footprints (Dogan et al., 2025). Harvesting and postharvest stages, in particular, represent critical points where resource efficiency and sustainability can be enhanced. Harvesting operations consume substantial fuel and energy, contribute to greenhouse gas (GHG) emissions, generate mechanical losses, and can degrade soil structure through compaction. Postharvest losses remain a major global challenge, particularly in developing countries where storage, transport, and packaging infrastructures are often limited. Reducing losses can be as impactful as increasing yields, because food that is not lost does not need to be produced, harvested, transported, and processed again.

Nano-harvest technologies are broadly defined as nano-enabled tools, materials, and systems that improve harvesting operations and postharvest processes. These technologies include nanosensors for harvest timing, nano-enabled machine coatings that reduce friction and wear, nano-based lubricants that reduce fuel consumption, nano-encapsulated formulations that improve crop quality before harvest,

Nanotechnology in Photosynthesis
https://doi.org/10.1016/B978-0-443-44376-3.00014-8

and nano-enabled packaging materials that extend shelf-life. Their adoption is being promoted in the context of precision agriculture, smart farming, and sustainable intensification. However, nano-enabled systems also introduce new environmental concerns, such as energy-intensive nanomaterial synthesis, chemical reagent use, potential nanoparticle release, and uncertain toxicity effects in soil and aquatic environments.

Life Cycle Assessment (LCA) provides a systematic and standardized approach to quantify environmental impacts across the full life cycle of products and technologies. LCA has been extensively used for conventional agricultural systems, fertilizers, pesticides, machinery, packaging, and food supply chains. Yet, applying LCA to nano-enabled technologies requires additional methodological considerations. Engineered nanomaterials (ENMs) may behave differently from bulk materials, and their environmental fate, exposure pathways, and toxicity profiles are not fully represented in traditional LCA impact models.

This chapter addresses these challenges by presenting a complete and detailed LCA framework for nano-harvest technologies, integrating environmental LCA with Life Cycle Costing (LCC) for economic sustainability. The chapter aims to support researchers, policy makers, and industry stakeholders in evaluating nano-harvest technologies in a transparent, science-based manner.

2. Nano-harvest technologies: Concepts, categories, and applications

2.1 Concept

Nano-harvest technologies represent an integrated and rapidly developing class of nano-enabled innovations designed to enhance harvesting efficiency, reduce postharvest losses, and preserve crop quality across the entire harvest value chain. Importantly, the term "nano-harvest" should not be interpreted narrowly as the use of nanomaterials only in harvesting machinery (Faizan et al., 2018, 2024, 2025). Rather, it reflects a systems-based concept in which nanotechnology intervenes at multiple stages that collectively determine the proportion of total biological yield that becomes marketable produce, the energy and labor required to harvest it, and the stability of harvested commodities during storage and distribution. In this broader framing, nano-harvest technologies encompass nanomaterials, nanosensors, nanocoatings, nano-enabled formulations, and nanocomposites that operate at three strongly interlinked levels: preharvest optimization, harvest-stage performance enhancement, and postharvest protection and monitoring. This multi-level view is consistent with the wider literature on agricultural nanotechnologies, which emphasizes that nano-enabled solutions often create value not only by increasing productivity but also by improving efficiency and reducing losses along the farm-to-fork chain (Parisi et al., 2015; Gogos et al., 2012; Kah et al., 2013).

At the preharvest optimization level, nano-harvest technologies influence harvest outcomes indirectly by improving crop uniformity, physiological maturity, and resilience before harvest. Harvest efficiency is strongly affected by plant health and synchronized ripening, because uneven maturity, lodging, drought stress, nutrient imbalance, or disease pressure can increase harvesting time, raise fuel consumption, and elevate field losses through shattering, bruising, or incomplete collection. Nano-enabled nutrient carriers and nano-formulated agro-inputs (including nano-silica, nano-chelated micronutrients, and controlled-release nano-encapsulated fertilizers) are increasingly investigated for improving nutrient use efficiency, stabilizing photosynthesis under stress, and strengthening plant tissues, thereby improving harvest ability and reducing preharvest quality decline (Kah et al., 2013; Gogos et al., 2012). In parallel, nanosensors and nano-biosensors deployed in precision agriculture can support harvest decision-making by monitoring real-time indicators such as moisture content, chlorophyll degradation, ethylene emission, sugar accumulation, or pathogen biomarkers, which are particularly critical in high-value horticultural crops where harvesting at the correct maturity stage determines shelf-life and market value (Duncan, 2011; Cushen et al., 2012). Thus, even though these interventions occur before harvest, they directly shape harvesting efficiency by improving crop readiness, reducing heterogeneity, and minimizing physiological and mechanical vulnerabilities.

At the harvest-stage improvement level, nano-harvest technologies act more directly by enhancing the mechanical and operational performance of harvesting systems. Harvesting is an energy-intensive process involving cutting, threshing, separation, and conveying, and inefficiencies in these operations can cause significant losses and damage. Nanostructured coatings applied to cutting blades, threshing drums, conveyors, and storage bins can reduce friction, wear, and corrosion, which improves durability and may reduce the energy demand of machinery over time. Similarly, nanoparticle-based lubricant additives have been explored for improving tribological performance, lowering heat generation, and reducing engine wear, potentially lowering fuel consumption and extending machinery lifespan. While the agricultural-specific literature is still emerging, these applications align with the broader evidence that nanomaterials can deliver functional efficiency gains through surface engineering and material reinforcement (Hischier and Walser, 2012; Walser and Hellweg, 2010). Beyond mechanical improvements, nanosensor-driven "smart harvesting" can improve precision by guiding optimal cutting height, machine speed, and harvest timing based on crop density, moisture gradients, or ripeness distribution. This reduces mechanical damage, minimizes grain losses, and improves the fraction of harvested material that remains marketable, an outcome that is central to sustainability because losses represent wasted land, water, fertilizers, and energy embedded in production (FAO, 2019; Gustavsson et al., 2011).

At the postharvest technology level, nano-harvest technologies address one of the most critical sustainability challenges in global agriculture: postharvest deterioration and food loss. A large fraction of harvested produce is lost during storage, transport, and processing due to microbial spoilage, oxidation, moisture loss, and physiological respiration. Nano-enabled packaging and storage systems provide advanced mechanisms to slow these degradation pathways. Nanocomposite packaging materials can improve gas and moisture barrier properties, reduce oxygen penetration, and enhance mechanical strength, while antimicrobial nano-additives (e.g., silver, zinc oxide, titanium dioxide) can inhibit microbial growth and reduce spoilage without relying solely on conventional preservatives (Duncan, 2011; Rhim et al., 2013). In addition, smart nano-packaging can incorporate nanosensors capable of detecting spoilage gases, microbial metabolites, or pH changes, enabling real-time freshness monitoring and reducing unnecessary waste. This is particularly significant because, from a life cycle perspective, preventing food waste often yields greater environmental benefits than marginal reductions in packaging impacts, given the high footprint of food production itself (FAO, 2019; Gustavsson et al., 2011). However, the postharvest domain also raises key sustainability concerns, including potential nanoparticle migration from packaging into food and the implications for end-of-life management and recycling of nanocomposites (Cushen et al., 2012; Duncan, 2011).

Collectively, nano-harvest technologies therefore represent a unified sustainability framework that includes both direct nano-enhancements in harvesting equipment and indirect nano-based interventions that influence harvest readiness, product integrity, and post-harvest stability. Their overall promise lies in reducing energy intensity, improving harvest precision, minimizing mechanical and microbial losses, and increasing the proportion of produce that remains marketable across supply chains. However, the sustainability of nano-harvest technologies cannot be assumed solely from functional performance, because engineered nanomaterials may impose significant upstream environmental burdens during synthesis and manufacturing, and uncertainties remain regarding nanoparticle release, fate, and toxicity in the environment. This is why life cycle thinking and Life Cycle Assessment are essential for evaluating whether nano-harvest technologies deliver net environmental and economic gains when assessed from cradle-to-grave, rather than shifting burdens across life cycle stages (Arvidsson et al., 2011; Meyer et al., 2011).

2.2 Major categories

2.2.1 Nanosensors and smart monitoring

Nanosensors and nano-biosensors constitute one of the most transformative components of nano-harvest technologies because they enable real-time, high-sensitivity, and highly selective monitoring of crop maturity, physiological status, and postharvest quality indicators across the harvest continuum. In conventional harvesting systems, decisions regarding harvest timing and quality assessment are often based on visual observation, periodic sampling, or generalized maturity indices, which can be subjective, labor-intensive, and insufficiently responsive to rapid changes in crop physiology or environmental conditions. In contrast, nano-enabled sensing platforms exploit the unique electrical, optical, and catalytic properties of nanomaterials such as a large surface area-to-volume ratio, tunable electronic conductivity, quantum confinement effects, and high reactivity to detect extremely low concentrations of biological or chemical signals. This enhanced sensitivity is especially important in harvesting and postharvest systems, where early detection of ripening signals, moisture fluctuations, pathogen contamination, or stress-induced metabolites can prevent substantial quantitative and qualitative losses. The value of nanosensors lies not only in measurement accuracy but also in their ability to support precision harvesting, reduce unnecessary handling, and enable predictive quality management, thereby strengthening both economic returns and environmental sustainability through waste reduction (Duncan, 2011; Parisi et al., 2015).

A key application of nanosensors in nano-harvest systems is the detection of ripeness and maturity indicators, which determine optimal harvest time and directly influence shelf-life, nutritional quality, and consumer acceptance. In climacteric fruits such as banana, mango, apple, and tomato, ripening is strongly associated with ethylene production, respiration rate, and changes in sugar-acid balance, firmness, and pigment composition. Nano-enabled ethylene sensors are therefore widely explored as tools for harvest timing and postharvest management. Nanostructured metal oxide semiconductors (MOS) such as ZnO, SiO_2, TiO_2, and WO_3 can detect ethylene at very low concentrations due to surface adsorption and conductivity changes, while carbon nanomaterials such as carbon nanotubes (CNTs) and graphene can be functionalized to enhance ethylene selectivity and stability. These nanosensors can be integrated into field monitoring networks, storage facilities, or packaging systems to provide continuous ripeness tracking, enabling farmers and supply chain operators to harvest at peak maturity and prevent premature harvesting or overripening, both of which increase losses and reduce market value (Duncan, 2011; Cushen et al., 2012). Beyond ethylene, nanosensors can also detect volatile organic compounds (VOCs) associated with fruit aroma development, fermentation, or spoilage, offering a more comprehensive ripening and quality profile.

Another critical domain of nanosensor applications is moisture content monitoring, which is particularly important in cereal grains, oilseeds, and legumes. Grain moisture at harvest is one of the strongest determinants of threshing efficiency, mechanical breakage, storage stability, and fungal contamination. Harvesting grain at excessively high moisture increases drying energy requirements and spoilage risk, while harvesting too late at very low moisture increases shattering losses and breakage during threshing. Nanosensors capable of rapid moisture detection can support decisions on harvest timing, machine settings, and drying requirements. For example, nanostructured dielectric sensors and conductive polymer nanocomposites can detect changes in moisture through shifts in electrical resistance, capacitance, or impedance. When integrated into harvesting machinery, such sensors can provide real-time feedback, enabling dynamic adjustment of threshing speed and separation intensity to minimize losses. In storage environments, nanosensors can monitor moisture hotspots and prevent fungal growth by triggering ventilation or drying interventions before spoilage occurs.

Nanosensors also play a major role in the early detection of microbial contamination and spoilage, which is a leading cause of postharvest losses in fruits, vegetables, grains, and processed commodities. Traditional microbiological methods for detecting pathogens such as *Escherichia coli*, Salmonella, *Listeria monocytogenes*, and fungal contaminants often require laboratory analysis, skilled personnel, and time-consuming culturing procedures. Nano-biosensors, in contrast, can detect microbial cells, toxins, or nucleic acids rapidly and at low detection limits. Many nano-biosensor platforms employ antibody-functionalized nanoparticles, DNA aptamer-based recognition systems, or enzyme-linked nanoscale transducers. Gold nanoparticles (AuNPs), for example, are frequently used in colorimetric biosensors because they exhibit surface plasmon resonance, enabling visible color changes upon binding with microbial targets. Similarly, CNT- and graphene-based biosensors can detect pathogen binding events through changes in electrical conductivity, while quantum dot-based systems provide fluorescence-based detection with high sensitivity. Such technologies can be integrated into smart packaging, storage monitoring systems, or portable handheld devices for rapid quality screening, thereby reducing the spread of contamination, minimizing waste, and improving food safety outcomes (Cushen et al., 2012; Duncan, 2011).

Beyond ripeness and contamination, nano-harvest nanosensors can monitor stress metabolites and physiological biomarkers that signal crop health and harvest readiness. Abiotic stresses such as drought, heat, salinity, and heavy metal toxicity alter plant metabolism, leading to changes in reactive oxygen species (ROS), antioxidant enzyme activity, proline accumulation, malondialdehyde formation, and shifts in phytohormones such as abscisic acid (ABA), jasmonates, and salicylic acid (Mohiuddin et al., 2023; Shareen et al., 2023; Wahid et al., 2025). Although many of these biomarkers are traditionally measured through laboratory-based biochemical assays, nano-biosensors are increasingly explored for detecting key metabolites or stress-related VOCs in real time. For example, electrochemical nanosensors based on metal nanoparticles or CNT composites can detect hydrogen peroxide (H_2O_2), which is a central oxidative stress marker, while nano-enabled enzymatic biosensors can quantify glucose, lactate, or phenolic compounds associated with ripening and stress responses. The ability to detect such markers in the field could enable more accurate harvest timing, particularly under climate change conditions where stress-induced premature senescence or rapid ripening can occur unexpectedly. This aspect is highly relevant for nano-harvest sustainability because stress-related losses reduce the efficiency of agricultural inputs and increase the environmental footprint per unit of marketable yield.

From a materials perspective, nano-harvest sensing platforms rely on several advanced nanomaterial classes. Graphene-based sensors are highly valued due to graphene's exceptional electrical conductivity, high carrier mobility, mechanical flexibility, and strong surface adsorption capacity. These features allow graphene sensors to detect gases, VOCs, and biomolecules with high sensitivity. Carbon nanotube-based biosensors offer similar advantages and can be functionalized with enzymes, antibodies, or aptamers to enhance selectivity for specific targets. CNT networks can also be integrated into flexible substrates, enabling wearable or packaging-integrated sensors. Metal oxide nanostructures (ZnO, SnO_2, TiO_2, and WO_3) are widely used for gas sensing due to their semiconducting behavior and strong surface reactivity, making them suitable for ethylene, ammonia, and other VOC detection. Meanwhile, quantum dot systems provide powerful optical sensing capabilities because their fluorescence properties can be tuned by particle size and surface chemistry. Quantum dots can be used in biosensing assays for pathogen detection, toxin monitoring, or freshness indicators, particularly in packaging-based applications (Duncan, 2011; Rhim et al., 2013).

A major advantage of nanosensors in nano-harvest systems is their compatibility with smart monitoring and digital agriculture frameworks. Nanosensors can be integrated with wireless sensor networks, Internet of Things (IoT) platforms, and data analytics tools to generate continuous streams of information. In harvesting operations, this enables predictive decision-making where harvest timing, labor allocation, and machinery scheduling can be optimized based on real-time maturity and weather forecasts. In postharvest chains, smart monitoring allows storage conditions to be dynamically controlled (temperature, humidity, ventilation), reducing spoilage and extending shelf-life. These digital integrations strengthen sustainability by reducing unnecessary energy use, preventing losses, and enabling targeted interventions rather than blanket treatments. However, from an LCA perspective, it is important to note that nanosensors-based systems may introduce additional upstream impacts associated with electronics, battery production, and e-waste disposal. Therefore, sustainability evaluation must consider whether the environmental benefits from loss reduction outweigh the burdens associated with sensor manufacturing and end-of-life management (Meyer et al., 2011; Hischier and Walser, 2012).

Despite their strong potential, several challenges remain for nanosensors in nano-harvest applications. Sensor performance in real-world agricultural environments can be affected by dust, humidity, temperature fluctuations, and complex mixtures of gases and VOCs that interfere with selectivity. Long-term stability and calibration drift are also major issues, especially for field-deployed sensors. In addition, the cost of nano-enabled sensors and their integration into farm systems can limit adoption in low-resource settings. Finally, end-of-life disposal and recycling of nano-enabled electronics must be addressed to prevent environmental release of nanomaterials. These challenges reinforce the need for integrated sustainability assessment frameworks that combine functional performance evaluation with life cycle thinking and safe-by-design principles (Arvidsson et al., 2011; Som et al., 2010).

2.2.2 *Nano-coatings for harvest machinery*

Nano-coatings are applied to cutting blades, threshers, conveyors, and storage bins to reduce friction, corrosion, and wear. This can reduce energy consumption and maintenance costs.

Examples include TiO_2, SiO_2, Al_2O_3, and carbon-based coatings.

2.2.3 *Nano-lubricants and fuel additives*

Nanoparticles are sometimes added to lubricants or fuels to improve combustion efficiency and reduce engine wear. Although widely explored in mechanical engineering, their application in agricultural machinery is emerging.

2.2.4 *Nano-encapsulated agrochemicals and biostimulants*

Controlled-release nanoformulations of fertilizers or growth regulators can enhance crop uniformity and quality, thereby improving harvest ability. However, such applications raise concerns regarding nanoparticle release into soils and plant tissues.

2.2.5 *Nano-enabled packaging and post-harvest preservation*

Nano-packaging is one of the most commercially advanced and scientifically mature domains within nano-harvest technologies, because it directly targets the major sustainability bottleneck in global agri-food systems: postharvest deterioration and food loss. Conventional packaging primarily serves as a passive barrier against moisture, oxygen, and physical damage. In contrast, nano-enabled packaging systems are engineered as functional materials that actively regulate gas exchange, inhibit microbial growth, delay oxidation, and in some cases monitor freshness in real time. This is achieved through the incorporation of nanomaterials into polymer matrices or coating layers, creating nanocomposites that deliver improved performance even at very low additive concentrations. Among the most widely studied nano-packaging materials are polymer–clay nanocomposites, silver nanoparticles (AgNPs), zinc oxide nanoparticles (ZnO NPs), titanium dioxide nanoparticles (TiO_2 NPs), silica nanoparticles, and carbon-based nanofillers. These nano-additives enhance packaging functionality through three main pathways: improved barrier properties, antimicrobial activity, and mechanical reinforcement. Because shelf-life extension reduces spoilage and lowers the need for premature disposal of food, nano-packaging has strong potential to reduce the environmental footprint per unit of consumed food, provided that nanoparticle release, toxicity risks, and end-of-life management are adequately addressed (Duncan, 2011; Cushen et al., 2012; Rhim et al., 2013).

A central advantage of nano-packaging is its ability to improve gas and moisture barrier properties, which directly slows physiological and biochemical degradation processes in stored produce. Oxygen penetration into packaging accelerates lipid oxidation, enzymatic browning, vitamin degradation, and aerobic microbial growth, while water vapor transmission influences dehydration, texture loss, and fungal spoilage. Polymer–clay nanocomposites are particularly effective in this regard. Layered silicate clays (e.g., montmorillonite) dispersed at the nanoscale create a "tortuous path" for gas molecules, meaning that oxygen and water vapor must travel around plate-like nanoparticles rather than passing directly through the polymer matrix. This significantly reduces permeability without requiring thicker packaging, thereby maintaining lightweight designs while improving performance. Such barrier enhancement is critical for fresh-cut fruits, vegetables, dairy products, and meat, where oxygen exposure strongly influences shelf-life. Improved barrier properties also allow modified atmosphere packaging (MAP) systems to maintain stable internal gas compositions for longer periods, which is especially important for respiration-sensitive commodities such as leafy greens and climacteric fruits (Rhim et al., 2013; Duncan, 2011).

In addition to barrier enhancement, nano-packaging can provide strong antimicrobial activity, which is one of the most important functional benefits for food preservation.

Microbial spoilage accounts for a large fraction of postharvest losses, particularly in perishable commodities. Nanoparticles such as AgNPs, ZnO NPs, and TiO_2 NPs exhibit broad-spectrum antimicrobial effects through multiple mechanisms. Silver nanoparticles can release Ag^+ ions that disrupt microbial cell membranes, interfere with enzyme activity, and damage DNA, leading to cell death. ZnO nanoparticles generate ROS under certain conditions and can also disrupt microbial membrane integrity, while TiO_2 nanoparticles show photocatalytic antimicrobial action under UV or visible light exposure, producing ROS that degrade microbial cells. These antimicrobial nanocomposites are particularly valuable because they can reduce reliance on chemical preservatives and extend shelf-life without significantly altering sensory quality. However, the antimicrobial mechanism is also a sustainability concern because it raises questions about nanoparticle migration into food, consumer exposure, and potential ecological impacts if nanoparticles enter waste streams (Cushen et al., 2012; Duncan, 2011).

Nano-packaging also improves mechanical strength and durability, which is crucial for preventing physical damage during transport and storage. Mechanical injury such as bruising, puncturing, and crushing accelerates spoilage and increases the likelihood of microbial infection, especially in fruits and vegetables. Nanofillers such as clay nanoparticles, silica nanoparticles, and nanocellulose can reinforce polymer matrices, increasing tensile strength, stiffness, and resistance to tearing while maintaining flexibility. This allows packaging to better withstand handling stress, vibrations during transport, and stacking loads in supply chains. Enhanced mechanical stability also reduces packaging failure and leakage, which is particularly relevant for liquid or semisolid food products. From a sustainability perspective, stronger packaging can reduce material use because thinner films may achieve the same or better performance than conventional thicker plastics. This can potentially lower the packaging's material footprint, although the trade-off must be evaluated using life cycle assessment because nanomaterial production may add upstream burdens (Hischier and Walser, 2012; Meyer et al., 2011).

One of the most important sustainability benefits of nano-packaging is its ability to extend shelf-life and reduce food waste, which often dominates the environmental footprint of the food system. Life cycle studies consistently show that the environmental impacts of food production—land use, irrigation water, fertilizers, pesticides, energy use, and greenhouse gas emissions—are generally far greater than the impacts of packaging materials. Therefore, even small improvements in shelf-life that reduce food losses can yield large net environmental benefits. For example, preventing spoilage of high-impact commodities such as meat, dairy, and fresh produce can save substantial embedded emissions and resource inputs. This is why nano-packaging is frequently discussed not merely as a packaging innovation but as a waste-prevention strategy that improves the sustainability efficiency of the entire farm-to-fork chain (Gustavsson et al., 2011; FAO, 2019). In nano-harvest systems, where the aim is to maximize the proportion of harvested yield that remains marketable and consumable, nano-packaging becomes a critical downstream tool for achieving real sustainability gains.

However, despite these benefits, nano-packaging also introduces important risks and sustainability challenges that must be addressed. A major concern is the potential migration of nanoparticles from packaging into food, particularly for nanoparticles embedded in polymer matrices that may be exposed to heat, acidic conditions, fats, or long storage times. Studies on nanoparticle migration show that the extent depends on nanoparticle size, surface functionalization, polymer type, temperature, and food composition. While many nanocomposites are designed to immobilize nanoparticles within the polymer, imperfect dispersion, polymer degradation, or mechanical abrasion may increase release potential. This issue is not only a food safety concern but also an LCA and circular economy challenge because nanocomposite packaging can complicate recycling streams. Conventional plastic recycling systems are not designed to separate nanocomposites, and the presence of metal-based nanoparticles may affect recycled quality or introduce occupational exposure risks during processing (Cushen et al., 2012; Hischier and Walser, 2012).

From a life cycle assessment perspective, nano-packaging can create a complex trade-off: nanomaterial production can be energy-intensive and may involve hazardous chemicals, but the avoidance of food waste may deliver much larger environmental savings. Therefore, the sustainability of nano-packaging must be evaluated through cradle-to-grave LCA that includes raw material extraction, nanomaterial synthesis, polymer processing, packaging manufacturing, transport, use-phase benefits (shelf-life extension and waste reduction), and end-of-life pathways such as landfill, incineration, composting, or recycling. In many cases, LCA evidence suggests that packaging innovations are environmentally justified when they prevent losses of high-impact foods, but the conclusion depends strongly on assumptions about waste reduction rates, consumer behavior, and disposal scenarios (Meyer et al., 2011; Hischier and Walser, 2012). This reinforces the need for careful scenario-based LCA rather than general claims that nano-packaging is inherently sustainable.

3. Sustainability rationale: Why LCA is essential for nano-harvest systems

The sustainability rationale for applying Life Cycle Assessment (LCA) to nano-harvest systems is rooted in the fact that nano-enabled innovations are often promoted as inherently "green" because they improve harvesting efficiency, enhance shelf-life, and reduce postharvest losses; yet these functional benefits do not automatically translate into net environmental gains when the full life cycle is considered. Nano-harvest technologies frequently involve engineered nanomaterials whose production can be highly energy-intensive, chemically demanding, and dependent on upstream extraction of metals, solvents, and specialty reagents, meaning that a reduction in field-level fuel use or spoilage may be offset or even exceeded by emissions and resource burdens generated during nanoparticle synthesis and product manufacturing. For example, a nano-coating that lowers friction and reduces fuel consumption during harvesting may carry substantial cradle-stage impacts associated with nanomaterial production, purification, and coating processes, while nano-packaging that extends shelf-life and reduces food waste may create new downstream sustainability burdens if nanocomposite materials complicate recycling streams, increase persistence in landfills, or lead to nanoparticle release during disposal or incineration. LCA is therefore essential for nano-harvest systems because it systematically captures upstream burdens (raw material extraction, mining, synthesis, and manufacturing), quantifies use-phase benefits (fuel savings, yield stabilization, quality retention, and waste reduction), and identifies trade-offs across multiple impact categories such as climate change, energy demand, toxicity potential, water use, and resource depletion. Most importantly, LCA prevents "burden shifting," where environmental improvements in one stage (e.g., reduced post-harvest losses) are achieved at the expense of increased impacts elsewhere (e.g., higher embodied energy or toxicity risks from nanomaterial production), and it provides a decision-support framework for eco-design, safe-by-design innovation, and evidence-based policy regulation. In this context, the ISO 14040 and ISO 14044 standards offer internationally accepted methodological foundations for conducting LCA, and while these standards are not nano-specific, they can be adapted to nano-harvest technologies by incorporating nanomaterial-relevant inventory data, release scenarios, and uncertainty analysis to ensure scientifically robust sustainability evaluation (Arvidsson et al., 2011; Meyer et al., 2011; Hischier and Walser, 2012).

4. Life cycle assessment methodology for nano-harvest technologies

4.1 LCA phases

According to ISO standards, LCA includes:

1. Goal and scope definition
2. Life cycle inventory (LCI)
3. Life cycle impact assessment (LCIA)
4. Interpretation

Each phase requires careful tailoring for nano-harvest technologies.

4.2 Goal and scope definition

4.2.1 *Goal*

Typical goals include:

- Comparing nano-harvest systems with conventional alternatives
- Identifying environmental hotspots in nano-harvest supply chains
- Supporting design optimization and policy decisions
- Evaluating trade-offs between environmental and economic outcomes

4.2.2 *Functional unit*

The functional unit (FU) is critical. In agriculture, suitable FUs may include:

- **1 t of harvested crop** delivered to storage
- **1 ha harvested** per **season**
- **1 kg of marketable produce** after postharvest losses
- **1 day of harvesting operation** (less recommended)

For nano-packaging:

- **1 kg of packaged produce maintained at acceptable quality for X days**

The FU should reflect the real function: not simply producing nanomaterials, but delivering harvest services and marketable food.

4.2.3 *System boundaries*

Common boundaries include:

- **Cradle-to-gate**: raw material extraction → manufacturing
- **Cradle-to-farm gate**: includes use in harvesting
- **Cradle-to-grave**: includes end-of-life
- **Cradle-to-cradle**: includes recycling/circular loops

For nano-harvest technologies, cradle-to-grave is often recommended due to uncertainty in end-of-life nanoparticle release.

4.3 Life cycle inventory for nano-harvest technologies

LCI compiles all inputs and outputs:

- Raw materials (metals, polymers, solvents)
- Energy use (electricity, heat, fuel)
- Emissions (CO_2, NOx, VOCs, wastewater)
- Waste streams (hazardous waste, solid residues)

4.3.1 *Challenges in nano-LCI*

Key nano-specific issues include:

- Lack of industrial-scale data for emerging nanomaterials
- Proprietary manufacturing processes

- Inconsistent reporting of yields, reagents, and energy inputs
- Nanoparticle release during manufacturing and use

Studies highlight that inventory uncertainty is often the largest limitation in nano-LCA.

4.3.2 *Approaches for building LCI*

Researchers use:

- Process-based LCI from lab/pilot data
- Hybrid LCI combining process and economic input-output (EIO)
- Scenario modeling for scale-up
- Proxy datasets from similar bulk materials

4.4 Life cycle impact assessment

LCIA translates inventory flows into impact categories. Common methods include ReCiPe, CML, TRACI, and ILCD. Key categories:

- Global warming potential (GWP)
- Cumulative energy demand (CED)
- Fossil depletion
- Water use and scarcity
- Acidification and eutrophication
- Human toxicity and ecotoxicity
- Particulate matter formation
- Land use

4.4.1 *Nano-specific LCIA issues*

Standard LCIA models do not fully incorporate nanoparticle-specific fate and toxicity. For example:

- Nanoparticles have size-dependent transport and bioavailability
- Dissolution, aggregation, and surface chemistry alter toxicity
- Release to soil and water may be low but persistent

Some researchers propose integrating fate-exposure models such as USEtox adaptations for ENMs, but consensus remains incomplete.

4.5 Interpretation

Interpretation identifies:

- Hotspots (high-contribution stages)
- Trade-offs across categories
- Sensitivity to assumptions
- Improvement options

For nano-harvest technologies, interpretation must carefully address:

- Scale-up assumptions
- Nanomaterial release rates
- Energy mix (renewable vs. fossil)
- Recycling rates and end-of-life pathways

5. Environmental LCA of nano-harvest technologies

5.1 LCA of nano-coatings for harvest machinery

Nano-coatings applied to harvest equipment can reduce friction, corrosion, and wear. Potential benefits include:

- Reduced fuel use (lower engine load)
- Longer equipment lifetime
- Reduced maintenance and replacement frequency

However, the synthesis of coating nanoparticles (TiO_2, Al_2O_3, and SiO_2) can be energy-intensive. Hotspots often occur in:

- Precursor chemical production
- High-temperature calcination
- Solvent use and emissions

LCA studies on engineered nanomaterials frequently show that energy consumption during synthesis dominates GWP and cumulative energy demand.

5.2 LCA of nanosensors for precision harvesting

Nanosensors may reduce losses by improving harvest timing and quality control. Their LCA must include:

- Sensor material production (graphene, CNTs, and metal oxides)
- Electronics and battery components
- Data transmission infrastructure (if included)
- End-of-life e-waste disposal

The environmental impact per sensor may be high, but when distributed across large harvested yields, the impact per kg of crop can become small. Benefits become substantial when sensor-guided harvesting reduces losses by even 5%–10%.

5.3 LCA of nano-packaging for post-harvest preservation

Nano-packaging can reduce food waste by extending shelf life. The sustainability of nano-packaging depends on:

- Type of polymer matrix (PET, PLA, and PE)
- Nanofiller type (clay, AgNPs, ZnO, and TiO_2)
- Manufacturing process (extrusion, coating, and lamination)
- End-of-life (recycling, incineration, and landfill)

Nano-additives can complicate recycling due to contamination concerns. In many cases, the greatest environmental benefit arises not from reduced packaging impacts but from reduced food waste, because food production has a high footprint.

5.4 LCA of nano-encapsulated formulations linked to harvest quality

Nano-encapsulated fertilizers, micronutrients, and growth regulators can improve uniformity, yield, and harvest quality. Their LCA must include:

- Nanocarrier synthesis (chitosan, silica, and liposomes)
- Encapsulation energy and solvent use
- Field application emissions
- Potential release to soil and water

If nanoformulations significantly reduce fertilizer application rates while maintaining yield, they may reduce eutrophication and GHG emissions. However, if the nano-carrier has high upstream burdens, net sustainability may be uncertain.

6. Hotspot analysis: Where impacts concentrate

Across nano-harvest technologies, life cycle assessments consistently indicate that a limited set of recurring "hotspots" tends to dominate environmental impacts, particularly when systems are evaluated from cradle-to-grave rather than focusing only on use-phase benefits. The most prominent hotspot is typically nanomaterial synthesis, because many engineered nanomaterials require energy-intensive processing steps such as calcination, pyrolysis, plasma-based treatments, hydro/solvothermal synthesis, or chemical vapor deposition, which can contribute disproportionately to cumulative energy demand and global warming potential (GWP), especially when scaled beyond laboratory production. A second major hotspot arises from solvent and chemical consumption during nanoparticle synthesis, purification, surface functionalization, and dispersion into polymers or coatings, since organic solvents, reducing agents, stabilizers, and acidic/basic reagents can elevate toxicity-related impact categories and photochemical ozone formation, and may also generate hazardous waste streams if solvent recovery is limited. The electricity mix is another decisive driver of results, as upstream impacts are highly sensitive to whether manufacturing relies on fossil-dominated grids or low-carbon electricity; consequently, the same nano-harvest product can shift from being environmentally burdensome to comparatively favorable under renewable-powered production scenarios. Finally, end-of-life management represents an increasingly important hotspot for nano-packaging, nanocomposites, and nano-enabled sensing devices, because these materials may complicate recycling streams, introduce uncertainties in waste treatment, and pose potential nanoparticle release risks during mechanical recycling, incineration, or landfill degradation—an issue often excluded from LCAs due to limited fate and release data but potentially significant for long-term environmental relevance and circular economy performance (Arvidsson et al., 2011; Meyer et al., 2011; Hischier and Walser, 2012).

7. Economic sustainability: Life cycle costing of nano-harvest technologies

LCC is a complementary sustainability tool that evaluates the total economic cost of a technology across its entire life cycle rather than focusing only on the purchase price, and it typically includes capital costs (equipment purchase and installation), operating costs (fuel, electricity, labor, and consumables), maintenance and repair requirements, replacement frequency and downtime losses, and end-of-life costs associated with disposal, recycling, or compliance. This whole-life perspective is particularly important for nano-harvest technologies because many nano-enabled solutions, such as nanosensors, nano-coatings, and nanocomposite packaging, often involve higher initial investment due to specialized materials, fabrication processes, or quality control requirements, even when they promise long-term savings through efficiency gains and loss reduction. Integrating LCA with LCC provides a robust framework for assessing eco-efficiency by quantifying environmental benefits per unit cost, determining the cost-effectiveness of emission reduction, and identifying trade-offs between affordability and sustainability performance; for instance, nanosensor systems may reduce postharvest losses and improve product quality, yet increase upfront costs and require periodic calibration or replacement, so LCC becomes essential for estimating payback periods, profitability, and adoption feasibility under real market conditions. Economic case examples commonly discussed for nano-harvest systems include nano-coated harvesting blades that have a higher purchase cost but lower replacement frequency and reduced downtime, nano-lubricants that may slightly increase operating cost per liter but generate net savings through reduced fuel consumption and engine wear, and nano-enabled packaging that increases per-unit packaging cost but can deliver strong economic returns by extending shelf-life, reducing food waste, and improving market value through quality retention.

8. Uncertainty, sensitivity, and scenario analysis

Uncertainty is a defining challenge in the LCA of nano-harvest technologies because many systems are still emerging and rely on incomplete data, assumptions, and laboratory-scale performance metrics that may not translate directly to real-world conditions. Major sources of uncertainty include the use of lab-scale life cycle inventories that often overestimate energy and chemical consumption compared with optimized industrial-scale synthesis, the frequent need to assume nanoparticle release rates during application, washing, abrasion, or disposal due to limited empirical measurements, and strong regional variability in electricity mixes that can substantially change upstream impacts for the same nanomaterial. Additional uncertainty arises from agricultural variability itself, including differences in crop yield, harvesting efficiency, and postharvest loss rates across seasons, climates, management practices, and supply chains, as well as uncertainty in end-of-life scenarios where recycling rates, incineration conditions, landfill behavior, and nanocomposite separation are often modeled using simplified assumptions. Sensitivity analysis is therefore essential to identify which parameters most influence results, and key variables typically include nanomaterial synthesis yield and energy intensity, the degree of lifetime extension achieved by nano-coated machinery components, the percentage reduction in postharvest losses delivered by nano-packaging or nanosensors, and realistic recycling or disposal rates for nano-enabled materials. Beyond deterministic sensitivity testing, Monte Carlo simulation and probabilistic LCA approaches provide a more robust framework by treating uncertain inputs as distributions rather than fixed values, allowing analysts to quantify confidence intervals, assess the likelihood of net environmental benefits, and avoid over interpreting single-point results in systems where data gaps remain substantial (Arvidsson et al., 2011; Meyer et al., 2011; Hischier and Walser, 2012).

9. Nano-specific environmental risk considerations

LCA alone does not replace risk assessment, particularly for nano-harvest technologies where engineered nanomaterials may introduce exposure and hazard concerns that are not fully captured by conventional life cycle impact categories. While LCA is essential for quantifying resource use, emissions, and overall environmental trade-offs across the cradle-to-grave system, nano-harvest applications may also involve direct and indirect release pathways that require dedicated safety evaluation, including worker exposure during nanoparticle synthesis, formulation, coating, and packaging manufacturing; release into soil during field application of nano-enabled inputs; release into water bodies through runoff, washing, or waste disposal; and potential persistence, bioaccumulation, and trophic transfer within food chains. Because these risks depend strongly on particle properties (size, shape, surface charge, dissolution behavior), environmental fate and transport, and realistic exposure scenarios, a comprehensive sustainability assessment should integrate LCA with complementary tools such as risk assessment, exposure modeling, and toxicological studies. This combined approach supports more defensible conclusions by addressing both environmental burdens and safety implications, ensuring that nano-harvest technologies are evaluated not only for efficiency and climate benefits but also for human health protection and ecological integrity (Arvidsson et al., 2011; Som et al., 2010; Hischier and Walser, 2012).

10. Policy, standards, and regulatory dimensions

ISO (2006) 14040 and ISO 14044 provide the internationally accepted methodological foundation for Life Cycle Assessment, defining core principles, system boundaries, functional units, inventory analysis, impact assessment, and interpretation; however, while these standards are fully applicable to nano-harvest technologies, they are not nano-specific, and practical guidance on issues such as nanoparticle release, fate, characterization factors, and data uncertainty is still evolving as nanotechnology applications expand. In parallel, regulatory governance for nano-enabled products remains fragmented across regions, with significant differences in definitions, reporting requirements, and risk evaluation frameworks: for example, the European Union has explicitly incorporated "nanoforms" into the REACH regulation, requiring nano-relevant registration and safety information for substances placed on the market, while the OECD has played a central role in developing harmonized testing guidelines and international collaboration for nanomaterial safety evaluation; additionally, national agencies in multiple jurisdictions regulate nano-enabled food contact materials, particularly where nanocomposites or antimicrobial nanoparticles may migrate into food, creating regulatory scrutiny for nano-packaging applications. Because nano-harvest technologies sit at the intersection of agriculture, food systems, and advanced materials, effective policy frameworks must balance the need to support innovation and sustainability gains (such as reduced food loss and improved resource efficiency) with precautionary environmental and human health protection, ensuring transparency, traceability, and responsible lifecycle management of engineered nanomaterials across production, use, and end-of-life stages.

11. Future research directions

Key research and policy priorities for advancing the sustainability evaluation of nano-harvest technologies include the development of nano-specific life cycle impact assessment (LCIA) characterization factors, the creation of harmonized and transparent nano–life cycle inventory (LCI) databases, and the improvement of release, fate, and exposure models that realistically represent nanoparticle behavior across agricultural soils, storage environments, and waste-management systems, alongside emerging needs to integrate environmental LCA with social life cycle assessment (S-LCA) and to validate laboratory-based assumptions through field-scale testing and real-world monitoring. In conclusion, nano-harvest technologies offer substantial potential to improve agricultural efficiency, extend shelf-life, and reduce postharvest losses, but their true sustainability depends on whether downstream benefits outweigh the upstream environmental burdens associated with nanomaterial synthesis, processing, and end-of-life challenges; in this context, LCA—particularly when integrated with LCC—provides a robust framework for quantifying trade-offs and preventing burden shifting, and the most sustainable nano-harvest systems are those that rely on low-impact synthesis routes, deliver measurable reductions in energy use and food waste, and embed circular economy principles such as recyclability, safe disposal, and material recovery, while continued methodological refinement, improved datasets, and nano-specific impact models remain essential to ensure that nano-harvest technologies contribute meaningfully and responsibly to sustainable agriculture.

References

Alam, P., Faizan, M., Sultan, H., Albalawi, T., 2025. Silicon oxide nanoparticles boost rice resilience to salinity by enhancing antioxidant defenses and stress regulation. Plant Sci. 359, 112588.

Arvidsson, R., Molander, S., Sandén, B.A., Hassellöv, M., 2011. Challenges in life cycle assessment of nanomaterials: a review. Environ. Sci. Technol. 45, 612–619.

Cushen, M., Kerry, J., Morris, M., Cruz-Romero, M., Cummins, E., 2012. Migration and exposure assessment of nanoparticles from food packaging into food. Trends Food Sci. Technol. 24, 136–146.

Dogan, Y., Alam, P., Sultan, H., Sharma, R., Soysal, S., Baran, M.F., Faizan, M., 2025. Zinc oxide nanoparticles for sustainable agriculture: a tool to combat salinity stress in rice (Oryza sativa) by modulating the nutritional profile and redox homeostasis mechanisms. J. Agric. Food Res. 19, 101598.

Duncan, T.V., 2011. Applications of nanotechnology in food packaging and food safety: barrier materials, antimicrobials and sensors. J. Colloid Interface Sci. 363, 1–24.

Faizan, M., Ashoka, H., Karabulut, F., Alam, P., Ahmed, S. M., Khan, I., Soysal, S., Hayat, S., Ahmed, N., 2024. Annexins: a family of calcium binding proteins with variety of roles in plant development and abiotic stress tolerance. Plant Stress 14, 100573.

Faizan, M., Faraz, A., Hayat, S., 2018. Dose dependent response of epibrassinolide on the growth, photosynthesis and antioxidant system of tomato plants. Indian Hortic. J. 8 (2), 68–76.

Faizan, M., Gangadharappa, B.S., Alam, P., Tonny, S.H., Maruthi, K.R., Hayat, S., 2025. Untapped potential of calcium and nano-calcium to develop abiotic stress resilience in photosynthetic machinery: the primary source of plant food and fuels. Plant Stress 15, 100718.

FAO, 2019. The State of Food and Agriculture: Moving Forward on Food Loss and Waste Reduction. Food and Agriculture Organization of the United Nations.

Gogos, A., Knauer, K., Bucheli, T.D., 2012. Nanomaterials in plant protection and fertilizers: current state, foreseen applications, and research priorities. J. Agric. Food Chem. 60, 9781–9792.

Gustavsson, J., Cederberg, C., Sonesson, U., et al., 2011. Global Food Losses and Food Waste. FAO Report.

Hischier, R., Walser, T., 2012. Life cycle assessment of engineered nanomaterials: state of the art and strategies to overcome existing gaps. Sci. Total Environ. 425, 271–282.

ISO, 2006. ISO 14040: Environmental Management—Life Cycle Assessment—Principles and Framework. International Organization for Standardization.

Kah, M., Beulke, S., Tiede, K., Hofmann, T., 2013. Nanopesticides: state of knowledge, environmental fate, and exposure modeling. Crit. Rev. Environ. Sci. Technol. 43, 1823–1867.

Meyer, D.E., Curran, M.A., Gonzalez, M.A., 2011. An examination of existing LCA studies of nanomaterials: state of the art and best practices. Environ. Sci. Technol. 45, 743–752.

Mohiuddin, M., Muntha, S., Ali, A., Faizan, M., Samrana, S., 2023. The ecology of reactive oxygen species signalling. In: Faizan, M., Hayat, S., Ahmed, S.M. (Eds.), Reactive Oxygen Species: Prospects in Plant Metabolism. Springer, Nature Switzerland.

Parisi, C., Vigani, M., Rodríguez-Cerezo, E., 2015. Agricultural nanotechnologies: what are the current possibilities? Nano Today 10, 124–127.

Rhim, J.W., Park, H.M., Ha, C.S., 2013. Bio-nanocomposites for food packaging applications. Prog. Polym. Sci. 38, 1629–1652.

Shareen, Faraz, A., Faizan, M., 2023. Physiological impact of reactive oxygen species on leaf. In: Faizan, M., Hayat, S., Ahmed, S.M. (Eds.), Reactive Oxygen Species: Prospects in Plant Metabolism. Springer, Nature Switzerland.

Som, C., Berges, M., Chaudhry, Q., et al., 2010. The importance of life cycle concepts for the development of safe nanoproducts. Toxicology 269, 160–169.

Wahid, A., Faizan, M., Arif, Y., Alam, P., Hayat, S., 2025. Dose-dependent response of potassium on antioxidant enzymes, photosynthetic efficiency, and carbohydrate metabolism in Indian mustard (Brassica juncea L.) under NaCl stress. In: Discover Plants, 2, p. 51.

Walser, T., Hellweg, S., 2010. Life cycle assessment of nanomaterials. Environ, Sci. Technol. 44, 450–457.

CHAPTER

15

Nanotechnology for climate-resilient photosynthesis under abiotic stresses

Radwan Khalil, Amina Gamal and Nesma Elsayed

Botany and Microbiology Department, Faculty of Science, Benha University, Benha, Egypt

1. Introduction

One of the most significant issues facing farmers worldwide in the 21st century is climate change. Rising atmospheric temperatures, altered rainfall patterns, an increase in extreme weather events, and soil degradation have all made abiotic stressors on plant growth and productivity worse globally. Because they directly impact the physiological and biochemical processes that plants require to survive and thrive, drought, salinity, heat, cold, and heavy metal toxicity are some of the worst stresses (Mittler, 2006; Lesk et al., 2016). Since photosynthesis is the primary mechanism by which plants and crops develop, it is extremely vulnerable to abiotic effects brought on by climate change.

Abiotic stressors reduce the efficiency of photosynthesis by restricting both stomatal and nonstomatal processes. Stomatal conductance is reduced by drought and high salinity, making it more difficult for CO_2 to enter the leaf. Enzyme activity, membrane integrity, and photosystem stability are all impacted by thermal stress, both hot and cold (Flexas et al., 2012; Allakhverdiev et al., 2008). According to Shahid et al. (2015), heavy metals such as cadmium, lead, and mercury interfere with the synthesis of chlorophyll, electron flow, and the equilibrium of redox processes, making photosynthesis even more difficult. Together, these pressures reduce the efficiency of carbon uptake, photosynthesis, and ultimately crop output. Finding novel strategies to preserve photosynthesis under stress has emerged as a key research priority due to the anticipated increase in the global need for food.

Traditional breeding and genetic engineering are two antiquated methods of improving plants' stress tolerance. They have made significant progress, but it still requires a significant investment of time and resources, and they are frequently constrained by the complexity of the interactions between genotypes and environments. In this regard, nanotechnology has become a promising multidisciplinary tool that may be able to assist in resolving novel issues in plant physiology and agriculture. The study of materials at the nanoscale (1–100 nm) and their design, production, and applications is known as nanotechnology. These materials can be employed in biological applications because

Nanotechnology in Photosynthesis
https://doi.org/10.1016/B978-0-443-44376-3.00015-X

of their special physicochemical characteristics, which include a large surface area, enhanced reactivity, and changeable functionality (Roco et al., 2011; Khalil et al., 2021a).

In recent years, nanomaterials (NMs) have attracted a lot of attention due to their potential to improve nutrient utilization, regulate physiological processes, and increase plant resistance to nonliving stimuli. Nanomaterials (NMs) can alter metabolic pathways at the molecular level, penetrate cellular barriers, and interact with plant tissues more effectively than bulk materials (Kah et al., 2018). Because photosynthetic machinery is highly sensitive to environmental changes, these characteristics make nanotechnology particularly helpful for targeting it.

According to recent studies, even in harsh environments, nanomaterials can modify, protect, and enhance photosynthetic pathways. When applied in the proper quantities, certain metal-based and metal oxide nanoparticles, such as zinc oxide (ZnO), titanium dioxide (TiO_2), iron oxide (FeO_4), and cerium oxide (CeO_2), have been demonstrated to boost chlorophyll, stabilize photosystems, and improve light harvesting efficiency (Tripathi et al., 2017; Rico et al., 2015). Carbon-based nanomaterials, including carbon nanotubes, graphene derivatives, and fullerenes, have been shown to facilitate electron mobility, increase the amount of CO_2 that chloroplasts can absorb, and shield them from oxidative stress damage (Giraldo et al., 2014; Khalil et al., 2021a).

Reactive oxygen species (ROS) management is one of the primary ways by which nanomaterials improve photosynthesis in stressed plants. Abiotic stress frequently causes chloroplasts to produce excessive amounts of reactive oxygen species (ROS), which can damage thylakoid membranes, photosystem II (PSII), and photosynthetic enzymes. Certain nanomaterials have inherent antioxidant qualities or can support plants' antioxidant defense mechanisms. This preserves photosynthetic structures and the redox balance (Das et al., 2018). Cerium oxide nanoparticles, for example, have the ability to transition between Ce^{3+} and Ce^{4+}. This enables them to eliminate ROS and maintain photosynthesis in the face of adversity. By reducing oxidative stress and facilitating the access and utilization of nutrients, nanomaterials can also support photosynthesis. Trace minerals, including iron, magnesium, and manganese, are necessary for the production of chlorophyll and the transfer of electrons during photosynthesis. Because nano-fertilizers are more effective than conventional fertilizers at delivering these nutrients to plants, they will continue to be present even when the plants are under stress, and the nutrients are out of balance (Dimkpa and Bindraban 2016). Improved nutritional equilibrium leads to increased photosynthesis, Rubisco activity, and chlorophyll.

Additionally, nanomaterials can regulate signaling pathways and gene expression that are critical for photosynthesis and stress response. According to new research, certain nanoparticles can alter the function of genes involved in carbon metabolism, antioxidant defense, and photosystem repair. This makes plants more resilient to abiotic stress (Zhao et al., 2020). Because of its capacity for regulation, nanotechnology is both an active modulator of networks that react to stress and photosynthesis and a passive protective tool (Khalil et al., 2021a). The relationship between photosynthetic systems and nanomaterials is still intricate. The type, size, concentration, plant species, and environmental circumstances of the nanoparticles all play a major role. Nanomaterials can help plants deal with stress and photosynthesize more effectively, while excessive or poorly constructed nanomaterials can harm plants. This demonstrates the significance of using nanomaterials sensibly and cautiously (Ma et al., 2010).

In this instance, the current work examines how nanomaterials can enhance, modify, and shield photosynthetic processes from abiotic stress. Using recent developments in plant physiology and nanotechnology to enhance sustainable crop production in a changing environment, this work seeks to elucidate the strategic application of nanomaterials to lessen stress-induced photosynthetic inhibition. In order to develop safe and efficient nano-enabled farming techniques in the future, we must understand how these interactions function at the physiological, biochemical, and molecular levels.

2. Abiotic stress and photosynthesis

Abiotic stress has an impact on photosynthesis, a physiological function that is extremely sensitive. This is due to the fact that it absorbs signals from the external environment and combines them with intricate chemical and physical processes that take place in various cell components. Photosynthesis is affected both directly and indirectly by abiotic stressors such as heat, salt, drought, heavy metal toxicity, and cold. As a result, plants can produce fewer crops, absorb less carbon, and struggle to convert energy into food. These stressors frequently occur simultaneously in the field, which exacerbates their detrimental effects on the photosynthetic system (Khalil et al., 2021b).

2.1 Drought and heat stress effects on photosynthesis

The main way that drought stress inhibits photosynthesis is by closing the stomata, which prevents CO_2 from entering the leaf mesophyll. There is less CO_2 between cells when stomatal conductance decreases. According to Flexas et al. (2012), this directly reduces the net photosynthetic rate and the efficiency of Rubisco carboxylation (Fig. 15.1). When there is a mild to moderate drought, stomatal restriction is the main issue. However, nonstomatal limits occur when there is an extended period of dryness or insufficient water. These include damage to photosystem II (PSII), decreased Rubisco activity, and decreased regeneration of ribulose-1,5-bisphosphate (RuBP) (Madany and Khalil, 2017).

Photosynthetic inhibition is exacerbated when heat stress and drought coexist. Elevated temperatures cause thylakoid membranes to become unstable, change how proteins interact in photosystems, and hasten the denaturation of important Calvin cycle enzymes (Allakhverdiev et al., 2008). The heat-sensitive enzyme known as Rubisco activase is required to maintain Rubisco's active state. CO_2 fixation sharply decreases as the temperature rises (Khalil et al., 2021b). Heat stress and drought both produce excessive amounts of reactive oxygen species (ROS), including singlet oxygen (O_2O_2), superoxide radicals (O_2^{-}), and hydrogen peroxide (H_2O_2). These ROS are mostly produced in chloroplasts when CO_2 absorption is limited because the photosynthetic electron transport chain is excessively reduced (Mittler, 2017). Lipid peroxidation, protein oxidation, chlorophyll breakdown, and photoinhibition are all brought on by elevated ROS levels. Photosynthesis is less effective if antioxidant defenses are weak (Doneva et al., 2021).

2.2 Salinity stress and photosynthetic dysfunction

Salinity stress affects both ions and osmosis, which makes photosynthesis more difficult. When the soil is highly salinized, it is more difficult for plants to obtain water (Fig. 15.1). The plant experiences osmotic stress as a result, which has drought-like effects on the stomata's function and the water content of the leaves. However, salinity can also render some ions poisonous. The main reason for this is that there are insufficient potassium ions (K^+) and excess sodium ions (Na^+) (Khalil et al., 2022a, 2022b). Elevated Na^+ levels disrupt ion homeostasis by competing with K^+ for binding and absorption sites. This hinders the production of

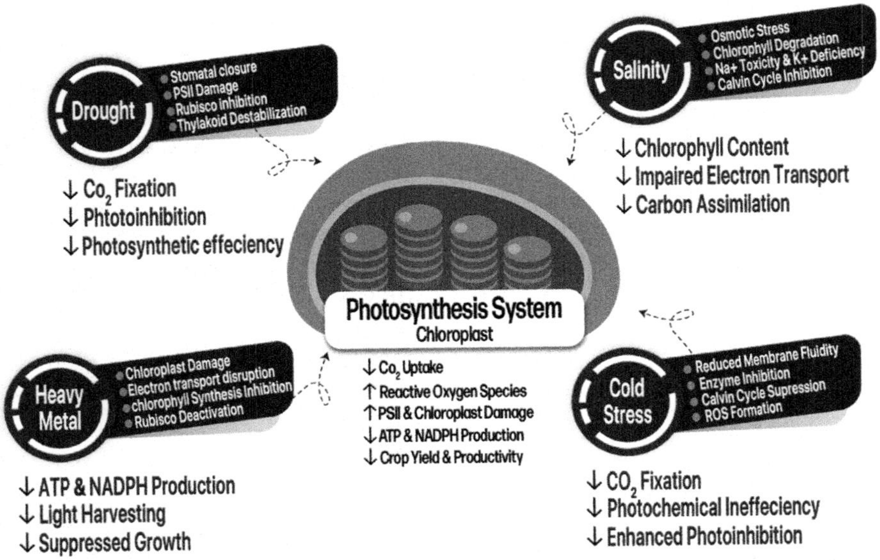

FIGURE 15.1 Effect of abiotic stress on photosynthesis in plants.

proteins, the function of enzymes, and the regulation of stomata (Munns and Tester, 2008). Stomata function, photosynthetic metabolism, and osmotic equilibrium all depend on potassium. This implies that CO_2 assimilation will decrease in salty environments if there is insufficient K^+.

Additionally, the production and duration of chlorophyll are altered by salinity stress. When there is a lot of Na^+ in the soil, plants find it more difficult to absorb magnesium. An essential component of chlorophyll is magnesium. Plants find it more difficult to receive light as a result of the breakdown of chlorophyll (Parida and Das, 2005). Thylakoid membranes and photosynthetic pigments are also harmed by salt-induced oxidative stress, which further reduces the efficiency of photosystems. Salinity modifies the metabolism of carbohydrates and inhibits important Calvin cycle enzymes. As a result, the source-sink interactions are disrupted, and less photosynthate is produced. When combined, these factors significantly impede photosynthesis and plant growth in salty environments (Khalil et al., 2022b).

2.3 Heavy metal stress and photosynthetic inhibition

Even at low concentrations, heavy metals, including cadmium (Cd), lead (Pb), mercury (Hg), and arsenic (As) pose a significant abiotic barrier to photosynthesis (Fig. 15.1). Unlike micronutrients, which are essential for life, these metals have no known biological use. They are harmful in a variety of ways. By altering the ultrastructure of chloroplasts, electron transport, and enzyme function, heavy metals obstruct photosynthesis.

The photosynthetic electron transport chain is significantly impacted by heavy metal stress. Lead and cadmium can take the place of crucial metal cofactors in photosystem proteins, which slows down the synthesis of ATP and NADPH and prevents electrons from moving from PSII to PSI (Shahid et al., 2015). ROS and oxidative damage occur when electron transport is disrupted because electrons are more likely to leak to oxygen. By degrading pigments and obstructing enzymes involved in the tetrapyrrole biosynthesis pathway, heavy metals also prevent the production of chlorophyll. Several studies have shown that heavy metals can damage chloroplast membranes, causing grana to become disorganized and thylakoids to expand. Due to these structural alterations, photosynthesis is less effective since it is more difficult to absorb light and transfer energy. Additionally, by attaching to sulfhydryl groups and altering the structure of proteins, heavy metals harm Calvin cycle enzymes, including Rubisco. Photosynthesis and plant development significantly decrease when electron transport is disrupted, pigments are lost, and enzymes malfunction (Khalil et al., 2017a).

2.4 Cold stress and photosynthetic limitations

Enzyme kinetics, membrane fluidity, and the metabolic synchronization of light and dark reactions are all affected by cold stress. Lower temperatures cause chloroplast membranes to become less fluid, which slows electron flow and reduces the efficiency of photosynthetic protein complexes (Allen and Ort, 2001). A less flexible membrane makes it more difficult for plastoquinone and other electron carriers to move about. Photochemical reactions are slowed down as a result (Fig. 15.1).

The enzymes involved in the Calvin cycle perform poorly in cold conditions. Enzymes that perform poorly at low temperatures include sedoheptulose-1,7-bisphosphatase, fructose-1,6-bisphosphatase, and rubisco. As a result, CO_2 fixation decreases (Sage and Kubien, 2007). The excess energy accumulates when light processes continue to produce excitation energy in the cold but are unable to absorb carbon. This increases the likelihood of photo-inhibition and ROS production. Cold-induced oxidative stress further damages photosynthetic components, particularly PSII's response centers. Plants that thrive in cold climates frequently enhance their photoprotective mechanisms, such as antioxidant enzyme activity and nonphotochemical quenching, to mitigate these impacts. However, cold-sensitive plants exhibit reduced photosynthetic activity when under cold stress (Majláth et al., 2020).

2.5 Integrated impact of abiotic stress on photosynthesis

Different kinds of abiotic stress affect photosynthesis in different ways (Fig. 15.1), but they all have the same effects, like less CO_2 uptake, oxidative stress, and damage to photosynthetic structures. The sensitivity of photosynthesis to environmental stress is demonstrated by stomatal limits, metabolic inhibition, and ROS accumulation. Understanding these processes provides us with a biological foundation for developing novel strategies, such as nanotechnology-based therapies, to safeguard and enhance photosynthesis in adverse settings (Janda et al., 2018).

3. Role of nanotechnology in climate-resilient photosynthesis

The demand for innovative methods to maintain photosynthesis in the absence of living organisms is growing as climate change has an increasing impact on agricultural systems. We can accurately manage the physiological, biochemical, and molecular processes that occur during photosynthesis thanks to nanotechnology, a powerful tool that is applicable to many different domains. Because of their small size and unique physicochemical properties, nanomaterials can directly interact with plant tissues and biological components. This provides us with new strategies to strengthen photosynthesis in the face of heat, salt, drought, and other stressors (Kah et al., 2018; Janda et al., 2019).

3.1 Nanofertilizers and controlled nutrient delivery

Abiotic stress frequently makes it more difficult for plants to get and utilize nutrients. This may result in inadequacies that directly damage the machinery involved in photosynthesis. By enabling the progressive and controlled release of vital macronutrients, including nitrogen (N), phosphorus (P), and potassium (K), nanofertilizers have been developed to address these problems. Nutrient-based nanoparticles facilitate the access and absorption of nutrients, particularly when roots are under stress and malfunctioning (Dimkpa and Bindraban, 2016). This is due to their large surface area and superior dissolution. During drought or salt stress, nitrogen nanoparticles help plants absorb light and fix carbon by aiding in the synthesis of Rubisco proteins and chlorophyll. Phosphorus nanoparticles aid in the production of ATP and NADPH, which are essential for the Calvin cycle to function. By controlling stomatal conductance and enzyme activation, potassium nanoparticles improve electron transport efficiency and maximize photosynthesis. According to (Khalil et al., 2017a, 2017b), nanofertilizers cooperate to maintain nutritional balance and shield photosynthesis from inclement weather (Fig. 15.2).

3.2 Nanoparticles as reactive oxygen species scavengers

The main cause of photosynthesis halting under abiotic stress is oxidative damage. The primary cause of this is the overproduction of reactive oxygen species (ROS) by chloroplasts (Fig. 15.2). Certain nanomaterials, such as metal and metal oxide nanoparticles, have inherent antioxidant qualities that enable them to eliminate ROS and safeguard photosynthetic structures (Mittler, 2017). Because of their remarkable redox properties, cerium oxide (CeO_2) nanoparticles can imitate antioxidant enzymes like catalase and superoxide dismutase by switching between Ce^{3+} and Ce_4^+ states (Das et al., 2018). It has been shown that zinc oxide (ZnO) and titanium dioxide (TiO_2) nanoparticles increase antioxidant enzyme activity and decrease lipid peroxidation in thylakoid membranes (Tripathi et al., 2017). By lowering photo-oxidative damage, these nanoparticles preserve membranes, preserve photosystem II (PSII) efficiency, and prevent chlorophyll degradation during periods of high stress.

3.3 Nanocarriers for stress-alleviating compounds

Using nano-encapsulation methods, nanotechnology also enables the rapid and simple delivery of stress-relieving substances (Fig. 15.2). Phytohormones like salicylic acid (SA), jasmonates, and abscisic acid (ABA) regulate the function of antioxidants, stomata, and stress signals. However, when employed in the typical manner, they frequently decompose rapidly. Hormones function better in the body when they are stabilized and released in a controlled manner thanks to nanocarriers (Zhao et al., 2020). By closing their stomata at the appropriate time without making it too difficult for them to absorb CO_2, controlled ABA administration helps plants use water more effectively during drought stress. The body uses salicylic acid and jasmones to combat free radicals and shield photosynthetic enzymes from oxidative stress. In order to maintain the stability of proteins, membranes, and the ultrastructure of chloroplasts, nanocarriers can also transport osmo protectants such as proline and glycine betaine. This maintains the machinery responsible for photosynthesis in the face of ionic and osmotic stress (Khalil et al., 2021c; Rico et al., 2015).

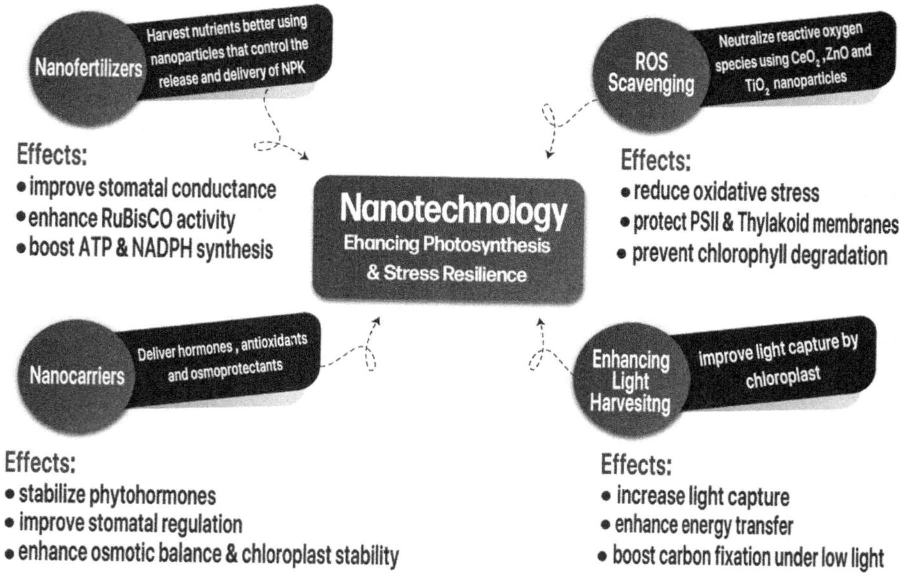

FIGURE 15.2 Role of nanotechnology in climate-resilient photosynthesis.

3.4 Enhancement of light harvesting efficiency

Nanotechnology presents new ways to increase photosynthesis's light utilization efficiency. Plasmonic nanoparticles, especially those made of gold (Au) and silver (Ag), exhibit localized surface plasmon resonance. This characteristic helps photosynthetic pigments absorb more light and strengthens local electromagnetic fields (Kim and Kim, 2021). Due to improved energy transfer, photochemical reactions function better in low-light or stressful environments. Another type of nanomaterial that appears promising is quantum dots, whose optical characteristics vary according to their size (Fig. 15.2). These substances have the ability to absorb light at a variety of wavelengths and return it to the photosynthetically active radiation range. As a result, more light may be utilized for photosynthesis (Giraldo et al., 2014). Although these approaches are still primarily experimental, they demonstrate how nanotechnology could directly improve the photophysical processes involved in photosynthesis (Khalil et al., 2021b).

3.5 Nano-enabled genetic regulation of photosynthesis

Recent developments in nanobiotechnology have made it possible to introduce genetic tools like small interfering RNA (siRNA) and CRISPR/Cas systems into plant cells without using traditional transformation methods. The ability of nanoparticles to transport nucleic acids allows them to regulate genes crucial for photosynthesis and stress tolerance (Demirer et al., 2019). You may alter the pathways that govern stomatal conductance, photoprotection, antioxidant defense, and Rubisco activity via nano-enabled gene regulation. Using fewer genetically modified technologies, this approach is a wonderful way to improve photosynthesis and increase plants' stress tolerance (Fig. 15.2).

4. Case studies and experimental evidence supporting nano-enabled photosynthetic resilience

The use of nanotechnology to lessen stress-induced photosynthetic damage in crops is supported by an increasing amount of experimental data. Nanomaterials improve chlorophyll content, preserve photosystem integrity, and fortify antioxidant defense systems, according to studies involving a variety of species and stress situations (Khalil et al., 2021a).

4.1 Titanium dioxide nanoparticles and drought stress in wheat

Wheat photosynthesis is severely hampered by drought stress because it reduces CO_2 availability and causes oxidative damage. In situations where there is a shortage of water, the application of TiO_2 nanoparticles has demonstrated an increase in chlorophyll content and net photosynthetic rate (Yang et al., 2006). While enhanced Rubisco activity facilitates continuous CO_2 fixation, higher chlorophyll stability enhances light absorption. Higher Fv/Fm values indicate that TiO_2 nanoparticles increase PSII's photochemical efficiency. They also lessen oxidative stress by triggering antioxidant enzymes like catalase and superoxide dismutase. During drought stress, their joint actions preserve the thylakoid membrane and maintain effective electron transport (Rico et al., 2015).

4.2 Zinc oxide nanoparticles and salinity tolerance in rice

Salinity stress disrupts the ion balance and results in oxidative damage, which reduces the efficiency of photosynthesis in rice. By increasing PSII stability and restoring chlorophyll levels, ZnO nanoparticles lessen salt stress (Singh et al., 2022). Increased zinc facilitates the synthesis of chlorophyll and allows enzymes to function in salty settings. ZnO nanoparticle treatment increases antioxidant enzyme activity and reduces ROS production. This shields photosynthetic membranes from oxidative stress-induced damage. Additionally, ZnO nanoparticles aid in maintaining the Na^+/K^+ balance, which indirectly enhances CO_2 absorption and stomatal function. When plants are under saline stress, this results in increased net photosynthetic rates.

4.3 Cerium oxide nanoparticles and heat stress in maize

Heat stress causes photosynthetic proteins in maize to become less stable and produces more reactive oxygen species (ROS). Because of their regenerative antioxidant properties, cerium oxide nanoparticles efficiently lower the build-up of reactive oxygen species (ROS), safeguarding PSII reaction centers and preserving the integrity of thylakoid membranes at high temperatures (Wu et al., 2017). When exposed to heat stress, plants treated with CeO_2 nanoparticles exhibit increased amounts of chlorophyll, improved photosynthetic efficiency, and prolonged activation of Calvin cycle enzymes. These impacts enhance photosynthetic equipment's physiological performance and increase its heat tolerance.

4.4 Integrative insights

A recurring pattern is seen in these case studies: by lowering oxidative stress, stabilizing photosystems, and maintaining chlorophyll integrity, nanomaterials lessen photosynthetic inhibition. Nanomaterials' diverse modes of action combine to regulate redox and safeguard photosystems, demonstrating their potential utility for climate-resilient agriculture (Khalil et al., 2021a).

5. Challenges and considerations in nano-enabled photosynthetic enhancement

Before nanotechnology is widely applied in agriculture, a number of scientific, environmental, and societal challenges need to be carefully resolved, even if it may help improve photosynthesis and increase plants' resistance to abiotic stress. Numerous issues include potential nanotoxicity and bioaccumulation, impacts on soil microbial communities, constraints on cost-effectiveness and scalability, and ambiguous environmental biosafety and regulatory frameworks (Khalil et al., 2021a).

5.1 Potential nanotoxicity and bioaccumulation

The potential for nanomaterials to be detrimental to plants, soil organisms, and higher trophic levels is a significant issue with their use in agriculture. At low concentrations, many nanomaterials promote photosynthesis and plant growth, but prolonged exposure can have detrimental effects on plants, including oxidative stress, membrane damage, slowed enzyme function, and disruption of cellular homeostasis (Ma et al., 2010). The tight margin between positive and negative effects in nano-plant interactions is shown by this dose-dependent response (Al-Musawi et al., 2022).

Nanoparticles can be absorbed by plant roots or leaves and subsequently transported to aerial tissues like grains and edible organs. Concerns regarding their integration into the food chain and possible long-term effects on human and animal health are raised by the presence of metal-based nanoparticles in plant tissues, such as ZnO, TiO_2, and CeO_2 (Rico et al., 2015). Furthermore, frequent use of nanomaterials may cause them to build up in agricultural soils, where it is yet unclear how long they will last and how they will change.

Nanoparticles' physicochemical characteristics, such as their size, shape, surface charge, and coatings, have a significant impact on their toxicity and biological interactions. Before utilizing nanoparticles in the field, it is crucial to have consistent nanoparticle characterization and safe dosage limits because the heterogeneity in these properties makes risk assessment more difficult (Tripathi et al., 2017).

5.2 Effects on soil microbiome and rhizosphere processes

In order to break down organic matter, cycle nutrients, and assist plants in coping with stress, the soil microbiome is crucial. The types and roles of soil microorganisms may alter when nanomaterials are added to soil ecosystems, which could have an impact on the soil's health and plant growth. Numerous studies demonstrate that some nanoparticles, particularly at higher dosages, can alter the variety of bacteria, slow down enzyme activity, or prevent microbes from proliferating (Kah et al., 2018).

Although useful in some situations, metal and metal oxide nanoparticles' antibacterial qualities may inadvertently harm beneficial rhizosphere microbes, such as mycorrhizal fungi and nitrogen-fixing bacteria. The beneficial benefits of nanomaterials on photosynthesis may ultimately be undermined by such disruptions, which may hinder nutrient availability and plant–microbe interactions (Khalil et al., 2017a). The antibacterial qualities of metal and metal oxide nanoparticles may be useful in certain situations, but they may also inadvertently damage beneficial rhizosphere microbes such mycorrhizal fungi and nitrogen-fixing bacteria. The beneficial impacts of nanomaterials on photosynthesis could be offset by these kinds of disruptions, which could make it more difficult for plants to obtain nutrients and for microorganisms to interact with them (Dawwam et al., 2023; Al-Sissi et al., 2025).

5.3 Scalability and cost-effectiveness in agriculture

The difficulty of scaling up nanotechnology is one of the main issues with its application in agriculture. Numerous studies demonstrate that stress can increase the efficiency of photosynthesis. Very small quantities of highly customized nanomaterials are used in these investigations, which are conducted in controlled laboratory or greenhouse environments. In order to make widespread use of these findings, we must consider their manufacturing costs, potential applications, and profitability. Nanomaterials are less economical than conventional fertilizers and agrochemicals since they frequently require complex processes, specialized machinery, and a lot of energy (Dimkpa and Bindraban, 2016).

Many farmers, particularly those in underdeveloped areas, might not be able to obtain nano-enabled products if there are no affordable, environmentally friendly ways to produce them and efficient ways to deliver them to farmers. Furthermore, the stability, mobility, and efficacy of nanoparticles may be impacted by environmental variability in field circumstances, such as soil heterogeneity, precipitation patterns, and temperature variations. These drawbacks highlight the necessity of conducting thorough field tests and cost-benefit assessments to assess the usefulness of agricultural inputs based on nanotechnology.

5.4 Environmental biosafety and regulatory challenges

An essential component of adopting nanotechnology in agriculture is environmental biosafety. Insects, soil animals, aquatic creatures, and nearby flora are examples of nontarget organisms with which nanomaterials released into the environment may interact. The transport and transformation properties of nanoparticles may differ significantly from those of conventional compounds because of their tiny size and strong reactivity (Kah et al., 2018). Nanomaterials are frequently categorized by current chemical or fertilizer standards, which do not sufficiently address features unique to the nanoscale. It is uncertain how to assess dangers, label, and obtain authorization for agricultural goods enabled by nanotechnology due to the absence of regulations. Additionally, efforts are currently underway to develop standard techniques for assessing the toxicity of nanoparticles, their environmental half-lives, and their long-term impacts on ecosystems. People's perceptions and societal acceptance of nanotechnology have a big impact on its utilization in food production. Explaining the advantages, dangers, and safety regulations in detail is crucial to fostering responsible innovation and fostering public trust.

5.5 Toward responsible and sustainable nano-agriculture

Future research must use a multidisciplinary approach that incorporates plant physiology, nanoscience, environmental toxicity, and policy analysis in order to fully utilize the advantages of nanotechnology for improving photosynthesis while minimizing dangers. Sustainable nano-agriculture requires long-term field experiments, life-cycle analyses, and the creation of biodegradable or plant-derived nanomaterials (Zhao et al., 2020). Although nanotechnology provides effective tools for improving photosynthetic tolerance to climate stress, its successful incorporation into agricultural systems requires careful risk assessment, viability from an economic standpoint, and strict regulatory control.

6. Future perspectives in nano-enabled photosynthetic enhancement

Abiotic factors that impact agricultural yield are getting worse due to climate change. In order to develop sophisticated, clever, and durable nano-enabled techniques for enhancing photosynthesis, future research should go beyond proof-of-concept investigations. Recent developments in synthetic biology, artificial intelligence, and nanotechnology present significant chances to improve carbon fixation and strengthen plant resilience in the face of increasingly unpredictable weather.

6.1 Smart nanomaterials responsive to stress signals

The creation of intelligent nanomaterials that can react dynamically to plant stress signals is one possible future path. Smart nanoparticles are made to detect certain physiological or environmental signals, including pH shifts, reactive oxygen species (ROS) levels, or ionic imbalances, in contrast to conventional nanomaterials that operate passively. When they do, they respond to stimuli by releasing their active components in a regulated manner (Kah et al., 2018).

Abiotic stresses frequently alter the pH, ion concentrations, and accumulation of reactive oxygen species (ROS) in plant tissues, particularly in the apoplast and chloroplasts. The release of antioxidants during oxidative stress or the delivery of nutrients and signaling molecules at the exact time when the photosynthetic system is most susceptible might be precisely triggered by nanomaterials designed to react to these signals. Antioxidants or protective chemicals may only be released by ROS-responsive nanoparticles when ROS levels surpass a specific threshold. By doing this, they reduce the chance of injury and prevent needless exposure (Zhao et al., 2020). By releasing potassium or calcium ions to restore ionic balance and maintain the stability of photosynthetic enzymes, ion-responsive nanocarriers may help lessen Na^+ toxicity brought on by elevated salt levels. This degree of distribution accuracy will significantly increase biosafety and efficiency while lowering the quantity of materials required and the amount of pollution that accumulates in the environment.

6.2 Integration with artificial intelligence and precision agriculture

The real-time, adaptive management of photosynthetic stress has advanced significantly with the integration of nanotechnology, artificial intelligence (AI), and precision agriculture. Precision agriculture technology, such as remote sensing, hyperspectral imaging, and Internet of Things (IoT)-based sensors, enables farmers to continuously monitor the health of their crops. According to Liakos et al. (2018), integrating these systems with nano-enabled delivery platforms may facilitate prompt and precise adjustments based on the plants' current requirements.

Large volumes of data from environmental sensors and plant photos can be analyzed by AI-powered data analytics to forecast stress before any overt symptoms appear. The use of specific nanomaterials, such as antioxidant nanoparticles in situations of thermal stress or nano-fertilizers in situations of nutrient deficiency, might then be initiated by these predictive algorithms. This would reduce waste and environmental damage while maintaining optimal photosynthetic performance. By transforming nano-agriculture from a static, input-focused model to a dynamic, feedback-based approach, these integrated systems would increase its productivity and sustainability. To achieve this integration, interdisciplinary collaboration between plant physiologists, data scientists, engineers, and agronomists is required.

6.3 Nano-enabled synthetic and semi-synthetic photosynthesis

Nanotechnology not only improves natural photosynthesis but also opens the door to the creation of nano-enabled synthetic or semisynthetic photosynthesis systems that are intended to boost the efficiency of carbon fixation. Enzymatic inefficiencies, photorespiration, and energy dissipation are the fundamental limitations of natural photosynthesis. By improving the processes of light harvesting, charge separation, and carbon conversion, nano-engineered systems seek to improve biological photosynthesis (Blankenship et al., 2011).

Quantum dots, plasmonic nanoparticles, and semiconductor nanocomposites are examples of nanostructured materials that can be designed to efficiently transfer energy to biological or artificial catalytic systems and absorb a broader range of solar radiation. In experimental settings, hybrid bio-nano interfaces that integrate nanomaterials with chloroplasts or photosynthetic proteins have demonstrated improved photochemical efficiency and superior electron transport (Giraldo et al., 2014). Artificial photosynthesis systems enabled by nanotechnology may eventually work alongside crops or in controlled agricultural settings to enhance atmospheric CO_2 sequestration and convert it into useful compounds. These approaches show great promise for addressing issues related to food security and climate mitigation, even if they are still in the early stages of development.

6.4 Sustainability, safety, and translational research priorities

Sustainability and biosafety should be given top priority in future advancements in nano-enabled photosynthetic enhancement. To address ecological issues, research projects should focus on developing biodegradable, plant-based, or environmentally safe nanomaterials. For evaluating environmental destiny, persistence, and cumulative consequences in real-world farming situations, life-cycle studies and extended field research are essential (Kah et al., 2018).

Equally important is the conversion of laboratory results into useful technology. This calls for affordable formulations, robust delivery systems that complement modern farming methods, and scalable synthesis procedures. To encourage safe innovation and global acceptability, international collaboration and uniform regulatory frameworks will be crucial.

6.5 Concluding outlook

The combination of digital farming, synthetic biology, and intelligent materials holds the key to the future of nano-enhanced photosynthesis. Through precise, sensitive, and long-lasting interventions, nanotechnology has the potential to revolutionize the protection and improvement of photosynthesis under climate stress. The full potential of nanotechnology in creating climate-resilient agricultural systems will require ongoing interdisciplinary study, responsible innovation, and evidence-based regulation.

7. Conclusion

Climate change is a unique threat to global agricultural productivity because it makes abiotic factors worse, which directly affects photosynthetic efficiency and crop yield. Since photosynthesis is the primary process by which plants produce food and thrive, it is crucial for initiatives aimed at strengthening plants' resistance to harsh environmental circumstances. In this regard, nanotechnology has developed into a potent and adaptable instrument for creating photosynthesis that can adapt to climate change. Enhancing nutrient delivery, strengthening antioxidant defense, stabilizing photosynthetic apparatus, optimizing light absorption, and precisely modulating stress-responsive pathways are just a few of the ways that integrating nanomaterials into plant systems can mitigate stress-induced photosynthetic inhibition.

Appropriately tailored nanomaterials can efficiently protect photosystems, preserve carbon fixation, and enhance physiological performance under drought, salinity, heat, and other abiotic stressors, according to experimental evidence from a variety of crops and stress situations. The successful application of nano-enabled photosynthetic enhancement in sustainable agriculture requires responsible development and utilization. Strict regulatory frameworks, substantial environmental research, and comprehensive biosafety assessments are necessary to lower the hazards of nanotoxicity and ecological disturbance. When paired with advancements in artificial intelligence, precision agriculture, and the creation of environmentally friendly nanomaterials, nanotechnology holds great promise for ensuring food production in the future and for making agriculture more climate-smart in a rapidly changing world.

References

Allakhverdiev, S.I., Kreslavski, V.D., Klimov, V.V., Los, D. A., Carpentier, R., Mohanty, P., 2008. Heat stress: an overview of molecular responses in photosynthesis. Photosynth. Res. 98 (1), 541–550.

Allen, D.J., Ort, D.R., 2001. Impacts of chilling temperatures on photosynthesis in warm-climate plants. Trends Plant Sci. 6 (1), 36–42.

Al-Musawi, H.A.H., Khalil, R., Mehasen, S.A.S., 2022. Effect of nano silica on bread wheat under water stress. Benha J. Appl. Sci. 7 (12), 49–52. https://doi.org/10.21608/bjas.2022.306490.

Al-Sissi, N., Yassin, M.H., Khalil, R., Gamal, A., Attia, M.S., Hashem, A.H., 2025. Innovative fungal bioagents: producing siderophores, IAA, and HCN to support plants under salinity stress and combat microbial plant pathogens. Microb. Cell Fact. https://doi.org/10.1186/s12934-025-02862-2.

Blankenship, R.E., Tiede, D.M., Barber, J., Brudvig, G.W., Fleming, G., Ghirardi, M., Sayre, R.T., 2011. Comparing photosynthetic and photovoltaic efficiencies and recognizing the potential for improvement. Science 332 (6031), 805–809. https://doi.org/10.1126/science.1200165.

Das, S., Dowding, J.M., Klump, K.E., McGinnis, J.F., Self, W., Seal, S., 2018. Cerium oxide nanoparticles: applications and prospects in plant stress tolerance. Environ. Sci. Nano 5, 270–290.

Dawwam, G.E., Fathy, M.H., Emam, T.M., AbdelMonem, M. O., Khalil, R., Nagah, A., 2023. Production of the phytohormone indole acetic acid by some rhizospheric bacteria associated with the Egyptian flora. J. Basic Environ. Sci. 10 (4), 130–137. https://doi.org/10.21608/jbes.2023.372265.

Demirer, G.S., Zhang, H., Goh, N.S., González-Grandío, E., Landry, M.P., 2019. Carbon nanotube–mediated DNA delivery without transgene integration in intact plants. Nat. Protoc. 14 (10), 2954–2971.

Dimkpa, C.O., Bindraban, P.S., 2016. Nanofertilizers: new products for the fertilizer industry? J. Agric. Food Chem. 64, 9794–9804doi. https://doi.org/10.1007/s11120-008-9331-0.

Doneva, D., Pál, M., Brankova, L., Szalai, G., Tajti, J., Khalil, R., Peeva, V., 2021. The effects of putrescine pre-treatment on osmotic stress responses in drought-tolerant and drought-sensitive wheat seedlings. Physiol. Plantarum 171 (2), 200–216. https://doi.org/10.1111/ppl.13150.

Flexas, J., Barbour, M.M., Brendel, O., Cabrera, H.M., Carriquí, M., Díaz-Espejo, A., Warren, C.R., 2012. Mesophyll diffusion conductance to CO2: an unappreciated central player in photosynthesis. Plant Sci. 193, 70–84.

Giraldo, J.P., Landry, M.P., Faltermeier, S.M., McNicholas, T. P., Iverson, N.M., Boghossian, A.A., Strano, M.S., 2014. Plant nanobionics approach to augment photosynthesis and biochemical sensing. Nat. Mater. 13 (4), 400–408.

Janda, T., Khalil, R., Tajti, J., Pál, M., Szalai, G., Rudnóy, S., Darko, É., 2018. The newly synthesized plant growth regulator S-methylmethionine salicylate may provide protection against high salinity in wheat. Plant Growth Regul. 85 (2), 305–315. https://doi.org/10.1007/s10725-018-0398-0.

Janda, T., Khalil, R., Tajti, J., Pál, M., Darkó, É., 2019. Responses of young wheat plants to moderate heat stress. Acta Physiol. Plant. 41 (8), 137. https://doi.org/10.1007/s11738-019-2930-x.

Kah, M., Kookana, R.S., Gogos, A., Bucheli, T.D., 2018. A critical evaluation of nanopesticides and nanofertilizers against their conventional analogues. Nat. Nanotechnol. 13 (8), 677–684.

Khalil, R.R., Moustafa, A.N., Bassuony, F.M., Haroun, S.A., 2017a. Kinetin and/or calcium affect growth of Phaseolus vulgaris L. plant grown under heavy metals stress. J. Environ. Sci. 46 (2), 103–120.

Khalil, R.R., Galal, H.A., Darwisch, W.B., 2017b. Role of bio-fertilizer treatments in alleviating the adverse effect of water stress in Mangifera indica. Egypt. J. Bot. 56 (2), 471–488. https://doi.org/10.21608/ejbo.2017.1146.

Khalil, R., ElSayed, N., Hashem, H.A., 2021a. Nanoparticles as a new promising tool to increase plant Immunity against abiotic stress. In: Faizan, M., Hayat, S., Yu, F. (Eds.), Sustainable Agriculture Reviews, 53. Springer, Cham. https://doi.org/10.1007/978-3-030-86876-5_3.

Khalil, R., Tajti, J., Hamow, K.A., Gondor, K.O., Darko, E., Elsayed, N., Majláth, I., 2021b. How does moderate drought affect quantum yield and the regulation of sugar metabolism at low temperature in durum wheat (Triticum durum L.)? Photosynthetica 59 (2), 313–326. https://doi.org/10.32615/ps.2021.030.

Khalil, R., Haroun, S., Bassyoini, F., Nagah, A., Yusuf, M., 2021c. Salicylic acid in combination with kinetin or calcium ameliorates heavy metal stress in Phaseolus vulgaris plant. J. Agric. Food Res. 5, 100182. https://doi.org/10.1016/j.jafr.2021.100182.

Khalil, R., Elsayed, N., Khan, T.A., Yusuf, M., 2022a. Potassium: a potent modulator of plant responses under changing environment. In: Role of Potassium in Abiotic Stress. Springer Nature Singapore, Singapore, pp. 221–247. https://doi.org/10.1007/978-981-16-4461-0_11.

Khalil, R., Yusuf, M., Bassuony, F., Haroun, S., Gamal, A., 2022b. Alpha-tocopherol reinforce selenium efficiency to ameliorates salt stress in maize plants through carbon metabolism, enhanced photosynthetic pigments and ion uptake. South Afr. J. Bot. 144, 1–9. https://doi.org/10.1016/j.sajb.2021.08.033.

Kim, D., Kim, Y., 2021. Extinction effect of gold nanocatalysts on photocatalytic activities under plasmonic excitation. Catalysts 11 (4), 413.

Lesk, C., Rowhani, P., Ramankutty, N., 2016. Influence of extreme weather disasters on global crop production. Nature 529 (7584), 84–87.

Liakos, K.G., Busato, P., Moshou, D., Pearson, S., Bochtis, D., 2018. Machine learning in agriculture: a review. Sensors 18 (8), 2674. https://doi.org/10.3390/s18082674.

Ma, X., Geiser-Lee, J., Deng, Y., Kolmakov, A., 2010. Interactions between engineered nanoparticles (ENPs) and plants: phytotoxicity, uptake and accumulation. Sci. Tot. Environ. 408 (16), 3053–3061. https://doi.org/10.1016/j.scitotenv.2010.03.031.

Madany, M., Khalil, R., 2017. Seed priming with ascorbic acid or calcium chloride mitigates the adverse effects of drought stress in sunflower (Helianthus annuus L.) seedlings. Egypt J. Exp. Biol. (Bot) 13 (1), 119–133. https://doi.org/10.5455/egyjebb.20170409090612.

Majláth, I., Éva, C., Tajti, J., Khalil, R., Elsayed, N., Darko, E., Janda, T., 2020. Exogenous methylglyoxal enhances the reactive aldehyde detoxification capability and frost-hardiness of wheat. Plant Physiol. Biochem. 149, 75–85. https://doi.org/10.1016/j.plaphy.2020.02.003.

Mittler, R., 2006. Abiotic stress, the field environment and stress combination. Trends Plant Sci. 11 (1), 15–19. https://doi.org/10.1016/j.tplants.2005.11.002.

Mittler, R., 2017. ROS are good. Trends Plant Sci. 22 (1), 11–19. https://doi.org/10.1016/j.tplants.2016.08.002.

Munns, R., Tester, M., 2008. Mechanisms of salinity tolerance. Annu. Rev. Plant Biol. 59, 651–681. https://doi.org/10.1146/annurev.arplant.59.032607.092911.

Parida, A.K., Das, A.B., 2005. Salt tolerance and salinity effects on plants: a review. Ecotoxicol. Environ. Saf. 60 (3), 324–349.

Rico, C.M., Majumdar, S., Duarte-Gardea, M., Peralta-Videa, J.R., Gardea-Torresdey, J.L., 2015. Interaction of nanoparticles with edible plants and their possible implications in the food chain. J. Agric. Food Chem. 63, 9274–9289.

Roco, M.C., Mirkin, C.A., Hersam, M.C., 2011. Nanotechnology Research Directions for Societal Needs in 2020: Retrospective and Outlook.

Sage, R.F., Kubien, D.S., 2007. The temperature response of C3 and C4 photosynthesis. Plant Cell Environ. 30 (9), 1086–1106.

Shahid, M., Khalid, S., Abbas, G., Shahid, N., Nadeem, M., Sabir, M., Dumat, C., 2015. Heavy metal stress and crop productivity. In: Crop Production and Global Environmental Issues. Springer International Publishing, Cham, pp. 1–25.

Singh, A., Sengar, R.S., Rajput, V.D., Minkina, T., Singh, R. K., 2022. Zinc oxide nanoparticles improve salt tolerance in rice seedlings by improving physiological and biochemical indices. Agriculture 12 (7), 1014.

Tripathi, D.K., Singh, S., Singh, S., Pandey, R., Singh, V.P., Sharma, N.C., Prasad, S.M., Dubey, N.K., Chauhan, D. K., 2017. An overview on manufactured nanoparticles in plants: uptake, translocation, accumulation, and phytotoxicity. Plant Physiol. Biochem. 110, 2–12. https://doi.org/10.1016/j.plaphy.2016.07.030.

Wu, H., Tito, N., Giraldo, J.P., 2017. Anionic cerium oxide nanoparticles protect plant photosynthesis from abiotic stress by scavenging reactive oxygen species. ACS Nano 11 (11), 11283–11297. https://doi.org/10.1021/acsnano.7b05723.

Yang, F., Hong, F., You, W., Liu, C., Gao, F., Wu, C., Yang, P., 2006. Influence of nano-anatase TiO2 on the nitrogen metabolism of growing spinach. Biol. Trace Elem. Res. 110 (2), 179–190. https://doi.org/10.1385/BTER:110:2:179.

Zhao, L., Hu, Q., Huang, Y., Fulton, A.N., Hannah-Bick, C., Adeleye, A.S., Keller, A.A., 2020. Nanotechnology in agriculture: new opportunities and perspectives. Trends Plant Sci. 25, 517–531.

CHAPTER

16

Integrating nanotechnology with smart farming and precision photosynthetic management

Meghna Sarkar[1], *Nilotpal Das*[2], *Atin Kumar*[3], *Rida Batool*[4], *Sharad Sachan*[5] *and Mohammad Faizan*[6]

[1]Department of Soil Science, Assam Agricultural University, Jorhat, Assam, India; [2]Research Scholar, Faculty of Agricultural Sciences, GLA University, Mathura, Uttar Pradesh, India; [3]School of Agriculture, Uttaranchal University, Dehradun, Uttarakhand, India; [4]Department of Plant Sciences, Faculty of Biological Sciences, Quaid-i-Azam University, Islamabad, Pakistan; [5]Department of Agricultural Economics and Extension, School of Agriculture, Lovely Professional University, Phagwara, Punjab, India; [6]Botany Section, School of Sciences, Maulana Azad National Urdu University, Hyderabad, Telangana, India

1. Introduction

Combining nanotechnology, smart farming, and precision photosynthetic control is proving to be an effective answer to increasing global demands regarding agriculture, in which population growth, climate change, soil erosion, and water scarcity challenge the ability of the existing food systems to produce enough and consistent production. The world agribusiness needs to both produce more, with fewer inputs per unit of production, and reduce the emission of greenhouse gases as temperatures rise, droughts and floods become more frequent, and arable land and fresh water are becoming limited (Wang et al., 2024; Mizik, 2021). Under such conditions, the traditional farming systems, based on blanket applications of water, fertilizers, and pesticides, are becoming more and more inadequate: they tend to be poorly and inefficiently utilized, spatially and temporally homogeneous in their management, and slow to respond to stress because crop status is visually inferred. Traditional nutrient management specifically is limited by low resolution of dosing, loss through leaching and volatilization, as well as lack of synchrony between nutrient release and crop requirements, which combine to decrease production and cause environmental pollution. Smart farming and

Nanotechnology in Photosynthesis
https://doi.org/10.1016/B978-0-443-44376-3.00016-1

precision agriculture, in contrast, uses sensors, Internet of Things (IoT) networks, remote and proximal sensing, robotics, and artificial intelligence to monitor real-time soil, plant, and atmosphere and apply water, nutrients, and agrochemicals at the optimal dose, location, effectively using resources and enhance their resilience (Martinho and Guiné, 2021; Wang et al., 2024). Agro nanotechnology can offer a complementary dimension of material-level control within this digital change: nanofertilizers, nanopesticides, nano-biosensors, and nano-facilitated delivery systems can offer controlled and targeted delivery, better nutrient uptake, prevention of active ingredient degradation, and stress resilience and directly tackle the inefficiencies and externalities of traditional nutrient management (Codreanu et al., 2025). Recent literature points to the possibility of nanoformulations and bio-based nanoparticles to achieve climate-smart agriculture, stress-resilient agriculture by enhancing soil quality, providing opportunities to support smart fertilisation, and decreasing the necessity of overuse of inputs (Codreanu et al., 2025). However, the majority of present-day smart farming technologies are optimised at the field or plant level and not at the biological machine of yield: photosynthesis. Spatially discrete control and monitoring of photosynthetic processes, light interception, electron transport, carbon fixation, and associated stomatal and nutrient statistics under changing environmental conditions using integrated sensing, modeling, and actuation to optimise carbon gain per unit of water, nutrient, and energy input could be described as precision photosynthetic management. High-resolution estimation of canopy vigor, chlorophyll content, and stress status, hyperspectral and multispectral measurements can serve as a proxy of photosynthetic performance and early stress, and machine learning and remote sensing technologies already allow making site-specific management decisions (Wang et al., 2024; Guebsi et al., 2024). The combination of these features with nano-enabled inputs and nano-encapsulated fertiliser, the release of which depends on soil moisture or root exudates, or nano-biosensors on a leaf, and informing the farmer about the real-time photosynthesis restrictions as well as smart farming infrastructures (IoT sensor networks, autonomous machinery, digital twins) allows the junction of closed-loop control of photosynthesis at the plant and canopy scale. The objectives of this chapter, in this context, are fourfold, namely: (i) synthesizing the global problems of crop yield, resource shortage, and the climate crisis that drive the need to move beyond traditional farming and nutrient management; (ii) reviewing the advent of nanotechnology and smart farming as radically new, and more convergent, strategies toward enhancing resource-use efficiency and sustainability in agriculture; (iii) defining the term precision photosynthetic management, iv) demystifying its biophysical structure, and arguing that it should be the center of attention in a more convergent manner. Through this articulation of integration, it is expected that the chapter will offer a conceptual and technological roadmap that describes how inputs-centered precision agriculture can be transitioned to photosynthesis-centered precision agriculture that can achieve food and climate objectives (Codreanu et al., 2025; Wang et al., 2024).

2. Fundamentals of photosynthesis and its agricultural relevance

2.1 Overview of photosynthetic processes (light reactions and carbon fixation)

Photosynthesis in chloroplasts transforms light energy, water, and CO_2 into carbohydrates and oxygen, and this forms the biochemical foundation of crop productivity. During the light reactions, photosynthetically active radiation (PAR) is absorbed by pigments present in thylakoid membranes and leads to the electron transport with the participation of photosystems I and II to produce ATP and NADPH by photophosphorylation and splitting of water (Livadariu et al., 2023; Yang et al., 2025). In the calvin cycle of carbon fixation, CO_2 is reduced in the stroma to triose phosphates and finally to sugars, which are then used to support growth, respiration, and storage. The net CO_2 assimilation and consequently the crop performance are controlled by the balance between light harvesting, electron transportation, and calvin cycle activity (Yang et al., 2025; Zhang et al., 2025).

2.2 Important considerations that have an impact on photosynthetic efficiency of crops

The photosynthetic efficiency is regulated by the amount and quality of light, CO_2 level, temperature, water, nutrient availability, and canopy structure. Enhanced CO_2 enrichment in confined systems considerably enhances photosynthetic velocity, water use efficiency, and yield, although responses by species and setting vary (Villagran et al., 2025). The spectral distribution and intensity of light control photosynthesis and morphology of the plant; the control of the wavelength of LED lighting can increase the activity of chlorophyll, biomass, and the production of bioactive substances (Livadariu et al., 2023). The form and availability of nitrogen present a strong effect on photosynthesis since nitrate and ammonium assimilation is a drain on photosynthetic energy; mixed nutrition of nitrate-ammonium enhances electron transport, Rubisco performance, and carbon assimilation in a variety of crops (Yang et al., 2025).

2.3 The correlation between photosynthesis, biomass accretion, and yield

The rate of fixation of carbon into carbohydrates that form biomass and grain or fruit yield is regulated by net photosynthesis. High CO_2 and light control promote carbon assimilation and enhance dry matter and high productivity in greenhouse and vertical farming systems (Villagran et al., 2025). The artificial interventions in synthetic biology, which improve the electron transport, kinetics of Rubisco, or photorespiration in rice and tobacco species has improved photosynthesis, biomass, and grain value, indicating the close association between photosynthetic activity and agricultural productivity (Zhang et al., 2025) (Fig. 16.1).

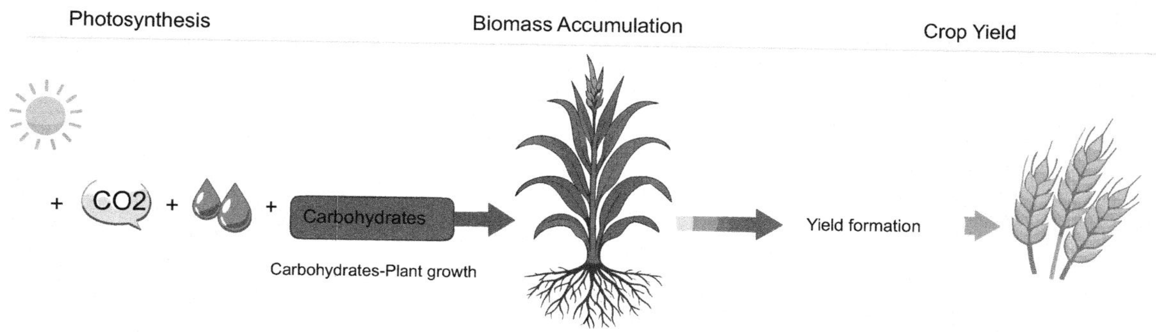

FIGURE 16.1 Conceptual relationship between photosynthesis, biomass accumulation, and crop yield. Photosynthesis converts light energy, carbon dioxide, and water into carbohydrates that support plant growth and biomass formation, which ultimately determines agricultural yield.

2.4 Photosynthetic stress during the abiotic and biotic stress

During drought, salinity, heat, and other abiotic stresses, photosynthesis is limited by stomatal closure, decreased CO_2 diffusion, pigment degradation, and damage to photosystems I and II, which results in a decrease in carbon fixation and loss of yield (Wahab et al., 2022; Roşca et al., 2023). The impact of drought is decreased chlorophyll content, dysfunction of PSII (via chlorophyll fluorescence), and Calvin cycle enzymes, leading to the long-term reduction of CO_2 assimilation and biomass (Wahab et al., 2022). Salinity also impairs photosynthesis and growth of crops like tomato in all the stages of growth and development, and leads to serious yield loss (Roşca et al., 2023). These stresses mostly work together with nutrient and light conditions, and therefore, there is a need to have coordinated management of these factors in order to sustain the photosynthetic efficiency of the field.

3. Nanotechnology in agriculture: Concepts and applications

3.1 Definition and classification of agricultural nanomaterials

As structures, agricultural nanomaterials have a minimum size of 1–100 nm along at least one dimension and are engineered to respond to plants, soils, or agrochemicals to increase productivity and sustainability. They can be categorised by dimensions 0D (nanoparticles, quantum dots), 1D (nanotubes, nanofibers), 2D (nanosheets, nanolayers), and 3D (nanocomposites, porous networks) and by composites: organic (polymeric carriers, lipid nanoparticles), carbon-based (fullerenes, graphene), and inorganic (metal/metal oxides, silica, phosphates, and other ceramics). Their surface area is very high and their surface chemistry can be tuned, giving them special optical, catalytic and adsorption properties on which agricultural applications are based (Codreanu et al., 2025) (Table 16.1).

3.2 Types of nano-inputs: Nano-fertilizers, nano-pesticides, nano-bio-stimulants

The nutrient-carrying nanoparticles can be nutrients contained in biopolymer nanocarriers (examples of which are chitosan and starch) that serve as slow-release systems, improving nutrient uptake and decreasing leaching and volatilization (Codreanu et al., 2025; Dhanapal et al., 2024). Nano-pesticides contain active substances that are contained in polymeric or inorganic nanocarriers (silica, metal oxides, clays) or even metallic nanoparticles, such as Ag, Cu, ZnO, so that they can achieve higher stability,

TABLE 16.1 Representative nano-inputs in agriculture.

Category	Main forms and functions	Typical benefits in crops	Citations
Nano-fertilizers	Nutrients in nano-chelates, nano-encapsulated, or nano-coated on carriers	Controlled/slow release, higher nutrient-use efficiency, improved growth under stress	Codreanu et al. (2025), Dhanapal et al. (2024), Chandrashekar et al. (2023)
Nano-pesticides	Nano-emulsions, polymer/inorganic nanocarriers, metallic/oxide NPs with active ingredients	Better stability, targeted delivery to pests, reduced dose, and residues	Deka et al. (2021), Luneja and Mkindi (2025), Sati et al. (2025)
Nano-bio-stimulants	Nanoformulations of microbial/plant extracts, signaling molecules, and nano-enzymes	Enhanced root growth, photosynthesis, stress tolerance, improved soil–plant interactions	Codreanu et al. (2025), Chandrashekar et al. (2023)

adhesion, penetration, and controlled release against insect pests, pathogens, and weeds (Deka et al., 2021; Luneja and Mkindi, 2025; Sati et al., 2025). The nano-bio-stimulants involve nano-formulations of useful microorganisms, algal or plant extracts, signal molecules, and nano-enzymes to cause growth, nutrient uptake, photosynthesis, and drought or other resilience (Codreanu et al., 2025; Chandrashekar et al., 2023).

3.3 The uptake, translocation, and action of nanoparticles in plants

The nanoparticles gain entry into plants by roots (apoplast and symplast with the help of the pores, ion channels, transporter interactions) or leaves (stomata, cuticle nano-defects), followed by transportation into the shoots and the reproductive organs by the xylem and phloem (Chandrashekar et al., 2023; Pokharel et al., 2024). They penetrate cell walls and membranes, occasionally by endocytosis, and are also located in cell walls, cytoplasm, chloroplasts, and vacuoles, due to their small size. Nano-fertilizers release nutrients in areas of the root or chloroplast, and nano-pesticides and nano-bio-stimulants release active molecules into particular cells or organelles, which may increase growth and drought tolerance because of their ability to modulate photosynthesis, antioxidant networks, hormone signaling, and stress-responsive genes (Codreanu et al., 2025; Chandrashekar et al., 2023) (Fig. 16.2).

3.4 Nanotechnology has benefits over the conventional agro-input

Nanotechnology is more efficient in terms of use, delivery, and loading environment compared to bulk fertilizers and pesticides. The use of nano-carriers facilitates a regulated release according to the needs of the plants and reduces their application rate and the losses to the soil and water (Codreanu et al., 2025; Dhanapal et al., 2024). Nano-pesticides enhance efficacy and persistence and may impose reduced doses and smarter formulations on the reduction of operator exposure and off-target effects (Deka et al., 2021; Luneja and Mkindi, 2025; Sati et al., 2025). Nanoparticle treatments often enhance water and nutrient absorption, photosynthesis, and antioxidant defense, which partially helps to offset yield losses, especially in times of stress (e.g., drought) (Chandrashekar et al., 2023). Together, these attributes make agri-nanotechnology an important solution to climate-smart, resource-efficient agriculture.

4. Smart farming and precision agriculture technologies

Smart farming and precision agriculture combine digital technologies to enhance the efficiency, locality, and data-driven nature of crop production and change the manner in which farmers observe fields and apply inputs.

4.1 Digital agriculture and smart farming concepts

Digital agriculture (also known as Agriculture 4.0) is the application of integrated sensors, robotics, and communication networks and data analytics throughout the entire farm system, including soil and weather monitoring as well as machinery control and market decisions (Abbasi et al., 2022; Araújo et al., 2023). One of its central elements is precision agriculture, which aims to control the spatial and temporal fluctuations, that is, to treat various areas of the field in different ways, depending on the measured conditions to maximize yield, consumption, and environmental outcomes (Sharma et al., 2021; Yousaf et al., 2023). Smart farming is an expansion of the concept of precision agriculture, which implements automation,

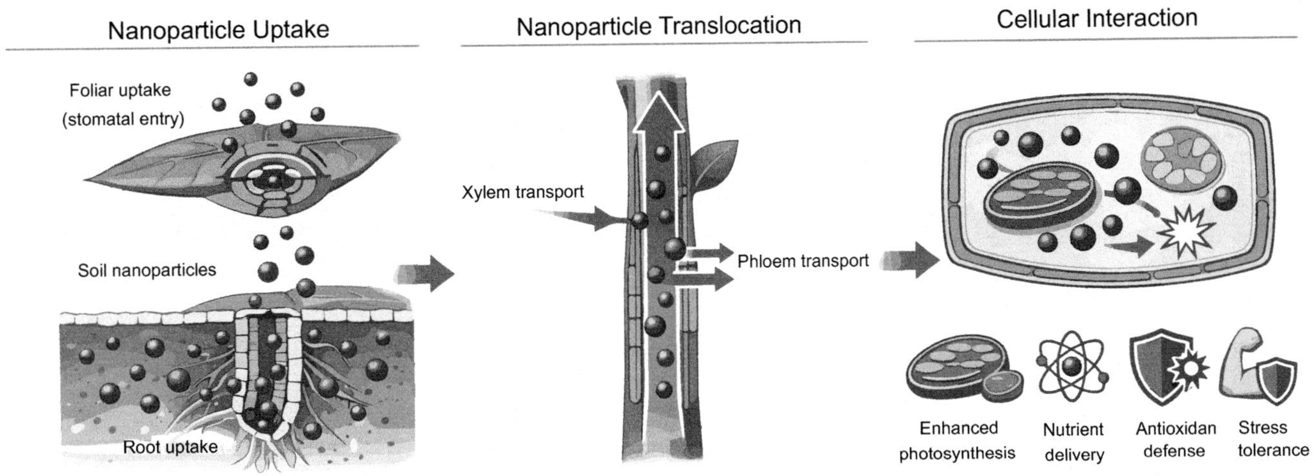

FIGURE 16.2 Schematic representation of nanoparticle uptake, translocation, and physiological action in plants. Nanoparticles can enter plants through root absorption from soil or through foliar uptake via stomata. After entry, they are transported through the xylem and phloem to various plant tissues where they interact with cellular structures such as chloroplasts, influencing photosynthesis, nutrient utilization, and stress tolerance.

artificial intelligence, and feedback loops between computer models and physical land into its practice, allowing remote monitoring, autonomous devices, and automatic feedback (Abbasi et al., 2022; Araújo et al., 2023).

4.2 Role of IoT, AI, machine learning, and big data in crop management

Internet of Things (IoT) refers to the links between the distributed field devices (soil moisture sensors, weather stations, fertilizer controllers, and machinery) and either the local or the cloud platform, offering real-time data streams of the crop, soil, and microclimate conditions (Sharma et al., 2021; Mughele et al., 2024). Such large datasets are processed by artificial intelligence (AI) and machine learning (ML) to generate actionable information that can be used in various works, including yield prediction, disease and weed detection, crop growth modeling, and scheduling irrigation or fertilization. Regression (used to predict soil and yield), support vectors and random forests (used to classify, e.g., disease or stress mapping), and deep learning (CNNs) (used to diagnose using images of UAV and field cameras) are common ML approaches in agriculture (Sharma et al., 2021; Miller et al., 2025). Big data and cloud computing make it possible to store and process heterogeneous sensor, drone, satellite, and machinery log data to support advanced analytics and scalable decision support services, but issues concerning data quality, interoperability, and architecture of agricultural big data systems are present (Araújo et al., 2023; Cravero et al., 2022).

4.3 Sensors, drones, satellites, and decision-support systems

Smart farming is based on a stratified sensing infrastructure. The ground sensors are used to measure soil moisture, temperature, electrical conductivity, nutrient status, and microclimate, plant sensors, and proximal imaging to measure the canopy temperature, chlorophyll indices, and biomass. High-resolution maps of crop vigor, water stress, disease foci, and nutrient deficiencies are provided by drones (UAVs) armed with either RGB, multispectral, hyperspectral, thermal, or LiDAR sensors, more prominently used even in smallholder systems to monitor crops and fertilise them (Gokool et al., 2023; Amarasingam et al., 2022) (Fig. 16.3). Satellite remote sensing provides high-frequency and large-scale coverage of the land to track irrigated areas, crop development, and productivity with vegetation indices and time series analysis, and usually with ML to classify the type of crop or determine biomass (Wang et al., 2024; Darra et al., 2023; Yousaf et al., 2023). Such data streams are used to feed decision-support systems (DSS) where sensor and remote sensing data are combined with predictive models and agronomic rules they use to produce site-specific irrigation, fertilizer, pest control, and harvesting information using dashboards or mobile applications (Miller et al., 2025; Araújo et al., 2023). Soon, digital twins will attempt to match real-time data and simulations in virtual replications of disciplines to enable scenario testing and optimum interventions (Awais et al., 2025).

4.4 Precision nutrient, water, and crop stress management strategies

Variable-rate technology and ML-based models are applied in precision nutrient management, where soil maps, yield history, and spectral data are used to adjust the type of fertilizer and the dose applied on the field to enhance nutrient use efficiency and minimize losses. Precision irrigation is an irrigation method that unites soil moisture sensors, weather predictions, and crop models with IoT-enabled pumps or valves to schedule and apply water according

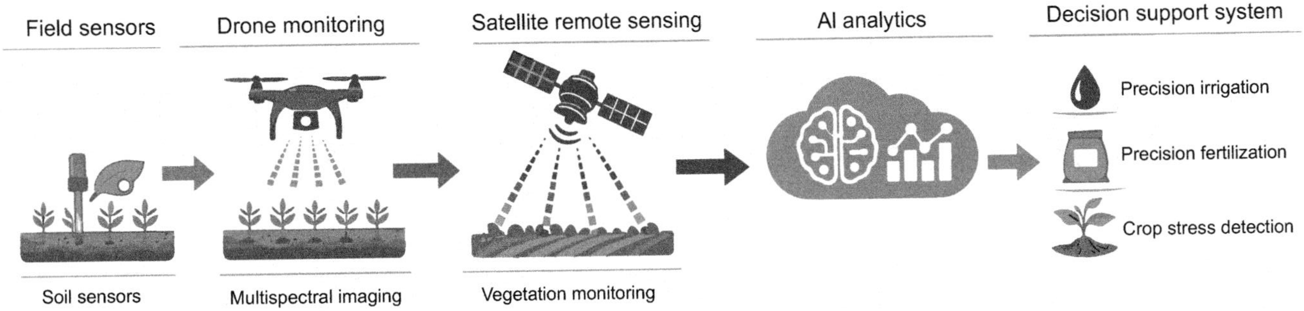

FIGURE 16.3 **Integration of sensors, drones, satellites, and decision-support systems in smart farming.** Field sensors, UAV imaging, and satellite remote sensing collect real-time agricultural data, which are analyzed using artificial intelligence and machine learning algorithms. These systems support precision agriculture practices such as variable-rate fertilization, precision irrigation, and crop stress monitoring.

to real-time crop water conditions, and smart irrigation has been extensively reported to save both water and energy consumption and maintain yields (Sharma et al., 2021; Abdelmoneim et al., 2025; Yousaf et al., 2023). In crop stress management, ML and IoT systems can detect and identify stress types and severity, respectively, using multispectral and thermal drone and satellite images, as well as ground sensors to identify water stress, nutrient deficiencies, disease, and weed patches in time to apply specific pesticides and local irrigation or any other mitigation measures (Miller et al., 2025; Adli et al., 2023; Guebsi et al., 2024). In these areas, the combination of IoT, AI, and remote sensing can be used to manage nutrients, water, and stresses more precisely, automatically, and sustainably, but the high costs, connection losses, and the problem of data control restrict the use at a large scale, particularly in developing areas (Abbasi et al., 2022).

5. Precision photosynthetic management: Concept and approaches

Precision photosynthetic management aims at measuring and controlling photosynthesis in order to achieve maximum carbon uptake, resource-use efficiency, and productivity in fluctuating environments. It uses intricate sensing of photosynthetic characters, distant and near-infrared instruments, and feedback controls of microclimate (light, CO_2, temperature, and water) and agronomic procedures to ensure that canopies perform close to the physiological optimal (Villagran et al., 2025; Wang et al., 2024; Zandi et al., 2025) (Table 16.2).

5.1 Definition and principles of precision photosynthesis management

Photosynthetic management processes at leaf, canopy, or system scale can be optimized accurately in real time with precision photosynthesis management, based on sensing, modeling, and automated control. The main principles consist of: (i) constant monitoring of the main photosynthetic indicators (chlorophyll status, fluorescence, gas exchange); (ii) a combination of such data with environmental measurements (light, CO_2, temperature, humidity) and crop models; and (iii) the active control of inputs (CO_2 enrichment, light, irrigation, ventilation) to maintain high photosynthetic efficiency without triggering metabolic saturation and stress. In laboratory settings, CO_2 (usually 550–650 μmol mol −1), light intensity, and spectrum, as well as temperature, have been accurately controlled to substantially enhance photosynthesis, water use efficiency, and biomass, assuming carbohydrate accumulation and other feedback constraints are controlled (Villagran et al., 2025).

5.2 Monitoring photosynthetic traits (chlorophyll content, fluorescence, gas exchange)

Chlorophyll content, which is commonly estimated with optical indices or handheld meters, represents the N status of the leaf and possible photosynthetic ability and can be used to support nutrient and canopy management decisions. The parameters of chlorophyll fluorescence, including Fv/Fm (maximum PSII efficiency) and NPQ (non-photochemical quenching), can be used to measure the proportion of absorbed light to photochemical versus heat dissipation, and early abiotic stress and optimal light or CO_2 conditions can be observed before visual symptoms develop. The indices of multi-band fluorescence (e.g., CFR, CCI, PRI, XI) are used to describe the composition of pigments, photoprotection, and dynamic light use efficiency with fine-grained control of light regimes. Parameters of net photosynthesis and transpiration at a leaf or canopy level are parameterized and tested with direct gas exchange

TABLE 16.2 Key photosynthetic metrics, sensors, and uses.

Photosynthetic trait/ index	Typical measurement approach	The main purpose of precision management	Citations
Leaf chlorophyll content/ pigment indices (e.g., CCI)	SPAD meters; hyperspectral or multispectral reflectance in red–NIR; pigment-sensitive indices	Diagnose N status, canopy vigor, and long-term photosynthetic capacity trends; guide fertilization and canopy management	Wang et al. (2024), Zandi et al. (2025), Marques et al. (2024)
Chlorophyll fluorescence (e.g., Fv/Fm, NPQ, CFR, CCI, PRI, XI)	Pulse-amplitude modulated fluorometry; active/passive fluorescence imaging; hyperspectral fluorescence bands (e.g., 520–750 nm)	Assess PSII efficiency, photochemical versus non-photochemical energy dissipation, early stress detection, and dynamic light-use efficiency	Zandi et al. (2025)
Leaf/canopy gas exchange	Infrared gas analyzers, open/ closed photosynthesis systems; whole-canopy chambers	Direct quantification of CO_2 assimilation and transpiration; calibration of photosynthesis models and CO_2/light control algorithms	Villagran et al. (2025), Zandi et al. (2025)
Canopy spectral indices for photosynthesis (e.g., NDVI, EVI, PRI, SIF-derived indices)	UAV, satellite, or proximal multispectral/hyperspectral imaging	Spatial mapping of canopy greenness, LUE proxies, and stress, forming inputs to precision control and decision-support tools	Wang et al. (2024), Muruganantham et al. (2022), Zandi et al. (2025)

measurements and then run in real time to modify CO_2 enrichment or microclimate in greenhouses and vertical farms (Villagran et al., 2025; Maraveas et al., 2023).

5.3 Remote sensing and spectral indices for photosynthesis assessment

Remote sensing obtains reflectance of the canopy in visible, NIR, and SWIR bands to estimate canopy structure, pigment content, and stress on a large scale (Wang et al., 2024; Muruganantham et al., 2022; Marques et al., 2024). Multispectral and, in particular, hyperspectral sensors can give plant phenotyping, crop index, and stress monitoring with finer spectral signatures of plant characteristics associated with photosynthesis. Popular vegetation indices (e.g., NDVI, EVI) monitor the canopy greenness and biomass, and indices of pigment selectivity (e.g., photochemical reflectance index (PRI)) and other pigment active indices are more directly correlated with the photosynthetic light use efficiency and xanthophyll cycle operation. Recent multi-mode analytics methods combine spectral imaging, image-based phenotyping, and advanced machine learning to enhance plant stress measurement, specifically improving the evaluation of photosynthetic performance and stress responses on chlorophyll fluorescence signals and pigment indices that are pigment-related indices.

5.4 Site-specific and real-time photosynthetic optimisation

Precision processor Photo synthesis management in a protected agricultural system applies sensor-model-control loops to the control of CO_2, light, temperature, and humidity to ensure the optimization of net photosynthesis and carbon use efficiency. The CO_2 sensor, light sensor, and microclimate sensor are wireless sensor networks that transmit data to the control platforms, which are optimized by computational fluid dynamics, machine learning, or other optimization techniques to optimize photosynthesis with minimal energy consumption and carbon footprint (Villagran et al., 2025; Maraveas et al., 2023). In fields, UAVs and satellites can be used to obtain spatially explicit photosynthesis related index images (e.g., chlorophyll, fluorescence proxies, stress indices) that can be used with machine learning to produce variable rate irrigation, shading or nutrient prescriptions, targeting the regions with low photosynthetic performance (Wang et al., 2024; Muruganantham et al., 2022; Zandi et al., 2025). Multimodal analytics combining spectrals, thermal, and structural measurements have additional benefits in ensuring early stress detection and enabling early interventions to conserve photosynthetic capability and prevent yield losses before they grow.

6. Role of nanotechnology in enhancing photosynthetic efficiency

Nanotechnology can take several paths in order to increase photosynthetic efficiency through improvements in chlorophyll synthesis and enzyme regulation, greater light collection and energy utilization, and reduced oxidative damage to allow stomata and carbon fixation to better respond to stress.

6.1 Nano fertilizers: The synthesis of chlorophyll and enzyme

Metal (e.g., Fe, Zn, Ti, Si) and biopolymer (e.g., chitosan NPs) nano fertilizers enhance the bioavailability of nutrients and can often enhance chlorophyll content and photosynthetic enzyme activity (Fig. 16.4). In a drought, TiO_2 and ZnO nanoparticles enhanced the chlorophyll *a* and *b*, K and P uptake, and photosynthetic rate and biomass of crops like wheat, maize, and linseed. The total chlorophyll, net photosynthesis, and antioxidant enzyme (CAT, APX) activities of periwinkle were coordinately increased by chitosan nanoparticles, which suggested that pigment synthesis and metabolic enzymes are coordinately activated. ZnO NPs sustained the net photosynthetic rate by conserving pigments and supplementing metabolically important carbohydrate-metabolizing enzymes (UDP glucose pyrophosphorylase, phosphoglucoisomerase, cytoplasmic invertase), which facilitate effective exploitation of fixed carbon in maize (Chandrashekar et al., 2023).

6.2 Nanomaterials advancing light absorption and energy transfer

Carbon in the form of graphene and carbon nanotubes is also shown to be able to boost chlorophyll content and electron transport, thereby increasing light harvesting and photosynthesis in certain species (Chandrashekar et al., 2023). Moving to a higher level of materials, nanostructuring has been demonstrated to be useful in increasing light absorption and energy transfer due to increased surface area, bandgap optimization, and plasmonic effects; for example, plasmonic metals (Au, Ag) and doped nanostructures have been found to significantly enhance light capture in photocatalytic systems (Asghar et al., 2024). Other applications that are more recent in application are similar principles applied to engineer light to manage nanomaterials in leaf surfaces or in nutrient formulations to maximize the use of the canopy light.

6.3 Nano-enabled oxidative stress and photo-inhibition mitigation

A significant number of nanoparticles have an indirect effect on photosynthesis by alleviating oxidative stress that would otherwise cause pigment damage, thylakoid membranes, and photosystem damage. Various NPs (Fe_3O_4, TiO_2, ZnO, Se, chitosan) in drought conditions constantly enhance enzyme and non-enzyme

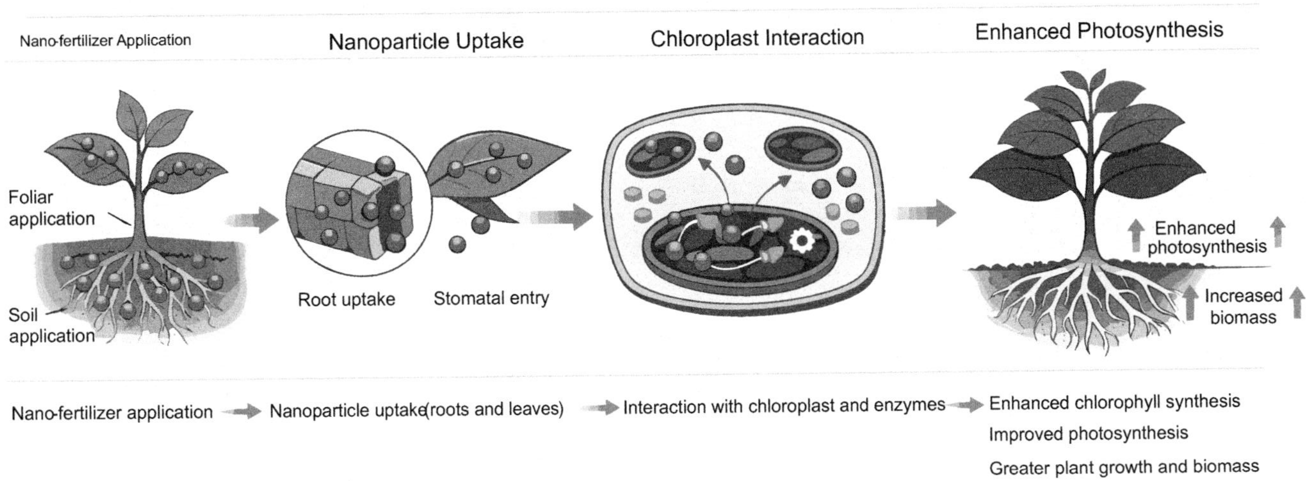

FIGURE 16.4 **Mechanistic role of nano-fertilizers in enhancing photosynthetic performance in plants.** Nano-fertilizers applied through soil or foliar routes are absorbed by plants and transported to cellular sites such as chloroplasts. These nanoparticles enhance chlorophyll synthesis, activate photosynthetic enzymes, and improve nutrient availability, leading to increased photosynthetic efficiency and plant biomass.

antioxidant activities, decrease ROS and malondialdehyde, and enhance osmolytes (proline), protection of chloroplast structures, and photoinhibition (Chandrashekar et al., 2023). Nanomaterials can increase photosynthesis in microalgae and cyanobacteria, regulate the level of ROS, increase pigmentation, and production of metabolites (Codreanu et al., 2025). In a broader sense, nanozymes' superoxide dismutase-like activity exemplifies how nanomaterials might be designed to catalyze the elimination of superoxide radicals, which is a conceptual approach to protecting photosynthetic tissues against superoxide radical-induced damage (Zhao et al., 2021).

6.4 Effects on carbon assimilation, stomatal conductance, and water use efficiency

The common positive effects of NP treatments under water deficit include relative water content, stomatal conductance, and net photosynthesis, which lead to an increase in biomass and yield (Chandrashekar et al., 2023). These mechanisms are the changes in the properties of the cell walls and root pore structure that facilitate the inflow of water and nutrients and the changes in hormones (e.g., ABA) that fine-tune stomatal behavior (Chandrashekar et al., 2023). ZnO NPs treated maize lowered pigment degradation, ensured a higher net CO_2 assimilation, and enhanced water use efficiency by controlling stomata movement and ROS concentration (Chandrashekar et al., 2023). Nanotechnology has also been used in photosynthetic microorganisms to enhance CO_2 fixation performance and development through the stimulation of photosynthesis and regulation of ROS, highlighting the general applicability of the use of nano-enabled strategies to enhance carbon assimilation in the context of biological systems (Codreanu et al., 2025).

7. Integration of nanotechnology with smart farming systems

Smart farming and digital agriculture are becoming increasingly linked with nanotechnology to allow real-time monitoring, optimization of nano inputs, and specific improvement of photosynthesis and plant behaviors.

7.1 Nano-enabled sensors for real-time physiological monitoring of plants

Nano biosensors take advantage of high surface area and adjustable chemistry of nanomaterials to identify ions, metabolites, pathogens, or pesticide residues at extremely low concentrations and can be incorporated into soils, leaves, or irrigation systems. These nano sensors can be incorporated into IoT networks, enabling this type of continuous data on soil nutrients, plant health, or contaminant loads, which can be sent to farm management platforms and can be used to support site specific management and early stress detection in relation to photosynthetic performance. These nano biosensors are also under investigation to be used in precision farming in order to forecast soil disease and track microbial activity and nutrient dynamics to use water, fertilizers, and pesticides more efficiently (Codreanu et al., 2025; Srinivasan et al., 2024).

7.2 Nano-input application decision-making AI-driven

The smart farming architectures are based on IoT sensors, drones, and remote sensing to produce massive datasets of crop conditions and the environment; machine learning and AI transform these data into recommendations of the input timing, rate, and spatial targeting (Araújo et al., 2023). Similar frameworks can be applied to nano fertilizers and nano stimulants, where AI models combine sensor and weather data to maximize nano input application to yield maximum and minimally risky to the environment (Fue et al., 2025; Adli et al., 2023). Fields or farms Digital twins are virtual replicas operated by real-time sensor and UAV data, which are a potent means to simulate alternative strategies of nano applications before implementation and assist in adaptive management (Awais et al., 2025).

7.3 Nano-scale precision delivery systems of nano-fertilizer and nano-stimulant

Precision agriculture is already based on the use of GNSS-controlled machines, variable rate application, UAVs, and robotics to target inputs into specific locations (Sharma et al., 2021; Gokool et al., 2023; Guebsi et al., 2024). These systems are inherent vectors of nano fertilizers and nano biostimulants, which allow the spatial patterns and fine-scale dosage to be controlled, which is comparable to the field variation in soil fertility, canopy vitality, or stress in sensor and remote sensing (Wang et al., 2024; Cai et al., 2025). Some proposed Smart delivery ideas are nano-encapsulated biofertilizers or pesticides with sensor-informed variable rate maps and drone-based foliar spraying guided by high-resolution crop status maps (Wang et al., 2024; Gokool et al., 2023).

7.4 Interaction effects of nano-inputs with digital technologies on photosynthesis

A combination of nano inputs that optimize nutrient utilization, stress resistance, or pigment stability with digital technologies that constantly monitor the health of plants and optimize inputs has a high potential of increasing photosynthesis and reducing input consumption and emissions. Precision fertilization (AI-driven) has already been found to enhance the efficiency of nutrient use and yields and minimize environmental effects; incorporation of nanoformulations into these digitally controlled regimes can further enhance uptake efficiency and canopy photosynthesis (Fue et al., 2025; Cai et al., 2025; Petrović et al., 2024). The IoT (AI + IoT) systems of the future will be expected to apply multi sensor fusion (optical, soil, microclimate, perhaps nano biosensors) and edge AI to offer real time, localised control over nano input application and establish a closely coupled feedback loop to keep photosynthesis more closely in line with its physiological optimum under variable field conditions (Miller et al., 2025; Adli et al., 2023).

8. Nanotechnology-based tools for stress-responsive photosynthetic management

Nanotechnology is actively being developed to stabilize and improve photosynthesis in abiotic stress conditions, increasing with climate change, primarily through better water and nutrient interactions, antioxidant protection of photosystems, and alteration of stress-induced signaling.

8.1 Nano-interventions in the drought, salinity, heat, and nutrient stress

Various carbon-based nanomaterials (CNTs, graphene, fullerene), metal nanomaterials (ZnO, TiO_2, Fe, and Cu) are shown to have a beneficial effect on photosynthetic activity in the face of drought conditions: they can enhance water and nutrient uptake, regulate stomatal conductance, increase the pigment content and net CO_2 assimilation (Table 16.3). These nanoparticles enhance antioxidant systems and

TABLE 16.3 Representative nano-interventions and photosynthetic outcomes.

Stress type	Nano-formulation/ system	Key photosynthetic/stress-responsive effects	Citations
Drought (various crops)	Metal (ZnO, TiO_2, Fe, Cu) and carbon-based NPs	↑ chlorophyll, ↑ photosynthetic rate, improved stomatal conductance, better water and nutrient uptake, ↓ ROS and MDA, ↑ osmolytes and antioxidant enzymes, modulation of drought-responsive genes and phytohormones (e.g., ABA)	Chandrashekar et al. (2023)
Drought (monocots and dicots)	NP-based fertilizers and foliar treatments	Enhanced germination, biomass, and physio-biochemical traits; improved water relations and photosynthetic activity through regulated stomatal closure and nutrient use	Chandrashekar et al. (2023)
Salinity (pearl millet)	Green-synthesized Ag NPs	Improved growth, chlorophyll content, osmotic regulation, and $O_2\bullet^-/H_2O_2$ scavenging; mitigation of oxidative damage and better photosynthetic performance under NaCl stress	Codreanu et al. (2025)
Salinity (barley)	Biogenic Ag and Cu NPs	↑ gas exchange, ↑ photosynthetic pigments, improved osmoregulation; ↓ MDA and H_2O_2; enhanced antioxidant enzyme activity and reduced phenolic overaccumulation	Codreanu et al. (2025)
Salinity (various crops with biofertilizers)	Graphene, graphene oxide, CNTs + microbial inoculants	Synergistic mitigation of growth inhibition at high salinity; improved overall plant performance likely through combined nutrient, ROS, and hormonal regulation that indirectly supports photosynthesis	Codreanu et al. (2025)
Climate-driven combined stresses (woody plants)	Conceptual nano-tools supporting NSC homeostasis	Stabilization of NSC pools to buffer energy demands, maintain cellular homeostasis, and support recovery of photosynthetic capacity under recurring drought, salinity, and heat.	Nawaz et al. (2025)

decrease ROS and malondialdehyde and stimulate the accumulation of osmolytes, thus protecting chloroplast membranes and photosystems (Chandrashekar et al., 2023). The green-synthesized Ag and Cu nanoparticles are nano-biostimulants that enhance chlorophyll content, gas exchange, osmoregulation, and antioxidant enzyme activity and reduce H_2O_2 and MDA, which directly promotes the photosynthetic activity of salt-stressed cereals and barley under salinity (Codreanu et al., 2025). The same mechanisms, namely osmotic adjustment, ROS, and protection of pigments, are also likely to underlie the benefits of heat and nutrient stress, though they are not as well documented.

8.2 Improving photosynthetic resilience in climate change conditions

Climate change is expected to exacerbate droughts, heat waves, and salinity spikes, and thus, stability will be pegged on stress-resilient photosynthesis. Nanoparticles aid in the process of photosynthesis by preserving water status, transpiration output, and non-structural carbohydrate (NSC) equilibrium, which equalises energy needs and recovers the stress. The nano-interventions can enhance the stomatal conductance, pigment stability, and ROS homeostasis and improve chronic photo-inhibition to maintain carbon gain during extreme events (Chandrashekar et al., 2023). Woody vegetation, especially those that are long-lived, may be helped by nano-enabled technologies to stabilize pools of NSC and shield photosynthetic tissues, making them resilient to frequent stresses triggered by climate change (Nawaz et al., 2025).

8.3 Nano-antioxidants and nano-silicon and stress tolerance

A variety of metal and metal-oxide nanoparticles are nano-antioxidants or antioxidant inducers. They stimulate SOD, CAT, APX, and other antioxidant enzymes, non-enzymatic antioxidants, and osmolytes activities under drought conditions, which all limit oxidative stress in photosynthetic pigments and thylakoid membranes. Salinity-biogenically synthesized Ag and Cu nanoparticles also effectively increase antioxidant potential and decrease MDA and H_2O_2, and oxidative stress is directly suppressed in chloroplasts (Codreanu et al., 2025). Nano-silicon is also reported extensively elsewhere, but not in these papers, to strengthen cell walls, enhance water relations, efficacy of ROS-scavenging systems, and thus protect photosystems II and I in situations of drought, salinity, and heat. Theoretically, nano-Si can be included in the more general category of nano-antioxidant interventions that enhance structural and biochemical defenses to the photosynthetic apparatus (Chandrashekar et al., 2023).

9. Environmental, safety, and regulatory considerations

9.1 Environmental fate and behavior of agricultural nanomaterials

When applied, drifted, washed off, and disposed of, nanoparticles in smart farming eventually find their way into soils and waters as nanofertilizers, nano-enabled sensors, nano-enabled nanofertilizers, nano-enabled nano-pesticides, and nano-enabled nano-sensors. Nanomaterials dissolve, become aggregated, and interact primarily with organic matter and minerals and determine their mobility, persistence, and bioavailability when released (Gambardella and Pinsino, 2022; Asghar et al., 2024). The systematic reviews of ecotoxicology point to the fact that the environmental fate heavily relies on particle size, coating and matrix: transformed or hetero-aggregated particles may behave in a much different way compared to pristine lab-grade materials, meaning that idealised particles in tests may not reveal how it would act in the field (Pradel et al., 2023). In the case of nanoparticles, such as nanoparticle-decorated biochar (NPs@BC), the situation is initially complex due to the sorption to biochar surfaces and involvement in biogeochemical cycling between biomass and mineral phases in soil, and design frameworks are currently explicitly integrated between the choice of material and environmental behavior and sustainability metrics in soil remediation and agriculture (Zheng et al., 2023).

9.2 Food safety concerns and ecotoxicological

Wide-ranging toxicological data indicate that most engineered nanoparticles (e.g., ZnO, TiO_2,

Ag, CNTs) are capable of inducing oxidative stress, DNA damage, apoptosis, and inflammatory responses in cell lines, invertebrates, fish, and plants. A scoping review has identified that the majority of the studies are conducted in the human format or in vitro, with a very minor proportion of the studies investigating the impacts of the environment, of which there is a considerable gap in knowledge about the effects at the ecosystem level and long-term exposure (Kumah et al., 2023; Gambardella and Pinsino, 2022). Nanoparticles have the potential to disrupt earthworm survival, growth, and reproduction in a dose- and material-dependent way in soils; some materials (e.g., TiO_2 at particular levels) have minimal long-term effects (Zheng et al., 2023). Plant ingestion of metallic nanoparticles can potentially change plant yield and biochemical composition to cause transfer to edible tissues as a source of food safety, though the existing data is very disparate and largely short-term (Kumah et al., 2023). This lack of certainty, together with the signs of nanoparticles deposition and trophic passage in aquatic and terrestrial organisms, is the basis of the concern about the residues of nano-enabled agrochemicals and smart-farming inputs finding their way into the food chain.

9.3 Risk evaluation and regulatory standards

The most common pesticide and fertilizer risk-assessment schemes are not adequately applicable to nanoformulations since the properties of nanoparticles (size, shape, surface charge, coatings) vary in terms of toxicity, persistence, and exposure patterns compared to bulk chemicals. The reviews of nanopesticides underscore that the existing regulatory procedures were optimized to test conventional active constituents and might not be able to measure nano-specific behavior, necessitating modified test guidelines, characterisation criteria, and testing exposure measures (Deka et al., 2021). On a broader scale, the arguments of ecotoxicology syntheses would be to adopt a safer-by-design strategy: (1) target the development of less hazardous nanomaterials, (2) assess risks throughout the entire life cycle and use phase, and (3) reduce releases and waste during production and use. Materials such as NPs@BC are currently being developed as frameworks that incorporate the hazard, environmental performance, and economic consideration to assist in the decision on whether a particular nano-enabled product is acceptable to be introduced into the soil to enhance soil remediation and agriculture (Zheng et al., 2023). However, the authors frequently mention the absence of harmonized regulations, a lack of chronic and field data, and a lack of research that would connect environmental exposure, the response of organisms, and the outcomes on a population level or on an ecosystem level (Kumah et al., 2023; Gambardella and Pinsino, 2022; Deka et al., 2021).

9.4 Smart farming nanotechnology sustainability and responsibility

Sustainable nano-agriculture demands that the increase in productivity should be matched with ecological and health protection. The biogenic production of nanomaterials with photosynthetic microorganisms, plant residues, or other bioresources is encouraged to reduce energy consumption, eliminate toxic reagents, and incorporate nanotechnology into the circular bioeconomy. Meanwhile, nano-enabled remediation (e.g., NPs@BC on contaminated soils) and nano-sensors have the potential to mitigate legacy pollution and maximize the use of inputs, although these strategies will only succeed when the life-cycle impact of these technologies is properly controlled in terms of safety-by-design, controlled release, and end-of-life (Zheng et al., 2023). Smart farming architectures (IoT, AI, precision application) may also assist with responsible usage by reducing doses, hitting hotspots, and reducing overall environmental load through the reduction of off-target emissions of nano-inputs (Musa et al., 2022). The issue of interdisciplinary research, transparent risk communication, governance that includes the farmers, and adaptive regulation that adjusts to the new evidence of environmental fate, ecotoxicology, and food safety are a priority in the current reviews to move to truly climate-smart and sustainable systems.

10. Case studies and field-level applications

10.1 Successful integration of nano-inputs with precision agriculture tools

Applications in the field and on the farm are beginning to demand the integration of nano-inputs (nanofertilizers, nano-pesticides, nano-biostimulants) with precision agriculture systems using sensors, remote sensors, drones, and machine learning, to provide inputs at the correct dose and place and time (Fig. 16.5). Nano-biosensors and nanosensors are also under test in precision farming designs to track soil diseases, micro-organisms, and nutrient levels, inputting real-time information into the decision-support system to optimize the application of variable rates of water, fertilizers, and crop protection products (Codreanu et al., 2025; Sharma et al., 2021). Simultaneously, machine learning combined with remote sensing and the UAV (drone)-based imaging will enable locating the site-specific occurrence of crop stress and variability, which can be overlaid with the maps of nano-input response in intelligent farming systems (Wang et al., 2024; Araújo et al., 2023). This combined nano + digital strategy will help decrease blanket usage of agrochemicals and take advantage of the controlled-release and increased efficiency of nano-formulations.

10.2 Crop-specific examples (cereals, pulses, oilseeds, horticultural crops)

The majority of published case studies are on an experimental or pilot scale but cover large crop groups. Nanofertilizers have increased nutrient use efficacy, output, and stress tolerance in relation to conventional fertilisers in various production conditions in cereals like rice and wheat, suggesting a possibility of use with site-specific nitrogen and water management in precise systems (Onyango et al., 2021; Pudhuvai et al., 2025). Drought-stressed or nutrient-stressed pulses and oilseeds have been shown to grow better and yield more when subjected to micronutrient nano-formulations (e.g., soil and canopy sensors drive a variable-rate foliar program with ZnO, CuO-based micronutrient nano-bits in soybean), generally. Some of the best nano-field connections are provided by horticultural crops: nano-bio-stimulant sprays made of chitosan nanoparticles loaded with microalgal protein hydrolyzate enhanced tomato plant height (up to 50%) and fruit phytonutrient profile, which is evidence of a foliar nano-input that can be scheduled based on sensor- or model-based growth diagnostics. In the case of climacteric fruits, edible nanoparticle-coated surfaces of fruits such as ZnO, Ag, or chitosan nanoparticles have been used to delay the ripening process, reduce moisture loss, and enhance the shelf life of fruits, a type of nano-application precision-based on supply-chain logistics and not field variability (Odetayo et al., 2022).

10.3 Yield, resource-use efficiency, and economic outcomes

In crops, systematic reviews indicate that nanofertilizers always have higher yields and

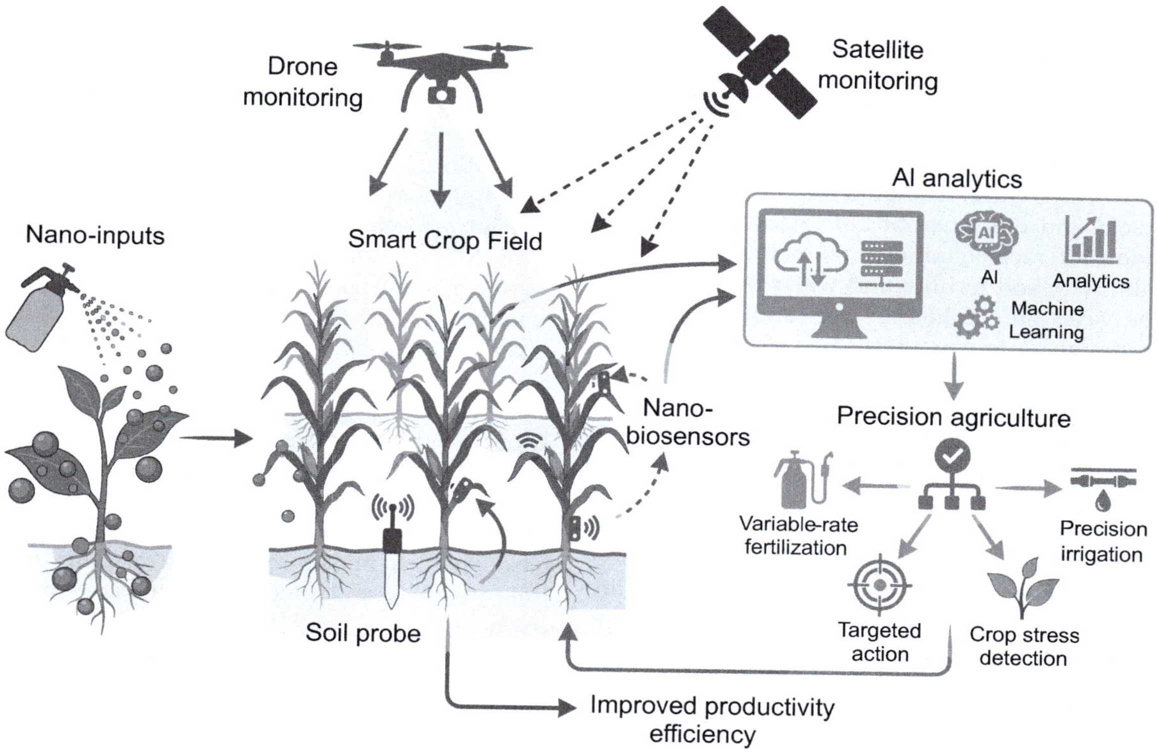

FIGURE 16.5 Integration of nano-inputs with precision agriculture technologies in smart farming systems. Nano-fertilizers, nano-pesticides, and nano-biosensors interact with crop systems while field sensors, drones, and satellite monitoring collect real-time data. These data are analyzed using artificial intelligence and digital farming platforms to support precision agriculture practices such as variable-rate fertilization, precision irrigation, and targeted crop management.

use efficiency (NUE) of the nutrient as compared to conventional fertilizers, which is mostly because of a sustained nutrient release, decreased leaching, and enhanced root uptake. It is translated into the increased productivity per unit of N, P, and K and possible reductions in the cost of fertilizers, particularly when paired with accurate nutrient control technologies (soil/plant sensors, satellite imagery, crop models) to avoid over-applying fertilizer (Onyango et al., 2021; Pudhuvai et al., 2025). More generally, in precision agriculture, machine learning, IoT sensors, and remote sensing have been integrated and are known to bring significant increases in resource-use efficiency, especially with water and agrochemicals, due to the ability to make site-specific decisions. These benefits in the digital domain are complementary to nano-inputs: when applied correctly in terms of timing and location, controlled-release nanofertilizers and nanopesticides could be applied with less input and fewer applications, leading to better input-output ratios and potentially farm profitability (Martinho and Guiné, 2021; Pudhuvai et al., 2025). Nevertheless, the majority of nano-specific economic studies are of a short-term and plot-oriented nature; the extensive cost-benefit information across several seasons and at a commercial scale is still scarce.

10.4 Learners in experimental and on-farm trials

Field experiment evidence and on-farm precision trial evidence indicate a number of learning points for scaling nano-enabled smart farming. To begin with, it should be site-specific: the yield and efficiency of both nanofertilizer and tools are most significant when the spatial variability of soil fertility and water is high, as can be seen in multi-site rice and smallholder nutrient-management studies where profitable and yield-enhancing precision and tailored fertilizer control were achieved (Mizik, 2023; Onyango et al., 2021). Second, it might be complex and expensive to adopt technology, particularly among the smallholders; systematic reviews point out that affordable precision technologies, modular and low in costs, shared services, and high-quality advisory support are required to render precision practices, such as the adoption of nano-inputs, viable in small farms. Third, the majority of technologies in digital agriculture (IoT, drones, hyperspectral imaging, AI-based decision support) are in their prototype or initial deployment phases, and their combination with nano-inputs is a prospect, as opposed to a commercial reality (Ram et al., 2024; Abbasi et al., 2022). Lastly, among the lessons of integrated-smart agriculture, it should be stressed that the achievement of success in the adoption process demands the harmonization of nano and digital innovations with the sustainability, strong risk assessment, farmer training, and inclusive governance, such that the increase in yield and efficiency is not at the cost of environmental or socio-economic resilience (Martinho and Guiné, 2021).

11. Challenges, limitations, and knowledge gaps

The combination of nano-inputs, sensing, and data-driven decision tools through smart farming is constrained interlocking by technical, social, and scientific limitations that restrict its usage and create knowledge gaps. Smart agriculture is based on dense networks of sensors, wireless communications, cloud or edge computing and frequently machine learning and digital-twin architectures on the technical and infrastructural level; in many agricultural areas these requirements are compromised by low connectivity, unreliable power, high hardware prices, and unstandardized data protocols and thus interoperable systems are challenging to deploy and maintain at scale (ElBeheiry and Balog, 2023; Awais et al., 2025). Precision and AIoT-based tools also require high-quality, timely data in large volumes; however, agricultural data are mostly seasonal and heterogeneous and device-, vendor-, and platform-based, which makes integration of data, real-time analytics, and effective model performance in changing field conditions challenging. Interoperability of data, in particular, is a special bottleneck: the devices and software are not compatible in terms of formats and communication protocols, and combining soil, crop, climate, and management data streams would not be as easy, and the computational requirements of the advanced simulations or digital twins may be even greater than the resources of small and medium farms. Such infrastructural and data challenges are related to farmer adoption and scalability challenges. The web of AI, IoT, and ML in agriculture is regularly described as reviews of both factors mentioning low levels of digital literacy, distrust to the new technology, perceived complexity, and high investment requirements as main reasons that many farmers, in particular, smallholders are reluctant to implement smart systems, even though they are technically viable. The lack of access to training, advisory services, and local technical assistance further increases the disparity between the early adopters and the majority of farmers and makes the reliance on foreign specialists to work and gain knowledge of digital tools. Layered on top of this digital stack when using nano-enabled inputs, the barrier to adoption may be even greater due to the unfamiliarity with the nano-scale materials, uncertainty about the long-term safety, and lack of obvious economic support at scale. Scientifically, gaps in knowledge between nano plant environment interactions under field circumstances are still large. Although laboratory experiments have investigated plant uptake and short-term physiological impacts of different nanomaterials, much less is known about their behavior in heterogeneous soils seasonally and their interactions with native microbiomes and local management practices, as well as the effect on yield stability, ecosystem services, and food safety over a period of years. Meanwhile, beyond the technical issues of data integration and interoperability, governance and ethics will also be of concern in integrating nanosensors, agronomic, and remote-sensing data on shared platforms: the question of data ownership, privacy, cybersecurity, and equitable access will become a concern as digital-twin and IoT systems access and utilize large amounts of sensitive information about farms. Combining all these constraints and gaps, this suggests that improving nano-enabled smart farming will not be limited to improved hardware and algorithms, but also an investment in rural digital infrastructure, open and interoperable data standards, user-focused design and training of farmers, and specific research on long-term nano plant soil environment interactions so that decision-support systems are not based on short-term or idealistic experiments.

12. Future prospects and research directions

The integration of high-risk nano-formulations, climate-smart, regenerative agriculture, and high-level digital models requires enabling policy and interdisciplinary research. The new generation nano-formulations are going beyond nutrient carriers to nanobionic and biogenic materials that are optimally designed to maximize photosynthesis, stress tolerance, and use of resources, such as nanoparticles biosynthesized by photosynthetic microorganisms (cyanobacteria, microalgae) that can deliver nutrients and bioactive molecules in a controlled fashion and serve as intelligent inputs to sustainable systems. In principle, these nano-inputs can be aligned with climate-smart and regenerative agriculture by reducing the synthetic fertilizer and pesticide burden, increasing nutrient use efficiency, and aligning with the objectives of the circular economy (i. e., production of bio-based nanoparticles using waste streams, in combination with soil health and biodiversity objectives). Climate-smart frameworks put forward simultaneous productivity, adaptation, and mitigation gains, and reviews of climate-smart technologies and integrated-smart agriculture note that digital, nanotechnological, and ecological innovations should be utilized in system-level approaches, not as isolated, using smart irrigation, conservation practices, and better nutrient management to reduce greenhouse-gas emissions and create resilience (Martinho and Guiné, 2021; Musa and Lim, 2025; Tadesse and Ahmed, 2023). In this new design, predictive photosynthesis modeling and digital twins should be at the center stage: digital twins can combine IoT sensors, remote sensing, and AI/ML, which means that dynamic virtual replicas of crops, soils, and greenhouses can be created, and light use, canopy structure, CO_2 assimilation, and nutrient delivery can be simulated and optimized in real-time. The overall state of digital twins in agriculture and greenhouse horticulture is according to reviews a fast but still immature development, with the current applications being climate control, energy management, growth prediction and nutrient concentration modeling, and future applications being climate resilience and GHG reduction as well as the connection to blockchain, robotics and multi-farm cooperative platforms (Awais et al., 2025; Slob and Hurst, 2022; Silva et al., 2023; Ariesen-Verschuur et al., 2022). Remote sensing, vegetation indices, and environmental data are also being integrated through machine learning and deep learning to predict yields and detect stress and provide the database required to train predictive photosynthesis and crop-response models, which could be linked to nano-formulation scheduling and dosing. These opportunities have to be unlocked by means of powerful policy support, capacity building, and interdisciplinary research. Systematic reviews of Agriculture 4.0 and integrated-smart agriculture emphasise that the majority of advanced digital technologies (big data, cyber-physical systems, digital twins) are currently at prototype stage where few are adopted because of cost, complexity and absence of standards, and recommends policies to encourage interoperability, data security, open platforms and incentives to adopt sustainable technology instead of input intensification. The extension services, vocational training, and farmer-centered design will also make climate-smart and regenerative transitions, so that nano-digital tools can be accessible to the smallholders, respond to local climate risks, and be in line with soil, biodiversity, and socio-economic objectives (Martinho and Guiné, 2021; Musa and Lim, 2025; Tadesse and Ahmed, 2023). Continued interdisciplinary cooperation by plant physiologists, nanotechnologists, soil scientists, data scientists, economists, and social scientists is once again proposed to be necessary to bridge knowledge gaps about nano plant environment interactions, strongly measure the long-term sustainability successes of regenerative and climate-smart systems, and co-develop digital twin and AI technology that is scientifically robust, ethically controlled, and practically utilized in actual farms.

13. Conclusion

This chapter brings out the importance of the realisation of nanotechnology that can be combined with smart agriculture and precise photosynthetic management as a groundbreaking outcome of contemporary agriculture. Nano-enabled inputs and sensors improve the efficiency of nutrient utilization, synthesis of chlorophyll, and photosynthetic activity, and smart farming technologies, such as IoT, remote sensing, and AI, also allow decisions to be made in real-time and site-specifically. These technologies are combined to maximize the consumption of resources, enhance stress resilience, and lower the environmental footprints. The nano-smart farming strategy is essential in attaining an accurate control of photosynthetic processes with a resultant high productivity of crops in chameleonic climatic and soil environments. Notably, such convergence helps in sustainable intensification as it enhances crop production without correspondingly raising the level of external inputs, hence saving natural resources and preserving soil and environmental conditions. On the whole, the interdependence of nanotechnology and digital agriculture is a future perspective that increases photosynthetic efficiency, crop performance, and the profitability of the farm, as well as promotes climate-smart and sustainable agriculture.

References

Abbasi, R., Martinez, P., Ahmad, R., 2022. The digitization of agricultural industry – a systematic literature review on agriculture 4.0. Smart Agric. Technol. 2, 100042. https://doi.org/10.1016/j.atech.2022.100042.

Abdelmoneim, A.A., Kimaita, H.N., Al Kalaany, C.M., Derardja, B., Dragonetti, G., Khadra, R., 2025. IoT sensing for advanced irrigation management: a systematic review of trends, challenges, and future prospects. Sensors 25 (7), 2291. https://doi.org/10.3390/s25072291.

Adli, H.K., Remli, M.A., Wan, K.N.S., Wong, S., Alina Ismail, N., González-Briones, A., Corchado, J.M., Saberi Mohamad, M., 2023. Recent advancements and challenges of AIoT application in smart agriculture: a review. Sensors 23 (7), 3752. https://doi.org/10.3390/s23073752.

Amarasingam, N., Ashan Salgadoe, A.S., Powell, K., Gonzalez, L.F., Natarajan, S., 2022. A review of UAV platforms, sensors, and applications for monitoring of sugarcane crops. Remote Sens. Appl.: Soc. Environ. 26. https://doi.org/10.1016/j.rsase.2022.100712. Available from: http://www.journals.elsevier.com/remote-sensing-applications-society-and-environment/.

Araújo, S.O., Silva Peres, R., Cochicho Ramalho, J., Lidon, F., Barata, J., 2023. Machine learning applications in agriculture: current trends, challenges, and future perspectives. Agronomy 13 (12), 2976. https://doi.org/10.3390/agronomy13122976.

Ariesen-Verschuur, N., Verdouw, C., Tekinerdogan, B., 2022. Digital Twins in greenhouse horticulture: a review. Comput. Electron. Agric. 199, 107183. https://doi.org/10.1016/j.compag.2022.107183.

Asghar, N., Hussain, A., Nguyen, D.A., Ali, S., Hussain, I., Junejo, A., Ali, A., 2024. Advancement in nanomaterials for environmental pollutants remediation: a systematic review on bibliometrics analysis, material types, synthesis pathways, and related mechanisms. J. Nanobiotechnol. 22 (1). https://doi.org/10.1186/s12951-023-02151-3.

Awais, M., Wang, X., Hussain, S., Aziz, F., Mahmood, M.Q., 2025. Advancing precision agriculture through digital twins and smart farming technologies: a review. AgriEngineering 7 (5), 137. https://doi.org/10.3390/agriengineering7050137.

Cai, B., Shi, F., Geremew, B.A., Addis, A.K., Abate, M.C., Dessie, W., Bayu, T., 2025. Precision agriculture techniques for optimizing chemical fertilizer use and environmental sustainability: a systematic review. Front. Agronomy 7. https://doi.org/10.3389/fagro.2025.1665444. Available from: www.frontiersin.org/journals/agronomy#.

Chandrashekar, H.K., Singh, G., Kaniyassery, A., Ashok Thorat, S., Nayak, R., Murali, T.S., Muthusamy, A., 2023. Nanoparticle-mediated amelioration of drought stress in plants: a systematic review. 3 Biotech 13 (10). https://doi.org/10.1007/s13205-023-03751-4.

Codreanu, S., Cepoi, L., Rudi, L., Chiriac, T., 2025. Photosynthetic microorganisms and biogenic synthesis of nanomaterials for sustainable agriculture. Nanomaterials 15 (13), 990. https://doi.org/10.3390/nano15130990.

Cravero, A., Pardo, S., Sepúlveda, S., Muñoz, L., 2022. Challenges to use machine learning in agricultural big data: a systematic literature review. Agronomy 12 (3), 748. https://doi.org/10.3390/agronomy12030748.

Darra, N., Anastasiou, E., Kriezi, O., Lazarou, E., Kalivas, D., Fountas, S., 2023. Can yield prediction Be fully digitilized? A systematic review. Agronomy 13 (9), 2441. https://doi.org/10.3390/agronomy13092441.

Deka, B., Babu, A., Baruah, C., Barthakur, M., 2021. Nanopesticides: a systematic review of their prospects with special reference to tea pest management. Front. Nutr. 8. https://doi.org/10.3389/fnut.2021.686131.

Dhanapal, A.R., Thiruvengadam, M., Vairavanathan, J., Venkidasamy, B., Easwaran, M., Ghorbanpour, M., 2024. Nanotechnology approaches for the remediation of agricultural polluted soils. ACS Omega 9 (12), 13522–13533. https://doi.org/10.1021/acsomega.3c09776.

ElBeheiry, N., Balog, R.S., 2023. Technologies driving the shift to smart farming: a review. IEEE Sens. J. 23 (3), 1752–1769. https://doi.org/10.1109/jsen.2022.3225183.

Fue, K.G., Prudence Baitu, G., Jokonya, O., Banwart, S., Korsten, L., 2025. Digitalization of precision fertilization in East Africa: adoption, benefits and losses. Front. Sustain. Food Syst. 9. https://doi.org/10.3389/fsufs.2025.1497577.

Gambardella, C., Pinsino, A., 2022. Nanomaterial ecotoxicology in the terrestrial and aquatic environment: a systematic review. Toxics 10 (7), 393. https://doi.org/10.3390/toxics10070393.

Gokool, S., Mahomed, M., Kunz, R., Clulow, A., Sibanda, M., Naiken, V., Chetty, K., Mabhaudhi, T., 2023. Crop monitoring in smallholder farms using unmanned aerial vehicles to facilitate precision agriculture practices: a scoping review and bibliometric analysis. Sustainability 15 (4), 3557. https://doi.org/10.3390/su15043557.

Guebsi, R., Mami, S., Chokmani, K., 2024. Drones in precision agriculture: a comprehensive review of applications, technologies, and challenges. Drones 8 (11), 686. https://doi.org/10.3390/drones8110686.

Kumah, E.A., Djou Fopa, R., Harati, S., Boadu, P., Vida Zohoori, F., Pak, T., 2023. Human and environmental impacts of nanoparticles: a scoping review of the current literature. BMC Public Health 23 (1). https://doi.org/10.1186/s12889-023-15958-4.

Livadariu, O., Maximilian, C., Rahmanifar, B., Petruta Cornea, C., 2023. LED technology applied to plant development for promoting the accumulation of bioactive compounds: a review. Plants 12 (5), 1075. https://doi.org/10.3390/plants12051075.

Luneja, R.L., Mkindi, A.G., 2025. Advances in botanical-based nanoformulations for sustainable cotton insect pest management in developing countries. Front. Agronomy 7. https://doi.org/10.3389/fagro.2025.1558395.

Maraveas, C., Karavas, C.-S., Loukatos, D., Bartzanas, T., Arvanitis, K.G., Symeonaki, E., 2023. Agricultural greenhouses: resource management technologies and perspectives for zero greenhouse gas emissions. Agriculture 13 (7), 1464. https://doi.org/10.3390/agriculture13071464.

Marques, P., Pádua, L., Sousa, J.J., Fernandes-Silva, A., 2024. Advancements in remote sensing imagery applications for precision management in olive growing: a systematic review. Remote Sens. 16 (8), 1324. https://doi.org/10.3390/rs16081324.

Martinho, V.J.P.D., Guiné, R.de P.F., 2021. Integrated-smart agriculture: contexts and assumptions for a broader concept. Agronomy 11 (8), 1568. https://doi.org/10.3390/agronomy11081568.

Miller, T., Mikiciuk, G., Durlik, I., Mikiciuk, M., Łobodzińska, A., Śnieg, M., 2025. The IoT and AI in agriculture: the time is now—a systematic review of smart sensing technologies. Sensors 25 (12), 3583. https://doi.org/10.3390/s25123583.

Mizik, T., 2023. How can precision farming work on a small scale? A systematic literature review. Precis. Agric. 24 (1), 384–406. https://doi.org/10.1007/s11119-022-09934-y.

Mizik, T., 2021. Climate-smart agriculture on small-scale farms: a systematic literature review. Agronomy 11 (6), 1096. https://doi.org/10.3390/agronomy11061096.

Mughele, E.S., Okuyade, O.S., Abdullahi, I.M., Maliki, D., Duada, I.A., 2024. Roles of Iot, big data and machine learning in precision agriculture: a systematic review. Sci. Afr. 23 (1), 33–44. https://doi.org/10.4314/sa.v23i1.3.

Muruganantham, P., Wibowo, S., Grandhi, S., Hoque Samrat, N., Islam, N., 2022. A systematic literature review on crop yield prediction with deep learning and remote sensing. Remote Sens. 14 (9), 1990. https://doi.org/10.3390/rs14091990.

Musa, S.F.P.D., Lim, S.A., 2025. Revitalising agriculture through climate change mitigation: a systematic literature review on smart technologies and sustainable practices. Int. J. Clim. Change Strateg. Manag. 17 (1), 481–499. https://doi.org/10.1108/ijccsm-05-2024-0071.

Musa, S.F.P.D., Hidayatullah Basir, K., Luah, E., 2022. The role of smart farming in sustainable development. Int. J. Asian Bus. Inf. Manag. 13 (2), 1–12. https://doi.org/10.4018/ijabim.20220701.oa5.

Nawaz, A.F., Gargiulo, S., Pichierri, A., Casolo, V., 2025. Exploring the role of non-structural carbohydrates (NSCs) under abiotic stresses on woody plants: a comprehensive review. Plants 14 (3), 328. https://doi.org/10.3390/plants14030328.

Odetayo, T., Tesfay, S., Ziphorah Ngobese, N., 2022. Nanotechnology-enhanced edible coating application on climacteric fruits. Food Sci. Nutr. 10 (7), 2149–2167. https://doi.org/10.1002/fsn3.2557.

Onyango, C.M., Nyaga, J.M., Wetterlind, J., Söderström, M., Piikki, K., 2021. Precision agriculture for resource use efficiency in smallholder farming systems in Sub-Saharan Africa: a systematic review. Sustainability 13 (3), 1158. https://doi.org/10.3390/su13031158.

Petrović, B., Bumbálek, R., Zoubek, T., Kuneš, R., Smutný, L., Bartoš, P., 2024. Application of precision agriculture technologies in Central Europe-review. J. Agric. Food Res. 15, 101048. https://doi.org/10.1016/j.jafr.2024.101048.

Pokharel, B.R., Sheri, V., Kumar, M., Zhang, Z., Zhang, B., 2024. The update and transport of aluminum nanoparticles in plants and their biochemical and molecular phototoxicity on plant growth and development: a systematic review. Environ. Pollut. 340 (Pt 1), 122875. https://doi.org/10.1016/j.envpol.2023.122875.

Pradel, A., Catrouillet, C., Gigault, J., 2023. The environmental fate of nanoplastics: what we know and what we need to know about aggregation. NanoImpact 29, 100453. https://doi.org/10.1016/j.impact.2023.100453.

Pudhuvai, B., Koul, B., Das, R., Shah, M.P., 2025. Nanofertilizers (NFs) for resurgence in nutrient use efficiency (NUE): a sustainable agricultural strategy. Curr. Pollut. Rep. 11 (1). https://doi.org/10.1007/s40726-024-00331-9.

Ram, B.G., Oduor, P., Igathinathane, C., Howatt, K., Sun, X., 2024. A systematic review of hyperspectral imaging in precision agriculture: analysis of its current state and future prospects. Comput. Electron. Agric. 222, 109037. https://doi.org/10.1016/j.compag.2024.109037.

Roşca, M., Mihalache, G., Stoleru, V., 2023. Tomato responses to salinity stress: from morphological traits to genetic changes. Front. Plant Sci. 14. https://doi.org/10.3389/fpls.2023.1118383.

Sati, A., Ranade, T.N., Mali, S.N., Ahmad Yasin, H.K., Pratap, A., 2025. Silver nanoparticles (AgNPs): comprehensive insights into bio/synthesis, Key influencing factors, multifaceted applications, and toxicity—A 2024 update. ACS Omega 10 (8), 7549–7582. https://doi.org/10.1021/acsomega.4c11045.

Sharma, A., Jain, A., Gupta, P., Chowdary, V., 2021. Machine learning applications for precision agriculture: a comprehensive review. IEEE Access 9, 4843–4873. https://doi.org/10.1109/access.2020.3048415.

Silva, L., Rodríguez-Sedano, F., Baptista, P., Coelho, J.P., 2023. The digital twin paradigm applied to soil quality assessment: a systematic literature review. Sensors 23 (2), 1007. https://doi.org/10.3390/s23021007.

Slob, N., Hurst, W., 2022. Digital twins and industry 4.0 technologies for agricultural greenhouses. Smart Cities 5 (3), 1179–1192. https://doi.org/10.3390/smartcities5030059.

Srinivasan, S., Raajasubramanian, D., Ashokkumar, N., Vinothkumar, V., Paramaguru, N., Selvaraj, P.,

Kanagalakshimi, A., Narendra, K., Shanmuga Sundaram, C.K., Murali, R., 2024. Nanobiosensors based on on-site detection approaches for rapid pesticide sensing in the agricultural arena: a systematic review of the current status and perspectives. Biotechnol. Bioeng. 121 (9), 2585–2603. https://doi.org/10.1002/bit.28764. Available from: http://onlinelibrary.wiley.com/journal/10.1002/(ISSN)1097-0290.

Tadesse, B., Ahmed, M., 2023. Impact of adoption of climate smart agricultural practices to minimize production risk in Ethiopia: a systematic review. J. Agric. Food Res. 13, 100655. https://doi.org/10.1016/j.jafr.2023.100655.

Villagran, E., Javier Espitia, J., Amado, G., Rodriguez, J., Gomez, L., Flores Velasquez, J., Gil, R., Baeza, E., Aguilar, C.E., Akrami, M., Arias, L.A., 2025. CO2 enrichment in protected agriculture: a systematic review of greenhouses, controlled environment systems, and vertical farms—Part 2. Sustainability 17 (7), 2809. https://doi.org/10.3390/su17072809.

Wahab, A., Abdi, G., Saleem, M.H., Ali, B., Ullah, S., Shah, W., Mumtaz, S., Yasin, G., Muresan, C.C., Marc, R.A., 2022. Plants' physio-biochemical and phyto-hormonal responses to alleviate the adverse effects of drought stress: a comprehensive review. Plants 11 (13). https://doi.org/10.3390/plants11131620. Available from: https://www.mdpi.com/2223-7747/11/13/1620/pdf?version=1655777519.

Wang, J., Wang, Y., Li, G., Qi, Z., 2024. Integration of remote sensing and machine learning for precision agriculture: a comprehensive perspective on applications. Agronomy 14 (9), 1975. https://doi.org/10.3390/agronomy14091975.

Yang, Z., Yan, H., Liu, H., Yang, L., Mi, G., Wang, P., 2025. Enhancing crop nitrogen efficiency: the role of mixed nitrate and ammonium supply in plant growth and development. Biology 14 (5), 546. https://doi.org/10.3390/biology14050546.

Yousaf, A., Kayvanfar, V., Mazzoni, A., Elomri, A., 2023. Artificial intelligence-based decision support systems in smart agriculture: bibliometric analysis for operational insights and future directions. Front. Sustain. Food Syst. 6. https://doi.org/10.3389/fsufs.2022.1053921.

Zandi, A., Hosseinirad, S., Kashani Zadeh, H., Tavakolian, K., Cho, B.K., Vasefi, F., Kim, M.S., Tavakolian, P., 2025. A systematic review of multi-mode analytics for enhanced plant stress evaluation. Front. Plant Sci. 16. https://doi.org/10.3389/fpls.2025.1545025. Available from: www.frontiersin.org/journals/plant-science.

Zhang, D., Xu, F., Wang, F., Le, L., Pu, L., 2025. Synthetic biology and artificial intelligence in crop improvement. Plant Commun. 6 (2). https://doi.org/10.1016/j.xplc.2024.101220. Available from: https://www.sciencedirect.com/science/journal/25903462.

Zhao, H., Zhang, R., Yan, X., Fan, K., 2021. Superoxide dismutase nanozymes: an emerging star for anti-oxidation. J. Mater. Chem. B 9 (35), 6939–6957. https://doi.org/10.1039/d1tb00720c.

Zheng, T., Ouyang, S., Zhou, Q., 2023. Synthesis, characterization, safety design, and application of NPs@BC for contaminated soil remediation and sustainable agriculture. Biochar 5 (1). https://doi.org/10.1007/s42773-022-00198-3.

CHAPTER

17

Policy prespectives: Regulatory pathways for nanomaterials in agriculture

Rajashree Krishna Bharadwaj[1], *Sudhir Kumar Rajpoot*[2], *Animekh Hazarika*[3] and *Mohammad Faizan*[4]

[1]Department of Agronomy (Agroforestry), Institute of Agricultural Sciences, Banaras Hindu University, Varanasi, Uttar Pradesh, India; [2]Department of Agronomy, Institute of Agricultural Sciences, Banaras Hindu University, Varanasi, Uttar Pradesh, India; [3]Department of Ecology and Environmental Science, Assam University, Silchar, Assam, India; [4]Botany Section, School of Sciences, Maulana Azad National Urdu University, Hyderabad, Telangana, India

1. Introduction

Nanotechnology has emerged as a transformative scientific domain with profound implications for modern agriculture. The application of nanomaterials in crop production systems offers opportunities for enhancing nutrient use efficiency, improving pest and disease management, enabling precision farming, and minimizing environmental pollution. Nanomaterials are generally defined as materials with at least one dimension in the range of 1–100 nm, exhibiting unique physicochemical properties such as high surface area, tunable reactivity, and enhanced solubility (Kah, 2018). Agricultural nanotechnology has witnessed rapid growth over the past decade, with increasing commercialization of nano-fertilizers, nano-pesticides, nano-sensors, and nano-encapsulation systems. While these innovations promise substantial agronomic and environmental benefits, their deployment has raised concerns regarding environmental persistence, ecotoxicity, bioaccumulation, and food safety. Traditional regulatory frameworks for chemicals and agro-inputs were not designed to address the distinct properties of nanoscale materials, creating regulatory gaps and governance challenges (FAO & WHO, 2019). This chapter critically examines the global regulatory landscape governing nanomaterials in agriculture, highlights major challenges in risk assessment and policy implementation, and proposes pathways toward harmonized and adaptive governance.

2. Nanomaterials in agriculture: Applications and scope

The application of nanotechnology in agriculture spans multiple domains of crop production and resource management.

https://doi.org/10.1016/B978-0-443-44376-3.00017-3

2.1 Nano-fertilizers

Nano-fertilizers are designed to deliver nutrients in a controlled and targeted manner, enhancing nutrient uptake efficiency and reducing losses through leaching, volatilization, and runoff (Raliya, 2018). Nano-encapsulation and nano-chelation technologies improve the bioavailability of micronutrients such as zinc, iron, and copper.

2.2 Nano-pesticides

Nano-pesticides include nano-formulated insecticides, fungicides, and herbicides that provide improved stability, controlled release, and site-specific delivery. These formulations can reduce the total quantity of active ingredients required while improving pest control efficacy (Kookana, 2014).

2.3 Nano-sensors and precision agriculture

Nano-sensors enable real-time monitoring of soil moisture, nutrient status, plant health, and pathogen detection, facilitating precision agriculture and data-driven farm management (Parisi et al., 2015).

2.4 Soil and water management

Nanomaterials are increasingly used in soil conditioners, water purification systems, and moisture retention technologies, contributing to climate-resilient and resource-efficient agriculture (Gogos et al., 2012; Fig. 17.1).

FIGURE 17.1 Regulatory roadmap for nanomaterials.

3. Regulatory challenges for agricultural nanomaterials

3.1 Definitional and classification challenges

One of the foremost regulatory challenges lies in defining what constitutes a nanomaterial. Jurisdictions differ in their definitional thresholds and criteria. The European Commission defines nanomaterials as materials with 50% or more of particles in the size range of 1–100 nm (EC, 2011), whereas other countries adopt broader or functional definitions.

Classification is further complicated by particle agglomeration, surface coatings, and transformations in soil and water matrices.

3.2 Risk assessment complexity

Conventional chemical risk assessment paradigms are inadequate for nanomaterials due to their size-dependent properties, surface reactivity, and dynamic environmental behavior. Nanoparticles may display toxicological profiles distinct from their bulk counterparts, necessitating new exposure models, toxicokinetic studies, and ecotoxicological endpoints (Handy, 2018).

Key concerns include:

Uptake and translocation in plants.
Interaction with soil microbial communities.
Bioaccumulation in food chains.
Long-term environmental persistence.

3.3 Life-cycle risk management

A comprehensive regulatory approach must incorporate a life-cycle perspective, encompassing manufacturing, formulation, transport, application, degradation, and disposal stages. Life-cycle assessment (LCA) has emerged as a critical tool for evaluating cumulative environmental and human health impacts of nanomaterials (Grieger, 2019; Table 17.1).

4. International regulatory frameworks

4.1 European Union

The European Union (EU) has developed one of the most comprehensive regulatory systems for nanomaterials. The Registration, Evaluation, Authorization and Restriction of Chemicals (REACH) regulation incorporates nano-specific provisions requiring registrants to provide detailed physicochemical and toxicological data.

Nano-pesticides are regulated under Regulation (EC) No 1107/2009, while nano-fertilizers fall under the Fertilising Products Regulation (EU) 2019/1009. The European Food Safety Authority (EFSA) has issued guidance on risk assessment of nanomaterials in food and agriculture (EFSA, 2021).

4.2 United States

In the United States, nanomaterials are regulated under existing statutes such as the Federal Insecticide, Fungicide, and Rodenticide Act, Toxic Substances Control Act, and the Federal Food, Drug, and Cosmetic Act. The Environmental Protection Agency (EPA) evaluates nano-pesticides on a case-by-case basis, while the Food and Drug Administration (FDA) provides guidance on nano-enabled food and agricultural products (EPA, 2017; FDA, 2014).

4.3 FAO, WHO, and Codex Alimentarius

FAO and WHO jointly provide international guidance on nanotechnology in food and agriculture through Codex Alimentarius. Their framework emphasizes transparency, standardized testing methodologies, and international cooperation (FAO & WHO, 2019).

TABLE 17.1 Global regulatory frameworks for nanomaterials in agriculture.

Region/ country	Regulatory authority	Governing legislation/ Framework	Scope of regulation	Current status
European Union	European Chemicals Agency (ECHA); European Food Safety Authority (EFSA)	REACH regulation; regulation (EC) No 1107/2009; Fertilising products regulation (EU) 2019/ 1009	Nano-pesticides, nano-fertilizers, food contact materials	Nano-specific provisions integrated into chemical and food safety laws
United States	Environmental Protection Agency (EPA); Food and Drug Administration (FDA)	FIFRA; TSCA; FDCA	Nano-pesticides, food additives, agricultural inputs	Regulated under existing chemical laws with nano-guidance
India	DST; DBT; MoEFCC; FSSAI	Environment protection Act; insecticides Act; food safety Act	Agrochemicals, fertilizers, food products	No nano-specific law; governed under existing statutes
China	Ministry of Agriculture and Rural Affairs; NSTC	Agricultural chemicals regulation; national nano standards	Pesticides, fertilizers, food additives	Mandatory nano-product registration
Brazil	ANVISA; Ministry of Agriculture	Chemical safety law; agrochemical registration Act	Agro-inputs, food safety	Developing nano-regulatory framework
South Africa	Department of Science and Innovation; Department of Agriculture	National nanotechnology Strategy; environmental Acts	Agricultural inputs, environmental protection	Policy framework under development
OECD/ Codex	OECD; FAO-WHO Codex Alimentarius	OECD nano safety Program; Codex guidelines	International standards and testing methods	Voluntary harmonization frameworks

5. Regulatory landscape in emerging economies

5.1 India

India's nanotechnology governance is coordinated through agencies such as the Department of Science and Technology (DST), Department of Biotechnology (DBT), Ministry of Environment, Forest and Climate Change (MoEFCC), and the Food Safety and Standards Authority of India. Although India lacks nano-specific agricultural legislation, nanomaterials are regulated under existing environmental and chemical safety laws. The Nano Mission of India promotes responsible innovation and regulatory capacity building (DST, 2020).

5.2 China

China has established extensive nanotechnology standards through the National Nanotechnology Standardization Technical Committee. Nano-enabled agrochemicals require registration under the Ministry of Agriculture and Rural Affairs.

5.3 Brazil and Africa

Brazil and South Africa are developing nanotechnology policy frameworks aligned with OECD guidelines. However, limited regulatory infrastructure and technical capacity remain key challenges (Table 17.2).

TABLE 17.2 Key regulatory challenges and policy gaps in agricultural nanotechnology.

Regulatory dimension	Major challenges	Policy gaps	Recommended actions
Definition & classification	Lack of uniform nanomaterial definition; dynamic particle behavior	Inconsistent size thresholds across jurisdictions	International harmonization of nano-definitions
Risk assessment	Inadequate toxicological models; limited exposure data	Absence of nano-specific testing guidelines	Develop standardized nano-risk assessment protocols
Environmental fate	Unknown persistence and bioaccumulation	Lack of long-term ecological studies	Mandate life-cycle and environmental fate assessments
Human health	Uncertain food chain transfer	Insufficient dietary exposure data	Strengthen food safety nano-surveillance
Regulatory capacity	Limited infrastructure in emerging economies	Weak enforcement mechanisms	Capacity building and regulatory training
Public perception	Low awareness; ethical concerns	Limited stakeholder engagement	Public consultation and transparency
Trade & harmonization	Divergent national regulations	Barriers to international trade	Align national policies with OECD/Codex standards

6. Ethical, socioeconomic, and sustainability considerations

Nanotechnology regulation extends beyond technical risk assessment to encompass ethical, social, and economic dimensions. Key issues include:

Farmer access and affordability
Intellectual property rights
Technology transfer to smallholders
Environmental justice
Public perception and trust

Public engagement and participatory governance are essential for fostering societal acceptance and responsible innovation (Macnaghten, 2014).

7. Toward harmonized and adaptive governance

Effective governance of agricultural nanotechnology requires:

Harmonization of international standards
Development of nano-specific risk assessment protocols
Investment in regulatory science and infrastructure
Adaptive governance mechanisms responsive to innovation
Stakeholder participation and transparency

OECD and ISO initiatives provide valuable platforms for regulatory convergence and best-practice sharing.

8. Conclusion

Nanotechnology offers unprecedented opportunities to transform global agriculture by enhancing productivity, sustainability, and climate resilience. However, these benefits can only be realized through robust, transparent, and science-based regulation. The current patchwork of regulatory frameworks reflects divergent national capacities and priorities, underscoring the need for international cooperation and harmonization. A life-cycle approach, coupled with adaptive governance and ethical foresight, will be critical to ensuring that agricultural nanomaterials contribute responsibly to food security and sustainable development.

References

DST, 2020. Nano Mission: Policy Framework. Department of Science and Technology, Government of India.

EC, 2011. Commission Recommendation on the Definition of Nanomaterial (2011/696/EU). European Commission, Brussels.

EFSA, 2021. Guidance on risk assessment of nanomaterials in food and feed. EFSA J. 19 (8), e06768.

EPA, 2017. Nanotechnology White Paper. U.S. Environmental Protection Agency, Washington, DC.

FAO & WHO, 2019. Safety aspects of nanotechnologies in food and agriculture. In: FAO Food and Nutrition Paper. Rome.

FDA, 2014. Guidance for Industry: Considering whether an FDA-Regulated Product Involves the Application of Nanotechnology. U.S. Food and Drug Administration.

Gogos, A., Knauer, K., Bucheli, T.D., 2012. Nanomaterials in plant protection and fertilization: current state, foreseen applications, and research priorities. J. Agric. Food Chem. 60, 9781–9792.

Grieger, K.D., 2019. Life cycle assessment of nanomaterials. Environ. Sci. Nano 6, 2049–2060.

Handy, R.D., 2018. Ecotoxicity and bioaccumulation of nanomaterials. Ecotoxicology 27, 123–135.

Kah, M., 2018. Nanopesticides: state of knowledge, risks and opportunities. Crit. Rev. Environ. Sci. Technol. 48 (13), 1199–1237.

Kookana, R.S., 2014. Nanopesticides: guiding principles for regulatory evaluation. J. Agric. Food Chem. 62, 4227–4240.

Macnaghten, P., 2014. Responsible innovation across borders. J. Respons. Innov. 1 (2), 191–199.

Parisi, C., Vigani, M., Rodríguez-Cerezo, E., 2015. Agricultural nanotechnologies: what are the current possibilities? Nano Today 10 (2), 124–127.

Raliya, R., 2018. Nano-enabled fertilizers for precision agriculture. Front. Plant Sci. 9, 1–12.

Further reading

ISO, 2018. Nanotechnologies—Vocabulary and Characterization. International Organization for Standardization, Geneva.

Mukherjee, A., et al., 2016. Emerging trends in agricultural nanotechnology. J. Nanoparticle Res. 18, 1–16.

OECD, 2016. Nanomaterials in Agriculture: Risk Assessment and Regulatory Perspectives. OECD Publishing, Paris.

Sekhon, B.S., 2014. Nanotechnology in agri-food production. Nanotechnol. Sci. Appl. 7, 31–53.

Sharma, P., 2020. Nanotechnology in agriculture: current status and future prospects. Environ. Chem. Lett. 18, 73–88.

Singh, R., 2017. Nano-fertilizers and their role in sustainable agriculture. J. Plant Nutr. 40 (1), 1–14.

CHAPTER

18

Risks and concerns in adopting nanotechnology to tackle hunger: Environmental, health, and ethical assessments

Sadia Afzal[1], *Sumera Javad*[1], *Saima Bashir*[2], *Amina Tariq*[1] *and Saiqa Ilyas*[3]

[1]Department of Botany, Lahore College for Women University, Lahore, Pakistan; [2]Department of Botany, University of Gujrat, Gujrat, Pakistan; [3]Department of Biotechnology, Lahore College for Women University, Lahore, Pakistan

1. Introduction

Emerging trends of nanotechnology in food and agriculture can offer a groundbreaking approach to tackle hunger of growing population that is estimated to reach 9 billion by 2050 (Duro et al., 2020). This technology revolutionized the agricultural sector by production of nano sized (1–100 nm) formulations with the aim of increasing crop productivity in a sustainable manner. European commission recognized nanotechnology amongst one of six key enabling technologies (KETs) (Wessendorf et al., 2025). Agricultural nanotechnology encompasses nano fertilizers, pesticides, biostimulants, seed coating, hydrogels to increase seed vigor and germination by protecting them from drought and pathogen attack. Engineered nano-formulations have specific physicochemical, biological and optical properties that enabled the targeted and controlled delivery of nutrients, pesticides and genetic material (DNA or RNA) to plants (Santos et al., 2025). Similarly, engineered nano particles (NPs) can modify soil properties like texture, water retention and fertility for better plant growth. Moreover, integration of nano sensors with internet of things (IoTs) enabled automated crop monitoring (Bhat et al., 2026).

Beyond direct applications to plants, nanoformulations are also extended to food industry. Use of nano sensors, nano particles and nano coatings during food packaging made it easier to post harvest monitoring and control of

Nanotechnology in Photosynthesis
https://doi.org/10.1016/B978-0-443-44376-3.00018-5

microbial activities and spoilage over time (Li et al., 2021). Despite all these advantages, some phytotoxicity, environmental and health risks are also associated with uncontrolled use of nano-formulations. The small size in the range of nanometers enables easy penetration and accumulation of nano-formulations (nano fertilizers and nano pesticides) into plant tissues and organelles. The excess amount of these formulations directly or indirectly can cause oxidative damage to cellular membranes, organelles and genetic material (DNA and RNA) leading to impaired photosynthesis and yield reduction (Racuciu et al., 2025). Long-term exposure can cause their persistence and accumulation in ecosystems, which facilitate their entry into trophic levels and food webs. The health of humans especially infants, pregnant women and patients is also at risk due to inhalation and ingestion of NPs (Xuan et al., 2023). Recently, scientists are emphasizing for production of cost-effective biodegradable nano-formulations with strict risk assessment before their market release.

Integration of nanotechnology in agriculture needs intention to formulate a set of separate uniform international regulations for testing and labeling of nano enabled products. This chapter will provide useful insights into applications of nanotechnology from tackling hunger to its possible toxicity risks of soil persistence, bioaccumulation, and trophic level transfer to human health. This study also highlighted existing regulatory networks, challenges in risk assessment and need of public awareness for safer and responsible use of nano-formulations Fig. 18.1.

2. Nano enabled pathways to tackle hunger across food system

2.1 Production and applications

In agricultural context, nanotechnology enabled production of vast range of nanomaterials (1–100 nm size) in the form of nano fertilizers, nano pesticides, nano bio-stimulants and soil modifications (Table 18.1).

2.1.1 *Nano fertilizers*

These are nano sized encapsulated formulations of one or more nutrients that are used to increase plant growth and yield (Table 18.1). Research data shows that application of nano fertilizers causes 25%–30% increase in plant yields. The tiny size enabled controlled and targeted delivery in plant systems, so their applications cause almost 40% reduction of the need for conventional fertilizers. The controlled release of nutrients also declines soil degradation, greenhouse gas emission, heavy metal accumulation, nutrient leaching and eutrophication. Due to dual aspects (crop health improvement and environmentally friendly); nano fertilizers are gaining popularity among agriculturists (Khundi et al., 2025).

2.1.2 *Nano pesticides*

These are nano formulations encapsulated with pesticides or active compounds against insects, fungal or plant pathogens (Table 18.1). These are used in the form of nano sized capsules, particles, tubes or metal oxides (Al_2O_3, CuO and ZnO etc.). Like nano fertilizers, nano pesticides also help to increase crop yield by efficiently attacking plant pathogens. The controlled release at optimum conditions (e.g., pH, humidity and temperature) reduced the dependency on conventional pesticides. The property of targeted delivery enabled minimum impact on non-target organisms and environment (Khan et al., 2023; Pagano et al., 2025).

2.1.3 *Nano bio-stimulants*

These are the encapsulated nano sized formulations that are used to improve activity of phytohormones along with biotic (microbes, fungi, insects, weeds etc.) and abiotic (temperature, salt, drought, water logging, heavy metals, nutrient deficiency) stress resistance of plants.

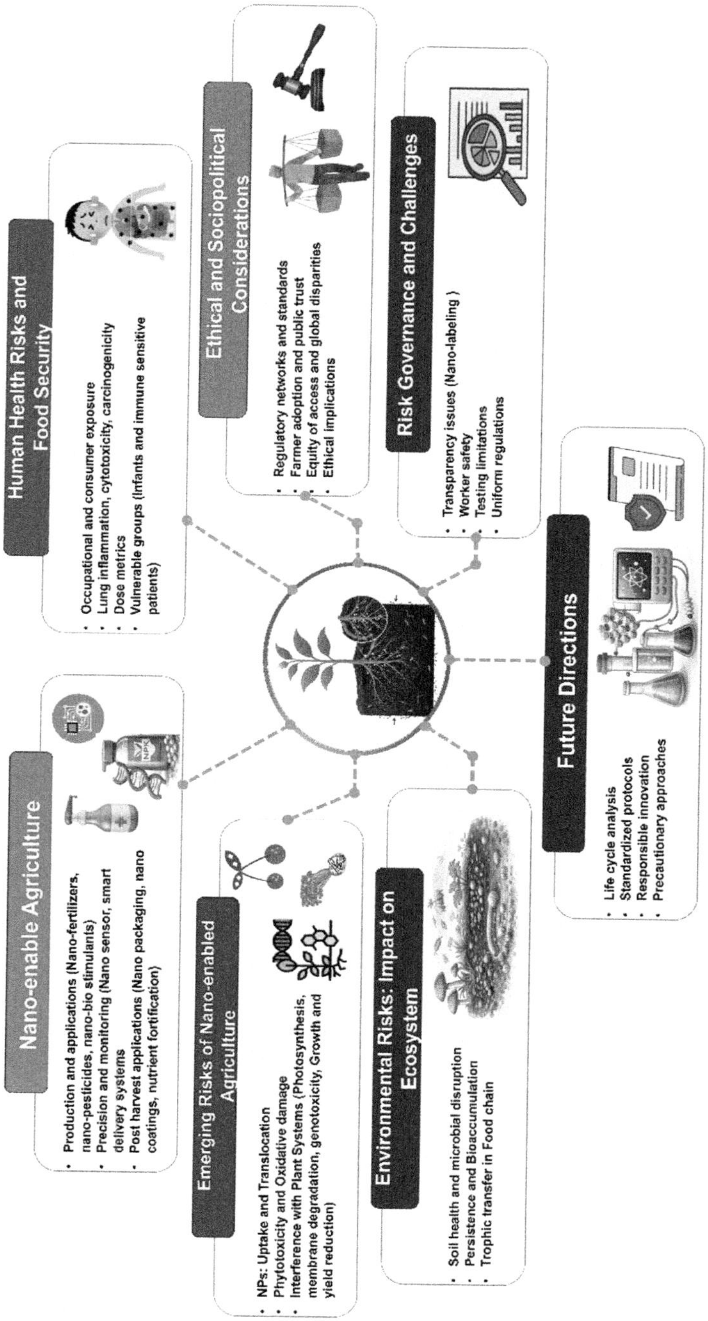

FIGURE 18.1 Graphical abstract of risks and concerns associated with use of nanotechnology.

TABLE 18.1 Different classes of nano-formulations and their effects on plants.

Class	Example	Effect	Plant specie	Reference
Non-metal NPs	5HT-Se NPs	Drought resistance	*Solanum tuberosum*	AL-Amri (2026)
	SiO_2 NPs	Reduction of G-quantum dots stress	*Zea mays*	Yan et al. (2024)
Metal NPs	Cu NPs, Ag NPs	Drought resistance	*Triticum aestivum*	Ahmed et al. (2021)
Metal oxide NPs	ZnO NPs, SiO_2 NPs	Drought resistance	*Solanum lycopersicum*	Seleiman et al. (2023)
	ZnO NPs	Drought resistance	*Oryza sativa*	Waqas Mazhar et al. (2022)
Nano composites	Cu-Ag NPs	Pathogen resistance	*Oryza sativa*	Chowdhury et al. (2024)
Nano polymers	GE-NPCH	Growth improvement	*Triticum aestivum*	Mondéjar-López et al. (2024)
Nano tubes	Graphene	Salinity tolerance	*Gossypium hirsutum*	Pandey et al. (2019)
Hydrogels	Nano DAP-hydrogel	Drought resistance	*Solanum lycopersicum*	Jayasudha et al. (2025)
	Nano silica	Drought resistance	*Oryza sativa*	El-Aziz et al. (2025)
Herbicide	Chitosan based NPs	Weed resistance	*Triticum aestivum*	Khan et al. (2023)
Nano coatings	Chitosan nano TiO_2	Enhanced shelf life	*Vaccinium corymbosum*	Li et al. (2021)
Nano emulsions	Farnesol	Pest resistance	*Capsicum annum*	Martín et al. (2024)
	Mustard oil	Weed resistance	*Vicia faba*	El-Wakeel and Zaki (2025)
Quantum dots	C-quantum dots	Improved photosynthesis	*Oryza sativa*	Tan et al. (2021)
	C-quantum dots	Drought resistance	*Glycine max*	Farhangi-Abriz et al. (2025)
	ZnO-quantum dots	Biomass improvement	*Lactuca sativa*	Liang et al. (2021)

Nano bio-stimulants are used in the form of nanoparticles (e.g., CuO, ZnO, AgO and FeO), nanocarriers (e.g., liposomes, micelles and dendrimers), or nanostructured materials with high surface area to volume ratio. These help to improve plant growth and yield by increasing nutrient absorption from roots, stress resistance, photosynthetic capability and modulations of phytohormones (e.g., auxins, ABA, gibberellins, ethylene and cytokinin etc.) (AL-Amri, 2026; Magnabosco et al., 2023).

2.1.4 *Soil modifications*

It includes mixing of nano sized formulations in soil to improve its texture, fertility and quality by modifying water retention capability, microbial interactions, nutrient profile, porosity and pH. Currently, most commonly used nano formulations for soil amendments include nano clays (e.g., montmorillonite), silica, hydro gels, fertilizers (e.g., encapsulated N, P and K) and materials (e.g., nano zero valent iron: nZVI) for heavy metals' sequestration. All these nano formulations provide controlled environment to increase plant growth and yield (Weisany, 2025).

2.2 Precision and monitoring

2.2.1 *Nano sensors*

These are highly sensitive nanosized devices that are designed to monitor various external and internal parameters of plants (Fig. 18.2).

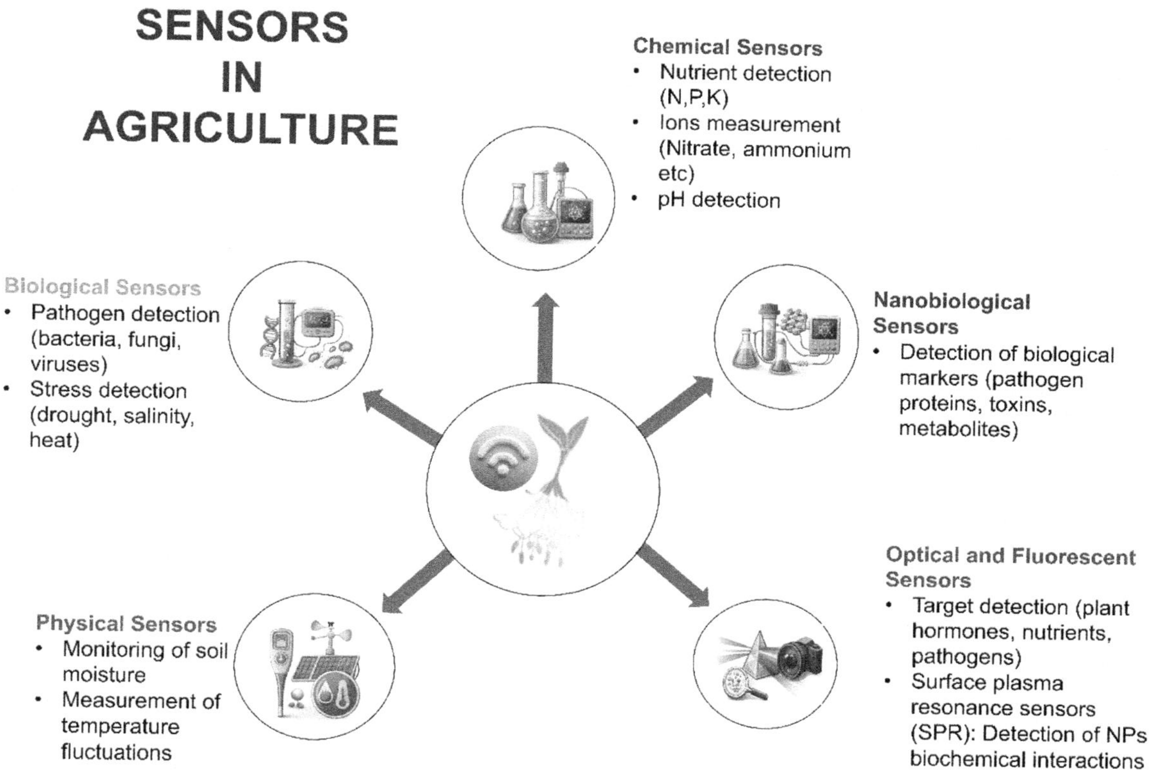

FIGURE 18.2 Different types of sensors and their applications in agriculture.

These tiny devices are embedded on plant tissues or soil systems to detect specific signals related to plant nutrients, moisture content, heat, cold, pH and pathogens (e.g., microbes, fungi, insects, caterpillars and weeds etc.). Nano sensors enabled accurate data collection by real time monitoring of plant systems for various types of biotic and abiotic stresses. For instance, nano sensor devices can help in early detection of pathogens or stressors (e.g., temperature, drought, salts, heavy metals and nutrients level) before their potential damage to crops. These can also monitor level of phytohormones, metabolites and gene expression patterns under normal and stress conditions. Integration of nano sensors with internet of things (IoTs) and cloud computing systems can couple data collection and analysis for early decision making and necessary precautions to maintain crop yield (Zeng et al., 2025).

2.2.2 *Smart delivery systems*

These are modern techniques to provide nutrients, pesticides and bio-stimulants in plant systems by ensuring their localized delivery and gradual release in sustainable manner. These techniques include nano encapsulations, hydrogels, polymers, nutrient particles, biodegradable carriers and enzyme-based systems. Nano encapsulations provide protection to pesticides and regulate their gradual release over time, which reduces the frequent need for conventional pesticides. Similarly, nanoparticles and nano polymers (Table 18.1) provide regulated nutrients (N, P, K etc.) to the plants, while hydrogels are made up of water absorbents that can encapsulate nutrients. These gel based nano systems can release water and nutrients under specific conditions such as drought stress; so, provide drought resistance as well as nutrients to soil or plant tissues (El-Aziz et al., 2025). Biodegradable carriers are made up of organic materials encapsulating phytohormones or other agrochemicals. These systems provide controlled amount of desired agrochemicals or phytohormones to the plants and afterward their automatic degradation is also safe for environment. Moreover, certain delivery systems are coupled with enzymes that can break coatings of encapsulated nano formulations (e.g., nano particles, pesticides, and polymers); to trigger their activity under optimal conditions of pH and temperature. Together, all these delivery systems help to provide efficient resistance against pathogens, drought, acidity/alkalinity, salt, cold, heat and heavy metals (Pagano et al., 2025).

2.3 Post harvest applications

2.3.1 *Nano packaging*

Integration of nano sensors and antimicrobials with packaging of post-harvest materials (e.g., fruits) can help to monitor and control quality, microbial activity and spoilage of post-harvest. Implanting nano sensors in packaging can help real time monitoring of fruit quality to prevent post-harvest damage by spoilage. However, more research is needed to explore new ways for integration of nano sensors with packaging materials (Dar and Nayik, 2022).

2.3.2 *Shelf-life enhancement*

Nanocoating can help to create a barrier between microbial growth and shelf life of post-harvest goods (Table 18.1). These coatings cover the surface of edibles; therefore, reduce exposed area for microbial attack and oxidation that ultimately enhance shelf life of post-harvest (Li et al., 2021; Chang et al., 2025).

2.3.3 *Nutrient fortification*

Nano coatings of postharvest edibles can help to prevent dehydration and nutrient loss, which ultimately increases food quality for a longer time. More locked nutrients can reduce consumption of edibles and act as key to tackling hunger of growing population (Chaudhary et al., 2025).

3. Emerging risks of nano enabled agriculture

3.1 Nanoparticles: Uptake and translocation

Nano particles (NPs) are nano sized (1–100 nm) formulations that are designed to contain nutrient elements, pesticide ingredients or nucleic acids (DNA or RNA) and act as carriers to transfer these materials at target cites of plants. Uptake and translocation of NPs depend upon size, chemical composition, shape, aggregation and dissolution properties (Li et al., 2022). It also depends upon plant morphologies, environmental conditions and modes of application. Applications of NPs in agriculture normally involve two basic pathways, (a) assisted delivery and (b) passive delivery. As the name indicates, assisted delivery involves external force for applications of NPs in plants and for their translocation to target cites. It involves techniques like gene-gun (Rajkumari et al., 2021), sonoporation (Zolghadrnasab et al., 2021), magnetic field, microinjection and electroporation. These techniques forcefully introduce nanoparticles into callus cultures, protoplasts or embryo transfers. However, controlled environment is needed for assisted delivery, as environmental conditions such as pH, temperature, humidity may interfere with efficacy of NPs translocation (Ahmad et al., 2022).

Passive delivery involves natural pathways for uptake and translocation of NPs either through plant roots or leaves. In this case NPs are simply applied to plants in solution form via soil or by foliar spraying. In soil applications, plants absorb NPs as other nutrients with water and may enter xylem tissues by apoplast or symplast pathways and finally reach to aerial parts. In some cases, Ag NPs are found to enter in plant cells by making small pores in cell walls (Rastogi et al., 2019). In foliar applications, NPs can enter to leave tissues by stomata, cuticle or epidermis and then enter to phloem tissues from where these can be translocated to storage sites with photosynthate (Gao et al., 2023). Moreover, some other methods of NPs applications involve seed priming and root dipping in which seeds or plant roots are soaked into NPs solution for a specific time interval and then transplanted to soil. Some studies suggest that foliar applications assist faster uptake and translocation of NPs as compared with soil mixing (Lian et al., 2020). Slow uptake from soil may be due to presence of mucilage covering on root surface and very low pore size of cell walls (5–20 nm) (Zhong and Ma, 2022). However, in some cases NPs of 50 nm size are found in cytoplasm in certain plant species (Djanaguiraman et al., 2018). Whatever the method NPs adopt, once they get entry into plants, they can spread throughout plant systems by diffusion, endocytosis or through ion channels and transporters (Wang et al., 2023). NPs can also accumulate in different tissues depending upon their size and dissolution properties (Li et al., 2022; Candan et al., 2024). Upon reaching their target cites, NPs release nutrients or genetic material to improve plant growth, yield or stress tolerance ability against drought, salinity, heavy metals, nutrients or thermal variability.

3.2 Phytotoxicity and oxidative stress

Although encapsulated nano-formulations have great potential to improve crop yield with minimum environmental risks, their unregulated delivery can cause toxicity and damage in plant systems (Racuciu et al., 2025). Recently research is ongoing to validate the impact of different concentrations of nano particles on crop plants (Table 18.2); and it suggests that excessive use of every type of nano particles can cause phytotoxicity (Mathur et al., 2023). For instance, optimum concentration of titanium dioxide nano particles (TiO_2 NPs) is found to be useful for plant species but their repeated delivery also has some toxic effects (Hu et al., 2020). Which can lead to hormonal imbalances, growth and developmental impairments.

A major consequence of nano-formulations based phytotoxicity is elevated oxidative stress in plant systems (Zhang et al., 2025). Their unchecked entry into plant tissues cause mass production of reactive oxygen species (ROS) such as Oxide radicals (O), super oxides (O^{2-}), hydroxyl radicals (OH) and hydrogen per oxide (H_2O_2). Under normal conditions ROS stimulate the production of enzymatic and non-enzymatic antioxidants (SOD, CAT, POD, APX, vit. C, E, flavonoids and glutathione) to trigger stress responsive signaling pathways (Fallah et al., 2024). However, their massive amount is harmful for cellular carbohydrates, proteins, lipids, DNA and RNA (Racuciu et al., 2025). The tiny size nano-formulations provides large surface area for redox reactions that can cause burst of ROS and oxidative stress pathways (Banerjee et al., 2025); leading to cellular damages. So, optimized low concentration of nano-formulations is useful for plants but higher and frequent delivery can cause oxidative damage to plant tissues (Rizwan et al., 2025; Wang et al., 2020).

TABLE 18.2 Applications of nano-formulations and phytotoxic effects on plants.

Nano-formulation Type	Size	Concentration	Phytotoxicity	Plant specie	Reference
ZnO-MWCNTs (Nanocomposite)	90:60 nm	50:500 mg/L	Increased ROS, reduced growth	*Arabidopsis thaliana*	Yang et al. (2023)
TMA-IONPs (Nanoparticles)	10.78 nm	>45.6 mg/L	Oxidative stress, genotoxicity	*Zea mays*	Racuciu et al. (2025)
Nano urea	56.79 nm	4% N	Yield reduction	*Triticum aestivum*	Sikka et al. (2025)
Ag (Nanoparticles)	<10 nm	2.5 mg/L	Reduction of fresh weight in callus culture	*Populus nigra*	Iori et al. (2023)
TiO_2 (Nanoparticles)	30 nm	20 mg/L	Oxidative stress	*Triticum aestivum*	Dogaroglu and Koleli (2017)
VO_2 (Nanoparticles)	20–30 nm	>10 mg/L	Reduction in Seedling growth, hotosynthesis	*Pisum sativum*	Li et al. (2022)
Y_2O_3 (Nanoparticles)	20–30 nm	>20 mg/L	Reduction in root length and chlorophyl content	*Oryza sativa*	Zhao et al. (2021)
TiO_2 (Nanoparticles)	–	50–400 mg/L	Decrease in nutritional content	*Lactuca sativa*	Hu et al. (2020)
Ag (Nanoparticles)	<100 nm	5 mM	Oxidative damage to chloroplast	*Triticum aestivum*	Hu et al. (2020)
ZnO (Nanoparticles)	30 nm	100 mg/L	Reduced activity of growth hormones	*Arabidopsis thaliana*	Vankova et al. (2017)
CuO (Nanoparticles)	40 nm	500 mg/L	Reduction in seedling growth	*Oryza sativa*	Wang et al. (2020)
ZnO (Nanoparticles)	<50 nm	50 mg/L	Growth inhibition	*Oryza sativa*	Chen et al. (2018)
Ag (Nanoparticles)	8–28 nm	>30 μg/mL	Enhance antioxidant activity	*Citrus reticulata*	Hussain et al. (2018)
Fe_2O_3 (Nanoparticles)	20 nm	100 mg/L	Chlorophyl content reduction	*Citrus maxima*	Hu et al. (2017)
Ag (Nanoparticles)		50 ppm	Chloroplast disruptions	*Pinus sylvestris*	Aleksandrowicz-Trzcinska et al. (2018)
Fe_3O_4 (Nanoparticles)	11.4 nm	>294 μg/mL	Inhibition of seedling growth	*Lavandula angustifolia*	Racuciu (2019)

3.3 Interference with plant systems (physiology and genetics)

3.3.1 *Photosynthesis and chloroplast*

Frequent applications of nano-formulations can cause oxidative damage to chloroplasts by mass production of ROS. As chloroplasts disruptions are directly linked to photosynthetic efficiency of plants; so, it affects plant growth and metabolic processes. Under normal conditions, some ROS can be produced by splitting water molecules during light reactions of photosynthesis. However, excessive nano-formulations (Table 18.2) can double the process of ROS production, which can cause oxidative damage to thylakoid membrane (Rastogi et al., 2019), vital proteins and enzymes of photosystems (PSI and PSII) such as RuBisCo. Hydroxyl radicals (OH) and oxygen species (O, O^{2-}) can interact with D1 proteins of PSII resulting in reduced efficiency of electron transfer to ETC and PSI, leading to impaired production of NADPH and ATP. These reactive species, when interact with RuBisCO enzyme, cause reduction of carbon fixation and impair the efficiency of Calvin cycle for production of glucose.

Silver nanoparticles (Ag NPs) are well studied for their antimicrobial activities in plants; but their excessive applications caused reduction in chlorophyl and carotenoid contents in *Oryza sativa* resulting in stunted plant growth and yield losses (Muzammil et al., 2023). Similar results for excessive applications of metal oxide nanoparticles (Fe_2O_3 and ZnO NPs) are also observed in different plant species (Hu et al., 2017; Islam, 2025). The reduction in chlorophyl and carotenoid contents is the indication of possible damage to chloroplast and thylakoid membrane; that might be due to direct interference of nanoparticles or ROS with light capturing ability of plants. So long term exposure of plants to NPs can cause reduction in photosynthesis that impacts overall growth and yield.

3.3.2 *Membrane degradation*

Excessive ROS due to heavy applications of nano-formulations can cause lipid peroxidation of cell and organellar membranes (Table 18.2). The free radicals (O, OH) of ROS remove electrons from lipid bilayer and contribute to making it more rigid, prone to leakage of cytosolic content and results in cell lysis. These damages are not restricted to plasma membranes but also disrupts intracellular organelles (Muzammil et al., 2023). Instead of oxidative damage, NPs of less than 10 nm size can easily penetrate (Racuciu et al., 2025) and directly interact with cellular membranes. Where they interfere with normal membrane functions by effecting its fluidity. In this way NPs cause double facet damage by oxidative stress and physical disruptions to cell membranes (Zhang et al., 2025). Ultimately these damages cause leakage to extracellular environment, impaired ion balance and homeostasis that can lead to apoptosis or necrosis of cells and impacts overall physiological processes of plants (Fallah et al., 2024).

3.3.3 *Genotoxicity*

Higher amounts of nano-formulations can indirectly or directly cause genotoxicity (Table 18.2). Plants can overcome short term stress of nano-formulations by activating stress mitigating genes related to the production of antioxidants, hormones and stress related proteins. However, prolonged exposure to metal NPs can cause oxidative stress by production of ROS. ROS can cause damage to DNA by breaking the strands or by modifications of nitrogenous bases (Racuciu et al., 2025). On the other hand, being small in size, nano-formulations can also directly bind to DNA molecules, chromosomes and spindle fibers. These changes can cause DNA fragmentation, mutations, chromatin thickness, chromosomal abrasions and alteration in expression patterns of stress regulatory genes (Muzammil et al., 2023;

Yang et al., 2023). Which can harm cell division machinery leading to cell cycle arrest. For instance, tungsten dioxide NPs cause DNA and chromosomal damage in root cells of *Allium cepa* (Banerjee et al., 2025). The same genotoxic impairments are also found in different other plant species when exposed to higher or prolonged doses of metal NPs (FeO, AgO, TiO_2 and ZnO) (Banerjee et al., 2025; Racuciu et al., 2025).

3.3.4 *Growth and yield reduction*

Frequent doses of nano-formulations cause oxidative and genotoxic effects at cellular level. The oxidative damage to organelles like chloroplasts (Rastogi et al., 2019), is directly linked to reduction in photosynthesis and growth. Similarly, genotoxicity leads to altered expression patterns of genes and hormonal imbalances that can ultimately interfere with overall plant growth and yield. Primarily NPs are used in agriculture to enhance crop yield, but their prolonged exposure can produce opposite results (Table 18.2). Research suggested that high exposure of ZnO and CuO NPs can retard seed germination and root elongation (Chen et al., 2018; Wang et al., 2020). Similarly prolonged exposure to Ag NPs reduced plant biomass and yield (Mathur et al., 2023) in (Khattak et al., 2024; Yang et al., 2018). Overall, uncontrolled applications of nano-formulations for prolonged time can cause physiological and genetic disruptions at cellular level that can ultimately cause yield losses.

4. Environmental risks: Impact on ecosystem

4.1 Soil health and microbial disruption

Agricultural use of encapsulated nano-formulations (nano particles, nano fertilizers and nao pesticides) can impact soil health by damaging its texture and by interacting with soil microbes. However, damage intensity to microbes depends upon concentration, number of doses, particle size, solubility, exposure duration and soil conditions (pH, EC, porosity, mineral and organic contents). For instance, in some cases concentration of ZnO NPs (250 mg/kg soil) is not harmful for soil and microbes but their higher concentrations can be fatal (Strekalovskaya et al., 2024). Similarly small sized nano particles (ZnO and Ag NPs) have antimicrobial properties and can easily penetrate into microbes to cause oxidative damage (Kandhil et al., 2025).

Sometimes soil conditions can also interfere with normal properties of NPs. For instance, abundant organic matter can hamper release of nutrients from NPs and reduce their bioavailability. In contrast higher porosity, low pH and organic content can increase mobility and bioavailability of NPs (Strekalovskaya et al., 2024). Moreover, damage intensity also depends upon type of microbes as some microbes cannot tolerate even low concentrations of nanoparticles. Such as Ag NPs in the range of 1 mg/kg soil can be toxic for soil microbes (Mishra and Yang, 2025). In contrast 20 mg/Kg soil of Ag NPs is reported to reduce biomass of arbuscular mycorrhizal fungi (Zhao et al., 2024). Higher doses of Ag NPs reduced the symbiotic ectomycorrhizal association of fungi with *Pinus sylvestris* (Aleksandrowicz-Trzcinska et al., 2018). Symbiotic fungi releases Glomalin related soil protein (GRSP) that is involved in soil aggregation and carbon sequestration. High amount of Ag NPs harms symbiotic fungi and therefore contributes to disrupting soil texture making it unfavorable for carbon sequestration (Zhao et al., 2024).

4.2 Persistence and bioaccumulation

Being very small in size (1–100 nm), nano-formulations are harder to decompose so can remain in soil for years depending upon nature of soil. For instance, Ag NPs can remain in upper layers of soil for more than 1 year. Its presence in upper layers indicates low leaching and higher persistence (Zhang et al., 2023). Moreover, mixing topsoil with biochar and organic matter can coat Ag NPs and their bioavailability to plants is reduced. Less leaching rate and more organic content of soil also contribute to their physical aggregation and longer persistence. Repeated applications can cause accumulation of NPs in soil, where they do not remain static but transform or become dissolved into free metal ions (Zn^{2+}, Cu^{2+}); which sometimes become more toxic than original molecules. In sulfur rich soils metal NPs (Ag NPs) can react to form silver sulfide over time; which reduced Ag toxicity by making it less bioavailable (Kraas et al., 2017). Further, biochar and organic content in soil also help to reduce the toxicity of Ag NPs (Zhang et al., 2023).

Persistence and transformation of NPs also cause their bioaccumulation in living organisms. Rotational crops can absorb NPs by their roots from exposed soil followed by translocation and accumulation (Table 18.2) in different plant tissues (Iori et al., 2023). Bioaccumulation is also linked to dissolution properties of NPs (Chen et al., 2018). In some crops roots act as barrier to filter nano particles (e.g., Ag NPs) by accumulating them and only small fractions can pass to distal parts like shoots and leaves (Kolesnikov et al., 2021). Other non-target plants and organisms like soil dwelling organisms (i.e., nematodes and detrivores like earthworms and sprigtails etc.), insects and birds that eat plants or feed on plant roots also consume the NPs. These NPs can accumulate in their bodies and may transform to cause toxicity and imbalances in cellular environments. The degree of nano particles bioaccumulation in different organisms varied from each other. For instance, garden snails that fed on Ag and TiO_2 NPs containing lettuce can pass more Ag NPs via its feces but retain 70% of ingested TiO_2 NPs in its tissues (Wu et al., 2021). Overall, different types of NPs behave differently depending upon their environmental conditions and dissolution properties.

4.3 Trophic transfer in food chains

Persistence and bioaccumulation potential of NPs also contribute to their transfer from consumers to higher trophic levels and food chains. One example of trophic transfer is lettuce and garden snail, where more concentration of Ag and TiO_2 NPs are found in garden snail than lettuce. Bioaccumulation of these metal NPs in garden snails caused neurological and digestive issues under experimental conditions; with more severe effect upon exposure to combined treatments of Ag NPs + TiO_2 NPs (Wu et al., 2021). Another example is transfer of NPs from algae to higher trophic levels across food chain; suggest that NPs pass from algae to herbivores and then to higher order predators. Passing through different trophic levels, when NPs reach in the bodies of birds and mammals their bioaccumulation can cause severe metabolic disruptions. Bioaccumulation and transfer to higher trophic levels also suggest risk of more concentration of NPs in top predators than producers. At consumer levels the metal NPs can cause oxidative stress and may transform to other forms like metal ions, which can react with other cellular substances to form new toxic complexes at each trophic level (Dang et al., 2023). However, all NPs do not behave constantly at each trophic level, as some NPs can excrete out from bodies of consumers without causing any toxicity. Even long-term transformations of NPs or complexes with other organic matter can reduce their bioavailability and toxicity over time (Zhang et al., 2023). Overall, we should take precautionary measures and ensure controlled use of NPs to get their benefits and to avoid their potential risks.

5. Human health risks and food security

5.1 Occupational and consumer exposure

Researchers, employees and farm workers face direct contact with nano-formulations (nano fertilizers, nano pesticides, nano particles, nano sensors or nano additives) during their synthesis, evaluation or mixing and loading for field applications (Iavicoli et al., 2017). Some fractions of NPs can persist in atmosphere (either from topsoil by wind or during field applications of spraying). Farmers, workers or other consumers may also inhale NPs from their surroundings or by dermal contact with plants, soil or other nano-formulations containing goods. Long term exposures may cause untended effects, however, inhalation risks of nano-formulations in humans can be minimized by using masks and gloves during synthesis, handling and field applications. As no proper biomarkers are available to monitor the exposure of nano formulations (Iavicoli et al., 2017), so strict regulations and guidelines for their safe use can help to minimize the potential risks.

Moreover, NPs (e.g., Ag NPs and Cu NPs) from nano coatings of food stuff can penetrate into flesh or deeper layers under unregulated storage conditions such as high temperatures or during long term storage. The ingestion of NPs from food stuff can cause severe health issues. The small size (less than 100 nm) enables NPs to absorb into gastrointestinal canal and pass through cell membranes. NPs can spread in different locations at cellular and organismic levels to cause oxidative damage, genotoxicity and carcinogenic effects (Wang et al., 2025). However, improper use of nano supplements to improve solubility properties of vitamins, minerals, antioxidants and food nutrients can also raise health concerns (Altemimi et al., 2024). Research should emphasize In vivo **experiments** to find long term effects of nano-formulations in human body.

5.2 Mechanism of potential harm

NPs can cause toxicity in living organisms by various pathways such as oxidative stress, genotoxicity, barrier penetration and by microbial interactions. Accumulation of NPs in living tissues can cause mass production of ROS leading to oxidative damage to cell membranes, lipids, proteins and amino acids that may result into cell death or other mal-functions. High concentration of NPs can trigger immune responses by activating immunity related cells such as macrophages and cytokines leading to inflammation. For instance, prolonged exposure to Ni, Cr, CuO, ZnO and TiO_2 NPs caused inflammatory, fibrotic and tumorigenic issues in rodents (Xuan et al., 2023). Accumulation of NPs can cause genotoxicity by direct binding with DNA or indirectly by ROS production, which can cause breakdown of its strands or modifications in nitrogenous bases leading to mutations and abrasions. The oxidative damage to mitotic apparatus may lead to cell cycle arrest or programmed cell death and changes in gene expression patterns can cause carcinogenic and reproductive problems at organism level (Wang et al., 2025).

Moreover, human body possesses certain barriers such as, air blood barrier prevents movement of large unwanted molecules from lungs to blood stream, blood brain barrier from blood vessels to brain cells, placental barrier from placenta to fetus tissues, blood testes barrier from blood to testes. However, being small in size (1–100 nm), inhaled or ingested NPs can easily cross all these barriers and can cause oxidative damage to blood, brain or fetus tissues that may lead to genetic, neurological and reproductive disorders (Xuan et al., 2023). Furthermore, some useful bacteria (e.g., Akkermansia and Enterococcus) are present in digestive tract of humans and other animals to facilitate digestion. Ingested NPs due to their non-biodegradable nature can interact with these bacterial populations in digestive tract and may weaken them. On the other hand, metal NPs can promote growth of inflammatory bacteria leading to gut and tissue inflammations, which can ultimately cause digestive, metabolic and neurological disorders (Landsiedel et al., 2022).

5.3 Dose metrics problems

Unlike traditional fertilizers which can be measured by their weight, the NPs dose measurement required additional metrics. A specific mass of NPs contains more nutrient particles than same mass in compound form. Inhalation toxicology data suggests that lung inflammation intensity can better correlate with NPs surface area than mass. The reason is that greater available surface area can interact over large range of lung tissue and create more oxidative stress. Along with surface area, NPs number and shape also matter for toxicity (Schmid and Stoeger, 2016). Similarly, toxicity issues are also related to size, distribution and aggregation of NPs. Smaller NPs due to larger surface area to volume ratio can penetrate easily into tissues and cause more toxicity than larger NPs. Moreover, if NPs form aggregates, then their surface area and bioavailability can be changed compared to individual NPs. Environmental conditions also contribute to change nature (from individual particles to aggregates), dissolution or bioavailability parameters of NPs (Iavicoli et al., 2017). So, it is essential to consider size, distribution and aggregation characteristics of NPs for their exact dose determination in target tissues.

Dissolution properties of nano-formulations are another determinant of toxicity level. For instance, inert NPs show less toxicity issues than highly soluble and ionic forms. Metal based NPs such ZnO NPs can form Zn^{2+} ions, which can cause more toxicity than their particle form. Research suggests that the concentration of intact NPs, their active forms, degradable, inert and ionic forms, all should be considered for dose determination (Kah et al., 2021).

5.4 Vulnerable groups and ethics of uncertainty

Nano formulations do not affect all individuals equally; some age groups are more vulnerable to their exposure. For instance, early age group individuals (infants and young children), pregnant females and immune sensitive patients are at more risk with exposure to NPs than healthy young adults. NPs contaminated food can cause severe inflammation in digestive canals of infants, because immature and weak immune system cannot cope with the toxic effects. Some baby products contain nanohydroxyapatite and nano TiO_2, but proper guidelines are limited to estimate their potential risks. Similarly, pregnant females are also susceptible for nano-formulations, especially those working at farms are dealing directly with nano materials. The small sized NPs can easily cross placental barriers and can cause genetic or physical impairments in developing fetus. NPs can also interact with immune functions and can reduce or cause their overactivation, so patients with chronic diseases are more susceptible to these effects (Xuan et al., 2023).

Beyond vulnerability, there remains some ethical concerns as many people are unaware about the presence of NPs in their food or surroundings. People living near agricultural farms, where nano-formulations are used for crop improvement may unwillingly inhale them. This raises ethical concern about informed consent as lack of awareness can cause severe health issues. Some research groups urge the need to label the NPs containing food or other products. Awareness and informed consent for the use of nano-formulations can reduce the risks of uncertainty and will build trust in communities about their safety (Throne-Holst and Rip, 2011).

6. Ethical and sociopolitical considerations

6.1 Regulatory networks and standards

Integration of nanotechnology with agriculture raised the concerns to device proper regulatory networks and standards for the safe use of nano-formulations. In most states same regulations are followed for the use of nano-formulations and traditional agrochemicals. For instance, European union (EU) follows "REACH" regulations and general laws without any modifications for the use of nano formulations. In parallel, EPA in United states (US) follows same regulatory laws "FIFRA" and "TSCA" for the use of conventional chemicals and nano-formulations. Similarly, FAO and WHO also follow ordinary regulations for risk analysis of nano-formulations (Zeng et al., 2025). Research data suggest that no proper guidelines are present for labeling of nano enabled products. Lack of proper regulations at international levels makes it difficult to evaluate safety issues and approval of nanomaterials (Gottardo et al., 2021). Large ambiguities in regulatory networks created a barrier for safety concerns and commercialization of nanotechnology in food and agricultural sector. Formulation of uniform international and national regulations for labeling and safe use of agricultural nano-formulations remains a serious issue to be addressed.

6.2 Farmer adoption and public trust

The victory of any emerging technology heavily relies on its efficacy, safety and public trust. Use of nano-formulation (nano fertilizers, nano pesticides, NPs, nano coatings and nano sensors) can efficiently reduce the dependency on bulk conventional agrochemicals with minimum environmental risks. Research data suggested that stakeholders showed more interest toward nanotechnology products in agriculture than GM crops. Because controlled use of nano-formulations can enhance crop productivity without altering natural processes (Horgan et al., 2025). However, there are some safety and toxicity issues for uncontrolled use of nano-formulations. Some public groups are afraid about safety issues and unintended long-term effects of NPs and nano pesticides in agriculture. Similarly, lack of clear labeling also makes it difficult for consumers to select nano-enabled food stuff. Interestingly it is easy for farmers and consumers to accept use of nano sensors in soil and nano packaging materials of food stuff than direct applications of NPs, nano pesticides and nano coatings. So, researchers and stakeholders should provide actual data about benefits and potential risks of nano-formulations to gain public trust. Programs and campaigns for giving proper awareness to farmers and consumers about controlled use of nano-formulations will help to increase their trust on nano-enabled agricultural products.

6.3 Equity of access and global disparities

A major concern for the use of nano-formulations in food and agriculture is whether it will benefit small farm holders of developing countries equally as well-developed countries. And whether the price of nano-formulations will be low enough to be affordable for small farmers. Because high cost of nano-formulations will create global disparities for division of food sources among rich and poor. Research groups of some countries are developing nano-formulations with the aim of cost reduction like production of nano urea that is much cheaper than bulk urea fertilizer (Midde et al., 2022). Currently research is ongoing on green synthesis of NPs by using waste materials like fruit peels, husk, wheat and rice straw to make them cost effective, environment friendly and to reduce patent related issues. However, most advanced techniques of precision farming and nano sensors need more sensitive equipment and knowledge for their proper handling. Moreover, these advanced techniques also need broadband systems for their connectivity. Which can be unaffordable for small farmers and can create worst situations rather than potential advantages. Equity of access should also be considered for testing regulations of nano-formulations at international level. Because improper use of untested nano materials can cause localized health and environmental issues in low-income regions. So, policymakers should focus on making policies to make agricultural nano-formulations more consumer friendly and affordable for small farmers.

6.4 Ethical implications and unintended consequences

Internationally, proper regulations and guidelines for the labeling of nano-formulations and nano-enabled products are limited. European union instructed to specify nano materials with food labels but didn't take strict action for its compliance. Some NGOs also raised the issue for labeling and public awareness about nano-formulations. For instance, FDA (2019) encouraged label nano materials and their potential risks on nanomedicine (de Vlieger et al., 2019). Ethically, it is the responsibility of all stakeholders to provide honest information about all known and unknown aspects of nano ingredients in their products. Experts should emphasize taking precautionary measures for the use of nano-formulations and regularly update the guidelines according to new aspects of nanotechnology.

7. Risk governance

7.1 Challenges in risk assessment methods

Risk assessment of nano-formulations is more challenging than conventional agrochemicals. This is due to their small size (1–100 nm), as risk assessment for dose of such tiny formulations need standardized methods and techniques such as inductively coupled plasma mass spectrometry (ICP-MS), electron microscopy etc. to cover all parameters like surface area, particle number, size, shape, dissolution and aggregation capabilities. So, it is difficult for risk assessors to decide which parameter can best define the biological activity of nano-formulations.

Unavailability of long-term field data for applications of nan-formulations is another challenge for risk assessment. Most of the research experiments provide short term laboratory-based data that relies upon applications of high doses of NPs in controlled conditions. As natural soil conditions vary greatly in composition and microbial populations. NPs can show different interaction with microbial communities under natural field conditions than laboratory based experimental conditions (Islam, 2025). So, experimental data can only give an idea about beneficial and toxic doses of nano-formulations, but this data is insufficient to provide complete insights to define beneficial or toxic faces of nanotechnology in agriculture.

Uniformity of risk assessment protocols is also challenging. Most of the protocols rely upon plant biomass calculations to find toxic doses of NPs, without reaching to the changes in gene expression patterns. Moreover, existing risk assessment protocols for same parameter vary greatly among different regions depending upon equipment availability. Protocols for assessment of transformation rate of NPs over time and their accumulation capability in soil and living systems are also lacking. Scientists are emphasizing new regulations and integration of latest techniques for testing nano-formulations. For instance, use of electron microscopy can analyze effects of NPs dissolution patterns and accumulation rates in different tissues over multiple generations of crops (Zeng et al., 2025). Currently different NGOs like

OECD and ISO are working for proper test regulations. Integration of modern computational techniques like nano QSAR modeling is also in progress to predict risks depending upon physical and chemical attributes of nano-formulations before their large-scale commercialization Fig. 18.3.

7.2 Future directions (green nanotechnology and responsible innovation)

Synthesis of nano-formulations mainly relies upon inorganic methods for production of NPs, that leads to their non-biodegradability and dissolution issues. So, research focus should be the synthesis of biodegradable NPs. Recent research data suggest that scientists are focusing on producing NPs with biodegradable coatings by using organic polymers and edibles (Ben Zineb et al., 2025). Biodegradable materials can decompose easily after delivering NPs at target areas, so reduce the chances of toxicity issues. Similarly, research focus is to increase germination rate, drought resistance and seed vigor by priming with NPs and nano hydrogels (Alam et al., 2024). Research is also ongoing for synthesis of multipurpose and hybrid nano-formulations by fusing nano fertilizers with biocidal coatings or nano clays with microbial populations for simultaneous applications of nutrients and pesticides (Luo et al., 2025). Future focus should be paid on determination of effective dose and risk assessment of biodegradable coated and hybrid nano-formulations on soil, microbiota, crops, and consumers under natural field conditions.

Moreover, integration of nano sensors with IoT and data analytics can enable automated real time crop monitoring, that can help to predict exact time for applications of nano-formulations depending upon crop needs (Khattak et al., 2024). Protection of more

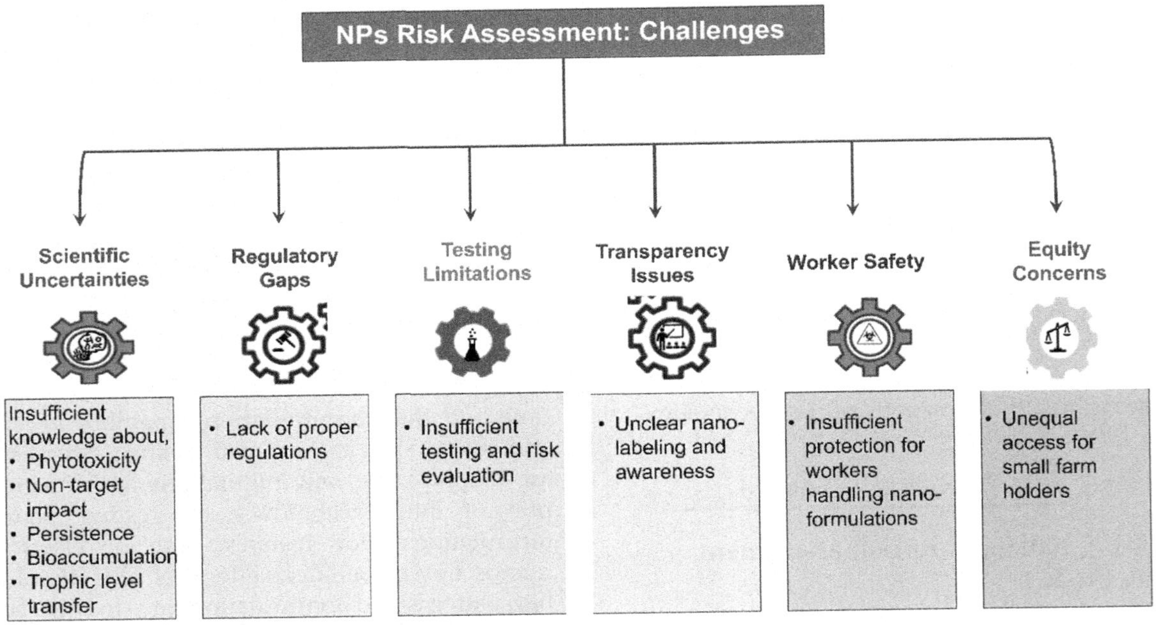

FIGURE 18.3 Challenges in risk assessment of NPs in food and agriculture.

susceptible individuals (infants, pregnant females, immunological patients and farm-workers) is a critical issue to be addressed. So, it is responsibility of stakeholders, researchers and experts to make proper regulations and guidelines for the use of nano-formulations. governing bodies should facilitate NGOs for proper awareness to general public about benefits as well as safety concerns of nano-formulations.

8. Conclusion

In modern era of emerging nanotechnology, various nano-formulation such as nano sized fertilizers, pesticides, bio-stimulants and soil amendment techniques provides plants with controlled environment for growth. Nano sensors and smart delivery systems enable real time monitoring of plants under normal and stress conditions that can aid early treatments before a potential risk happening into reality. Moreover, nano packaging with nano sensors and antimicrobials can help to improve quality of post-harvest edibles and nanocoating enhances shelf life of fruits. Together, all these techniques are used to tackle hunger challenges of growing population. Besides these benefits, some risks are also associated with uncontrolled use of nano-formulations. Excessive number of NPs can cause phytotoxicity and genotoxicity by oxidative stress or by direct interactions of NPs with DNA molecules leading to DNA fragmentation, mutations and cell cycle arrest in plants. As far as environmental risks are concerned, NPs can disrupt soil texture and microbial communities. Different NPs behave differently depending upon available soil conditions (i.e., pH, texture, composition etc.), with different antimicrobial and antifungal activities. Under some cases very low amounts of NPs can become damaging both for soil texture and symbiotic associations. Moreover, the small size and undecomposable nature is responsible for persistence of NPs in soil and bioaccumulation in non-target organisms. Soil mixing with organic matter can reduce NPs toxicity but not fully affective in all cases. NPs can also transfer to other trophic levels via different food webs (Wu et al., 2021). Humans can inhale NPs from their surrounding environment or can ingest from edibles, and bioaccumulation in humans can be lethal by causing lung inflammation, genotoxicity and carcinogenicity.

So, careful and strict regulations are needed for use of nano-formulations. Food and drug administration (FDA) and European food safety authority (EFSA) have started practices to evaluate bioaccumulation and toxicity parameters of nano formulations (Altemimi et al., 2024). But these are insufficient to meet risk assessment needs. To get sustainable advantages of green nanotechnology, proper risk assessment related to dose optimization and toxicity effects should be honestly communicated to farmers and consumers. Determination of nano-formulations in relation to target areas is a complex task that needs expert opinion related to number, surface area and dissolution properties to avoid their toxic effects. Protection of more susceptible individuals (infants, pregnant females, immunological patients and farm-workers) is a critical issue to be addressed. It is responsibility of researchers, stakeholders and governing bodies to make strict regulations for synthesis and testing of nano-formulations. Clear labeling of nano-enabled products along with their potential risks and cautions will help to increase trust of consumers in nanotechnology. More research focus is needed to unveil long term effects of nano-formulations at organism, population and ecosystem level.

AI disclosure

During the preparation of this work the authors used ChatGPT, Gemini etc. in order to collect literature. After using this tool/service, the authors reviewed and edited the content as needed and takes full responsibility for the content of the publication.

References

Ahmad, A., Hashmi, S.S., Palma, J.M., Corpas, F.J., 2022. Influence of metallic, metallic oxide, and organic nanoparticles on plant physiology. Chemosphere 290. https://doi.org/10.1016/j.chemosphere.2021.133329.

Ahmed, F., Javed, B., Razzaq, A., Mashwani, Z.U.R., 2021. Applications of copper and silver nanoparticles on wheat plants to induce drought tolerance and increase yield. IET Nanobiotechnol. 15 (1), 68–78. https://doi.org/10.1049/nbt2.12002.

Alam, M.W., Junaid, P.M., Gulzar, Y., Abebe, B., Awad, M., Quazi, S.A., 2024. Advancing agriculture with functional NM: "pathways to sustainable and smart farming technologies". Discover Nano 19 (1). https://doi.org/10.1186/s11671-024-04144-z.

AL-Amri, S.M., 2026. Synergistic application of biostimulant and selenium nanoparticles enhances drought tolerance in potato through integrated physiological and transcriptomic modulation. Plant Nano Biol. 15, 100244. https://doi.org/10.1016/j.plana.2025.100244.

Aleksandrowicz-Trzcinska, M., Szaniawski, A., Studnicki, M., Bederska-Blaszczyk, M., Olchowik, J., Urban, A., 2018. The effect of silver and copper nanoparticles on the growth and mycorrhizal colonisation of Scots pine (Pinus sylvestris L.) in a container nursery experiment. iFor. Biogeosci. For. 11 (5), 690–697. https://doi.org/10.3832/ifor2855-011.

Altemimi, A.B., Farag, H.A.M., Salih, T.H., Awlqadr, F.H., Al-Manhel, A.J.A., Vieira, I.R.S., Conte-Junior, C.A., 2024. Application of nanoparticles in human nutrition: a review. Nutrients 16 (5). https://doi.org/10.3390/nu16050636.

Banerjee, S., Rajput, P., Naz, I., Anand, V., Rajput, V.D., Sathyanarayana, E., Chauhan, A., Mandzhieva, S., Minkina, T., Sousa, J.R., 2025. Nanoparticles and plant DNA damage: a review of Genotoxic mechanisms and research outlook. Chemosphere 391. https://doi.org/10.1016/j.chemosphere.2025.144731.

Ben Zineb, A., Khalid, M.F., Gargouri, M., Alatalo, J.M., Ahmed, T., 2025. Green-synthesized nanoparticles: recent trends and innovations towards sustainable agriculture. Biocatal. Agric. Biotechnol. 67. https://doi.org/10.1016/j.bcab.2025.103667.

Bhat, U.H., Uddin, M., Khan, M.M.A., Islam, S., Khan, M.N., Chishti, A.S., Singh, S., Singh, S., Corpas, F.J., 2026. Harvesting the future: revolutionizing agriculture with engineered nanoparticles (NPs) – a comprehensive review. J. Plant Growth Regul. 45 (2), 821–832. https://doi.org/10.1007/s00344-025-11915-0.

Candan, F., Markushin, Y., Ozbay, G., 2024. Nanoparticle uptake and bioaccumulation in Pisum sativum L. (Green pea) analyzed via dark-field microscopy, infrared spectroscopy, and principal component analysis combined with machine learning. Agronomy 14 (7), 1473. https://doi.org/10.3390/agronomy14071473.

Chang, Y., Ferreira, M.D., Correa, D.S., Teodoro, K.B.R., Procopio, F.R., Brexó, R.P., Sarkhosh, A., Brecht, J.K., 2025. Advances in postharvest nanotechnology: enhancing fresh produce shelf life and quality to reduce losses and waste. Postharvest Biol. Technol. 222. https://doi.org/10.1016/j.postharvbio.2025.113397.

Chaudhary, S., Prasanna, D., Trivedi, P., Nandy, R., Prasad, S.S., Sunitha, S., Panigrahi, C.K., 2025. Nanotechnology applications in agriculture for crop protection and nutrient delivery: a review. Int. J. Plant Soil Sci. 37 (9), 106–119. https://doi.org/10.9734/ijpss/2025/v37i95690.

Chen, J., Dou, R., Yang, Z., You, T., Gao, X., Wang, L., 2018. Phytotoxicity and bioaccumulation of zinc oxide nanoparticles in rice (Oryza sativa L.). Plant Physiol. Biochem. 130, 604–612. https://doi.org/10.1016/j.plaphy.2018.08.019.

Chowdhury, A.R., Kumar, R., Mahanty, A., Mukherjee, K., Kumar, S., Tribhuvan, K.U., Sheel, R., Lenka, S., Singh, B.K., Chattopadhyay, C., Sharma, T.R., Bhadana, V.P., Sarkar, B., 2024. Inhibitory role of copper and silver nanocomposite on important bacterial and fungal pathogens in rice (Oryza sativa). Sci. Rep. 14 (1). https://doi.org/10.1038/s41598-023-49918-0.

Dang, F., Yuan, Y., Huang, Y., Wang, Y., Xing, B., 2023. Trophic transfer of nanomaterials and their effects on high-trophic-level predators. NanoImpact 32. https://doi.org/10.1016/j.impact.2023.100489.

Dar, A.H., Nayik, G.A., 2022. Nanotechnology Interventions in Food Packaging and Shelf Life. CRC Press. https://doi.org/10.1201/9781003207641.

de Vlieger, J.S.B., Crommelin, D.J.A., Tyner, K., Drummond, D.C., Jiang, W., McNeil, S.E., Neervannan, S., Crist, R.M., Shah, V.P., 2019. Report of the AAPS guidance forum on the FDA draft guidance for industry: "drug products, including biological products, that contain nanomaterials". AAPS J. 21 (4). https://doi.org/10.1208/s12248-019-0329-7.

Djanaguiraman, M., Belliraj, N., Bossmann, S.H., Prasad, P.V.V., 2018. High-temperature stress alleviation by selenium nanoparticle treatment in grain sorghum. ACS Omega 3 (3), 2479–2491. https://doi.org/10.1021/acsomega.7b01934.

Dogaroglu, Z., Koleli, N., 2017. Effects of TiO_2 and ZnO nanoparticles on germination and antioxidant system of wheat (Triticum aestivum L.). Appl. Ecol. Environ. Res. 15 (3).

Duro, J.A., Lauk, C., Kastner, T., Erb, K.H., Haberl, H., 2020. Global inequalities in food consumption, cropland demand and land-use efficiency: a decomposition analysis. Glob. Environ. Change 64. https://doi.org/10.1016/j.gloenvcha.2020.102124.

El-Aziz, M.A.A., Elbagory, M., Arafat, A.A., Aboelsoud, H. M., El-Nahrawy, S., Khalifa, T.H., Omara, A.E.D., 2025. Evaluating the impact of nano-silica and silica hydrogel amendments on soil water retention and crop yield in rice and clover under variable irrigation conditions. Agronomy 15 (3). https://doi.org/10.3390/agronomy15030652.

El-Wakeel, M.A., Zaki, F.S.A., 2025. Efficacy of rocket and mustard oils and their nano-emulsions as alternatives to chemical herbicides for controlling weeds associated faba bean. Sci. Rep. 15 (1). https://doi.org/10.1038/s41598-025-29915-1.

Fallah, S., Yusefi-Tanha, E., Peralta-Videa, J.R., 2024. Interaction of nanoparticles and reactive oxygen species and their impact on macromolecules and plant production. Plant Nano Biol. 10, 100105. https://doi.org/10.1016/j.plana.2024.100105.

Farhangi-Abriz, S., Ghassemi-Golezani, K., Torabian, S., Rahimzadeh, S., Osati, F., Safarpour, H., 2025. Response of soybean plants to the foliar application of carbon quantum dots under drought stress: a field study. J. Plant Growth Regul. 44 (2), 621–631. https://doi.org/10.1007/s00344-024-11464-y.

Gao, M., Chang, J., Wang, Z., Zhang, H., Wang, T., 2023. Advances in transport and toxicity of nanoparticles in plants. J. Nanobiotechnol. 21 (1). https://doi.org/10.1186/s12951-023-01830-5.

Gottardo, S., Mech, A., Drbohlavová, J., Małyska, A., Bøwadt, S., Riego Sintes, J., Rauscher, H., 2021. Towards safe and sustainable innovation in nanotechnology: state-of-play for smart nanomaterials. NanoImpact 21. https://doi.org/10.1016/j.impact.2021.100297.

Horgan, M.D., Cummings, C.L., Kuzma, J., Dahlstrom, M., Cimadori, I., Cuchiara, M., Larter, C., Loschin, N., Grieger, K.D., 2025. Beyond the hype: stakeholder perceptions of nanotechnology and genetic engineering for sustainable food production. Sustainability 17 (15), 6795. https://doi.org/10.3390/su17156795.

Hu, J., Wu, X., Wu, F., Chen, W., Zhang, X., White, J.C., Li, J., Wan, Y., Liu, J., Wang, X., 2020. TiO_2 nanoparticle exposure on lettuce (Lactuca sativa L.): dose-dependent deterioration of nutritional quality. Environ. Sci. Nano 7 (2), 501–513. https://doi.org/10.1039/c9en01215j.

Hu, J., Guo, H., Li, J., Gan, Q., Wang, Y., Xing, B., 2017. Comparative impacts of iron oxide nanoparticles and ferric ions on the growth of Citrus maxima. Environ. Pollut. 221, 199–208. https://doi.org/10.1016/j.envpol.2016.11.064.

Hussain, M., Raja, N.I., Iqbal, M., Ejaz, M., Aslam, S., Rehman, A.U., Javaid, U., 2018. Seed germination and biochemical profile of Citrus reticulata (Kinnow) exposed to green synthesised silver nanoparticles. IET Nanobiotechnol. 12 (5), 688–693. https://doi.org/10.1049/IET-NBT.2017.0303.

Iavicoli, I., Leso, V., Beezhold, D.H., Shvedova, A.A., 2017. Nanotechnology in agriculture: opportunities, toxicological implications, and occupational risks. Toxicol. Appl. Pharmacol. 329, 96–111. https://doi.org/10.1016/j.taap.2017.05.025.

Iori, V., Muzzini, V.G., Venditti, I., Casentini, B., Iannelli, M. A., 2023. Phytotoxic impact of bifunctionalized silver nanoparticles (AgNPs-Cit-L-Cys) and silver nitrate ($AgNO_3$) on chronically exposed callus cultures of Populus nigra L. Environ. Sci. Pollut. Control Ser. 30 (54), 116175–116185. https://doi.org/10.1007/s11356-023-30690-7.

Islam, S., 2025. Toxicity and transport of nanoparticles in agriculture: effects of size, coating, and aging. Front. Nanotechnol. 7. https://doi.org/10.3389/fnano.2025.1622228.

Jayasudha, M.N., Reddy, M.K., Madhavi, A., 2025. Enhancing drought resilience and sustainability in Solanum lycopersicum L cultivation: synergistic effects of nano-fertilizer and hydrogels. J. Appl. Nat. Sci. 17 (2), 753–759. https://doi.org/10.31018/jans.v17i2.6560.

Kah, M., Johnston, L.J., Kookana, R.S., Bruce, W., Haase, A., Ritz, V., Dinglasan, J., Doak, S., Garelick, H., Gubala, V., 2021. Comprehensive framework for human health risk assessment of nanopesticides. Nat. Nanotechnol. 16 (9), 955–964. https://doi.org/10.1038/s41565-021-00964-7.

Kandhil, U., Singh, G., Rani, A., Dang, A.S., Giri, S.K., Dhiman, S.S., Verma, N., Kumar, A., 2025. Soil-microbe-plant continuum under ZnO and TiO_2 nanoparticle stress: an insight into toxicological implications, risk evaluation and management strategies. Plant Nano Biol. 14. https://doi.org/10.1016/j.plana.2025.100201.

Khan, B.A., Nadeem, M.A., Alawadi, H.F.N., Javaid, M.M., Mahmood, A., Qamar, R., Iqbal, M., Mumtaz, A., Maqbool, R., Oraby, H., Elnaggar, N., 2023. Synthesis, characterization, and evaluation of nanoparticles of clodinofop propargyl and fenoxaprop-P-ethyl on weed control, growth, and yield of wheat (Triticum aestivum L.). Green Process. Synth. 12 (1). https://doi.org/10.1515/gps-2023-0105.

Khattak, W.A., Ullah, M.W., Manan, S., Islam, S.U., Khattak, W.A., Ul-Islam, M., 2024. Emerging applications and future trends of agri-nanotechnology. Nanotechnol. Life Sci. Part F3986, 429–458. https://doi.org/10.1007/978-3-031-76000-6_19.

Khundi, Q., Jiang, Y., Sun, Y., Rui, Y., 2025. Nanofertilizers for sustainable African agriculture: a global review of agronomic efficiency and environmental sustainability. Nanomaterials 15 (5), 390. https://doi.org/10.3390/nano15050390.

Kolesnikov, S., Tsepina, N., Minnikova, T., Kazeev, K., Mandzhieva, S., Sushkova, S., Minkina, T., Mazarji, M., Singh, R.K., Rajput, V.D., 2021. Influence of silver nanoparticles on the biological indicators of haplic chernozem. Plants 10 (5). https://doi.org/10.3390/plants10051022.

Kraas, M., Schlich, K., Knopf, B., Wege, F., Kägi, R., Terytze, K., Hund-Rinke, K., 2017. Long-term effects of sulfidized silver nanoparticles in sewage sludge on soil microflora. Environ. Toxicol. Chem. 36 (12), 3305–3313. https://doi.org/10.1002/etc.3904.

Landsiedel, R., Hahn, D., Ossig, R., Ritz, S., Sauer, L., Buesen, R., Rehm, S., Wohlleben, W., Groeters, S., Strauss, V., Sperber, S., Wami, H., Dobrindt, U., Prior, K., Harmsen, D., van Ravenzwaay, B., Schnekenburger, J., 2022. Gut microbiome and plasma metabolome changes in rats after oral gavage of nanoparticles: sensitive indicators of possible adverse health effects. Part. Fibre Toxicol. 19 (1). https://doi.org/10.1186/s12989-022-00459-w.

Li, Q., Hu, R., Chen, Z., Chen, L., Zhang, J., Wu, X., Li, J.B., Gao, Y., Yang, S.T., Wang, H., 2022. Phytotoxicity of VO_2 nanoparticles with different sizes to pea seedlings. Ecotoxicol. Environ. Saf. 242. https://doi.org/10.1016/j.ecoenv.2022.113885.

Li, Y., Rokayya, S., Jia, F., Nie, X., Xu, J., Elhakem, A., Almatrafi, M., Benajiba, N., Helal, M., 2021. Shelf-life, quality, safety evaluations of blueberry fruits coated with chitosan nano-material films. Sci. Rep. 11 (1). https://doi.org/10.1038/s41598-020-80056-z.

Lian, J., Zhao, L., Wu, J., Xiong, H., Bao, Y., Zeb, A., Tang, J., Liu, W., 2020. Foliar spray of TiO_2 nanoparticles prevails over root application in reducing Cd accumulation and mitigating Cd-induced phytotoxicity in maize (Zea mays L.). Chemosphere 239, 124794. https://doi.org/10.1016/j.chemosphere.2019.124794.

Liang, Z., Pan, X., Li, W., Kou, E., Kang, Y., Lei, B., Song, S., 2021. Dose-dependent effect of ZnO quantum dots for lettuce growth. ACS Omega 6 (15), 10141–10149. https://doi.org/10.1021/acsomega.1c00205.

Luo, X., Li, J., Guo, S., Yu, H., Zeng, X., Zhou, Z., Shangguan, Y., He, M., Ouyang, Y., Chen, K., Chen, Z., Qin, Y., 2025. Analysis of research status and trends on nano-agricultural application: a bibliometric study. Front. Plant Sci. 16. https://doi.org/10.3389/fpls.2025.1530629.

Magnabosco, P., Masi, A., Shukla, R., Bansal, V., Carletti, P., 2023. Advancing the impact of plant biostimulants to sustainable agriculture through nanotechnologies. Chem. Biol. Technol. Agric. 10 (1). https://doi.org/10.1186/s40538-023-00491-8.

Martín, F., Garzo, E., Guirao, P., Pascual-Villalobos, M.J., Fereres, A., Moreno, A., 2024. Persistence of nanoemulsions of bioactive volatiles and their impact on aphid feeding behaviour. J. Pest. Sci. 97 (4), 1–15. https://doi.org/10.1007/s10340-024-01746-5.

Mathur, P., Chakraborty, R., Aftab, T., Roy, S., 2023. Engineered nanoparticles in plant growth: phytotoxicity concerns and the strategies for their attenuation. Plant Physiol. Biochem. 199, 107721. https://doi.org/10.1016/j.plaphy.2023.107721.

Midde, S.K., Perumal, M.S., Murugan, G., Sudhagar, R., Mattepally, V., Badda, M., 2022. Evaluation of nano urea on growth and yield attributes of rice (Oryza Sativa L.). Chem. Sci. Rev. Lett. 11 (42), 211–214.

Mishra, S., Yang, X., 2025. How to safeguard soil health against silver nanoparticles through a microbial functional gene-based approach? Environ. Int. 202, 109680. https://doi.org/10.1016/j.envint.2025.109680.

Mondéjar-López, M., López-Jiménez, A.J., Gómez-Gómez, L., Ahrazem, O., García-Martínez, J.C., Niza, E., 2024. Field crop evaluation of polymeric nanoparticles of garlic extract–chitosan as biostimulant seed nano-priming in cereals and transcriptomic insights. Polymers 16 (23). https://doi.org/10.3390/polym16233385.

Muzammil, S., Ashraf, A., Siddique, M.H., Aslam, B., Rasul, I., Abbas, R., Afzal, M., Faisal, M., Hayat, S., 2023. A review on toxicity of nanomaterials in agriculture: current scenario and future prospects. Sci. Prog. 106 (4). https://doi.org/10.1177/00368504231221672.

Pagano, M., Lunetta, E., Belli, F., Mocarli, G., Cocozza, C., Cacciotti, I., 2025. Advancements in agricultural nanotechnology: an updated review. Plants 14 (18), 2939. https://doi.org/10.3390/plants14182939.

Pandey, K., Anas, M., Hicks, V.K., Green, M.J., Khodakovskaya, M.V., 2019. Improvement of commercially valuable traits of industrial crops by application of carbon-based nanomaterials. Sci. Rep. 9 (1). https://doi.org/10.1038/s41598-019-55903-3.

Racuciu, M., 2019. Iron oxide nanoparticles phytotoxicity on Lavandula angustifolia. Rom. J. Biophys. 29 (3).

Racuciu, M., Barbu-Tudoran, L., Oancea, S., 2025. Evaluation of phytotoxicity and genotoxicity of TMA-stabilized iron-oxide nanoparticle in corn (Zea mays) young plants. Sci. Rep. 15 (1). https://doi.org/10.1038/s41598-025-03872-1.

Rajkumari, N., Alex, S., Soni, K.B., Anith, K.N., Viji, M.M., Kiran, A.G., 2021. Silver nanoparticles for biolistic transformation in Nicotiana tabacum L. 3 Biotech 11 (12). https://doi.org/10.1007/s13205-021-03043-9.

Rastogi, A., Zivcak, M., Tripathi, D.K., Yadav, S., Kalaji, H. M., 2019. Phytotoxic effect of silver nanoparticles in Triticum aestivum: improper regulation of photosystem I activity as the reason for oxidative damage in the chloroplast. Photosynthetica 57 (1), 209–216. https://doi.org/10.32615/ps.2019.019.

Rizwan, H.M., Shafqat, U., Ishfaq, A., Batool, F., Mahmood, F., Su, Q., Yaseen, N., Raza, T., Altihani, F. A., 2025. Identifying the phytotoxicity of biosynthesized metal oxide nanoparticles and their impact on antioxidative enzymatic activity in maize under drought stress. Plants 14 (7). https://doi.org/10.3390/plants14071075.

Santos, P.A., Biraku, X., Nielsen, E., Ozketen, A.C., Ozketen, A.A., Hakki, E.E., 2025. Agricultural nanotechnology for a safe and sustainable future: current status, challenges, and beyond. J. Sci. Food Agric. 105 (6), 3159–3169. https://doi.org/10.1002/jsfa.13922.

Schmid, O., Stoeger, T., 2016. Surface area is the biologically most effective dose metric for acute nanoparticle toxicity in the lung. J. Aerosol Sci. 99, 133–143. https://doi.org/10.1016/j.jaerosci.2015.12.006.

Seleiman, M.F., Al-Selwey, W.A., Ibrahim, A.A., Shady, M., Alsadon, A.A., 2023. Foliar applications of ZnO and SiO_2 nanoparticles mitigate water deficit and enhance potato yield and quality traits. Agronomy 13 (2). https://doi.org/10.3390/agronomy13020466.

Sikka, R., Kalia, A., Ahuja, R., Sidhu, S.K., Chaitra, P., 2025. Substitution of soil urea fertilization to foliar nano urea fertilization decreases growth and yield of rice and wheat. Plant Soil 512 (1–2), 1475–1491. https://doi.org/10.1007/s11104-024-07157-w.

Strekalovskaya, E.I., Perfileva, A.I., Krutovsky, K.V., 2024. Zinc oxide nanoparticles in the "soil–bacterial community–plant" system: impact on the stability of soil ecosystems. Agronomy 14 (7). https://doi.org/10.3390/agronomy14071588.

Tan, T.L., Zulkifli, N.A., Zaman, A.S.K., Jusoh, M.b., Yaapar, M.N., Rashid, S.A., 2021. Impact of photoluminescent carbon quantum dots on photosynthesis efficiency of rice and corn crops. Plant Physiol. Biochem. 162, 737–751. https://doi.org/10.1016/j.plaphy.2021.03.031.

Throne-Holst, H., Rip, A., 2011. Complexities of labelling of nanoproducts on the consumer markets. Eur. J. Law Technol. 2 (3).

Vankova, R., Landa, P., Podlipna, R., Dobrev, P.I., Prerostova, S., Langhansova, L., Gaudinova, A., Motkova, K., Knirsch, V., Vanek, T., 2017. ZnO nanoparticle effects on hormonal pools in Arabidopsis thaliana. Sci. Total Environ. 593–594, 535–542. https://doi.org/10.1016/j.scitotenv.2017.03.160.

Wang, J., Liang, X., Hou, H., Deng, J., Lin, Q., Wen, L., Bai, J., 2025. A review on migration behavior and safety regulation of inorganic nanoparticles in food packaging: multiscale mechanisms, risk assessment, and innovative strategies. Food Control 178, 111490. https://doi.org/10.1016/j.foodcont.2025.111490.

Wang, W., Liu, J., Ren, Y., Zhang, L., Xue, Y., Zhang, L., He, J., 2020. Phytotoxicity assessment of copper oxide nanoparticles on the germination, early seedling growth, and physiological responses in Oryza sativa L. Bull. Environ. Contam. Toxicol. 104 (6), 770–777. https://doi.org/10.1007/s00128-020-02850-9.

Wang, X., Xie, H., Wang, P., Yin, H., 2023. Nanoparticles in plants: uptake, transport and physiological activity in leaf and root. Materials 16 (8), 3097. https://doi.org/10.3390/ma16083097.

Waqas Mazhar, M., Ishtiaq, M., Hussain, I., Parveen, A., Bhatti, K.H., Azeem, M., Thind, S., Ajaib, M., Maqbool, M., Sardar, T., Muzammil, K., Nasir, N., Shah, A.N., 2022. Seed nano-priming with Zinc Oxide nanoparticles in rice mitigates drought and enhances agronomic profile. PLoS One 17 (3), e0264967. https://doi.org/10.1371/journal.pone.0264967.

Weisany, W., 2025. Nanotechnology for sustainable water management in precision agriculture. In: Handbook of Nanotechnology in Agriculture. Springer.

Wessendorf, C., Kopka, A., Fornahl, D., 2025. Key enabling technologies (KETs) in the technological space: embeddedness and regional knowledge creation. Eur. Plan. Stud. 33 (2), 161–182. https://doi.org/10.1080/09654313.2024.2420857.

Wu, J., Bosker, T., Vijver, M.G., Peijnenburg, W.J.G.M., 2021. Trophic transfer and toxicity of (Mixtures of) Ag and TiO_2 nanoparticles in the lettuce–terrestrial snail food chain. Environ. Sci. Technol. 55 (24), 16563–16572. https://doi.org/10.1021/acs.est.1c05006.

Xuan, L., Ju, Z., Skonieczna, M., Zhou, P.-K., Huang, R., 2023. Nanoparticles-induced potential toxicity on human health: applications, toxicity mechanisms, and evaluation models. MedComm 4 (4). https://doi.org/10.1002/mco2.327.

Yan, R., Zhang, Y., Tian, H., Hao, Y., Sun, H., 2024. SiO_2 nanoparticle attenuates phytotoxicity of graphene quantum dots in Zea mays (L.) plants. Land Degrad. Dev. 35 (12), 3745–3755. https://doi.org/10.1002/ldr.5164.

Yang, J., Jiang, F., Ma, C., Rui, Y., Rui, M., Adeel, M., Cao, W., Xing, B., 2018. Alteration of crop yield and quality of wheat upon exposure to silver nanoparticles in a life cycle study. J. Agric. Food Chem. 66 (11), 2589–2597. https://doi.org/10.1021/acs.jafc.7b04904.

Yang, S., Yin, R., Wang, C., Yang, Y., Wang, J., Sun, W., 2023. Phytotoxicity of zinc oxide nanoparticles and multiwalled carbon nanotubes, alone or in combination, on Arabidopsis thaliana and their mutual effects on oxidative homeostasis. PLoS One 18 (2), e0281756. https://doi.org/10.1371/journal.pone.0281756.

Zeng, S., Shakoor, N., Rui, Y., 2025. Nanotechnology and agricultural sustainability: a review. Nanomaterials 15 (23), 1755. https://doi.org/10.3390/nano15231755.

Zhang, T., Wang, Q., Rui, Y., 2025. The impact of nanomaterials on plant health: a review of exposure, toxicity, and control. Environ. Sci. Nano 12 (6), 2965–2982. https://doi.org/10.1039/d5en00037h.

Zhang, X., McKee, M.S., Filser, J., 2023. Long-term effects of silver nanoparticles (NM-300K) and soil amendments on soil respiration and mesofauna in a semi-field experiment. Environ. Sci. Nano 10 (8), 2090–2101. https://doi.org/10.1039/d2en00730d.

Zhao, H., Liu, Z., Han, Y., Cao, J., 2024. Impact of silver nanoparticles on arbuscular mycorrhizal fungi and glomalin-related soil proteins in the rhizosphere of maize seedlings. Diversity 16 (5), 273. https://doi.org/10.3390/d16050273.

Zhao, X., Zhang, W., He, Y., Wang, L., Li, W., Yang, L., Xing, G., 2021. Phytotoxicity of Y_2O_3 nanoparticles and Y3+ ions on rice seedlings under hydroponic culture. Chemosphere 263, 127943. https://doi.org/10.1016/j.chemosphere.2020.127943.

Zhong, X., Ma, E., 2022. A novel approach for characterizing pore size distribution of wood cell wall using differential scanning calorimetry thermoporosimetry. Thermochim. Acta 718, 179380. https://doi.org/10.1016/j.tca.2022.179380.

Zolghadrnasab, M., Mousavi, A., Farmany, A., Arpanaei, A., 2021. Ultrasound-mediated gene delivery into suspended plant cells using polyethyleneimine-coated mesoporous silica nanoparticles. Ultrason. Sonochem. 73, 105507. https://doi.org/10.1016/j.ultsonch.2021.105507.

Index

'*Note*: Page numbers followed by "f" indicate figures and "t" indicate tables.'

D

E

S

T

U

V

W

X

Z

Printed and bound by CPI Group (UK) Ltd, Croydon, CR0 4YY
22/09/2026
02220555-0012